Electrical Machines
and Drive Systems

Electrical Machines
and Drive Systems

C. B. Gray
Lecturer in Electrical Engineering
University of Bradford

Copublished in the United States with
John Wiley & Son, Inc., New York

Longman Scientific & Technical,
Longman Group UK Limited,
Longman House, Burnt Mill, Harlow,
Essex CM20 2JE, England
and Associated Companies throughout the world.

Copublished in the United States with
John Wiley & Sons, Inc., 605 Third Avenue, New York, NY 10158

© Longman Group UK Limited 1989

All rights reserved; no part of this publication may be reproduced, stored in a retrieval system, or transmitted in any form or by any means, electronic, mechanical, photocopying, recording, or otherwise without either the prior written permission of the Publishers or a licence permitting restricted copying in the United Kingdom issued by the Copyright Licensing Agency Ltd, 33–34 Alfred Place, London, WC1E 7DP.

First published 1989

British Library Cataloguing in Publication Data
Gray, C. B.
 Electrical machines and drive systems.
 1. Electromechanical energy conversion
 I. Title
 621.31'3
ISBN 0-582-30540-3

Library of Congress Cataloging-in-Publication Data
88-8462
ISBN 0470-21319-1

Set in 10/11pt Times Roman

Produced by Longman Singapore Publishers (Pte) Ltd.
Printed in Singapore

'How great a being, Lord, is Thine which doth all beings keep!
Thy knowledge is the only line to sound so vast a deep.
Thou art a sea without a shore, a sun without a sphere;
Thy time is now and evermore, Thy place is everywhere.'

John Mason c. 1645–91

Contents

Preface xiii
Acknowledgments xv
List of principal symbols xvii

1 Current and power flow in electrical circuits

Electrical machines – what they are and what they do | 1
Electrical machine types | 2
Current flow in electric circuits | 3
Electric field of a conductor | 4
Review of electric circuit power | 9
Complex power flow in a.c. circuits | 14
Tutorial examples | 18

2 Basic electrostatic and electromagnetic field theory

Introduction | 20
The electrostatic field | 20
Boundary relationships for the electrostatic field | 27
Energy storage and force development in the electrostatic field | 28
The electromagnetic field | 31
Polarity, right-handedness | 33
Curl **H**, Ampère's law | 34
Magnetic vector potential **A** | 36
Magnetic field intensity due to current in a long straight wire | 38
Mechanical stresses in the magnetic field, force between current-carrying conductors | 39
Inductance, flux-linkage | 41
Electromagnetic induction | 42
The Lorentz equation | 42
Faraday's law, Lenz's law | 43
Energy storage and flow in the electromagnetic field | 45
The Poynting vector | 46
Power transfers at the surface of a conductor | 46
Tutorial examples | 49

viii   Electrical machines and drive systems

3 Force and torque development in magnetic circuits – properties of magnetic materials

Force on a current-carrying conductor located in a magnetic field	53
Force and torque development on current-carrying loops placed in a magnetic field	55
The properties of magnetic materials	57
The hysteresis loop	61
Eddy current loss	63
Boundary relationships for the magnetic field	64
Energy stored in the magnetic field – fringing	66
Maxwell's stresses in the magnetic field	69
Normal and tangential mechanical stresses at ion/air boundaries	72
Force development at iron/air boundaries appropriate to a linear actuator	74
Worked example	75
Rotary motion actuators, torque development	79
Torque development in rotating machines with cylindrical or salient iron/airgap surfaces	82
Effects of slotting	83
Rotary actuator with permanent-magnet polarisation	83
Direct-current moving-coil actuator	85
Faraday's rotator	86
Faraday's disc	87
Permanent magnetism	89
Worked example	94
Tutorial examples	96

4 The transformer

Introduction	100
Voltage and current relationships in an ideal two-winding transformer	101
Effect of core saturation	104
Leakage flux	105
Equivalent circuit and phasor diagram of a two-winding iron-cored transformer	106
Voltage regulation of a transformer	110
Magnetising current requirements and energy losses in transformer cores	112
Transformer efficiency	116
Transformer testing	118
Equivalence of winding resistance and leakage reactance values referred to the same number of turns	119
Power transfers across primary and secondary windings of a two-winding transformer	120
Polyphase (three-phase) transformer connections and core arrangements	121
Autotransformers	126
High-frequency transformers	128

Current transformers	128
Tutorial examples	130

5 Windings for rotating machines

Introduction	134
The magnetic field of an armature winding	136
The radial magnetic field of a.c. armature windings	140
Space-harmonic analysis of the radial field developed by a short-pitch coil	142
Winding distribution, distribution factor evaluation	145
Worked example	147
The tangential magnetic field of armature windings	151
General procedures for airgap radial and tangential field evaluation	153
Radial and tangential fields of a sinusoidal winding (sinusoidal current sheet)	156
Practical winding considerations	159
Resultant magnetic fields of polyphase windings (rotating field)	161
Principal magnetic field properties of armature windings (summary)	163
Electromotive force induction in armature windings	164
Coil interconnections within phase windings and their effect on induced e.m.f., distribution and pitch factors	169
Power flows relating to rotating machine windings	171
Tutorial examples	174

6 Direct-current machines

General arrangement of d.c. machines	178
Generation of e.m.f. between armature terminals of a rotating d.c. machine	179
Calculation of armature e.m.f.—The e.m.f. equation of a two-pole d.c. machine	187
The general e.m.f. equation of a d.c. machine	194
Torque development within a d.c. machine	194
The airgap field of a d.c. armature winding – armature reaction	195
Radial and tangential airgap field components of a d.c. armature current sheet	197
Tangential mechanical stress and torque development at the armature surface of a d.c. machine	201
Output coefficient and rating of a d.c. machine	204
Commutation of d.c. machines	205
Methods of stator pole energisation	211
Performance of d.c. machines	213
Direct-current motor characteristics	215
Worked example	220
Direct-current motor dynamics – speed control, starting and braking	221

Electronically controlled d.c. machine drive systems	225
Direct-current chopper drive systems	226
Worked example	232
Phase-controlled converter-fed drive systems	234
Single-phase, half-wave converter	236
Single-phase, full-wave thyristor bridge converter	238
Single-phase, half-controlled thyristor bridge converter	242
Polyphase a.c./d.c. converters	243
Overlap	247
Anti-parallel bridge converters for four-quadrant, controlled d.c. motor drives	250
Half-controlled, polyphase bridge converter	252
Closed-loop d.c. motor-control system	253
Permanent-magnet d.c. machines	254
Brushless permanent-magnet d.c. machines	255
Tutorial examples	258

7 Synchronous machines

Introduction	264
The synchronous machine – constructional details	266
The uniform-airgap synchronous machine – principle of operation	267
Radial magnetic field developed in a uniform airgap by a balanced three-phase winding energised with three-phase currents	270
Worked example	270
Radial airgap field development by sinusoidal current distribution	273
Tangential airgap field due to sinusoidally distributed current	276
Determination of resultant radial and tangential fields in a uniform airgap due to a balanced, three-phase winding energised with balanced, three-phase, sinusoidally distributed current	278
Torque development in a polyphase synchronous machine with a uniform airgap	279
Electromotive force equation of a polyphase synchronous machine with a uniform airgap	282
Equivalent circuit and phasor diagram of a polyphase synchronous machine with a uniform airgap	285
Evaluation of armature reaction reactance in terms of airgap parameters	286
Synchronous machine operation as an alternator – voltage regulation	287
Uniform-airgap synchronous machine operation on constant-voltage, constant-frequency busbars	290
Starting of synchronous machine, synchronisation and putting on load	292
Terminal power/load angle relationships for a uniform airgap synchronous machine on an infinite busbar	293
Dynamic response of a synchronous machine to changes in load	295

The natural frequency of oscillation of a synchronous machine	297
Dynamic response of a synchronous machine to load or system changes; Equal-area critierion	299
Synchronous machine with salient poles – two-reaction theory	302
The polyphase reluctance machine	303
The single-phase reluctance motor	309
The polyphase salient-pole synchronous machine – phasor diagram	310
Construction of a phasor diagram of a salient-pole synchronous machine given knowledge of the machine parameters and terminal power conditions	314
Power/load-angle relationship for the salient-pole machine on constant-voltage, constant-frequency busbars	316
Measurement of direct- and quadrature-axis synchronous reactances of a salient-pole synchronous machine	318
Sudden three-phase short circuit of a synchronous machine	318
Negative (phase-) sequence reactance of a polyphase synchronous machine	322
Zero (phase-) sequence reactance of a polyphase synchronous machine	323
Stepping motors	323
Stepping motor static torque characteristics	329
Stepping motor dynamic behaviour	330
Stepping motor drive circuit requirements	338
Switched-reluctance motor	339
Permanent-magnet a.c. synchronous machines	343
Power electronic drive systems for synchronous motors	345
Single-phase voltage-source bridge inverter	347
Quasi-square voltage waveform generation	350
Forced commutation of a bridge inverter supplied from a voltage source	350
Worked example	354
Worked example	359
Amplitude control of inverter output voltage	360
Worked example	362
Three-phase voltage-source bridge inverter	364
Worked example	367
Three-phase voltage-source inverter drive systems	370
Three-phase current-source inverter drive systems	372
Tutorial examples	377

8 Induction machines

Introduction and principle of operation	385
Construction of polyphase induction machines	389
Electromotive force and torque evaluation in polyphase induction machines	391
Maximum torque condition at constant radial flux	396
Equivalent circuit and phasor diagram of a polyphase induction machine	397
Worked example	402
Speed/torque characteristics of polyphase induction machines	404

Starting and speed control of polyphase induction motors	406
Stator current locus of polyphase induction machine	407
Doubly fed induction machines	408
Variable-speed induction-motor drives employing cage rotors	412
Harmonic effects on polyphase induction motor performance	415
The single-phase induction machine	417
Other single-phase motors	424
The polyphase linear-induction motor	427
Tutorial examples	431
Bibliography	436
Index	438

Preface

This book is presented as a comprehensive text identifying with undergraduate and similar courses concerned with the principles of electromechanical energy conversion, its utilisation within particular drive systems, its practical implementation via power electronic circuitry and its relevance to integrated power networks. The crowded curriculum of a modern course in electrical engineering places lecture time at a premium, whilst the appeal and apparent urgency of developments in the information system technologies tend to divert the attention and interest of students and businessmen away from the 'old-hat' technologies of high power and heavy current.

Perhaps because electrical machines have worked so well, so reliably and so efficiently for so long it is tempting to suppose that no intellectual challenges remain in this area and that an introductory course might legitimately be reduced to a routine with a few essential rules and formulae, with an equivalent circuit representation—this is hardly justified. The present book seeks to counter this tendency by establishing the foundations squarely on the basic principles of electromagnetics which should contribute core curriculum material for any course in electrical engineering which has academic credibility.

Thus the early chapters of the book incorporate the essential aspects of electrical conductive, electrostatic and magnetic fields as conventionally understood. It is the author's view that all electrical devices are in essence field-effect devices and that students of electrical engineering might reasonably be expected to become familiar with the essential concepts of field phenomena, such as may be identified with the electromagnetic transmission of power or signal via transmission line, waveguide, space, antenna, electronic device or electromechanical machine.

Force and torque on iron surfaces bounded by an airgap are generally evaluated in the text using the Maxwell concept of mechanical stress development within a magnetised medium. Such an approach emphasises the distributed nature of the forces developed on magnetised surfaces, unlike alternative global energy methods, and is of increasing relevance to the present-day designer as the field distribution of practical systems is increasingly capable of accurate computation. The Maxwell stress approach leads naturally to the recognition that tangential as well as radial magnetic field components exist in the airgap of rotating machines. The simplifying assumptions applied to the topology of electrical machine types subsequently studied in detail are such that the fundamental laws of electromagnetism are applicable without the need for mathematical dexterity, yet give rise to conventional performance equations and equivalent circuits, thus facilitating an appreciation of the inherent simplicity of the machine structures which these models represent.

One chapter has been devoted to rotating machine armature windings, primarily in the context of a.c. machines. By tradition, a study of machine windings engenders little enthusiasm and many texts avoid a detailed examination. Nevertheless the disposition of conducting material has such an important influence on the behaviour of a practical machine that no realistic analysis can ignore winding layout. The text has been designed, however, such that Chapter 5 may be omitted on a first reading or if the text is used as part of an introductory course.

No longer is it possible for an interest in electrical machines to be restricted to traditional d.c., a.c. synchronous and a.c. induction machines. The current demand for precision-built and precisely controlled machines in the aerospace, automobile, computer, industrial and domestic markets, including home video and audio equipment, has stimulated production of brushless d.c. machines, stepping motors and switched reluctance drives. Such developments are possible because of rapid parallel developments in control procedures and power electronic equipment.

A modern course in rotating electrical machines requires a complementary understanding of the power electronics circuitry by means of which the available power supplies are conditioned to suit the application. Consequently, the essential characteristics of representative converter and inverter circuits are included in the present text with descriptions of behaviour emphasising the features relating to energy transfers. Although it appears likely that control electrode turn-off devices like the GTO thyristor will increasingly displace the basic thyristor for controlled switching purposes, force-commutated inverter circuits are likely to remain of interest into the foreseeable future.

The final decade of the twentieth century marks a hundred years of development for the industrial electrical drive. With the continuing sophistication of electronic control and the prospect of further dramatic improvements in the properties of magnetic materials together with the discovery of high-temperature superconductors, the second century of progress in this field must hold the promise of equal fascination and opportunity.

Acknowledgements

I should like to express my gratitude to the many people who directly or indirectly have contributed to the content of this book and facilitated its publication: to those early philosophers, experimenters, engineers and industrialists who discerned and evaluated the logic of electrical science and first applied it to the service of man; to teachers, friends and mentors encountered personally in the early, formative years of a career and later colleagues, particularly Professor Denis O'Kelly, for many useful discussions on matters of common interest; to Drs Paul Acarnley, David Howe, John Edwards and Norman Fulton; and to Brian Richardson and various publishers for generously permitting the inclusion of material which may not yet have achieved the accolade or anonymity of 'common knowledge'.

Thanks also to several electrical equipment manufacturers who have kindly provided illustrative material: to the University of Bradford authorities, Dr Austin Hughes and the Engineering Council for allowing examination question material to be used (for which the responsibility regarding correctness of the answers given is entirely mine); to Professors William Shepherd and David Howson of the Department of Electrical Engineering at the University of Bradford for practical encouragement during the whole of the enterprise; to Marlyn Walsh, aided by Ricki Connors, for word-processing the script; and to David Jowett for tracing the line diagrams. Finally, thanks are expressed to the editorial and production staff at Longman Scientific and Technical.

Clifford B. Gray
Bradford, 1988

We are indebted to the following for permission to reproduce copyright material:

the author, P P Acarnley and Peter Peregrinus Ltd for figs 7.40, 7.43, 7.44 and 7.45 adapted from *Stepping Motors: A Guide To Modern Theory and Practice* by P P Acarnley (2nd Edn, 1984); the author, J D Edwards and Macmillan Education Ltd for figs 3.23 and 3.24 adapted from *Electrical Machines* by J D Edwards (2nd Edn); The Engineering Council Examinations Department for questions from past examinations papers; the author, D Howe for fig 7.49 adapted from the paper 'New design opportunities for permanent magnet excited machines' by D Howe, T Birch and I J Williams; McGraw-Hill Book Co (UK) Ltd for fig 2.22 from *Electromagnetics* by J D Kraus (3rd Edn, 1984); the author, B Richardson for an extract from his paper 'Transformer core losses' in *Electronics and Power* (May 1986); the University of Bradford for questions from past examination papers and tutorial sheets.

List of principal symbols

A	ampere	I, i	instantaneous current, A
A	linear current density, A m^{-1}	$I; I$	r.m.s. phasor; r.m.s., steady, mean current, A
A	specific electric loading, A m^{-1}	J	joule
$\mathbf{A}$	magnetic vector potential, Wb m^{-1}	J	moment of inertia, kg m^2
a	area, m^2	J	current density, A m^{-2}
a	conductor radius, m	j	90° operator
a_v	Fourier coefficient	K	a constant
b	radius of cylindrical surface bounding airgap	K_{pv}	coil span factor for vth space harmonic
b_v	Fourier coefficient	K_{dv}	distribution factor for vth space harmonic
$\mathbf{B}$	magnetic flux density, $\text{T} = \text{Wb m}^{-2}$	K_{wv}	$K_{pv}K_{dv}$, winding force for vth space harmonic
B_{mean}	specific magnetic loading, T	k	kilo = 10^3
C	coulomb	L	self inductance, H
C	capacitance, F	$l, \mathbf{l}$	length (vector, scalar), m
C	output coefficient	M	mutual inductance, H
$\mathbf{D}$	electric flux density, C m^{-2}	m	metre
d	distance (spacing), m	m	number of phases
$\delta l, \Delta l$	element of length (scalar), m	$\mathbf{m}$	magnetic dipole moment, A m^2
$\boldsymbol{\delta l}\ \boldsymbol{\Delta l}$	element of length (vector), m	m.m.f.	magnetomotive force, A
$\delta s, \Delta s$	element of surface (scalar), m^2	N	newton
$\boldsymbol{\delta s}, \boldsymbol{\Delta s}$	element of surface (vector), m^2	N	number of turns
Δv	element of volume, m^3	n	number of stacks (phases) of stepping motor
$\mathbf{E}$	electric field intensity, V m^{-1}	n	rotational speed, rev/s
$E, \mathbf{E},$ e.m.f., V_{oc}	electromotive force, V	n_0	synchronous speed
e		p	pressure, stress, Pa
F	farad	p	instantaneous power, W
$\mathbf{F}$	force, N	P	average power, W
f	frequency, Hz	P	permeance, H
g	airgap length, m	Pa	pascal = N m^{-2}
H	henry	Q	slots/pole/ph
$\mathbf{H}$	magnetic field intensity, A m^{-1}	Q	reactive volt-amperes, VA_r
Hz	hertz, cycles per second	Q, q	charge, C

Symbol	Meaning	Symbol	Meaning
Q'	charge/unit length, C m^{-1}	ε	absolute permittivity, F m^{-1}
R, r	resistance, Ω	ε	fundamental angle of short-pitch, °elec or rad
R	reluctance, H^{-1}		
r	radius, m	ε	elec (abbrev.)
rad	radian	ε_0	permittivity of free space $= 8.854 \times 10^{-12} \text{ F m}^{-1}$
S	Poynting vector, W m^{-2}		
S	volt-amperes, VA	η	efficiency, p.u.
S, s	surface area, m^2	η	impedivity (complex generalisation of resistivity) Ω m
S	siemen $= \Omega^{-1}$		
s	second (of time)		
s	Laplace operator	θ, θ'	(phase angle) °elec or rad, angular position
s	slip (p.u.)		
T	tesla $= \text{Wb m}^{-2}$	θ_m, θ'_m	angular position, °mech or rad
T	torque, N m		
t	time, s	Θ	slot angle, fundamental °elec or rad
t_r	number of rotor teeth on stepping motor		
		λ	flux linkage, Wb
t	stress, Pa	μ	absolute permeability, H m^{-1}
U	magnetic scalar potential, A		
		μ_0	permeability of free space $= 4\pi \times 10^{-7} \text{ H m}^{-1}$
V	volt		
V, v	instantaneous voltage, potential difference	ρ	resistivity Ω m
		ρ	electric volume charge density, C m^{-3}
$V; V$	r.m.s. phasor; r.m.s., steady, mean voltage, V	ρ_s	electric surface charge density, C m^{-2}
v	velocity, m s^{-1}	σ	electrical conductivity, S m^{-1}
v	harmonic number		
v, V	volume, m^3	σ	flux-leakage coefficient
W	watt	σ	phase spread °ε
W	energy, J	τ	time-constant, s
Wb	weber	ϕ	time-phase angle, power factor angle
w	energy density, J m^{-3}		
X	reactance	Φ	magnetic flux, Wb
$\hat{x}$	unit vector in direction $\text{O}x$	Ψ	electric flux, C
		ψ	space-phase angle, torque angle, power factor angle
Y	admittance, S		
$\hat{y}$	unit vector in direction $\text{O}y$		
		Ω	ohm
Z, Z, z	impedance, Ω	ω	angular frequency $(= 2\pi f)$, rad s^{-1}
Z_L, Z_L	load impedance		
$\hat{z}$	unit vector in direction $\text{O}z$	ω	synchronous speed, elec rad s^{-1}
α	phase angle, delay angle	ω_0	synchronous speed, mech rad s^{-1}
β	space-phase angle	ω_0, ω_n	resonant frequency, natural frequency, rad s^{-1}
γ	overlap angle		
γ	reciprocal of time-constant, s^{-1}	ω_m, ω_r	actual speed, mech rad s^{-1}
δ	load angle		
ε	voltage regulation, p.u.		

1 Current and power flow in electrical circuits

1.1 Electrical Machines – What They Are and What They Do

One definition of a machine states that it is a device which is capable of doing work. In the process energy is expended and may be converted from one form to another. On this basis a simple electrical resistance heater is a machine, converting electrical energy into heat. In most energetic processes, however, for example mechanical power transmission via the torque converter of an automatic gearbox installed in a motor car, heat is an unwanted consequence of conversion efficiencies below 100%. In this book we shall be concerned mainly with electromechanical machines which convert energy from the electrical form to the mechanical form or vice versa. We shall also consider devices and systems like the electrical transformer, or the a.c./d.c. converter, which do not involve a change in the energy state. Within such systems energy storage in a region of space is a common feature and this ability of space to store energy lies at the heart of electromechanical energy conversion processes; further, the parameters of the effective space involved predetermine the prospective performance characteristics.

Air or free space is able to store energy electromagnetically in two ways – due to either the electric field or to the magnetic field or to both. The energy *densities* w available are expressible in terms of the corresponding electric and magnetic field intensities E V m^{-1} and H A m^{-1}, respectively as

$$w = \tfrac{1}{2}\varepsilon_0 E^2, \quad \tfrac{1}{2}\mu_0 H^2 \text{ J m}^{-3}$$

where ε_0 and μ_0 are respectively the *permittivity* and *permeability* of air or free space, with values of 8.854×10^{-12} F m^{-1} and $4\pi \times 10^{-7}$ H m^{-1}, respectively. Whilst there appears to be no limit to the magnetic field intensity H tolerated by air, the values of magnetic flux densities B ($= \mu_0 H$) which may be readily established in the airgaps of electromagnetic machines having composite magnetic circuits comprising ferromagnetic material and air are limited to about 1.5 T. Electric field strengths in excess of around 2.5×10^6 V m^{-1} in air are likely to lead to breakdown by ionisation, giving a limiting energy density for the electric field of the order of

$$w_{max} = \tfrac{1}{2} \times 2.5^2 \times 10^{12} \times 8.854 \times 10^{-12} \simeq 25 \text{ J m}^{-3}.$$

This value compares unfavourably with that obtainable for energy storage in the magnetic field which is given by

$$w_{max} = \frac{1}{2}\frac{B^2}{\mu_0} = \frac{1.5^2}{2 \times 4\pi \times 10^{-7}} \simeq 10^6 \, \text{J m}^{-3}.$$

(Hydraulic systems may improve on this figure by a factor of 2 or more).

Superior energy storage capability is not the only reason tending to favour electromagnetic devices for the purpose of electromechanical energy conversion. That electric current establishes a magnetic field and that a *changing* magnetic field induces voltage, makes possible machine configurations which enable energy conversion to proceed continuously and smoothly, much more readily than any electric (electrostatic) system.

This book is therefore concerned essentially with electromagnetic machines. Nevertheless, some understanding of the characteristics of electrostatic fields as well as of magnetic fields will be required. This is because the *propagation*, or flow, of electrical power requires the confluence of both electric and magnetic field systems. Whereas in *mechanical* terms power results from the co-existence of linear motion and force, or angular velocity and torque, in electrical *circuit* terms power relates to a voltage–current product. With current flow is associated magnetic field development: voltage or potential difference relates to an electric field integrated between appropriate points. It is the *current* disposition of an electromechanical energy converter which is responsible for the force or torque development of the properties of the consequential magnetic field, whereas *motion* gives rise to the induced electric fields which account largely for the terminal voltage.

1.2 Electrical Machine Types

Electrical machines are therefore characterised by having interlinked electrical and magnetic circuits. It is the business of the machine designer to optimise the configuration of appropriate materials and structures to meet a required specification, but it will be our main concern to recognise why electrical machines behave in the way they do. At the onset it is appropriate to consider briefly the types in widespread use today, noting their significant features.

The first point worthy of observation relates to the number and scale of rotating electrical machines to be found in a modern technological society. There are literally millions of them, ranging in power rating from a few microwatts to a gigawatt, and with speeds ranging from less than $1 \, \text{rev s}^{-1}$ to $10000 \, \text{rev s}^{-1}$ or even more. Consider extreme examples – a stepping motor drive for a quartz analogue wristwatch, and a turbine generator supplying power to a national electricity grid. How many electromagnetic machines are installed in your home, or in your motor car? The answer is likely to be dozens, and more! As technological societies develop so does their need for motive power, provided cheaply, cleanly, efficiently, reliably and neatly by one or other type of electric motor. Nor is motion necessarily rotational. Electromagnetic linear motion devices find increasing use in automated production techniques, robotics and the peripheral equipment of data processing systems.

Electrical machines are frequently classified in relation to the time-variation of the electricity supply required, whether alternating or direct current. Alternating current supplies of nominally constant voltage amplitude and frequency are most readily available commercially, and rotating machines for use on a.c. supplies are generally more robust and simpler to construct than d.c. machine equivalents. Single-phase rotating a.c. machines suffer, however, from an inherent inability to develop a constant torque, the average torque developed being modified by a superimposed alternating component varying at twice the supply frequency, as shown in Section 1.4.2. Polyphase a.c. machines are more compact and can match instantaneous torque to average torque as shown in Section 1.4.5, but unless compensated they retain the disadvantage of operating with a *power factor* other than unity. This requires that the a.c. power lines carry a 'reactive' component of electrical power which alternates in the direction of flow at twice the supply frequency, superimposed upon the unidirectional flow of 'real' power. The simpler types of a.c. machine provide substantially constant speed drives, if the supply frequency is fixed. Direct current machines lend themselves more readily to variable speed applications.

1.3 Current Flow in Electric Circuits

All electrical machines constitute a part of an electric circuit. The magnetic field created by current flow in paths defined by machine windings is a feature in the development of torque. If the machine is a motor the electrical circuit will be completed through a *source* via connecting wires. The source is the means by which electrical energy is input to the circuit. Within the motor *load* electrical energy received via the transmission circuit is converted into useful mechanical energy and unwanted heat. Some energy is lost in transmission also, as a consequence of the resistance of the connecting wires. Power losses due to the resistance of the electrical conductors in the source, transmission circuit and motor load need to be minimised for efficient operation of the elementary power system described.

An electric current consists essentially of electric charges in motion. In electrolytic conductors positively and negatively charged ions move in opposite directions, as also do the electrons and holes responsible for current flow in semiconducting materials. In the metallic conductors which determine the current path in electrical machines and power transmission circuits, the moving charge is accounted for by negatively charged electrons which normally occupy the outer shell of the atomic structure and are present in abundance in a crystalline sample of the pure metal. These valency electrons are only loosely bound to the heavy, positively charged ion which constitutes the remainder of the atom and, in a bulk sample of the material, are in constant random motion with a mean velocity dependent upon the ambient temperature. Under the influence of an electric field the electrons achieve a *drift* velocity which is characterised by its snail's pace and proportionality to the applied electric field.

Now, an isolated electron subjected to an electric field in a vacuum, say, will experience a constant force proportional to the electric field intensity. By definition, the value of electric field at a point is equal to the force developed on a unit (positive) charge placed at that point. The electron in a vacuum which experiences this force will move with constant *acceleration* and consequent increasing speed. Thus it appears that the motion of a charge-carrying electron in good conductor material under the influence of an electric field is markedly different to that of an isolated charge carrier. One can model the motion of an electron in a conducting medium as analogous to that of a particle falling under constant gravitational force through a viscous liquid. After the release of the particle a constant terminal velocity is soon achieved.

In order to develop a good understanding of the energy flow processes involved in the operation of electrical machines we require an appreciation of the nature of electric and magnetic fields. We might usefully imagine an elementary electrical circuit to be an intricate, tortuous path of conducting material, subjected locally to intense electrical and magnetic fields in order to achieve some end purpose, which may be the rotation of the drum of a washing machine at a location far distant from a spinning steam turbine. Our main concern in this section is with the *electric field* which is accountable not only for the external electrostatic effects relating to displaced surface charges but also for the flow of current within conducting circuits.

We shall establish, therefore, how conducting (and insulating) materials respond when placed in an external electric field, and how the electric field *within* a conductor relates to the external field. We shall also consider how current flow in a closed circuit may be brought about by the generation of an electromotive force (e.m.f.) within the circuit and the consequential distribution of electric field over the circuit.

1.3.1 *Introduction of an isolated conductor to an external electric field*

Within an isolated conductor, perfect or imperfect, mobile charge carriers move under the influence of an external electric field. Thus a momentary *current* flows within the body of the material due to the displacement of electrons. A redistribution of charge takes place over the *surface* of the conductor, resulting in an induced electric field such as to reduce the total electric field at all points within the conductor to zero. The situation is depicted in Fig. 1.1. The external electric field in the vicinity of the conductor, and beyond, is modified by the induced field of the surface charge distribution. On the surface of the conductor the total electric field is zero.

1.3.2 *Electric field of a conductor comprising part of an electric circuit*

If the conductor is now regarded as constituting a part of a large circuit with a continuous current flow the situation is modified such

Figure 1.1 Electric field modification by isolated conductor: (a) initial conditions, (b) equilibrium conditions.

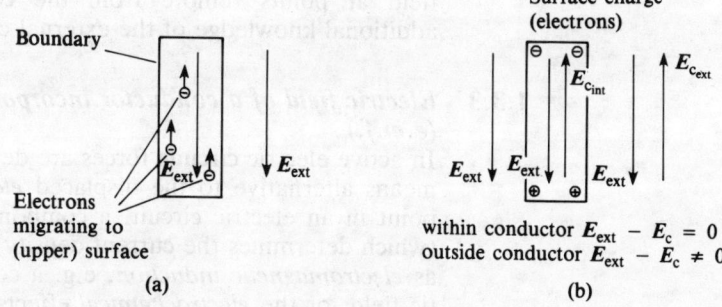

that the internal electric field is not zero but of an appropriate value to facilitate the continuous drift of electrons through the material. Thus the internal electric field is responsible for providing the force required by each electron as it moves through the conductor, constituting a part of the total current flow therein. The relationship between the current density and the electric field at a point defines the *conductivity* of the material; this itself is a property of the *mobility* of the charge carriers (electrons) and their *density*. This relationship is demonstrated as follows.

The current density J A m^{-2} at a section of conductor across which charge carriers of density ρ C m^{-3} move with a drift velocity v m s^{-1} is given by

$$J = \rho v.$$

The conductivity σ S m^{-1} of the material is related to the current density and the electric field intensity such that

$$J = \sigma E.$$

Hence

$$\sigma E = \rho v \quad \text{or} \quad \sigma = \rho(v/E). \qquad [1.1]$$

The quantity v/E is called the *mobility* of the charge carriers. In semiconductor materials, total current flow may be due to the simultaneous motion, in opposite directions, of oppositely charged electrons and holes having different mobilities.

The type of electric field associated with current flow in a resistive conductor is described as an *electrostatic* field since it is established by charges considered to be fixed in locations at the surface of the conductor and quite separate from those whose drift through the crystal lattice of the conductor constitutes the current. The surface charges have influence not solely within the conductor but also cause an electric field to exist in the external environment. An important property of electrostatic fields is that the *tangential* (field) component is continuous across a boundary (see S. 2.2.4). The electrostatic field responsible for current flow in the conductor (which may be likened to a long, thin, solid cylinder) is axially disposed and tangential to the boundary between the conductor and the insulating medium surrounding it, which may be air. Thus the electric field in the dielectric shown at the surface of the conductor in Fig. 1.2 is identical with the electric field within the conductor at the common boundary. Calculation of the electric

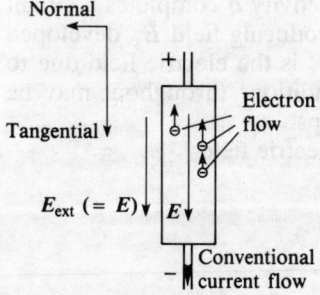

Figure 1.2 Surface and internal electric field conditions for a current-carrying conductor.

field at points remote from the conductor boundary requires additional knowledge of the external circumstances.

1.3.3 Electric field of a conductor incorporating electromotive force (e.m.f.)

In active electric circuits forces are developed on charge carriers by means alternative to the displaced *electrostatic* charge. Thus, at a point in an electric circuit, a component of the *total* electric field (which determines the current density) may result from such causes as *electromagnetic induction*, e.g. a conductor moving in a magnetic field, or the *electrochemical* effects of a battery. These electric-field components are described as *e.m.f.-producing* fields, and the line-integral of such a field between particular points in the circuit is said to be the *e.m.f.* (electromotive force) of the circuit element identified. In particular circuits several sources of e.m.f. may exist simultaneously but, if each is regarded in turn to be the sole effective source, the resulting current is conventionally considered to flow in the direction of the appropriate e.m.f.

If an isolated conductor has within it an e.m.f.-producing field, some electrons within the bulk of the material will undergo redistribution to provide fixed surface charges which enable the total electric field within the material to be zero at all points. The e.m.f.-producing field is balanced by the electrostatic field of the displaced charges. Since the tangential field due to charges is continuous across a boundary, the axial electric field at the surface of the conductor and effective externally will be equal and opposite to the responsible e.m.f.-produced field within the conductor.

If, subsequently, an electric circuit incorporating the conductor is completed through external resistance, so that continuous current can flow through the conductor, the resultant electric field therein must change to a value equal to the quotient of current density and conductivity. The direction of the current flow will be the same as that of the e.m.f.-producing field component. Thus the balancing electric field due to displaced charges must reduce, the lower value being mirrored by the electrostatic field in the insulating medium at its boundary with the conductor.

In order to distinguish further between e.m.f. and potential difference consider the simple circuit of Fig. 1.3. Here, a load conductor of length l, area a and conductivity σ completes a circuit having a uniformly distributed e.m.f.-producing field E_e developed in a source of infinite conductivity. If E_c is the electric field due to charges at any point in the circuit, conditions throughout may be summarised by the following relationships.

At any point in the circuit the total electric field

$$E = E_c + E_e.$$

Within the source of e.m.f.

$$E_c + E_e = 0.$$

Within the resistive load conductor

$$E_c = \frac{1}{\sigma} \cdot \frac{l}{a} \cdot \frac{I}{l} = \frac{1}{\sigma a} I = \frac{R}{l} I; \; E_e = 0,$$

where I is the load current and R the resistance of the load conductor.

The line integral of $\mathbf{E}$ around the circuit gives

$$\oint \mathbf{E}_\text{c} \cdot \mathbf{d}l + \oint \mathbf{E}_\text{e} \cdot \mathbf{d}l = I \oint \frac{R}{l} \mathrm{d}l.$$

(For comments on the scalar product of two vectors see Sect. 2.2.)
Now

$$\oint \mathbf{E}_\text{c} \cdot \mathbf{d}l = 0$$

(the line integral of an electric field due solely to charges is zero, see eqn [2.9], S. 2.2.2) hence

$$\left. \begin{array}{l} \oint \mathbf{E}_\text{e} \cdot \mathbf{d}l = I \int \dfrac{R}{l} \mathrm{d}l \\ \text{or} \\ E = IR. \end{array} \right\} \quad [1.2]$$

The above relationship is more generally stated as *Kirchhoff's voltage law*: 'The algebraic sum of the e.m.f.s around a closed circuit is equal to the algebraic sum of the resistive volt-drops around the circuit.' This law is strictly true under all circumstances, including those in which the closed circuit comprises one single closed mesh of an interconnected network. Under a.c. conditions its application is modified by incorporating some sources of e.m.f. within *impedance* as *inductive reactance*.

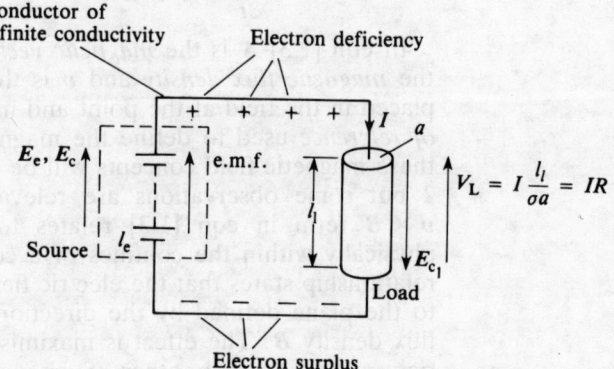

Figure 1.3 Electric circuit incorporating a source of e.m.f. and a load resistor.

With reference to the electric circuit represented by Fig. 1.3, certain points are worthy of emphasis.

- The redistribution of free charges which accounts for the development of the *electrostatic* field is brought about by the action of the source of e.m.f.
- The electrons responsible for the electrostatic field are distinguishable from those which drift to account for the circuit current. The former may be regarded as attached to the surfaces of the conducting material.
- The electric field due to charges extends beyond the confines of the circuit conductors into the space between the source and the load.

- If the source here represented by an electrolyte possesses resistivity, the electrostatic field will vary with load current, the e.m.f.-producing field remaining constant. The terminal voltage of the source, which is equal to the load voltage or the line integral of the electrostatic field (due to charges) in either source or load, will fall below the 'no-load' or 'open-circuit' value by the voltage drop across the source resistance.

1.3.4 *Electric field of a conductor due to electromagnetic induction*

Conductors placed in a time-invariant magnetic field or moving in a stationary magnetic field are subject to induced electric fields of the e.m.f.-producing type. Although electromagnetic induction will not be examined in depth until later chapters of this book, it is well to recognise at the start that electromagnetic induction does not require the presence of conducting material. Charges may be necessary at the point of interest in order that the effect of the induced electric field may be experienced as a force, but all that is required for the production of the electric field in a region of space is a magnetic field which changes with time. The presence of conducting material simply provides a dense population of electrons which will respond to the force exerted upon them by the induced electric field.

The e.m.f.-producing field E_e at a point due to electromagnetic induction has two components, in accordance with the *Lorentz equation*

$$E_e = -\frac{\partial A}{\partial t} + v \times B. \qquad [1.3]$$

In eqn [1.3] A is the *magnetic vector potential* at the point, B is the *magnetic flux density* and v is the velocity of a charge carrier placed in the field at the point and moving with respect to a *frame of reference* used to define the magnetic field. The significance of these magnetic field concepts will be examined in detail in Chapter 2 but some observations are relevant in this introduction. The $v \times B$ term in eqn [1.3] relates to charge carriers constrained physically within the confines of a conductor. The vector product relationship states that the electric field is developed at right-angles to the plane defined by the directions of velocity v and magnetic flux density B. The effect is maximised if v and B are directed at right-angles to each other. Such an electromagnetically induced electric field is confined to conducting material and behaves in the same way as the battery considered in Section 1.3.3 to develop an e.m.f. between the conductor terminals.

Further consideration of eqn [1.3] suggests that we can render the second term ineffective if our attention is restricted to a stationary circuit or, if the circuit is in motion, by defining the variation of A and B with regard to a reference frame which moves in synchronisation with the circuit. We may not dispense with the first term in eqn [1.3] when the magnetic field is varying with time.

It is evident that, since A is a field quantity, it is unlikely to be restricted to the area of space occupied by a conductor. Thus, in the region surrounding a conductor, an e.m.f.-producing field will be generated if the magnetic field of *A* external to the conductor changes. This has consequences for the electrostatic field component in the space adjacent to the conductor.

Consider an isolated conductor as depicted in Fig. 1.4 which is subjected to electromagnetic induction due to a time-changing magnetic field. Within that conductor the total electric field is zero, hence a complementary *electrostatic* field must be developed therein by the redistribution of charge from within the conductor on to its surfaces. Outside the conductor the electric field is supported electromagnetically to the extent that the line integral of $\partial A/\partial t$ over any path between the ends of the conductor balances the source e.m.f.

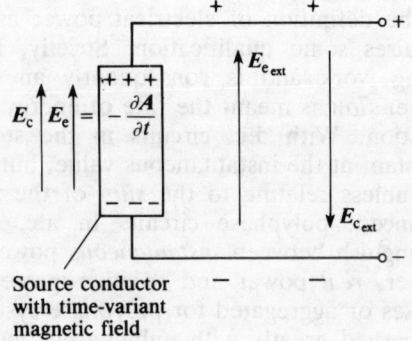

Figure 1.4 Electric field components within, and external to, a source of e.m.f.

Source conductor with time-variant magnetic field

In Section 2.6. it will be demonstrated that the electric and magnetic field conditions existing in a region of space between a source and a load in an electric circuit determine the propagation of energy over the system. In modelling networks, however, it is usual to consider the magnetic effects of a circuit element to be restricted to that element, so that all potential differences between the circuit nodes are accounted for electrostatically. If negligible stray capacitance exists between nodes in an equivalent circuit, negligible surface charge is required at each node to support any difference in potential.

Later chapters will be much concerned with the electric field at the surface of, and external to, a long conductor placed in a magnetic field environment. Some relevant observations drawn from the previous three sections may be summarised in the following statements.

- The total electric field within a conductor is given by the product of the current density and the resistivity, and equals the resultant of electric-field effects due to electromagnetic induction and electrostatic charges.
- The axial electric field at the surface of a conductor due to charges is continuous across the boundary between the conductor and the dielectric medium.

1.4 Review of Electric Circuit Power

Electric circuit theory involves the concepts of voltage, current and power as variables, with resistance, inductance, capacitance, impedance, voltage and current sources as circuit elements. Power supplied to the circuit by a source and absorbed by a load or 'sink', is associated with a voltage–current product.

In d.c. circuits the voltage and current variables are essentially unidirectional, although under transient conditions some circuits may give rise to an oscillatory response. In a.c. circuits the voltage and current vary with time, alternating in direction. Steady-state a.c. analysis of linear circuits is concerned with sinusoidally time-variant voltages and current. The sine wave is significant on account of the fact that its waveshape is basically unchanged when subjected to the mathematical procedures of addition, subtraction, multiplication, integration and differentiation.

The definition of electrical power as a voltage–current product requires some qualification. Strictly, power is the time rate of doing work and is consequently an *instantaneous* quantity. By power *flow* is meant the rate of energy transmission at a particular location. With d.c. circuits in the steady state, power remains constant at the instantaneous value, but with a.c. circuits this is not so, unless relating to the *sum* of the powers in each phase of a balanced, polyphase circuit. In a.c. circuits it is necessary to distinguish between *instantaneous* power, *average* power, *apparent* power, *real* power and *reactive* 'power', applicable to individual phases or aggregated for polyphase systems. This book will not be concerned greatly with unbalanced operation of power networks, including machines, as may occur under fault conditions, for example, and will take advantage of the fact that polyphase machines are (with few exceptions) designed to appear balanced, regarding the impedance presented by each phase to a connected a.c. power system. Thus balanced supply voltages will generally give rise to balanced load currents. Furthermore, the familiar $\sqrt{3}$ multiplier and 30° phase shift will apply as appropriate between line–line and phase–neutral voltages, or between line and phase currents, for three-phase star (wye) or delta (mesh) connected loads.

1.4.1 Power flow in a d.c. circuit

Consider the simple circuit shown in Fig. 1.5. A battery, of constant e.m.f. E and source resistance r, is connected to a load consisting of inductance L and resistance R in series. The terminal voltage v of the battery equals the sum of the voltages developed across the inductor and resistor load; it also equals the difference between the battery e.m.f. and the voltage drop across its internal (source) resistance.

Solution of the circuit after closure of the switch at $t = 0$, yields the following expression for the current $i(t)$, assuming that the initial current is zero,

$$i(t) = \frac{E}{R+r}\left[1 - \exp\left(-\frac{t}{\tau}\right)\right], \text{ where } \tau = \frac{L}{R+r}.$$

If the current is known at any instant the voltage and power distribution in the load may also be evaluated using such relationships as

$$v_R(t) = Ri(t),$$
$$v_L(t) = L\frac{di(t)}{dt},$$
$$p_L(t) = v_L(t)i(t); \quad p_R(t) = v_R(t)i(t).$$
[1.4]

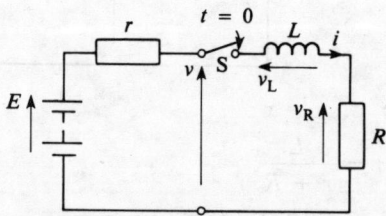

Figure 1.5 Simple d.c. circuit energised at instant $t = 0$.

The energy flow involved in the conversion of chemical energy to electrical energy within the battery is accounted for partially by the dissipation of heat due to the internal resistance of the battery and the external load resistance R. It also enables the energy stored in the magnetic field of the inductor L to increase. When sufficient time has elapsed, such that t is large compared with the *time constant* τ of the circuit, effectively steady conditions prevail and the corresponding current, voltages and powers become

$$I = \frac{E}{R + r};$$
$$V_R = IR = \frac{ER}{R + r};$$
$$V_L = 0;$$
$$P_L = 0;$$
$$P_R = I^2 R = \left[\frac{E}{R + r}\right]^2 R.$$

The above example serves to illustrate the notation in general use for electric circuit variables. Lower-case symbols indicate instantaneous, time-variant quantities, e.g. $v_L(t)$, abbreviated to v_L. Capital letters, e.g. E, V_R normally indicate steady d.c. quantities, or root-mean-square values of sinusoidally time-variant a.c. quantities. The r.m.s. value of a repetitive voltage or current is that measure which will dissipate power in a pure resistance at an *average* rate equal to that sustained by the same value of d.c. voltage or current.

1.4.2 *Power flow in an a.c. circuit*

Consider now the simple single-phase a.c. circuit of Fig. 1.6. A sinusoidally time-varying voltage source $\hat{E}\sin(\omega t + \alpha)$ replaces the battery of Fig. 1.5. Closure of the switch at time $t = 0$, at which instant $i = 0$, results in the flow of current i given by

$$i(t) = \frac{\hat{E}}{[(R + r)^2 + \omega^2 L^2]^{1/2}} \times$$
$$\left[\sin(\omega t + \alpha - \theta) - \sin(\alpha - \theta)\exp\left(\frac{-t}{\tau}\right)\right] \quad [1.5]$$

Figure 1.6 Simple a.c. circuit energised at instant $t = 0$.

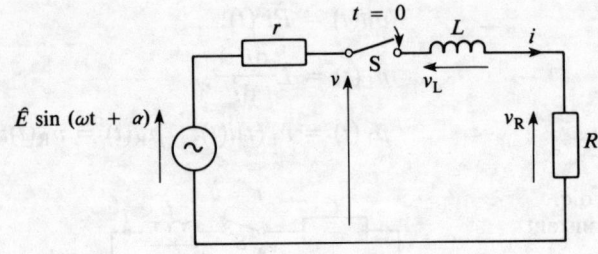

where

$$\tau = \frac{L}{r + R},$$

as before, and $\theta = \tan^{-1}\omega\tau$.

The circuit current is seen to be comprised of two components:

(i) a steady-state component, sinusoidally time-variant, lagging the source e.m.f. by phase angle θ and of magnitude equal to the quotient of source e.m.f. and modulus of circuit impedance; and

(ii) a transient component of initial magnitude such as to oblige the *total* current to equal the initial value zero, then decaying exponentially alone to zero with time constant τ.

Steady-state conditions exist when $t \gg \tau$ and the circuit current is given by only the sinusoidal component of eqn [1.5]. Under these conditions the instantaneous power dissipated in the load resistance R is given by

$$p_R = i^2 R = \frac{\hat{E}^2 R}{[(R + r)^2 + \omega^2 L^2]} \sin^2(\omega t + \alpha - \theta)$$

$$= R\frac{\hat{I}^2}{2}\{1 - \cos 2[\omega t + (\alpha - \theta)]\}, \quad [1.6]$$

where

$$\hat{I} = \frac{\hat{E}}{Z}$$

and $Z = [(R + r)^2 + \omega^2 L^2]^{1/2}$, this is the modulus of the impedance presented to the source, and is expressed symbolically as

$$\mathbf{Z} = R + r + j\omega L = Z\angle\theta.$$

Since

$$v_L = L\frac{di}{dt} = \omega L\hat{I}\cos(\omega t + \alpha - \theta)$$

after the transient current has died away, the instantaneous power absorbed by the load inductance L in the steady state is

$$p_L = v_L i = \omega L \hat{I}^2 \sin(\omega t + \alpha - \theta)\cos(\omega t + \alpha - \theta)$$

$$= \omega L\frac{\hat{I}^2}{2}[\sin 2(\omega t + \alpha - \theta)]. \quad [1.7]$$

Recognising that

$$\frac{\hat{I}}{\sqrt{2}}$$

is the r.m.s. value I of a sinusoidal current with peak value $\hat{I}$, it is clear from eqn [1.6] that the *instantaneous* power absorbed by the load *resistance* R in the steady state varies sinusoidally at twice the frequency of the source, is always positive and has an amplitude equal to its *average* value given by

$$\hat{p}_R = P_R = \frac{1}{2\pi}\int_0^{2\pi} p_R(\omega t)\,d\omega t = I^2 R. \qquad [1.8]$$

The waveform of p_R is shown in Fig. 1.7.

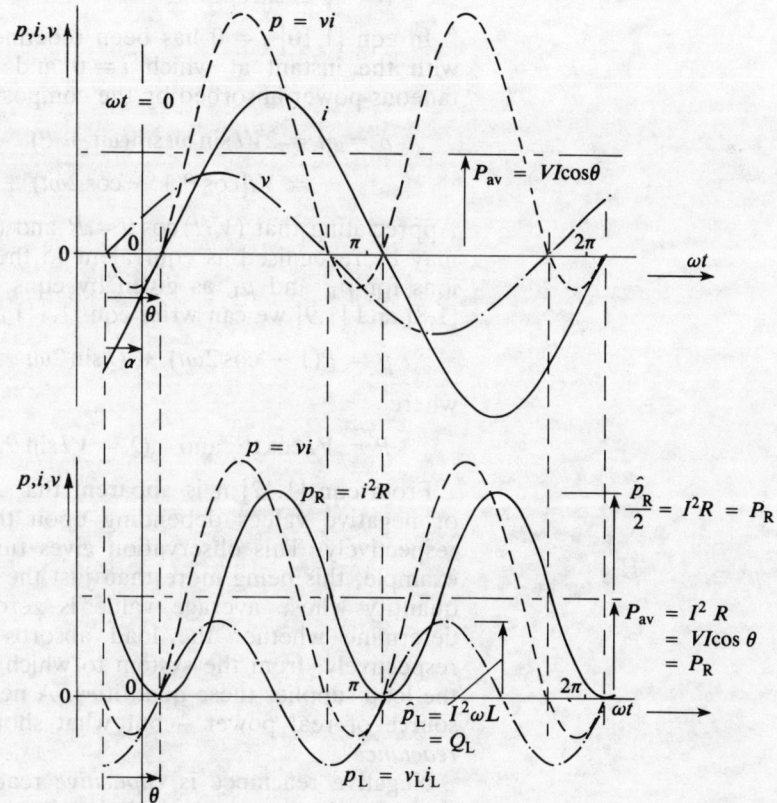

Figure 1.7 Steady-state waveforms of voltage, current and instantaneous power absorbed by resistive and inductive loads in series. Circuit of Fig. 1.6 with source resistance $r = 0$.

Considering the load inductance L it is apparent from eqn [1.7] that the *instantaneous* power absorbed in the steady state varies sinusoidally with twice the supply frequency about an average value of zero, with amplitude given by

$$\hat{p}_L = Q_L = I^2\omega L = I^2 X_L, \qquad [1.9]$$

where $X_L = \omega L$ is the *reactance* of the inductive load.

The waveform of p_L is also shown in Fig. 1.7. The reversals of polarity indicative of power flow direction are commensurate with the concept of energy interchange between the electric circuit and the magnetic field space of the inductor as its magnetic field alternates. The quantity Q_L, which is equal to the *peak* instantaneous power absorbed by an inductive load, is called the *reactive volt-amperes* (VA$_r$) of the load – by analogy with P_R, the *real power* of the load.

It is not necessary to subdivide a composite load impedance into resistive and reactive components in order to establish its steady-state real and reactive power properties. If the r.m.s. values of voltage across and current through the load impedance of Fig. 1.6 are known, together with the phase angle θ by which the current lags the voltage waveform, then

$$\left. \begin{array}{l} v = \sqrt{2}V\sin(\omega t + \theta) \\ i = \sqrt{2}I\sin \omega t. \end{array} \right\} \qquad [1.10]$$

In eqn [1.10] $t = 0$ has been redefined arbitrarily to correspond with the instant at which $i = 0$ and positive-going. The instantaneous power absorbed by the composite load is given by

$$p = vi = 2VI\sin \omega t \sin(\omega t + \theta)$$
$$= VI[\cos \theta (1 - \cos 2\omega t) + \sin \theta \sin 2\omega t]. \qquad [1.11]$$

Appreciating that $(V/I)\cos \theta = R$ and $(V/I)\sin \theta = \omega L$, eqn [1.11] may be recognised as equivalent to the sum of simplified expressions for p_R and p_L as given by eqns [1.6] and [1.7]. Using eqns [1.8] and [1.9] we can write eqn [1.11] as

$$p = P(1 - \cos 2\omega t) + Q \sin 2\omega t \qquad [1.12]$$

where

$$P = VI\cos \theta \quad \text{and} \quad Q = VI\sin \theta. \qquad [1.13]$$

From eqn [1.12] it is apparent that P and Q may have positive or negative values depending upon the sign of $\cos \theta$ and $\sin \theta$, respectively. This observation gives further significance to Q, for example, this being more than just the amplitude of a time-variant quantity whose average value is zero. The *signs* of P and Q determine whether the load absorbs real and reactive power, respectively, from the system to which it is connected, or whether the load supplies these quantities. A negative resistance is indeed a source of real power – but what should be said about negative *reactance*?

Negative reactance is *capacitive* reactance which, if it replaced the *inductive* reactance included in the circuit of Fig. 1.6 would cause the current i to lead the load voltage v, making the phase angle θ a negative angle (less than 90° if R is positive). Hence Q is negative, with certain consequences for the instantaneous power of the load given by eqn [1.12]. Now the polarity of the reactive component of p is *inverted* in comparison with that due to an inductive load, i.e. the phase of the double-frequency alteration is shifted by 180°.

1.4.3 *Complex power flow in a.c. circuits*

P and Q have been defined in Section 1.4.2 as the real power and reactive volt-amperes absorbed by a load comprising resistance and inductance operating under sinusoidal conditions. If the load is configured as a series combination of impedance $\mathbf{Z}_L = R + j\omega L = Z_L \angle \phi$, through which r.m.s. current I flows, the complex quantity S having P and Q as real and imaginary parts is known as the *complex power* or *apparent power*. The ratio

$$\frac{P}{S} = \frac{P}{[P^2 + Q^2]^{1/2}} = \frac{R}{[R^2 + \omega^2 L^2]^{1/2}} = \cos\phi$$

is called the *power factor* of the load, needing qualification to indicate whether the load current *leads* or *lags* the load voltage by less than 90°. Further

$$S = P + jQ = I^2 R + jI^2 \omega L = I^2(R + j\omega L). \qquad [1.14]$$

Since S must be entirely real if $L = 0$, and entirely imaginary with $R = 0$, it follows that the product I^2 in eqn [1.14] must be entirely real. If I is expressed as a phasor quantity leading an arbitrary reference phasor by angle α, say, i.e. $I = I\angle\alpha$, I^2 as a *real* number is achievable only if phasor quantity I is multiplied by the *conjugate* phasor quantity $I^* = I\angle-\alpha$ such that

$$\mathbf{II}^* = I^2\angle(\alpha - \alpha) = I^2\angle 0°.$$

Thus

$$S = \mathbf{II}^*(R + j\omega L) = I(R + j\omega L)I^*$$

i.e.

$$S = P + jQ = VI^*, \qquad [1.15]$$

where

$$V = I(R + j\omega L) = I\mathbf{Z}_L$$
$$= \text{phasor voltage across load impedance } \mathbf{Z}_L.$$

Hence, when complex power is expressed as a product of r.m.s. voltage and current phasors in an a.c. circuit, the correct relationship requires that the conjugate phasor of the current be used.

Equation [1.15] may also be used to indicate the power *flow* in electric circuits. Thus, in Fig. 1.8,

$$S_S = \text{Complex power into system from left-hand source}$$
$$= V_S I^*$$
$$S_R = \text{Complex power from system into right-hand load}$$
$$= V_R I^*$$
$$S_T = \text{Complex power absorbed by transmission medium}$$
$$= (V_S - V_R)I^*$$
$$= I^2 R + jI^2 X.$$

The relationships between the reference directions of power flow, voltage and current in Fig. 1.8. should be noted.

1.4.4 Real and reactive power balance in a.c. circuits

In a simple series a.c. circuit such as that shown in Fig. 1.8 it is apparent that the current is common to every element in the circuit

Figure 1.8 Power flows in a simple a.c. circuit.

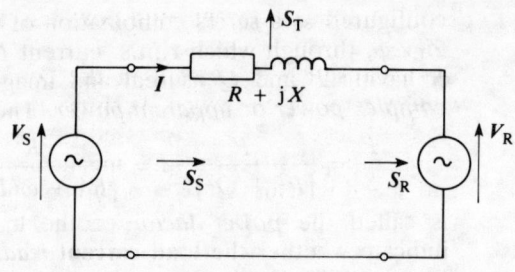

and that the algebraic sum of the voltages across each component is zero. It follows that the instantaneous total power of the circuit is also zero, i.e. $\Sigma vi = 0$. Consequently, the average (real) power supplied by the source equals the average (real) power absorbed by all the loads. It is also the case that the reactive volt-amperes supplied by all sources of reactive power must equal the reactive volt-amperes absorbed by sinks of reactive power. Convention has it that capacitors are regarded as sources of reactive power whereas inductors are regarded as corresponding sinks.

The same essential balance of real and reactive power in more general electric circuits is also true. The convention that an inductive load should absorb reactive power rather than supply it is arguably justified by the practice of electricity supply authorities to charge an increased tariff for industrial users who take power at low power factor, so increasing losses in the supply network and effectively de-rating the system plant. The low power-factor industrial demand is generally accounted for by uncompensated motor loads presenting inductive impedance to the supply. One can argue, therefore, that such a load absorbs a quantity that has to be supplied by the authority and hence may be charged for, by analogy with the real power requirement. Incidentally, it is commonly the case that power supply networks generate more reactive power via the transmission line capacitance, at times of light system loading, than meets the load requirements. Steps have then to be taken to reduce the number of lines or to absorb the surplus VA_r via installed shunt reactors to avoid the increased likelihood of operational problems through instability.

1.4.5 *Real and reactive power in polyphase circuits*

The simple circuits of Figs 1.6 and 1.8 may be regarded as constituting one phase of a balanced, polyphase system. For a balanced three-phase network the phase voltages and currents will differ in phase angle from each adjacent phase by $\pm 120°$. Evaluation of the *total* instantaneous power Σp leads to an interesting result. From eqn [1.12]

$$\Sigma p = p_a + p_b + p_c = P_{ph}\{[1 - \cos 2\omega t]$$
$$+ [1 - \cos 2(\omega t - 120°)]$$
$$+ [1 - \cos 2(\omega t - 240°)]\}$$
$$+ Q_{ph}\{\sin 2\omega t + \sin 2(\omega t - 120°)$$
$$+ \sin 2(\omega t - 240°)\}$$
$$= 3P_{ph},$$

i.e. the total instantaneous power is constant at three times the average power per phase!

This result is important in the context of an electromechanical energy conversion device. It shows that a steady rate of conversion of electrical power, expressed in terms of balanced sinusoidal voltages and currents, into mechanical power at constant torque and speed is achieved by a polyphase machine without the need for change in the total stored energy of any coupling medium. It should not be concluded, however, that such a polyphase machine has no reactive power requirement. Although the *total* reactive power evaluated as the sum of the rates of change of stored energy in each phase is zero at any instant, each phase has, in general, instantaneous reactive power balanced by that of the other phases. This reactive power is required to flow in the medium coupling the machine to the source. Polyphase machines, therefore, impose a reactive power requirement on the remainder of the power network when they operate at a non-unity power factor. By analogy with the consideration that total three-phase real power equals three times the value of the *average or real* power/phase, the total three-phase reactive volt-amperes, VA_r, are declared to be three times the value of the *reactive* power/phase, i.e.

$$\left. \begin{array}{l} P = 3\,P_{\text{ph}} \\ Q = 3\,Q_{\text{ph}}, \end{array} \right\} \quad [1.16]$$

where

$$P_{\text{ph}} = V_{\text{ph}} I_{\text{ph}} \cos\phi = I_{\text{ph}}^2 R_{\text{ph}}$$

$$Q_{\text{ph}} = V_{\text{ph}} I_{\text{ph}} \sin\phi = I_{\text{ph}}^2 X_{\text{ph}}$$

$$V_{\text{ph}} = I_{\text{ph}}[R_{\text{ph}}^2 + X_{\text{ph}}^2]^{1/2}; \quad \tan\phi = \frac{X_{\text{ph}}}{R_{\text{ph}}}.$$

Invoking the $\sqrt{3}$ relationship between line (i.e. line-to-line) and phase voltage magnitudes for a star-connected system, or between line and phase current magnitudes for a mesh-connected system, familiar expressions for total three-phase real and reactive power appropriate to a balanced, three-phase circuit are obtained, i.e.

$$\left. \begin{array}{l} P = \sqrt{3} V_L I_L \cos\phi \\ Q = \sqrt{3} V_L I_L \sin\phi \end{array} \right\} \quad [1.17]$$

1.4.6 *Single-phase equivalent circuits of polyphase machines*

Most polyphase electrical machines are designed for balanced operation. A three-phase induction motor, for example, should present identical impedances to each phase of the supply. Each set of phase voltages and currents will then be equal in magnitude and mutually displaced by a phase angle of 120°. Where an equivalent circuit is used to represent the impedance of a balanced polyphase machine, it is usual to employ the economies afforded by symmetry and display the circuit appropriate to one phase only.

In the absence of specific requirements to the contrary, circuit parameters are presented as appropriate to a *star* interconnection of the windings. One advantage of the star configuration over the delta alternative is that transmission-line series impedance may be incorporated directly in series with the machine winding impedance per phase.

1.5 Tutorial Examples

1.1 A circular disc of doped semiconductor material with a mobility of $0.135 \text{ m}^2\text{s}^{-1}\text{V}^{-1}$ has diameter 1 mm and thickness 0.1 mm. Metal contacts over each parallel face may be assumed to have infinite conductivity. Within the material free electrons with a density of 10^{20} m^{-3} constitute the charge-carrier population. Calculate the conductivity of the disc material and the resistance between contacts.

If a potential difference of 1.0 V is applied between the contact surfaces calculate (i) the current flow, (ii) the power dissipation, and (iii) the average electron drift velocity within the disc.

Note: The electronic charge $q_e = 1.602 \times 10^{-19}$ C.

[2.163 S m^{-1}, 58.87 Ω; 17.0 mA, 17.0 mW, 1.35 × 10^3 m s^{-1}]

1.2 An electrical circuit devised for investigating thermionic emission incorporates the series connection of lead-acid battery, thermionic valve and copper connecting wires. The *average* velocity of electrons accelerated from cathode to anode was evaluated at 100 m s^{-1}, whereas the average drift velocity of electrons in the connecting wires was 2×10^{-5} m s^{-1} and that for negatively charged sulphate ions within the cell 10^{-4} m s^{-1}. Explain how this state of affairs is compatible with the same value of current flowing at all points in the circuit.

1.3 A transmission line conductor of circular cross-section has diameter 2 mm and is made of copper having conductivity 57 MS m^{-1}. Calculate the d.c. resistance per metre length of conductor and determine the electric field therein when a current of 10 A flows.

If the electron mobility for copper is $0.003 \text{ m}^2\text{s}^{-1}\text{V}^{-1}$, determine the free-charge density in C m^{-3} for the copper conductor. Also establish the average electron drift velocity corresponding to the 10 A current.

[0.005 58 Ω m^{-1}, 0.055 84 V m^{-1}; 19 × 10^9, 1.675 × 10^{-4} m s^{-1}]

1.4 A pair of long conductors similar to that considered in **1.3** interconnects a source of d.c. e.m.f. and a load resistance. The source of e.m.f. is a similar copper conductor of length l m moving at constant speed v m s^{-1} at right-angles to a stationary magnetic field of constant flux density B T. If the load resistance is infinite calculate the electric field in (i) the source conductor, (ii) the dielectric adjacent to the source conductor, and (iii) the dielectric adjacent to the transmission conductor.

How are the external electrostatic fields established?

[(i) Zero, (ii) Bv V m^{-1}, axial (parallel to the axis of the conductor), (iii) radial (perpendicular to the axis of the transmission conductor) such that $\int E \cdot dx$ between transmission conductor pair $= Blv$ V. By free charges appropriately distributed over conductor surfaces.]

1.5 If the circuit of **1.4** is loaded such that a current of 10 A flows, and the surface charge in a particular region of transmission conductor is such as to establish a uniform radial electric field of 1.0 V m^{-1} at that surface, calculate the magnitude and direction of the resultant electric field in the dielectric just outside that surface. Calculate also the resultant electric field within and just outside the source conductor.

[1.0016∠3.2° V m^{-1} (to normal); 0.055 84 V m^{-1}, Bv − 0.055 84 V m^{-1}]

1.6 A balanced, three-phase, star-connected load of 40 + j30 Ω/phase is supplied from a balanced, three-phase, sinusoidal supply at 400 V (line, r.m.s.). Calculate (i) the (r.m.s.) line current, (ii) the average (real) power/phase, (iii) the reactive volt-amperes/phase, (iv) the power factor, and (v) the total volt-amperes.

[(i) 4.619 A, (ii) 853.33 W, (iii) 640 V A$_r$, (iv) 0.8 (lag), (v) 3200 V A]

1.7 If the current in phase a of the star-connected load of **1.6** is expressed in phasor form as $I_a = 4.619 + j0$ A determine compatible complex expressions for line currents I_b and I_c, and phase voltages V_a, V_b and V_c assuming phase sequence abc. Define reference directions on a labelled circuit diagram.

[−2.31 − j4.00 A, −2.31 + j4.00 A; 184.75 + j138.56 V, 27.71 − j229.32 V, −212.47 + j90.76 V]

1.8 Evaluate the instantaneous current in phases b and c of the balanced load of **1.6** when that in phase a is zero, positive going. Assume phase sequence abc and that the reference direction for each phase current is towards the common star point. Determine also the corresponding instantaneous power dissipated in each of the three phases and their instantaneous sum. Compare the latter with 3× the average power/phase calculated in **1.6** (ii).

[−5.657 A, 5.657 A; 0 W, 1280 W, 1280 W, 2560 W]

2 Basic electrostatic and electromagnetic field theory

2.1 Introduction

In Section 1.3 we came to appreciate that current flow in electric circuits was the consequence of electron motion in response to a net electric field within the conductor. At least a part of that field was due to static charges on the surface of the conductor adjusting to a value appropriate to the current density and the effect of any e.m.f.-producing field. We recognised that an e.m.f.-producing field exerted its influence solely within the conductor (or electrolyte in the case of a battery), whereas the effect of surface charges extended into the regions of space surrounding the conductor. It will be our concern in Sections 2.2 and 2.3 of the present chapter to review the properties of electric fields in dielectric media, particularly those which are relevant to the transfer of electrical energy through space. As we shall subsequently do likewise in respect of magnetic fields in Sections 2.4 and 2.5, we shall be able to establish in Section 2.6 the electromagnetic energy flows pertaining to power transmission systems and relevant also to such electromechanical devices as solenoids, transformers and rotating machines.

The magnetic field, electrostatic field and electric conduction field have many features in common and it is pertinent to consider these before identifying their distinguishing features. All produce a 'flux density' at a point in space as a result of a field 'force' or 'intensity'. Surface integrals of flux density give rise to the 'circuit' quantities of magnetic flux, electric flux or electric current, as appropriate; line integrals of field intensity give rise to m.m.f. (magnetomotive force), potential difference or e.m.f. Flux density and field intensity are related through parameters known respectively as *permeability*, *permittivity* and *conductivity*. These relationships are summarised in Table 2.1.

2.2 The Electrostatic Field

Electric fields arise in regions of space as a consequence of displaced stationary charge. Characteristically, the electric field exerts a force on any charged particle located in the space surrounding the fixed charges, but the medium appropriate to a study of the electrostatic field is ideally of zero conductivity so that free charges are not usually present. As a consequence of the

Table 2.1 A comparison of certain qualities of electric conduction, electrostatic and magnetic field systems

	Conduction field	Electrostatic field	Electromagnetic field
Field intensity	E	E_c	H
Flux density	$J = \sigma E$	$D = \varepsilon E_c$	$B = \mu H$
Potential difference	$\int_1^2 E \cdot dl = V_1 - V_2$ $\int E \cdot dl = 0$ $\int E_e \cdot dl =$ e.m.f.	$\int_1^2 E_c \cdot dl = V_1 - V_2$ $\oint E_c \cdot dl = 0$	$\int_1^2 H \cdot dl = U_1 - U_2$ $\oint H \cdot dl = I =$ m.m.f.
Flux	$\iint J \cdot ds = I$ $\oiint J \cdot ds = 0$	$\iint D \cdot ds = \Psi$ $\oiint D \cdot ds = Q$	$\iint B \cdot ds = \Phi = \int A \cdot dl$ $\oiint B \cdot ds = 0$

electric field the dielectric medium is under stress, both electrically and mechanically. The mechanical stress property accounts for the ability of a dielectric medium to store energy. Electric field strengths in practical media are limited, however, by the prospect of ionisation of the atoms comprising the structure of the dielectric. The liberated electrons and ions may then gain sufficient kinetic energy through their acceleration by the field to cause further ionisation by collision.

Each unit of electrostatic charge (coulomb) is considered in the SI system of units to account for one unit of electric flux. This relationship is incorporated in eqn [2.1], thus

$$\psi = Q = \oiint D \cdot ds = \oiint D_n \, ds \qquad [2.1]$$

where ψ is here the total electric flux emanating from charge Q.

D is the *electric flux density* crossing an elemental area ds constituting a part of a closed surface which embraces the charge. Since D and ds have a particular direction or orientation in space they need to be recognised as *vector* quantities. The orientation of a plane surface such as the elemental area ds (which is, strictly, the limit of a finite elemental area δs) is taken as the normal to that surface. The quantity $D \cdot ds$ is called the *scalar product* or '*dot*' product of the two vector quantities involved and is equal to the product of the magnitudes of two vector quantities multiplied by the cosine of the phase angle between their directions. The scalar product is so called because it is a *scalar* quantity – it has magnitude and polarity, but no unique direction. Electric charge and electric flux are such quantities, as also is *electric potential* which we shall consider in Section 2.2.2. The evaluation of a scalar product is always easy if the directions of the component vectors are the same. This is the case with the situation depicted in Fig. 2.1 if we consider the surface s enclosing the point charge Q to be a closed sphere of radius r with the charge at the centre. Equation [2.1] then becomes:

$$\psi = Q = \oint_s \mathbf{D} \cdot d\mathbf{s} = \int_s D \cos\theta \, ds = \int_s D_n \, ds = \int_s D \, ds$$

$$= D \int_s ds = D 4\pi r^2$$

giving

$$D = \frac{Q}{4\pi r^2} \text{ at radius } r. \tag{2.2}$$

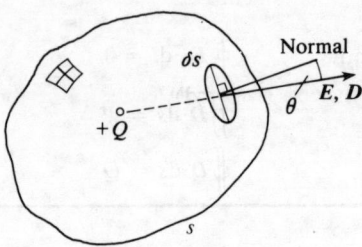

Figure 2.1 Electric charge enclosed by surface s.

Equation [2.2] shows that the electric flux density directed radially from a point charge varies in magnitude inversely as the radius squared. It follows as a special case of the more general *Gauss's law* which is stated mathematically in eqn [2.1] or, in words, as follows. 'The surface integral of the normal component D_n of the electric flux density $\mathbf{D}$ over any closed surface is equal to the charge enclosed.'

Unlike *magnetic* flux, the path of electric flux is not continuous but starts and finishes in different places defined by the locations of the responsible static charges, flux 'flowing' through the dielectric medium away from positive charge to terminate on negative charge. *En route* the electric flux density may change, but at all points in the dielectric, *electric field intensity* $\mathbf{E}$ and flux density $\mathbf{D}$ are related by the *constitutive equation* $D/E = \varepsilon$, where ε is the *permittivity* of the medium. If the medium is air or free space $\varepsilon = \varepsilon_0 = 8.854 \times 10^{-12}$ F m^{-1}. Insulating dielectric media have permittivity given by $\varepsilon = \varepsilon_r \varepsilon_0$, where the dimensionless *relative* permittivity ε_r has a value typically of the order of 2 or 3. The fact of low relative permittivity for insulating dielectric materials and their inability to withstand intense electric fields limits the value of electrostatic systems for energy storage or conversion purposes. Our interest in electrostatic fields is correspondingly muted but we shall need to appreciate the dielectric properties of the materials used to insulate machine windings secured in the slotted structures of a magnetic circuit. We shall see in Section 2.6 that the distribution of electric field in space has consequences for the distribution of electrical power during the transfer of energy over an electric circuit.

2.2.1 *Evaluation of electric field systems*

Electric fields are usually calculated on the basis of an assumed charge distribution and an application of Gauss's law stated in eqn [2.1]. Hopefully, considerations of symmetry and superposition will enable the flux density distribution to be simply stated. Use of

$D/E = \varepsilon$ should enable the distribution of electric field intensity E to be found, although a vector addition is implied when superimposing the effects of multiple charges. The field problem may then be regarded as solved, but in practice it will be considered desirable to extend the analysis to include knowledge of the distribution of electric potential and the identification of equipotential surfaces. As an example, consider *capacitance* which is equal to the ratio of responsible charge Q to the potential difference V which exists between equipotential surfaces at the locations of positive and negative charge, i.e.

$$Q = CV. \qquad [2.3]$$

Capacitance C (farad) is the important property of an electrostatic system which describes the relationship between current and voltage at its terminals. Thus, on differentiating eqn [2.3] with respect to time we get

$$i = C\frac{dV}{dt}. \qquad [2.4]$$

2.2.2 Electric field intensity E and electric (scalar) potential V

We have noted that a system of static charges is responsible for the production of an electric field in space. This field at a point in space may be expressed in terms of the *electric field intensity* vector E which is, by definition, equal to the force on a unit positive charge placed at that point. Since force has magnitude and direction and is properly represented by a vector quantity, the same will be true for E. If the field is due to a single point charge $+Q$ the field of E is directed radially outwards and is of value

$$\frac{Q}{\varepsilon 4\pi r^2}$$

at radius r. See eqn [2.2].

If the field is due to a long line of charge, distributed uniformly at $Q'\,\text{C}\,\text{m}^{-1}$, the field of E is readily shown to be again directed radially, uniform, and at radius r, of value

$$\frac{Q'}{\varepsilon 2\pi r}.$$

These relationships are readily proved using Gauss's law, eqn [2.1] applied to (i) spherical, and (ii) cylindrical surfaces with the charge at the centre and on the axis respectively. The *principle of superposition* is applicable to electrostatic fields so that the resultant field at a point is the vector sum of component fields at the same point due to individual charges – in other words, the permittivity ε of the medium is regarded as independent of electric flux density.

If a test charge moves under the influence of the electric field at its initial location *work* is involved. In the absence of other external influences a positive test charge will move always in the direction of E, i.e. in the direction of the force due to the system of charges. If the field is in air or free space the expectation is that the isolated test charge will accelerate in the direction of the field and gain kinetic energy.

If the test charge Q moves through an elemental distance δx *in the direction of the field* the work done by the field is given by

$$F\delta x = QE\delta x.$$

If the motion is constrained to be at right-angles to the field no work is done. If the motion is in the direction *opposite* to that of the field the work done *by* the field is given by $-QE\delta x$.

Generally, if the elemental movement is inclined at an angle θ to the direction of the field the work done by the field is

$$F\delta x = QE\cos\theta\,\delta x$$

The quantity $E\cos\theta\,\delta x$, in which both E and x are quantities having magnitude and direction and relative orientation of angle θ, is recognised as the *scalar product* or '*dot*' product of vector quantities $\boldsymbol{E}$ and $\delta\boldsymbol{x}$. That is

$$\boldsymbol{E}\cdot\boldsymbol{x} = E\cos\theta\,\delta x.$$

Thus, if the test charge Q moves through distance δx *in any direction* the work done by the field (a *scalar* quantity) is given by

$$F\delta x = Q\boldsymbol{E}\cdot\delta\boldsymbol{x} \qquad [2.5]$$

2.2.2.1 Absolute potential

Clearly a test charge located in an electric field has potential energy, available for conversion into kinetic energy, say, if the test charge is allowed to move under the influence of the field. This potential energy arises as a consequence of the *mechanical stressing* of the medium subjected to the field – a topic to be discussed in detail in the context of a magnetic field in Section 3.8. If we allow the electric field currently under consideration to be established by any system of charges separated by finite distances, the value of electric field at any point an infinite distance away will be zero. It is evident that the potential energy of a test charge at infinity is correspondingly zero and that an input or abstraction of energy must be associated with the movement of the test charge to any point where the field is other than zero. The *absolute (electric) potential* at any point in the field is defined as the *work done in moving unit positive charge from infinity to that point*.

The most obvious route to be considered in such a movement of charge is along a line of electric field intensity $\boldsymbol{E}$, but the definition above makes it clear that the actual path is immaterial.

2.2.2.2 Potential difference

It is also evident that electric potential is a *scalar* quantity, not a vector quantity, as it has no orientation in space. Different positions in the field will, in general, have different electric potentials – by definition, the *potential difference between two points is equal to the work done in moving unit positive charge from one location to the other*.

Thus if A and B are two separated points in the electric field depicted in Fig. 2.2, the electric potential difference $V_{AB} = V_A - V_B$ equals the work done in moving unit positive

Basic electrostatic and electromagnetic field theory 25

charge (one coulomb) from B to A.

Considering the arbitrary path shown in Fig. 2.2, the direction of the electric field over element δl is inclined at angle θ to the direction δl. The work to be done as an energy *input* in moving unit positive charge over the distance δl is correspondingly $-E_{\delta l}\delta l \cos\theta$, measured in joules per coulomb.

Using vector notation this may be rewritten as $-\boldsymbol{E}_{\delta l}\cdot\delta \boldsymbol{l}$. Integrating for all such elements at different values of l on the path shown in Fig. 2.2 we get

$$V_{AB} = V_A - V_B = -\int_B^A \boldsymbol{E}\cdot \mathrm{d}\boldsymbol{l}. \qquad [2.6]$$

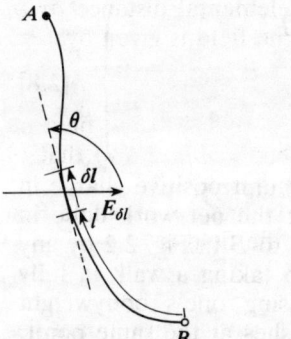

Figure 2.2 Determination of electric potential difference.

Electric potential difference has as its unit the *volt* which is equivalent to one joule of energy per coulomb of charge.

The quantity $\int_B^A \boldsymbol{E}\cdot \mathrm{d}\boldsymbol{l}$ is called the *line integral* of $\boldsymbol{E}$ along the path from B to A denoted by l. If we now consider the points A and B to be very close together and on a line of $\boldsymbol{E}$ such that A is distant from B by δl, the potential difference $V_A - V_B$ will be the incremental value δV such that

$$\delta V = -\int_B^{B+\delta l} \boldsymbol{E}_B \cdot \mathrm{d}\boldsymbol{l} = -\int_B^{B+\delta l} E_B \, \mathrm{d}l = -E_B \delta l$$

Hence

$$\frac{\delta V}{\delta l} = -E_B \qquad [2.7]$$

Equation [2.7] tells us that the voltage gradient *along a line of* $\boldsymbol{E}$ i.e. the incremental increase in potential for an incremental increase in distance *in the direction of* $\boldsymbol{E}$ is equal and opposite to the value of $\boldsymbol{E}$.

Further investigation of eqn [2.6] demonstrates that there is zero potential difference between points separated by incremental distances at right-angles to the direction of $\boldsymbol{E}$. Such points therefore define an *equipotential surface*. As a complementary observation, lines of $\boldsymbol{E}$ are always directed normally to an equipotential surface. Consequently, if we can define surfaces of equipotential in the 'field' of *scalar* electric potential (which may well vary in magnitude from place to place but has no direction) we may *deduce* the vector field of $\boldsymbol{E}$ as being everywhere at right-angles to the equipotential surface, directed in the sense of reducing electric potential, and equal in magnitude to the voltage gradient at that surface.

Casual interpretation of eqn [2.7] may appear to suggest that voltage gradient is a vector quantity, but it should be recalled that this equation was derived with a constraint on the direction of δl, being necessarily in the direction of $\boldsymbol{E}$.

Generally, scalar potential V will have different values in the different locations corresponding to incremental displacements in the three co-ordinate directions Ox, Oy and Oz about a fixed point. It is true to say that the potential gradients in the direction of each co-ordinate are equal and opposite to the *components* of the electric field $\boldsymbol{E}$ in the appropriate direction, i.e.

$$\frac{\partial V}{\partial x} = -E_x; \quad \frac{\partial V}{\partial y} = -E_y; \quad \frac{\partial V}{\partial z} = -E_z$$

or, in vector notation

$$\text{grad } V = -\mathbf{E}.$$

[2.8]

Appreciating that the path l taken by the unit positive charge in eqn [2.6] above is arbitrary, it follows that the net work done in moving a test charge from B to A and back to B in Fig. 2.2 by any route is zero. The situation is analogous to taking a walk in hilly countryside – the net work done in raising one's bodyweight against gravity is zero if one starts and finishes at the same point. It also follows, therefore, that *the line integral of the electric field intensity over any closed loop or circuit is zero*, i.e.

$$\int_c \mathbf{E} \cdot \mathrm{d}l = \oint \mathbf{E} \cdot \mathrm{d}l = 0.$$

[2.9]

The above result is noteworthy. The line integral of field intensity around a closed loop is not zero for the *magnetic* field to be studied later in this chapter (Sect. 2.4) but equal to the current enclosed by the path – the so-called *magnetomotive force* or *m.m.f.* Further, if we evaluate around a closed loop the line integral of a magnetic field quantity known as the *magnetic vector potential* $\mathbf{A}$ this is equal to the enclosed magnetic flux. As has been hinted in Section 1.3.4, if the magnetic vector potential at a location in space is varying with time an induced electric field is developed in consequence.

2.2.3 *The electrostatic field of a coaxial cable*

As a simple example, consider the electric field appropriate to a coaxial cable, in which the inner and outer conductors are regarded as concentric cylinders of radius r_1 and r_2 respectively, as shown in Fig. 2.3.

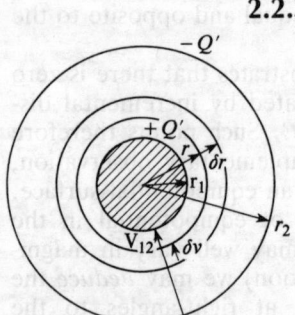

Figure 2.3 Cross-section of a coaxial cable.

Let the inner cylinder carry a distributed surface charge $+Q'$ C m^{-1}. Since the cylinders are concentric we would expect, from symmetry, that the charge distribution will be uniform over each surface, and more dense for the inner one. If we regard a cylindrical element of dielectric, with uniform permittivity ε, thickness δr and radius r concentric with the electrodes, considerations of symmetry and Gauss's law (eqn [2.1]) enable us to state that, over this element, the electric flux density will be uniform, radial and of value

$$D_r = \frac{Q'}{2\pi r} \text{ C m}^{-2}. \text{ Hence}$$

$$E_r = \frac{Q'}{\varepsilon 2\pi r} \text{ V m}^{-1}.$$

[2.10]

We thus conclude that the electric field intensity $\mathbf{E}$ is directed radially outward, of value inversely proportional to the distance from the centre of the system, and that the elementary surface at radius r is an equipotential surface. Using eqn [2.6] we can calculate the potential difference V_{12} between the inner and outer conductors as

$$V_{12} = -\int_{r_2}^{r_1} \boldsymbol{E}_r \cdot \mathrm{d}\boldsymbol{r} = -\int_{r_2}^{r_1} E_r\, \mathrm{d}r$$

$$= -\frac{Q'}{2\pi\varepsilon}\int_{r_2}^{r_1} \frac{1}{r}\,\mathrm{d}r$$

$$= \frac{Q'}{2\pi\varepsilon}\log_e\!\left(\frac{r_2}{r_1}\right). \qquad [2.11]$$

The *capacitance* of the cable,

$$C = \frac{Q'}{V_{12}} \text{ per metre length}$$

$$= \frac{2\pi\varepsilon}{\log_e(r_2/r_1)}. \qquad [2.12]$$

The electric field of a coaxial cable is depicted in Fig. 2.4 and serves to illustrate certain properties of electrostatic fields.

(i) The lines of $\boldsymbol{E}$ are directed at right-angles to the equipotential surfaces which are closer together in the region of higher flux density near the inner conductor. For this simple configuration the charge and flux distributions over particular equipotential surfaces are uniform, but this is not necessarily so in more general cases.

(ii) Over any path between equipotential surfaces the line integral of $\boldsymbol{E}$ is equal to the potential difference between those surfaces; over any *closed* path the line integral $\oint \boldsymbol{E} \cdot \mathrm{d}\boldsymbol{l} = 0$.

(iii) At any boundary between the dielectric and a conducting surface the *tangential* electric field component is zero.

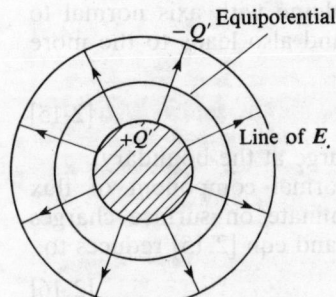

Figure 2.4 Electric field of a coaxial cable.

Point (iii) above requires some qualification as it is not generally applicable. In Section 1.3.1 we noted that the 'tangential electric field across a boundary (between a conductor and a dielectric) was *continuous*', not necessarily zero. We shall now prove the general relationship and then consider the special case where conducting material occupies one side of the boundary.

2.2.4 Boundary relationships for the electrostatic field

Consider initially that the two media having the common boundary indicated in Fig. 2.5 are perfect insulators so that the conductivities σ_1 and σ_2 are both zero. If the tangential electric field intensities at both sides of the boundary are respectively E_{t1} and E_{t2} the line integral $\oint \boldsymbol{E} \cdot \mathrm{d}\boldsymbol{l}$ over the closed rectangular path *abcda* indicated is zero, since the work done in moving a test charge from and back to *a* is zero. This relationship is true whether or not the route described encloses charge. If now the distance δx is made infinitesimally small the contribution of any *normal* electric field component to the line integral will be zero, so that $\oint \boldsymbol{E} \cdot \mathrm{d}\boldsymbol{l}$ over the indicated route reduces to

$$E_{t1}\delta y - E_{t2}\delta y = 0.$$

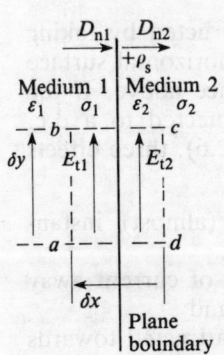

Figure 2.5 Electric field conditions at a plane boundary.

Hence

$$E_{t1} = E_{t2}. \qquad [2.13]$$

Equation [2.13] states that the tangential electric field intensity is continuous across the boundary.

If we now regard *medium 1*, say, to be a conductor with $\sigma_1 \neq 0$, E_{t1} will be zero (unless current is flowing in the conducting medium of finite conductivity in the direction of E_{t1}) to make

$$E_{t2} = E_{t1} = 0. \qquad [2.14]$$

If, finally, the conductivity of *medium 1* is finite and current flows therein, and/or an e.m.f.-inducing field exists in the conductor (both in the tangential direction of E_{t1}), eqn [2.13] holds.

Considering field components *normal* to the boundary it is almost self-evident that normal electric flux density D_n will be continuous across the boundary unless surface charge exists at the boundary. A simple application of Gauss's law, eqn [2.1], applied to an infinitesimally short cylindrical volume with axis normal to the boundary in Fig. 2.5 confirms this, and also leads to the more general result that

$$D_{n2} - D_{n1} = \rho_s, \qquad [2.15]$$

where ρ_s is the density of any surface charge at the boundary.

If medium 1 is a conductor, any normal component of flux density existing in medium 2 must terminate on surface charges located at the boundary with medium 1, and eqn [2.15] reduces to

$$D_{n2} = \rho_s. \qquad [2.16]$$

Equation [2.16] may be beneficially illustrated by visualising the electric flux lines emanating from a long, uniformly charged wire constituting the inner conductor of the coaxial cable of Figs. 2.3 and 2.4.

2.3 Energy Storage and Force Development in the Electrostatic Field

In this section we shall establish that energy is inherently stored in a region of dielectric subjected to an electric field. That this is so is apparent from simple experiment and intuitive reasoning as follows.

Suppose that a parallel-plate capacitor is constructed by taking two large identical flat plates, laying one upon a horizontal surface and suspending the other a small, uniform distance above. If the initially uncharged capacitor is subsequently connected to a d.c. voltage source via a switch and an ammeter (Fig. 2.6), three effects will be observed:

(i) the voltage across the capacitor will rise (almost) instantaneously to the value V_{oc} of the source e.m.f.,
(ii) the ammeter will indicate a momentary flow of current away from the positive pole of the voltage source, and
(iii) the suspended plate of the capacitor will be attracted towards the fixed plate.

Let us interpret the observations we have made. The surge of current suggests a movement of charge from the lower plate to the upper plate via the source — this is actually a displacement of *electrons* in the opposite direction. The rate of charge transference $dq/dt = i$, and the total charge transferred $Q = \int i \, dt$.

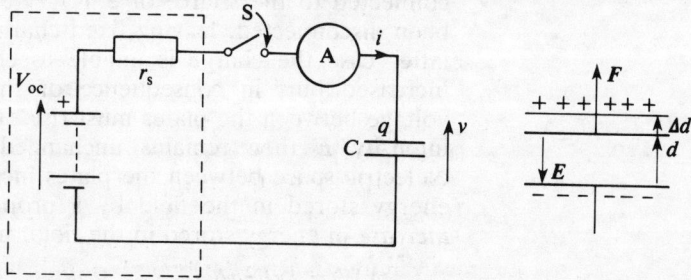

Figure 2.6 Parallel-plate capacitor connected to a voltage source.

During this transfer of charge power flows into the capacitor at the rate vi, hence the total *energy* absorbed by the capacitor is

$$W = \int vi \, dt = \int_{q=0}^{q=Q} \frac{q}{C} dq = \frac{Q^2}{2C} = \frac{1}{2}CV^2 \qquad [2.17]$$

with $V = V_{oc}$.

The energy absorbed by the charged capacitor must reside in the region between the plates subjected to the electric field. For the capacitor construction described one may readily envisage the field as being everywhere uniform at $E = V/d$, directed normally to the parallel plates. In so doing *end effects* are ignored (we shall consider these further in Section 2.3.2).

If the area of the plates is a m^2 the energy stored per unit volume is given by

$$w = \frac{W}{ad} = \frac{1}{2}\frac{CV^2}{ad} = \frac{1}{2}\varepsilon\frac{a}{d}\frac{V^2}{ad}$$

since

$$C = \frac{\varepsilon a}{d}$$

ignoring end effects, see Tutorial Example 2.4. Hence

$$w = \frac{1}{2}\varepsilon E^2 = \frac{1}{2}DE = \frac{1}{2}\frac{D^2}{\varepsilon} \text{ J m}^{-3}. \qquad [2.18]$$

2.3.1 Tensile mechanical stress in the electric field

We have yet to account for the force observed on the plates. We have assumed that the electric field E is directed normally to the plates which consitute equipotential surfaces and so it appears that the mechanical force is directed along the lines of electric field. This is true but is not the whole truth. Mechanical forces are also developed at right-angles to the lines of E but, in the symmetrical structure we have assumed, they have no net effect on the plates.

Consider the separation of the plates to be increased by an incremental distance Δd. The mechanical work to be done in achieving this is

$$W_m = F\Delta d,$$

where F is the mechanical force, distributed as a uniform *stress* over the plate surfaces and applied normally to these surfaces.

The way that the electrical system responds to this relative movement of the plates depends on whether the capacitor is still connected to the source of e.m.f. V_{oc}, or whether the source has been disconnected, leaving fixed charges $\pm Q$ on the plates. In the latter case the charge is unable to change as plate separation is increased but, in consequence of the reduced capacitance, the voltage between the plates must rise. The value of the electric field intensity E then remains unchanged whilst the *volume* of the dielectric space between the plates increases by $A\Delta d$ and the total energy stored in the field by a proportionate amount. Thus the *increase* in energy stored in the field, using eqn [2.18], is given by

$$\Delta W = \tfrac{1}{2}\varepsilon E^2 a\Delta d.$$

Now any energy supplied by the source (zero in this case) must equal the sum of any increase in energy stored in the electric field and the mechanical work done $F\Delta d$, where F is measured *in the direction of the displacement*. Hence

$$0 = \tfrac{1}{2}\varepsilon E^2 a\Delta d + F\Delta d,$$

giving for the mechanical stress developed on the plates:

$$\frac{F}{a} = -\frac{1}{2}\varepsilon E^2 = -\frac{1}{2}DE = -\frac{1}{2}\frac{D^2}{\varepsilon}. \qquad [2.19]$$

The similarity of eqns [2.18] and [2.19] should be noted.

The above virtual work exercise might alternatively have been carried out with the capacitor still connected to the voltage source. In this case the capacitor would be re-dimensioned with no change in voltage whilst the necessary charge *reduction* is associated with a flow of current in the circuit of Fig. 2.6 corresponding to power flow from the capacitor to the source.

The energy balance equation now has three non-zero components and the reader might usefully prove to his own satisfaction that, with $\Delta d \ll d$, one-half of any energy supplied by the source goes to increase the field energy whilst the remaining half accounts for the mechanical work done, the stress developed on the plates being given by eqn [2.19].

2.3.2 *Lateral mechanical stress in the electric field*

In Section 2.3.1 we established that the stress developed at the charge boundaries of the field was directed along the lines of E and tensile in nature. It is reasonable to assume that this stress is communicated through the dielectric and is not restricted to the charged surfaces, as the material must be in a state of mechanical force equilibrium. The situation is analogous to that of a structural tie-bar in tension. The mechanical stresses developed in a region of electrified space are not, however, entirely tensile as described. In fact, another system of stresses is also present, but is not always acknowledged, since its effects are not always apparent. We shall defer formal proof until we examine a similar effect created by a *magnetic* field (Sect. 3.8) and be content for the present with the statement that a second stress system associated with the electrostatic field is directed at right-angles to the field, of values

Basic electrostatic and electromagnetic field theory 31

$\frac{1}{2}\varepsilon E^2 = \frac{1}{2}DE$ equal to the magnitude of the longitudinal stress. The lateral stress acts in such a direction as to force adjacent lines of E apart. Thus we observe *fringing* of the electric field in the vicinity of the edges of the charged capacitor plates.

The complete stress systems across a section of capacitor shown in Fig. 2.7 correspond to two different degrees of alignment of the parallel plates. In Fig. 2.7(a) the plates are fully aligned and, due to symmetry, integration of the stresses over either plate surface gives rise to zero net horizontal force.

This is not so with case (b), however. Again, lateral stresses within the dielectric have no effect at the plate surfaces, but out-of-balance *longitudinal* stresses account for a force to the left on the upper plate and a complementary force to the right on the lower plate. The forces act in such a sense as to seek to achieve the aligned configuration of (a), which corresponds to maximum capacitance. This condition corresponds to *minimum* stored energy if the charge on the (isolated) plates is unchanged, but *maximum* stored energy if the plates are connected to a constant voltage source which allows the charge to increase in proportion to the increase in capacitance.

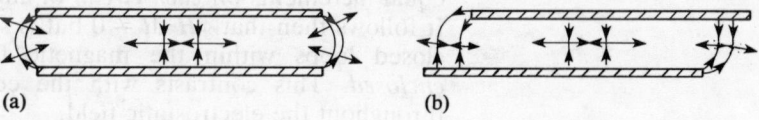

Figure 2.7 Mechanical stress distribution in charged parallel-plate capacitor (a) when the plates are aligned, and (b) when the plates are not aligned.

2.4 The Electromagnetic Field

Electromagnetic field theory is concerned with the effects of *moving* charges rather than of static charges. A flow of charge constitutes an electric current which the early experimenters showed to exhibit magnetic effects, *e.g.* a current flowing in a coil could cause a compass needle to deflect, as would a bar magnet.

A magnetic field system is proposed as the mechanism by which force at a distance may be exerted on an isolated magnetic pole in the same way as an electrostatic field is deemed responsible for force development on a test charge. Magnetic fields are capable of developing force on electrical charges also, and this accounts for the important phenomenon of *electromagnetic induction*. Thus an electric field, manifested as force capability on an isolated charge, may be brought about by a change in the magnetic state of the region as a result of changes in the system currents, or the dispositions of current-carrying conductors or permanent magnets.

The solution of magnetic field problems is generally more complicated than that of electrostatic field problems. An electric field problem is effectively solved when knowledge of the (scalar) electric potential distribution in space is established. In general it is difficult to deduce the *magnetic* field distribution in space due to a flow of current, unless simplifying assumptions are made, particularly when electromagnetic induction is involved in dynamic systems such as rotating machines.

Electrostatic and magnetic field systems have many features in common, as noted in Table 2.1, but there are also significant distinguishing features – notably that the 'flow' of magnetic flux is continuous in space whereas electric flux terminates on charges. These properties are illustrated in Fig. 2.8. In the electrostatic field the charges are regarded as responsible for electric flux 'flows' in space with the electric field intensity related to electric flux density through the permittivity of the medium. As noted in Section 2.2.2, E is readily calculable in complex charge configurations by the identification of equipotential surfaces defined through a knowledge of the electric potential distribution. Since electric potential is a scalar quantity no vector addition is necessary when superimposing the effects of charges at multiple locations.

We should like to be able to do the same for the magnetic field, i.e. calculate the force field H in terms of a scalar *magnetic potential* U, but we are in difficulty regarding the nature of the H field when selected closed paths enclose current. We can readily show experimentally, using a compass needle or iron filings, that the lines of H encircle a long conductor when it carries current, and therefore deduce that the potential energy of a fictitious isolated magnetic pole under the influence of H must change by equal increments *on each circuit* of any route enclosing the current. It follows then that $\oint H \cdot dl \neq 0$ but is some function of current over closed loops within the magnetic field space *when current is enclosed*. This contrasts with the equivalent identity $\oint E \cdot dl = 0$ throughout the electrostatic field.

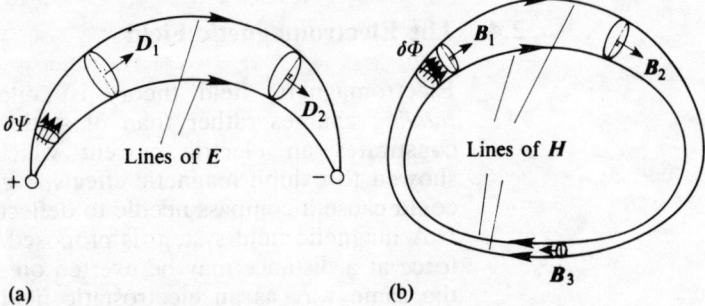

Figure 2.8 Tubes of (a) electric and (b) magnetic flux.

2.4.1 Field types

In fact, the use of a scalar potential quantity to avoid vector addition of field components resulting from multiple sources is directly applicable without ambiguity only to fields whose tubes of flux are discontinuous. Such a field type is described as *lamellar* and is characterised by having zero *curl* for its force-field vector and a non-zero *divergence* for its flux-density field vector. This latter property enables the sources responsible for the field to be identified. On the other hand, the electromagnetic field has the *divergence* of its flux-density field vector equal to zero and the *curl* of its force-field vector finite, enabling the current sources responsible to be located. Such are the characteristics of a *solenoidal* field.

The terms *curl* and *divergence* have been introduced in the above paragraph because they define neatly two rather complicated vector operations. These definitions are given below. The reader may rest content, however, that we shall have little need of their use explicitly in the electromagnetic theory to be further developed in this book.

2.4.2 *Definition of divergence*

'The *divergence* of a vector at a point in the field is a *scalar* quantity given by the integral of the normal component of flux density out of a closed surface surrounding the point divided by the volume Δv enclosed, as Δv tends to zero.'

The concept of divergence is an extension of Gauss's law, eqn [2.1], applied to infinitesimal volumes and, as applied to the electrostatic field, is zero-valued except at the locations of charge where the value of the divergence of the electric flux density D equals the value of the *volume* charge density, a scalar quantity ρ. For example

$$\operatorname{div} D = \left. \frac{\oint D_n \, ds}{\Delta v} \right]_{\Delta v \to 0} = \rho. \qquad [2.20]$$

The divergence of flux density in a three-dimensional field may be shown to equal the scalar sum of incremental changes in the components of flux density in each co-ordinate direction. Thus

$$\operatorname{div} D = \hat{x}\frac{\partial D_x}{\partial x} + \hat{y}\frac{\partial D_y}{\partial y} + \hat{z}\frac{\partial D_z}{\partial z} = \rho, \qquad [2.21]$$

where $\hat{x}$, $\hat{y}$ and $\hat{z}$ are unit vectors directed along rectangular co-ordinate axes Ox, Oy and Oz, respectively.

A vector field whose divergence is zero may have changes in the components of flux density applicable to a particular direction balanced by changes in components with other directions.

2.4.3 *Definition of curl*

'The *curl* of a vector at a point in the field is a *vector* quantity whose magnitude is equal to the line integral of the original vector around the boundary of an area Δs divided by the area Δs, as Δs tends to zero.

$$\text{e.g. } |\operatorname{curl} E| = \left. \frac{\oint E \cdot dl}{\Delta s} \right]_{\Delta s \to 0}. \qquad [2.22]$$

The direction of curl E is normal to the plane of the area Δs.'

The reader should readily perceive that, for the electrostatic field, curl E is everywhere zero. For the electromagnetic field, however, curl H is not zero at locations where current is enclosed, as within a conductor. A vector field whose curl is zero is constant along a line drawn at right-angles to the direction of the field; such a line is identified as an equipotential line.

2.4.4 *Polarity, right-handedness*

In our definition of curl we invoked the line integral over a closed loop of the scalar products of two vectors. In defining a scalar product, Section 2.2, we took into account the orientation of the vectors involved, e.g. electric field intensity E and elemental line length, dl, α being the angle between the two directions. The

34   Electrical machines and drive systems

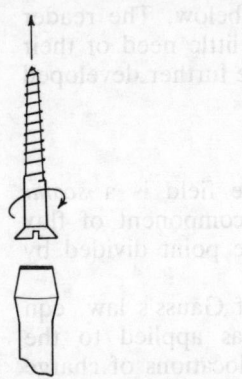

Figure 2.9 Right-hand screw rule.

sense of the integration of the scalar product over a line joining two points was significant in determining the *polarity* of the potential difference between the two points (eqn [2.6]).

If the line integral over a closed loop or 'circuit' has a finite value the polarity of that value is clearly of great importance and will depend upon the direction in which the route has been traversed. The direction of the route prescribes the direction of the elemental line length d*l* at any appropriate point in the field. *The universally adopted convention is that closed loops or contours should be traversed with a right-handed, or clockwise, rotation.* Further, if the resultant closed line integral is employed to formulate a *vector* quantity and yields a positive value, then the direction of the derived vector is that in which a right-handed screw would move. (See Fig. 2.9.)

2.4.5 Curl H, Ampères law

The curl of a three-dimensional field vector may be expressed in terms of components along rectilinear axes Ox, Oy and Oz, as follows, appreciating that the elemental area around which the line integral of the vector quantity is to be evaluated may be defined as the product of two of the three incremental distances Δx, Δy, Δz located at x, y and z.

Thus $\text{curl}_x \mathbf{H}$ = the x-axis component of curl $\mathbf{H}$

$$= \frac{\oint_{PQRSP} \mathbf{H} \cdot d\mathbf{l}}{\Delta y \, \Delta z}.$$

Referring to Fig. 2.10, integrating $\mathbf{H}$ around the periphery of the elemental area $\Delta y \, \Delta z$ right-handed about the direction Ox, we get

$$\oint_{PQRSP} \mathbf{H} \cdot d\mathbf{l} = \underbrace{H_y \Delta y}_{P \to Q} + \underbrace{\left(H_z + \frac{\partial H}{\partial y}\Delta y \right)\Delta z}_{Q \to R} - \underbrace{\left(H_y + \frac{\partial H}{\partial z}\Delta z \right)\Delta y}_{R \to S} - \underbrace{H_z \Delta z}_{S \to P}$$

$$= \left(\frac{\partial H_z}{\partial y} - \frac{\partial H_y}{\partial z} \right) \Delta x \, \Delta y$$

Therefore

$$\frac{\oint_{PQRSP} \mathbf{H} \cdot d\mathbf{l}}{\Delta x \, \Delta y} = \frac{\partial H_z}{\partial y} - \frac{\partial H_y}{\partial z} = \text{curl}_x \mathbf{H}, \text{ if } \Delta x \, \Delta y \to 0. \qquad [2.23]$$

If, as in the general case, curl $\mathbf{H}$ has components in directions Oy and Oz, complementary derivations give

$$\text{curl } \mathbf{H} = \hat{\mathbf{x}}\left(\frac{\partial H_z}{\partial y} - \frac{\partial H_y}{\partial z} \right) + \hat{\mathbf{y}}\left(\frac{\partial H_x}{\partial z} - \frac{\partial H_z}{\partial x} \right) + \hat{\mathbf{z}}\left(\frac{\partial H_y}{\partial x} - \frac{\partial H_x}{\partial y} \right). \qquad [2.24]$$

Relevant questions to ask at this point are: 'What is the nature of curl $\mathbf{H}$? What are its dimensions?' If we propose that $\mathbf{H}$ has the dimensions of A m^{-1} by analogy with the electric field $\mathbf{E}$ which is measured in V m^{-1}, and appreciating that $\mathbf{H}$ is set up by current, curl $\mathbf{H}$ will have the dimensions of current density. Also, as noted above in Section 2.4, $\oint \mathbf{H} \cdot d\mathbf{l} \neq 0$ for a magnetic field line which encloses current, therefore dimensional analysis suggests that

Figure 2.10 Pertaining to the evaluation of $\text{curl}_x \mathbf{H}$.

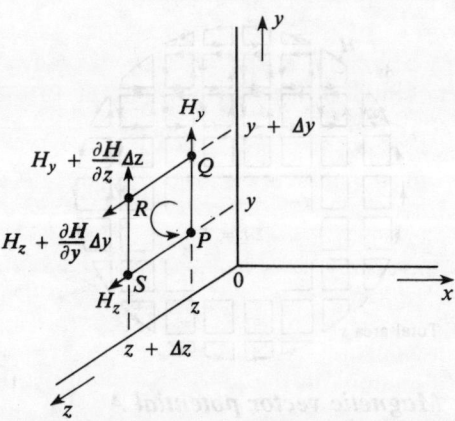

$\oint \mathbf{H} \cdot d\mathbf{l}$ will have the dimensions of current. It is reasonable to suggest therefore, that in an appropriate system of units curl $\mathbf{H}$ should equal the current density vector which may be oriented in any direction relative to the three co-ordinate axes; and additionally that $\oint \mathbf{H} \cdot d\mathbf{l}$ should equal the total current enclosed by the integration path. We shall now presume the former statement to be true and prove the latter result. Hence assume

$$\text{curl } \mathbf{H} = \mathbf{J}. \qquad [2.25]$$

Integrating both these quantities over the same surface area s yields

$$\int_s \text{curl } \mathbf{H} \cdot d\mathbf{s} = \int_s \mathbf{J} \cdot d\mathbf{s}. \qquad [2.26]$$

The area s may be subdivided into elemental areas of which Δs in Fig. 2.11 is typical. By $\int_s \text{curl } \mathbf{H} \cdot d\mathbf{s}$ we mean the sum of curl $\mathbf{H}$ and elemental area scalar products for all such elemental areas within s. Now

$$\text{curl } \mathbf{H} \cdot d\mathbf{s} = \oint \mathbf{H} \cdot d\mathbf{l}$$

around the periphery of each elemental area, and it is evident from Fig. 2.11 that the components of these line integrals on adjacent areas cancel, leaving as contributors to the integral only those on the periphery of the surface s. Mismatch of contour shape presents no problem as, in the limit, $\Delta s \to 0$. Hence the left-hand side of eqn [2.26] has a net value corresponding to $\oint_c \mathbf{H} \cdot d\mathbf{l}$ where c defines the periphery of the entire surface area s. This relationship is known as *Stokes' theorem* applied to the magnetic field.

Now, the right-hand side of eqn [2.26] is the sum I of currents flowing through all the elemental areas enclosed within the contour c and it is apparent that *the actual distribution of the total current is immaterial*. Hence we can write

$$\oint_c \mathbf{H} \cdot d\mathbf{l} = I, \qquad [2.27]$$

where I is the current enclosed within the contour c.

Equation [2.27] is known as *Ampère's law*. The relative directions of integration paths and positive current flow follow the right-handed convention of Section 2.4.4

Figure 2.11 Pertaining to Stokes' theorem applied to the magnetic field due to current.

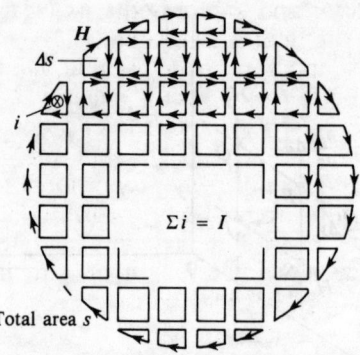

2.4.6 Magnetic vector potential A

We have noted above (Sect. 2.4) that there is no direct equivalent of the scalar quantity, electric potential, in the *magnetic* field due to its solenoid nature. This property is neatly expressed in the statement that the *divergence* of the field of *magnetic flux density* is everywhere zero, i.e.

$$\text{div } \boldsymbol{B} = 0. \quad [2.28]$$

It is a property of vectors that the divergence of the curl of any vector function is always zero. Therefore $\boldsymbol{B}$ may be expressed as the curl of some other vector function. To make this other vector quantity $\boldsymbol{A}$ unique we impose the constraint of solenoidal nature, i.e. we require the *divergence* of $\boldsymbol{A}$ to be zero. Hence the properties of vector $\boldsymbol{A}$ are given by

$$\left. \begin{array}{l} \text{curl } \boldsymbol{A} = \boldsymbol{B} \\ \text{div } \boldsymbol{A} = 0. \end{array} \right\} \quad [2.29]$$

Using eqn [2.25] and the *constitutive equation* for a medium of constant permeability, i.e.

$$\boldsymbol{B} = \mu \boldsymbol{H} \quad [2.30]$$

we write

$$\text{curl curl } \boldsymbol{A} = \text{curl } \boldsymbol{B} = \mu \boldsymbol{J}.$$

By use of another vector identity and the requirement of eqn [2.29] that div $\boldsymbol{A} = 0$ we obtain a further vector equation in which the operator ∇^2 (*del* squared, the *divergence* of the *gradient*) appears, i.e.

$$\nabla^2 \boldsymbol{A} = -\mu \boldsymbol{J}.$$

This may be expressed in rectangular co-ordinates as

$$\hat{\boldsymbol{x}} \nabla^2 A_x + \hat{\boldsymbol{y}} \nabla^2 A_y + \hat{\boldsymbol{z}} \nabla^2 A_z = -\mu (\hat{\boldsymbol{x}} J_x + \hat{\boldsymbol{y}} J_y + \hat{\boldsymbol{z}} J_z)$$

and presented as the vector sum of three scalar equations, each taking the form of a standard identity known as *Poisson's equation*. The solution for $\boldsymbol{A}$ is correspondingly

$$\boldsymbol{A} = \frac{\mu}{4\pi} \int_v \frac{\boldsymbol{J}}{r} \mathrm{d}v \quad [2.31]$$

where $\mathrm{d}v$ represents an elemental volume of current-carrying

Basic electrostatic and electromagnetic field theory 37

conductor and r represents its (scalar) distance from the point at which the field is evaluated.

It is apparent from eqn [2.31] that each element of current contributes to the vector quantity A which pervades all space. The *direction* of a component of A at any point corresponds to the direction of current flow in the contributing element, and its *magnitude* is inversely proportional to the separation distance.

If the current flow I corresponds to that in a thin wire of uniform cross-sectional area and for which dl is an elementary distance along the length of wire in the direction of the current flow, eqn [2.31] may be rewritten

$$A = \frac{\mu I}{4\pi} \int \frac{dl}{r}. \qquad [2.32]$$

The vector A is known as the *magnetic vector potential* and has value in solving complex magnetic field problems. We shall make use of it on account of its fundamental qualities and the compact mathematical formulations which its use permits. However, the derivation of A for simple magnetic circuit configurations is unjustifiably cumbersome, therefore we shall make use of it only by its application *in reverse*. Thus, having first calculated the magnetic field and flux distribution by a more simple technique we shall take advantage of a property which follows from the fundamental relationship that $B = \mathrm{curl}\, A$, namely

$$\Phi = \int_s B \cdot ds = \oint_s A \cdot dl \qquad [2.33]$$

where Φ is the magnetic flux through surface s bounded by contour c, illustrated in Fig. 2.12.

Thus A is the vector quantity whose line integral around a closed loop is equal to the enclosed magnetic flux.

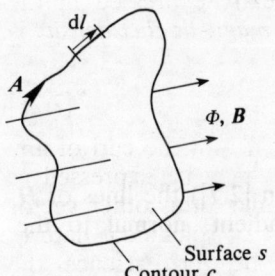

Figure 2.12 Pertaining to the relationship between magnetic vector potential A, flux Φ and flux density B.

2.4.7 Magnetic scalar potential U

We have now compiled two alternative procedures through which magnetic fields in media of constant permeability may be related to the responsible currents, e.g. Ampère's law eqn [2.27], and the use of magnetic vector potential A, eqn [2.31] via eqn [2.29]. It was suggested in Section 2.4 that we might be able to use a *scalar* magnetic potential concept analogous to electric scalar potential if we restrict any closed contours we consider to those embracing zero current. This we can do by introducing *barrier surfaces* defining 'no-go' areas for the paths of integration. Such a barrier surface will make contact with the current boundary and extend radially to infinity for the isolated, lossy conductor depicted in Fig. 2.13. From Ampère's law, eqn [2.27], we get

$$\oint_{PQRSP} H \cdot dl = 0.$$

An expression analogous to eqn [2.6] for the electrostatic field may be created, acknowledging that a magnetic scalar potential difference of value I must exist across the infinitely thin barrier, i.e.

$$\int_P^Q H \cdot dl = \int_S^R H \cdot dl = U_P - U_Q = I \qquad [2.34]$$

where U_P, U_Q are the scalar magnetic potentials at P and Q. In this context the barrier magnetic potential difference $U_P - U_Q$ is equal to the *magnetomotive force (m.m.f.)* of the magnetic circuit, equal to the current I and responsible for the magnetic field lines in the medium which terminate on the barrier. The barrier defines surfaces of magnetic equipotential at either side, with the potential difference I. From consideration of the uniformity of H over the integral path regions PQ and SR, we can produce other surfaces of equal magnetic potential, radiating from the location of the current like the spokes of a wheel. Such an equipotential is shown in Fig. 2.13 for $U = I/2$ relative to equipotential QR.

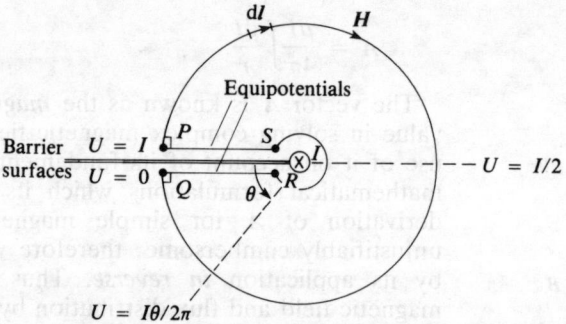

Figure 2.13 Isolated current-carrying conductor with a magnetic barrier.

By analogy with the electrostatic field (eqn [2.7]) the lines of H are directed down the scalar potential gradient, normal to the equipotential surfaces. Hence

$$H = -\text{grad } U \; (= -\nabla U \text{ in vector notation}). \qquad [2.35]$$

It may be deduced from Fig. 2.13 that the value of the magnetic potential of any equipotential surface, relative to that designated $U = 0$, is equal to the value of current enclosed by the angular displacement θ of the equipotential surface with respect to the reference. This property, and the principle of superposition, may be applied to evaluate the field distribution of multi-conductor systems *via* scalar magnetic potential.

2.4.8 *Magnetic field intensity due to current in a long, straight wire*

Figure 2.13 shows the concentric circular nature of the H-field lines external to the current located at the centre of the circular loci. Due to symmetry the value of H at all points on the circumference of a circle of radius r, with direction right-handed about that of the current I, is given by

$$H_r = \frac{I}{2\pi r} \qquad [2.36]$$

since, from eqn [2.34],

$$\int_P^Q \mathbf{H} \cdot d\mathbf{l} = H_r \int_P^Q dl = 2\pi r H_r = I.$$

Use of Ampère's law, eqn [2.27], gives the same result.

2.4.9 *Magnetic field intensity due to a twin-conductor transmission line*

Consider two long, thin parallel conductors, as in Fig. 2.14. sepa-

Figure 2.14 Magnetic equipotential surface due to two parallel conductors.

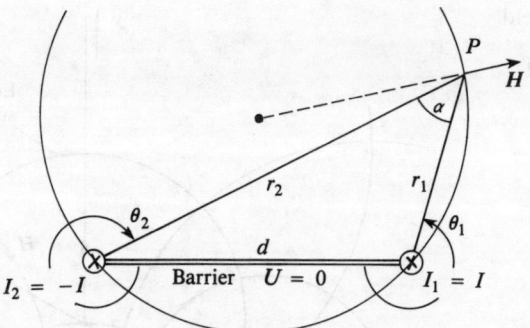

rated by distance d and carrying currents $I_1 = I$ and $I_2 = -I$.

The H field is most conveniently calculated using the concept of magnetic scalar potential to define equipotential surfaces. An appropriate plane barrier surface may be introduced in the field space between the currents. With respect to the lower surface of the barrier the magnetic scalar potential at point P may be stated as

$$U_P = I_1\left(\frac{\theta_1}{2\pi}\right) - I_2\left(\frac{\theta_2}{2\pi}\right) = I\frac{(\theta_1 + \theta_2)}{2\pi}.$$

Now $(\theta_1 + \theta_2) = \alpha + 3\pi$ and, if P traces a locus such as to keep α constant it will trace an equipotential surface. It is a property of a circle that the angle subtended by a chord at any point on the circle on the same side of the chord is fixed – hence the locus of P is *circular* where it defines an equipotential surface, the family of such circles having the barrier surface as a common chord. The equipotential circles are *co-axal* with centres on the Oy axis, and give rise to an *orthogonal* set of co-axal, Ox axis-centred circles defining the H field. Depicted in Fig. 2.15, these *magnetic field lines* mirror the *electrostatic equipotential surfaces* developed when complementary charges exist on the line conductors. Verification of the electrical field distribution is the subject of Tutorial Example 2.2.

2.4.10 *Mechanical stresses in the magnetic field, force between current-carrying conductors*

We shall leave a detailed discussion of (Maxwell's) stresses in the magnetic field until ferromagnetic materials are considered in Sections 3.5 and 3.8, but we will take a hypothetical view that such a system of mechanical stresses is likely to exist by analogy with the known properties of the electric field, Section 2.3. Further, using the idea of *duality*, we might expect the magnitude of the stress to be $\frac{1}{2}\mu H^2 = \frac{1}{2}BH$ N m^{-2}, both along and normal to the lines of field intensity, corresponding with $\frac{1}{2}\varepsilon E^2 = \frac{1}{2}DE$ for the electric field. If we are able to use such a stress system to confirm a definitive relationship between the magnitudes of current and force between two long, parallel conductors, separated by air or free space and carrying identical currents, then we should be able to validate both the stress system and the method of calculating the magnetic field. In the derivation of Ampère's law, eqn [2.27], use was made of an assumption stated in eqn [2.25]. Because of this

Figure 2.15 Magnetic field lines and equipotential surfaces for twin, parallel conductors carrying complementary currents.

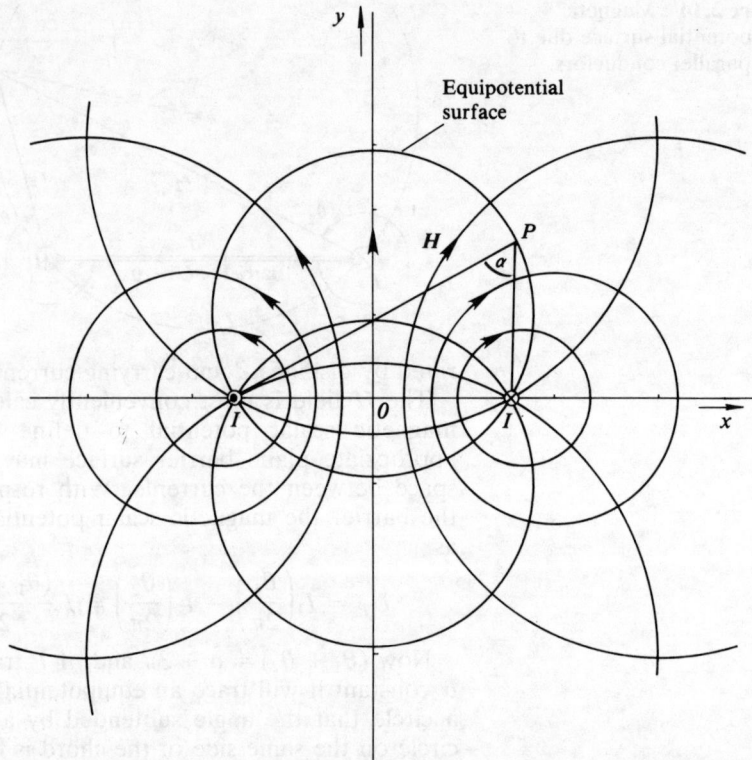

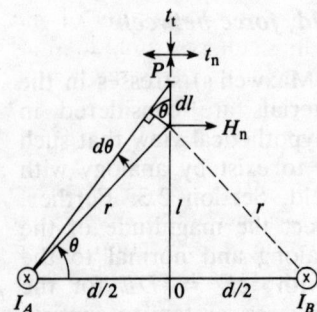

Figure 2.16 Calculation of mechanical stresses over a plane equidisant from twin, parallel conductors.

element of doubt we shall use Ampère's law to predict the field and consequential mechanical stresses over an appropriate area pertinent to evaluating the force between conductors.

Consider a one-metre length of parallel twin-conductor line with separation distance d, shown in Fig. 2.16. If P defines a line parallel to the conductors located on the plane surface which bisects d normally and extends to $\pm$ infinity, the normal stresses integrated over the whole of this surface will equal the force impressed on each conductor, tending to attract or repel according to the relative directions of current flow. If the currents I_A and I_B are equal and co-directional, as shown in Fig. 2.16, the resultant magnetic field at P, $H_P = H_A + H_B$ is directed normally to the plane and of value, from eqn [2.36]

$$H_P = \frac{I_A}{2\pi r}\cos\theta + \frac{I_B}{2\pi r}\cos\theta = \frac{I}{\pi r}\cos\theta,$$

where r is the radial distance of P from either current I.

We anticipate that such a field will give rise to a *tensile* mechanical stress at P normal to the plane of value

$$t_n = \tfrac{1}{2}\mu_0 H_P^2 \qquad [2.37]$$

and a *compressive* stress at P directed vertically, normal to the field, of the same value. Due to symmetry these latter stresses when integrated over the infinite surface of the plane sum to zero, whereas the total *normal* force over the plane extending from $\theta = -\pi/2$ to $\theta = +\pi/2$ for one metre length of line is given by

$$F = \int_{l=-\infty}^{l=\infty} t_n \, dl = \frac{1}{2}\mu_0 \frac{I^2}{\pi^2} \int_{l=-\infty}^{l=\infty} \frac{\cos^2\theta}{r^2} \, dl.$$

Now $r\,d\theta = dl\cos\theta$ and

$$\cos\theta = \frac{d}{2r}.$$

Hence

$$F = \frac{1}{2}\mu_0 \left(\frac{I}{\pi}\right)^2 \int_{-\pi/2}^{\pi/2} \frac{\cos^2\theta}{r^2} \frac{2r^2}{d} \, d\theta$$

$$= \frac{\mu_0}{2d} \frac{I^2}{\pi}. \qquad [2.38]$$

Substituting $\mu_0 = 4\pi \times 10^{-7}\,\mathrm{H\,m^{-1}}$, $d = 1$ m and $I = 1$ A we get

$$F = 2 \times 10^{-7}\,\mathrm{N}.$$

Now the unit of electric current, the ampere, is defined as *'that current which, if flowing in two infinitely long, parallel wires in vacuum separated by 1 m, produces a force of 200×10^{-9} N per metre of length.'* Hence the above result would appear to confirm both the validity of Ampère's law and our assumption regarding mechanical stress developed *along* the line of **H**. But what about the supposed component of stress *normal* to the direction of **H**? This is readily confirmed by the simple expedient of reversing the direction of flow of current in one of the conductors. Now the resultant field at P is directed *tangentially* to the infinite plane surface and of value

$$H_P = \frac{I}{\pi r}\sin\theta.$$

Due to symmetry the tensile stresses along the infinite plane now cancel, but the compressive stresses directed normally to the plane surface remain, directed so as to press adjacent field lines apart. The general field plot is shown in Fig. 2.15. It is left to the reader to prove that the stresses *tangential* to the direction of the field, if of value $\frac{1}{2}\mu_0 H_P^2$, when integrated over the infinite surface, give rise to an identical expression to eqn [2.38] for the magnitude of the force per metre between long, parallel conductors carrying identical currents in opposite directions. The direction of the force is now repulsive, instead of attractive as with co-directed currents.

2.4.11 *Inductance, flux-linkage*

As we have demonstrated in Section 2.4.10, that a medium subjected to a magnetic field is in a state of mechanical stress, it is likely that the magnetic field, like the electric field, Section 2.3, is capable of storing energy. Just as a *capacitor* may be regarded as a device which stores energy in an electric field and has a defined charge/voltage ratio, so an *inductor* is a device which can store energy in a magnetic field and has a defined *flux-linkage*/current ratio. *Inductance* is an important concept which significantly affects the behaviour of electric circuits under changing conditions – it relates a component of terminal voltage to the rate of current change.

The *flux-linkage* λ of a circuit is simply the magnetic flux embraced by the closed circuit. Where the circuit consists of coiled wire in which each turn links effectively the same value of flux, the flux-linkage is given simply as the product of the flux per turn and the number of turns on the coil. Such *concentrated* coils will be considered in Chapter 3 and 4 in the context of solenoid and transformer windings, generally incorporating ferromagnetic material within their magnetic circuits. For present considerations of inductance we shall restrict our attention to single loops, such as are provided by a single-turn coil or a length of twin, parallel conductor transmission line.

Where more than one electric circuit share the same magnetic field space, such that current flowing in one circuit affects flux linkage in another, *mutual inductance* exists. Not all the magnetic flux created by the energised circuit will create flux linkage with the other. The flux common to both is described as *mutual flux*; that which links the energised circuit but not the other is *leakage flux* associated with the energised circuit.

A clear distinction between the characteristics of the different designations of flux is important in the study of electromagnetic devices, due to the interlinking of the electric and magnetic circuits. The *total* flux linkage of a circuit gives rise to *self inductance*; that due to leakage flux gives rise to *leakage inductance*. *Mutual inductance* is that ratio of the flux linkage in one circuit due to the mutual flux divided by the current in the other circuit responsible for the magnetic field.

2.5 Electromagnetic Induction

Electromagnetic induction accounts for the ability of a magnetic field to drive current around a closed conducting circuit subject to its influence. The property of a magnetic field to impress force upon electrical charges is crucial to the electromagnetic energy transfer and conversion processes involved in electromagnetic transformers and electromechanical devices.

2.5.1 *The Lorentz equation*

The quantitative relationship between the variables involved in electromagnetic induction is most generally and neatly expressed by the *Lorentz equation* for the force on a unit charge at a point in the magnetic field, i.e. the electric field intensity E, which is stated thus

$$E = -\frac{\partial A}{\partial t} + v \times B \qquad [2.39]$$

where A is the magnetic vector potential, B is the magnetic flux density and v is the velocity of the charge at the point under examination. The charge will, in many circumstances, correspond to one of many loosely bound electrons associated with conducting material which itself provides the motion.

Each of the terms on the right-hand side of eqn [2.39] is important and worthy of close examination. The first term is evidently present whenever a changing magnetic field is to be found, irrespective of the presence or absence of conducting material. The use of a partial derivative suggests that A might be a function of *position* as well as time. The second term, involving a vector product, suggests that the conductor electrons through which the electric field is manifest may have a component of velocity at right-angles to the magnetic flux density, the force on the electrons being developed at right-angles to both. The effect will be maximised when the motion of the conductor is at right-angles to the B field, the force on the electrons being then along the axis of the conductor wire.

2.5.2 *Faraday's law, Lenz's law*

Many practical applications of electromagnetic induction involve the evaluation of e.m.f. around a closed circuit. This is obtained by the integration of the e.m.f.-inducing field given by eqn [2.39] over the circuit involved. Now we have already established that A is the vector quantity whose line integral around a closed loop equals the enclosed magnetic flux, eqn [2.33]. Integration of the $-\partial A/\partial t$ term in eqn [2.39] over a closed circuit gives rise to *Faraday's law* which states that the e.m.f. due to a time-changing magnetic flux is equal to the rate of change of flux enclosed by the circuit. Such an e.m.f. is called a 'transformer' e.m.f. when motion is not involved

$$E = \oint E \cdot dl = -\oint \frac{\partial A}{\partial t} \cdot dl = -\oint \frac{dA}{dt} \cdot dl$$

$$= -\frac{d\Phi}{dt}. \qquad [2.40]$$

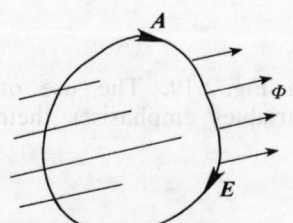

Figure 2.17 Spatial relationship between magnetic vector potential A, electric field intensity E and magnetic flux Φ.

E is *oppositely directed* to A which is right-handed about the direction of the flux Φ. The *reference* direction for E is the same, however, as that for A shown in Fig. 2.17, accounting from the minus sign in eqns [2.39] and [2.40]. If the circuit e.m.f E acting in the direction of E causes current to flow in the same direction, the magnetic field produced by the current (right-handed about the direction of the current) will act to oppose the change in flux linking the circuit which was responsible for E. This requirement is known as *Lenz's law*. The e.m.f. E should not be confused with the electric field intensity E. E is measured in volts, E in volts per metre. E, like Φ, is a scalar quantity; E and A are vectors.

Strictly speaking Φ in eqn [2.40] should be replaced by λ, the flux *linkage* of the circuit to allow for multiple encirclements of common flux Φ by identical winding turns connected in series, following the argument of Section 2.4.11. The relationship between flux linkage and current through inductance may also be invoked to relate the electrical circuit concepts of voltage and current, through inductance, as follows.

$$E = -\frac{d\lambda}{dt} = -\frac{d\lambda}{dI}\frac{dI}{dt} = -L\frac{dI}{dt} \qquad [2.41]$$

for a circuit possessing self inductance L, and

$$E_1 = -\frac{d\lambda_1}{dt} = -\frac{d\lambda_1}{dI_2}\cdot\frac{dI_2}{dt} = -M_{12}\frac{dI_2}{dt} \qquad [2.42]$$

$$E_2 = -\frac{d\lambda_2}{dt} = -\frac{d\lambda_2}{dI_1}\cdot\frac{dI_1}{dt} = -M_{21}\frac{dI_1}{dt} \qquad [2.43]$$

for circuits 1 and 2 possessing flux linkage resulting from current in circuits 2 and 1 respectively. Also

$$E_{11} = \frac{d\lambda_{11}}{dt} = -\frac{d\lambda_{11}}{dI_1}\cdot\frac{dI_1}{dt} = -L_{11}\frac{dI_1}{dt} \qquad [2.44]$$

for the contribution made to the e.m.f. of circuit 1 by its leakage flux.

The minus sign in eqns [2.41] to [2.44] is appropriate to the understanding that the circuit current and e.m.f have similar reference directions along the winding and that, where mutual coupling is employed, the induced e.m.f. and current flows are similarly directed with respect to *dots* appended to indicate winding sense (Fig. 2.18).

In conventional circuit analysis the minus sign inherent in eqns [2.47] to [2.50] is dispensed with the reference direction of the induced e.m.f. reversed. It is now *strictly* inappropriate to call this *induced voltage* an e.m.f. Expressions analogous to eqns [2.41]–[2.44] become

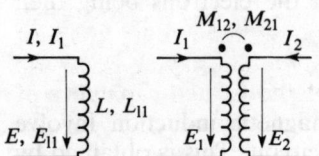

Figure 2.18 Electromotive force and current reference directions for self and mutual inductance.

$$\begin{aligned} v &= L\frac{di}{dt} \\ v_1 &= M_{12}\frac{di_2}{dt} \\ v_2 &= M_{21}\frac{di_1}{dt} \\ v_{11} &= L_{11}\frac{di_1}{dt} \end{aligned} \qquad [2.45]$$

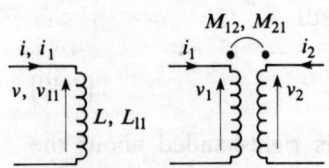

Figure 2.19 Reference directions of induced voltage and current for self and mutual inductance.

with the reference directions shown in Fig. 2.19. The use of lower-case symbols for the electrical variables emphasises their instantaneous nature.

2.5.3 Flux linkage or flux cutting?

All the relationships of eqns [2.45] are normally stated in the context of static circuits in which the changing flux linkage is not due to relative motion between circuits and flux. Such an influence on the e.m.f. induced is apparently accounted for by the $\boldsymbol{v} \times \boldsymbol{B}$ term in eqn [2.39]. Since the $\boldsymbol{A}$ and $\boldsymbol{B}$ fields are interrelated, being alternative descriptions of the same magnetic field, eqn [2.29], it seems likely that both terms on the right-hand side of eqn [2.39] may be alternative manifestations of the same phenomenon. Much discussion has centred around the relative merits of 'change in flux-linkage' and 'flux-cutting' with regard to electromagnetic induction; the key to the problem resting on the recognition that $\boldsymbol{v}$ represents the motion of the conductor of interest *relative to a reference frame* with respect to which the fields $\boldsymbol{A}$ and $\boldsymbol{B}$ are described. The most obvious reference frame is, perhaps, a stationary one – one in which $\boldsymbol{v}$ is the absolute velocity in space of the conductor. However, a more convenient frame of reference may be one which moves synchronously with the conductor so that the velocity term in eqn [2.39] disappears. It is important, then, to recognise that A will have its time dependency affected by the motion of the reference frame.

In our interpretation of eqn [2.39] in subsequent chapters we shall adopt whichever procedure is most appropriate to our purpose at the time.

2.6 Energy Storage and Flow in the Electromagnetic Field

Just as we have deferred a detailed consideration of mechanical stress development in the magnetic field until Chapter 3, so we shall postpone until later formal proof of the related capability of the magnetic field to store energy (Sect. 3.7). However, if for the present we accept that, by analogy with the electrostatic field, such energy storage facility exists, it is likely to have density $\frac{1}{2}BH = \frac{1}{2}\mu H^2$ J m^{-3} (c.f. eqn [2.18]).

If the magnetic field configuration is known it is feasible to calculate the energy distribution in space and the total field energy of the system, appropriate to current values. When currents are changing an interchange of energy takes place between the conductors which constitute the electric circuit and the field space. The power transfer accounting for the change in stored field energy external to the conductors takes place at the conductor surface. The instantaneous electrical power transfer for the whole of the effective field space is given by the product of the conductor current and the electric field induced in the conductor by the changing magnetic field, integrated along its length, i.e. the e.m.f. induced by the field.

As the concept of *inductance* is relevant in relating induced e.m.f. to current it can with advantage be used to account for the *global* energy storage in a magnetic field system. It cannot, however, account for the *distribution* of that energy within the field space. Derivation of the relationship between current, inductance and stored energy for a single-coil system parallels the argument employed in deriving eqn [2.17] for a capacitor. During the transfer of energy from the electric circuit to the magnetic field space (or vice versa) current flows with instantaneous value i, an e.m.f. is induced which has magnitude

$$E = \frac{d\lambda}{dt} = \frac{Nd\Phi}{dt} = L\frac{di}{dt}$$

and the instantaneous electrical *power* transfer

$$p = Li\frac{di}{dt}.$$

Hence the total *energy* transferred, corresponding to a change in current from zero to final value I, which equals the total final energy stored in the field, is

$$W = \int_{i=0}^{i=I} Li\frac{di}{dt} = \frac{1}{2}LI^2. \qquad [2.46]$$

It is left for the reader to prove that a two-coil system with self and mutual inductances L_1, L_2 and M, has total energy stored in the field given by

$$W = \tfrac{1}{2}L_1 I_1^2 + \tfrac{1}{2}L_2 I_2^2 \pm M I_1 I_2 \qquad [2.47]$$

with the $\pm$ sign dependent upon whether the fields due to the individual coil currents I_1 and I_2 support or oppose each other.

In the derivation of eqns [2.46] and [2.47] linearity of the magnetic circuit was assumed, i.e. that the inductance coefficients are constant. With magnetic circuits incorporating ferromagnetic materials, however, this is inappropriate. Evaluation of $\Sigma i d\lambda$, however, always gives the correct result for the *total* stored magnetic energy.

2.6.1 *The Poynting vector*

Recalling that the vector product of two vectors is another vector, it is of interest to evaluate the product $E \times H$, E and H being the electric and magnetic field intensity vectors at a point in space.

Since E is expressed in V m^{-1} and H in A m^{-1} the resulting vector has the dimensions of *power density* (W m^{-2}) and is directed at right-angles to the plane of EH. The vector product $E \times H$ is called the *Poynting vector*. When directed appropriately, the normal component of the Poynting vector, integrated over the surface of a closed volume, may be shown to yield the net electrical power crossing the boundary, either to undergo conversion into alternative forms of energy or to increase the stored energy within the enclosed volume. Thus, by definition

$$S = E \times H. \qquad [2.48]$$

The Poynting vector S is not unique in yielding this result on integration but is the most commonly used quantity to describe electrical power density over surfaces, partly on account of its simplicity. The Poynting vector provides a very clear visual image of the means whereby energy propagation is achieved within the electromagnetic field and, in simple field configurations, it leads to verifiable solutions without complicated algebra.

2.6.2 *Power transfer at the surface of a conductor*

For example, consider conditions at the surface of a long, thin conductor of circular cross-section and radius r, which is one of a pair of parallel conductors carrying equal currents $\pm I$ and for which the loop self inductance per unit length is L.

At the surface of the conductor the magnetic field intensity will be $H = I/2\pi r$, using Ampère's law, eqn [2.27]. H is directed circumferentially, right-handed about the direction of the current. If the current is increasing an electric field will be set up along the length of the conductor in such a direction as to oppose the increase in current and of value

$$E = |E| = \frac{L}{2}\frac{dI}{dt}$$

at the surface of the conductor.

The Poynting vector at the surface of the conductor, as shown in Fig. 2.20, is directed radially away from the conductor and is of uniform value

$$EH = \frac{EI}{2\pi r} \text{ W m}^{-2}.$$

Integration over the surface of one metre length of the conductor gives

$$2\pi r EH = EI \text{ W m}^{-1}$$

which represents the instantaneous power flow across the boundary between the electric circuit and the magnetic field space. This conductor is one of a pair, with energy crossing the boundary to increase the energy stored in the field at the rate

$$p = I\frac{L}{2}\frac{dI}{dt} = \frac{1}{2}\frac{d}{dt}\left(\frac{1}{2}LI^2\right) \text{ W m}^{-1}.$$

This is half the rate of increase of stored magnetic energy corresponding to a *loop* inductance of L H m^{-1}. The other half is, of course, transferred by the second conductor of the pair.

When the current *decreases* with increasing time, E reverses direction and energy is recovered by the electric circuit from the field space. As the E and H field distributions for more complex configurations are capable of mathematical formulation, the electrical power transfers over other defined surfaces in space are calculable, if not always with such relative ease.

Figure 2.20 Power flow at the surface of a circular conductor.

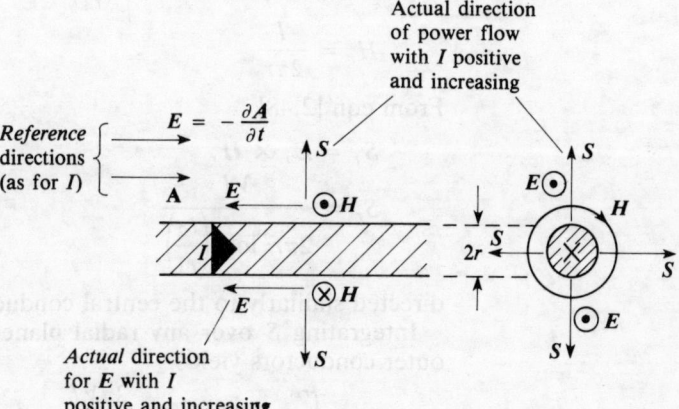

If we now consider the conductor to possess resistance and the uniform current flow to be in the direction shown in Fig. 2.20, the direction of the electric field *along* the conductor will be the same as that of the current and be of value $I/\sigma\pi r^2$ V m^{-1}. The H field at the conductor surface remains at $I/2\pi r$ A m^{-1}, directed as before, but the Poynting vector integrated over the surface now yields power *inflow* to the conductor from the field space at the rate

$$\frac{I^2}{\sigma 2\pi^2 r^3} \cdot 2\pi r = \frac{I^2}{\sigma\pi r^2} = I^2 R,$$

where R is the resistance of a one-metre length of conductor. The Poynting vector thus indicates the power dissipation in the conductor, i.e. the so-called 'copper loss'.

2.6.3 *Power flow over a coaxial d.c. cable*

The Poynting vector is also of use in describing the power flow over transmission networks. One of the simplest field configurations which serves to demonstrate this is that of a coaxial cable in which the electric field due to surface charges on the inner and outer conductors is radial, supporting the system voltage considered initially to be uniform at the constant value V throughout the loss-free cable. The magnetic field is directed tangentially at right-angles to any radius between the inner and outer conductors (set up by d.c. current I flowing in the inner conductor) and shown in Fig. 2.21.

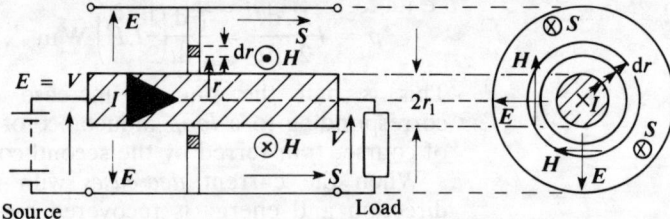

Figure 2.21 Power flow over a coaxial cable carrying direct current.

Thus from eqns [2.10], [2.12] and [2.36], at radius r

$$E_r = \frac{Q'}{\varepsilon 2\pi r} = \frac{1}{\varepsilon 2\pi r} \frac{V 2\pi \varepsilon}{\log_e\left(\frac{r_2}{r_1}\right)} = \frac{V}{r \log_e\left(\frac{r_2}{r_1}\right)}$$

$$H_r = \frac{I}{2\pi r}.$$

From eqn [2.48]

$$\mathbf{S}_r = \mathbf{E}_r \times \mathbf{H}_r$$

$$S_r = \frac{VI}{2\pi r^2 \log_e\left(\frac{r_2}{r_1}\right)}$$

directed similarly to the central conductor current flow.

Integrating S over any radial plane area between the inner and outer conductors yields

$$p = \int_{r_1}^{r_2} S_r 2\pi r \, dr = \frac{2\pi VI}{2\pi \log_e\left(\frac{r_2}{r_1}\right)} |\log_e r|_{r_1}^{r_2} = VI.$$

This interesting result demonstrates that the transmitted electrical power flow is everywhere along the length of cable accounted for within the field space between the conductors. The power *density* is greatest near the inner conductors where both electric and magnetic field strengths are greatest.

If now the cable conductors are considered to possess resistance the *radial* electric field will reduce progressively along the length of the cable and a longitudinal or *axial* component will develop at the surfaces of the inner conductor and sheath. Evaluation of the integral of the Poynting vector over the surface of the conductor under examination provides the power input necessary to satisfy the corresponding copper loss.

For example, if the inner conductor carries the current I uniformly over cross-sectional area πr_1^2 with conductivity σ, the copper loss per unit length within the inner conductor is given by

$$I^2 R = I^2 \frac{1}{\sigma \pi r_1^2} \text{ W m}^{-1}. \qquad [2.49]$$

The axial component of E at the inner conductor surface due to the flow of current through the resistive material is given by

$$E_a = I \frac{1}{\sigma \pi r_1^2}.$$

Hence S directed *inwards* at the surface of the inner conductor has the uniform magnitude

$$S = E_a H = \frac{I}{\sigma \pi r_1^2} \frac{I}{2\pi r_1}$$

which, on integration over the surface of a one-metre length of inner conductor gives

$$\int S \, ds = S 2\pi r_1 1 = \frac{I^2}{\sigma \pi r_1^2}.$$

This expression agrees with eqn [2.49]. A similar calculation carried out at the outer boundary of the dielectric accounts for the copper loss within the sheath.

It is apparent from the foregoing that, in general, the electric power density vector at the surface of a uniform conductor, expressed in terms of the Poynting vector, has components directed along, and at right-angles to, the conductor surface – i.e. both axially and radially. The first component identifies with the *power transmitted in the axial direction* by the transmission system due to the interaction between its magnetic field and the *radial* electric field between the conductors. The second relates to *power transfer between the conductor and the field space*, normal to their common boundary, which accounts for changes in stored electromagnetic field energy (due to conductor current changes) and conductor resistance losses. This power transfer is due to the interaction between the magnetic field and the *axial* electric field at the conductor surface. The feature common to both power flows is the *tangential* magnetic field due to the axially directed conductor current.

2.6.4 *Power transmission over a twin-conductor transmission line*

Figure 2.22 shows a map, reproduced from Krauss[1] by permission, of the axial power flow density over the field space surrounding an infinite twin-conductor transmission line for which the conductors are 25.4 mm in diameter and 76.2 mm spaced between centres. The contours indicate relative power density. Increasing the ratio of conductor spacing to conductor diameter will increase further the concentration of power density near the conductor surfaces.

2.7 Tutorial Examples

2.1 If the dielectric breakdown strength of air is 3 MV m^{-1} determine the

50   Electrical machines and drive systems

Figure 2.22 Power flow map for an infinite two-conductor lossless transmission line with a potential difference of 10 V r.m.s. The contours give the power density (Poynting vector) in watts per square metre. (Reproduced by permission from Krauss[1].)

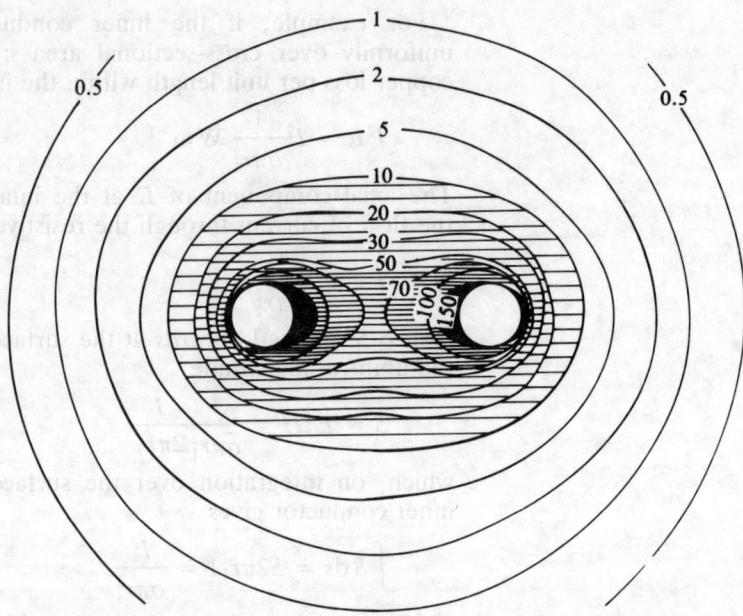

2.1 If the dielectric breakdown strength of air is $3\,\mathrm{MV\,m^{-1}}$ determine the minimum radius of curvature permissible for an isolated spherical conductor maintained at an absolute potential of 150 kV. Consider the complementary charge to be distributed over a concentric sphere of infinite radius.

[50 mm]

2.2 Show that the electrostatic field developed by a pair of complementarily charged, long, parallel conductors of circular cross-section, located at $x = \pm d/2$, $y = 0$, and of radius a gives rise to a family of cylindrical, co-axal equipotential surfaces with centre at

$$x = \pm \frac{d}{2}\left(\frac{K^2+1}{K^2-1}\right), \quad y = 0 \text{ and radius } \frac{dK}{K^2-1}$$

where K is the ratio of distances of point P from equivalent line charges.
Note: Consider initially the conductors of finite radius to be replaced by lines of charge having equal and opposite charge densities (C m^{-1}). Such theoretical line charges will not be at the centre of conductors of circular cross-section which establish the same field distribution, and the charge distribution over the surface of the latter will not, in general, be uniform.

The field distribution is illustrated by Fig. 2.18 with the field lines and equipotential surfaces interchanged.

2.3 Show that the capacitance/metre of a concentric cable with inner conductor of radius a, outer conductor of radius c, and mixed dielectrics with common boundary at radius b such that the relative permittivities are ε_1 for the inner dielectric and ε_2 for the outer dielectric is given by

$$C = 2\pi\varepsilon_0\left[\frac{1}{\varepsilon_1}\log_e\left(\frac{b}{a}\right) + \frac{1}{\varepsilon_2}\log_e\left(\frac{c}{b}\right)\right]^{-1} \mathrm{F\,m^{-1}}.$$

2.4 Show that the capacitance of a parallel-plate capacitor, neglecting fringing effects, is given by an expression of the form

$$C = \varepsilon_0\varepsilon_r a d^{-1}.$$

2.5 Calculate the capacitance of a mixed-dielectric parallel-plate capacitor in which dielectrics of (i) $\varepsilon_r = 4$, thickness 0.6 mm, (ii) $\varepsilon_r = 2$, thickness 0.4 mm, and common area 100 cm² are in contact with the plates and with each other. Calculate also the total energy stored within the electric field of the capacitor when charged to 1000 V.

[253 pF, 126.5 μJ]

2.6 A parallel-plate capacitor normally has a slab of solid dielectric with $\varepsilon_r = 2.5$ and thickness 2.5 mm separating the square plates of area 100 cm². If the capacitor is connected to a source of 10 000 V calculate the energy stored when the dielectric is half in, half out, after being displaced parallel to a pair of opposite edges. Calculate also the force on the dielectric tending to increase the capacitance of the system. State any assumptions made.

[3.1 mJ, 0.027 N]

2.7 The *Biot-Savart law* describing the magnetic field at point P distant r from conductor element dl carrying current I may be derived from eqns [2.29], [2.30], and [2.32] and expressed in integral form as

$$H = \frac{I}{4\pi} \int \frac{dl \times \mathbf{1}_r}{r^2}$$

where $\mathbf{1}_r$ is the unit vector in the direction r from dl to P. Show that an alternative formulation in differential form avoiding vector notation is

$$dH = \frac{1}{4\pi} \frac{I \, dl \sin \theta}{r^2}$$

and construct a diagram displaying the variables in a plane defined by the direction of the current along dl and the point P.

2.8 Use the Biot-Savart law to derive an expression for the magnetic field intensity produced at distance x from a long, thin straight conductor carrying current I. Compare with eqn [2.36].

2.9 Use the Biot-Savart law to calculate the magnitude flux density at the centre of a concentrated square coil in air, of dimensions 0.05 m × 0.05 m, having 100 turns and carrying 100 A.

[0.226 T]

2.10 Use the Biot-Savart law to show that the magnetic field along the axis of a concentrated circular coil with N turns, of radius a and carrying current I, at distance x from the centre of the coil is

$$H_x = \frac{NI}{2} \frac{a^2}{(a^2 + x^2)^{3/2}}$$

2.11 Calculate the H field at the centre of an infinitely long solenoid of circular cross-section carrying current I with n turns/m using (i) Ampère's law, and (ii) the Biot-Savart law. State any assumptions made.

[nI]

2.12 Utilise eqns [2.30], [2.33] and [2.36] to show that the A field at distance x from an infinitely long, thin, straight conductor carrying current I is directed parallel to the conductor and given by

$$A = -\frac{\mu_0 I}{2\pi} \log_e x + \text{constant}.$$

Hence show that the A field external to an infinitely long, straight conductor of radius a may be given by

$$A = -\frac{\mu_0 I}{2\pi} \log_e \left(\frac{x}{a}\right)$$

when the constant of integration is chosen to make A zero at the conductor surface.

Note: The same result may be arrived at with some difficulty by use of eqn [2.32].

Use the Lorentz equation [2.39] to determine the voltage/km induced in each of a pair of infinitely long, parallel conductors separated by x. Hence show that the mutual inductance per loop km between the conductors of a single-phase transmission line in air with conductors of radius a and separation d is given by

$$M = 2 \times 10^{-7} \log_e(d/a) \text{ H}.$$

Show further that the effective self inductance per metre *loop* of single-phase transmission line is given by

$$L = 4 \times 10^{-7} \log_e(d/a) \text{ H} \qquad [2.50]$$

ignoring conductor internal flux leakages.

2.13 Suggest how a pure inductor and a pair of switches might be employed to effect the efficient transfer of energy from a 12 V battery to a 6 V battery.

3 Force and torque development in magnetic circuits. Properties of magnetic materials

3.1 Force on a Current-carrying Conductor Located in a Magnetic Field

During our considerations of electromagnetic induction in the previous chapter (Sect. 2.5) we made use of the Lorentz equation for the force on a unit charge, placed in a changing magnetic field and having motion relative to the frame of reference in which the magnetic field is expressed

$$\frac{F}{q} = E = -\frac{\partial A}{\partial t} + v \times B. \qquad [3.1]$$

If we consider the field to be unchanging with time the only effective term is the latter one and the direction of the force on the charge is perpendicular to the plane of the motion and the flux density. If the motion is at right-angles to the axis of a conducting but *isolated* wire the force causes the redistribution of charge along the conductor, thus accounting for its electromagnetically induced e.m.f. Such a process involves no steady interchange of energy and no net force is developed on the *conductor*. If, however, an electric circuit is completed, enabling current to flow along the conductor – in, or against, the direction of the induced e.m.f. – the situation is changed. Electrical power will flow out of, or into, the conductor and mechanical force will be developed thereon, opposing or assisting its motion. The mechanical force results from the modification to the magnetic field brought about by the conductor current. The direction of the force may be deduced from eqn [3.1], substituting for v the motion *of the charge carriers* in the wire. We shall now evaluate the magnitude of the force assuming a uniform flux density, with conductor motion and current flow at right-angles to each other.

Figure 3.1 shows a conductor of length l in contact with parallel, stationary conducting rails with the circuit completed via resistance R. The conductor is moving with constant velocity v in a direction at right-angles to both the parallel rails and to a uniform magnetic field of flux density B. For conductor motion to the right and flux directed into the page, the direction of the induced electric field is such as to make current circulate in the clockwise direction and the upper rail potential positive with respect to the bottom rail. Applying eqn [3.1] to the charge carriers within the moving conductor we deduce that the force **F** developed on these charges is at right-angles to *their* motion and opposes the movement of the conductor.

Figure 3.1 Conductor moving in a uniform magnetic field.

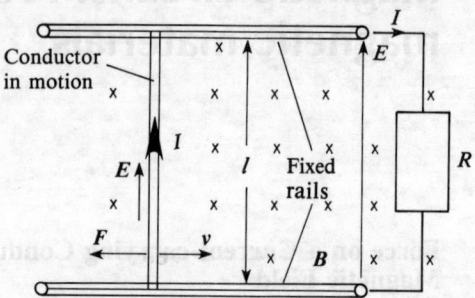

The simplest way to calculate the value of the force manifested on the moving conductor is to equate the mechanical power supplied to the system in moving the conductor at constant velocity v against constant force F to the electrical power dissipated in the load resistor. The validity of the method is not affected if the remainder of the circuit contributes a part of the total load resistance.

The e.m.f. induced in the moving conductor is given by

$$E = \int \mathbf{E} \cdot d\mathbf{l} = \int (\mathbf{v} \times \mathbf{B}) \cdot d\mathbf{l} = Blv.$$

Equating mechanical and electrical powers

$$\mathbf{F} \cdot \mathbf{v} = Fv = EI.$$

Hence

$$F = \frac{EI}{v} = BlI. \qquad [3.2]$$

Thus the force per unit length on a conductor carrying current I is given by the product BI. We need to appreciate also that the force is directed in such a sense that, if the current flow is in the direction of the electromagnetically induced e.m.f., the force opposes the motion. The law of energy conservation would require this, and the view is confirmed by a qualitative application of the principles of Maxwell's stresses to the magnetic field conditions in the vicinity of the moving conductor. Thus, if current is flowing in the moving conductor in the direction of $\mathbf{E}$, the direction of the magnetic field $\mathbf{H}$ due to current flow is such as to enhance the resultant $\mathbf{B}$ field – and consequently the lateral mechanical stress – on the side of the conductor conceding the motion. The resultant $\mathbf{B}$ field is reduced on the other side corresponding to reduced lateral stress at this side of the conductor.

It is important to recognise that the equation for the mechanical force developed on a conducting element requires, in general, a formulation in terms of a vector product. Equation [3.2] above is valid only when the flux density and current are directed at right-angles to each other, with the force perpendicular to both. The general expression for the force on a current element of length $d\mathbf{l}$ carrying current I in the direction $d\mathbf{l}$ is given by

$$dF = I(dl \times B). \qquad [3.3]$$

As confirmed by reference to Fig. 3.1 the direction of F is right-handed about the direction of dl (or I), turned into B. With I directed as shown in Fig. 3.1, F is necessarily directed oppositely to v. Reversal of the direction of the current by, for example, the inclusion of an additional source of e.m.f. in the circuit, will change the direction of dl and hence that of F. Electromechanical energy conversion is now from the electrical to the mechanical form.

3.2 Force and Torque Development on Current-carrying Loops Placed in a Magnetic Field

The early experimental work establishing the force effects between current-carrying circuits was carried out by Ampère during the early 1820s and continued by others including Oersted, Biot and Savart. As well as establishing quantitatively the relationship between force, distance apart and the current carried by parallel conductors as discussed in Section 2.4.10, Ampère found that pairs of circular coils when carrying current could exert forces and *torques* (couples) on each other which were similar to those experienced by an excited coil and a permanent magnet. Ampère thus deduced that the fields of permanent-magnet materials must be due to elementary current flows at the atomic level, and built up a theory of magnetism which provides the basis for modern explanations of ferromagnetic phenomena.

The force or torque developed on current loops placed in a magnetic field is utilized in many practical applications: examples are the loudspeaker transducer which employs a multiturn circular coil placed in a radial, permanent-magnet field to convert electrical current into sound pressure waves; and the permanent-magnet, moving-coil instrument for measuring current using a spring-restrained, pivoted pointer moving over a graduated scale. We shall require an appreciation of the force and torque developed on current loops in order to establish equilibrium states likely where current orientation with respect to an external field is capable of change. This is a necessary preliminary to the development of the theory of ferromagnetism which is pursued in Section 3.3.2.

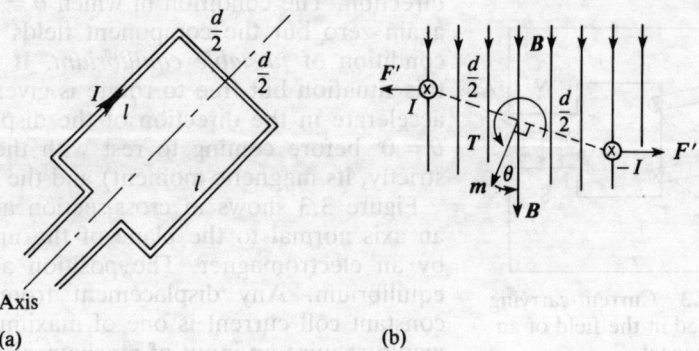

Figure 3.2 (a) Current-carrying loop situated in uniform magnetic field. (b) Sectional view.

Consider a current-carrying loop placed in a uniform field of magnetic flux density B. The loop shown in Fig. 3.2 lies in one plane, is symmetrical and, for simplicity, has rectangular form. It is pivoted about a central axis at right-angles to the plane of the coil. The sectional view of Fig. 3.2(b) has the current flow at right-angles to the plane of the paper and the external flux density in the plane of the paper. Application of eqn [3.3] shows that the forces F' developed on the two active coil sides of length l are oppositely directed in the plane of the paper and of value BIl. Due to the symmetry of the structure and the uniformity of the external magnetic flux density there is no net force acting on the coil. There is, however, in general, a torque or couple tending to rotate the coil about its axis. Referring to Fig. 3.2(b) the anti-clockwise torque is

$$T = 2F'\frac{d}{2} = BIld \sin \theta$$

$$= BIa \sin \theta \qquad [3.4]$$

where a is the area of the loop.

The product Ia is a property of the current loop which is identified as its *magnetic (dipole) moment, m*, hence

$$T = mB \sin \theta.$$

Now m may be given the status of a vector quantity $\mathbf{m}$, with direction normal to the plane of the loop, as defined by the direction of the H field appropriate to the flow of current in the loop – in accordance with the right-hand or 'corkscrew' rule. θ is then the angle between vectors $\mathbf{m}$ and $\mathbf{B}$ and the torque may be expressed as a vector product:

$$\mathbf{T} = \mathbf{m} \times \mathbf{B} \qquad [3.5]$$

The torque is directed in the plane of $\mathbf{m}$ and $\mathbf{B}$, at right-angles to the axis of rotation. The *actual* direction of the torque on the coil is always such as would cause $\mathbf{m}$ to align with $\mathbf{B}$, reducing θ to zero. This is known as the *alignment principle*, which states that *the torque developed on a current-carrying coil placed in an external magnetic field is such as to try to co-align the field of the coil with the external field*. The co-aligned condition at which θ is zero is a condition of *stable equilibrium*, to which state the excited coil will tend to return after being given an angular displacement in either direction. The condition in which $\theta = 180°$, for which the torque is again zero but the component fields are directed oppositely, is a condition of *unstable equilibrium*. If an energised coil initially in this situation but free to rotate is given a small displacement it will accelerate in the direction of the displacement and oscillate about $\theta = 0°$ before coming to rest with the field of the coil (or, more strictly, its magnetic moment) and the external field co-aligned.

Figure 3.3 shows in cross-section an excited coil pivoted about an axis normal to the plane of the uniform flux density produced by an electromagnet. The position adopted is the one of stable equilibrium. Any displacement from this position – which for constant coil current is one of maximum stored magnetic energy – would require an input of mechanical energy to the system.

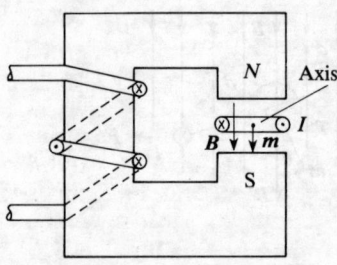

Figure 3.3 Current-carrying coil placed in the field of an electromagnet.

Force and torque development in magnetic circuits. Properties of magnetic materials

3.3 The Properties of Magnetic Materials

Following experimentation with the force and torque effects on permanent magnets and coils, Ampère observed that a current-carrying coil produced a magnetic field similar to that produced by a permanent magnet. Comparative field patterns are displayed in Fig. 3.4. Ampère deduced that, within the latter, the electrons orbiting around the nucleus of each atom developed numerous elementary magnetic dipoles.

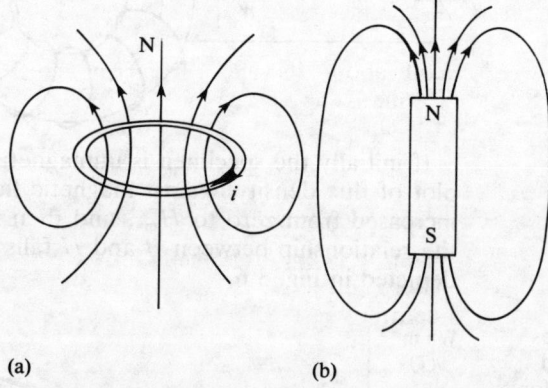

Figure 3.4 Magnetic fields of (a) a current loop, and (b) a permanent magnet.

Modern concepts of magnetism retain the essential model of Ampère with certain refinements, particularly a recognition that the *spin* of an electron about its own axis constitutes a moving charge able to give rise to a magnetic field. The complex atomic structures of the materials involve large numbers of electrons moving in many and varied orbits. The magnetic effects of individual electrons may either be cumulative or tend to cancel, accounting for the differing magnetic behaviour of different materials.

Most electromagnetic machine systems incorporate composite magnetic circuits which consist of a material of high permeability and an air space. These are usually configured in such a manner that the principal magnetic flux is common to both – this is a *series* magnetic circuit. The objective is generally to develop a high flux density in the air space with as little circuit m.m.f. as possible.

In Chapter 2 we considered the fundamental behaviour of magnetic fields in a medium of constant and uniform permeability for which B was a linear function of H, i.e. $B = \mu H$. We effectively identified the medium as free space, for which $\mu = \mu_0 = 4\pi \times 10^{-7}\,\mathrm{H\,m^{-1}}$, or air, which approximates very closely to free space in its magnetic behaviour. In fact, air has a permeability of $1.000\,000\,4 \times 4\pi \times 10^{-7}\,\mathrm{H\,m^{-1}}$, exhibiting a property known as *paramagnetism*. Iron and its compounds exhibit *ferromagnetism*, demonstrated by a B/H ratio around 1000 times greater than that of air but significantly non-linear. Most magnetic circuits in electrical machines employ ferromagnetic materials with, airgaps and so it is important that the characteristics of ferromagnetism are understood.

Prior to a discussion of how the various magnetic effects arise, consider the results of a laboratory test designed to evaluate the B/H relationship for samples of certain materials, made up into a toroidal shape and wrapped round with coiled wire connected via a switch to a controlled current source as shown in Fig. 3.5.

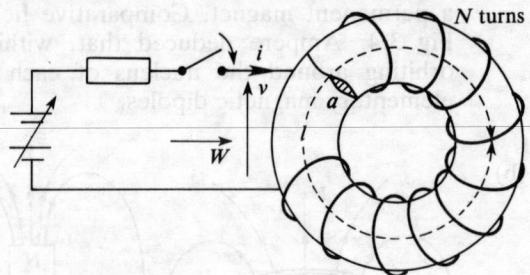

Figure 3.5 Toroidal winding with a magnetic core energised from a controlled current source.

If initially the specimen is unmagnetised, i.e. $B = 0$ for $H = 0$, a plot of flux density against magnetic field intensity, as the latter is increased from zero to H_{max} and then reduced to zero, shows that the relationship between B and H falls into one of three categories depicted in Fig. 3.6.

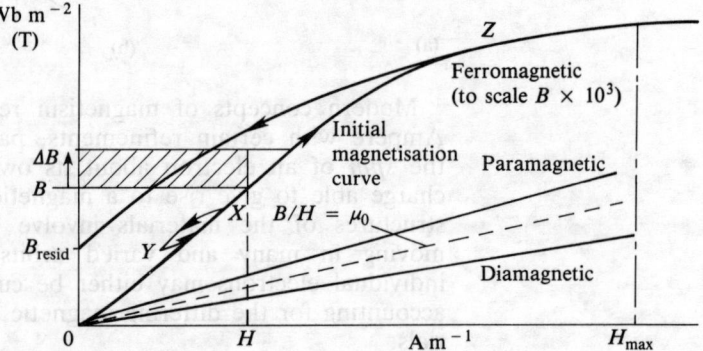

Figure 3.6 Relationships between flux density and magnetic field intensity for magnetic materials.

Diamagnetic and *paramagnetic* materials yield values of B which are linearly related to H via a permeability factor comparable to that of air or free space. A diamagnetic material actually has a value of *relative permeability* μ_r slightly less than unity when the constitutive equation for a magnetic material is presented in the form

$$B = \mu_0 \mu_r H. \qquad [3.6]$$

By contrast, the relationship between B and H for a *ferromagnetic* material has $\mu_r \sim 1000$, but variable. Furthermore the relationship between B and H when both are reducing is now markedly different from that when both are increasing. This *hysteresis* effect gives rise to a value of B when H (and B) is reducing greater than that at the same value of H when it is increasing. Additionally, evaluation of the energy flow W between source and coil shows that energy returned to the source whilst H is reducing is less than that taken from the source as the field is increasing. This is readily demonstrated with reference to Fig. 3.6 using the results of the analysis below, which show that the energy transferred between the source and the magnetic field is represented by the area between the B axis and the B/H curve between limits appropriate to the range of values for B which are of interest.

Thus, referring to Fig. 3.5, within the toroid

$$H = \frac{Ni}{l}$$

also, anticipating eqn [3.14],

$$v = N\frac{d\Phi}{dt} = Na\frac{dB}{dt},$$

with product al defining the volume of the torodial sample.

The increment of energy transferred from the source during an interval of time dt is given by

$$dW = vi\,dt = Na\frac{dB}{dt}\frac{Hl}{N}dt = alH dB$$

Hence,

$$W = al\int_{B_1}^{B_2} H\,dB \qquad [3.7]$$

where W is the total energy transferred to the magnetic field space as the flux density is increased from B_1 to B_2.

Reference to Fig. 3.6 demonstrates that the energy returned to the source when H is reduced from H_{max} to zero is less than that abstracted as the field is built up. The phenomenon of hysteresis is exhibited by ferromagnetic materials only and, whilst accounting for the useful feature of *permanent magnetism* discussed in Section 3.17, greatly complicates the study of magnetic field systems. The physical mechanism by which the observed phenomenon of ferromagnetism is explained is quite different from those accounting for diamagnetic and paramagnetic effects.

3.3.1 *Diamagnetism and paramagnetism*

For a general magnetic material eqn [3.6] may be rewritten as

$$B = \mu_0\left(\frac{M}{H}\right)H \qquad [3.8]$$

where M is called the *magnetisation*, the magnetic dipole moment per unit volume. The ratio M/H is called the *magnetic susceptibility*, representing the magnetic field enhancement in response to applied field H.

At the molecular level of material structure the resultant magnetic moment is due to the orbital motion of the many electrons around the nucleus and their intrinsic rotation (spin) about their own axis. Such components normally sum to zero in conditions of equilibrium for most materials. When, however, the material is placed in an external flux density, *each electron* experiences a consequential force which alters marginally its orbital motion. The *changes* in the electron orbits may be represented as a superimposed induced current whose magnetic field opposes that of the external flux density.

All substances exhibit the phenomenon of diamagnetism described above, which gives rise to negative M/H of the order of 10^{-5}. In addition, *paramagnetism* occurs in certain substances for which *each atom or molecule* generates a net magnetic dipole moment due to aggregated electron orbital motion or spin. In the presence of an external magnetic field the atomic dipoles experience a torque (eqn [3.5]) which encourages them to align with the external field. The effect of paramagnetism is temperature dependent, however, and exhibits saturation under intense external fields when alignment is complete.

3.3.2 *Ferromagnetism*

Paramagnetic effects are insignificant in comparison with the enhancement of magnetisation observable with ferromagnetic materials having indicated M/H ratios $\sim 10^3$. To account for this, Pierre Weiss proposed a modified model of the atomic structure which included an internal field, H_{int}, which is related to the magnetisation M by a constant λ – known as the *Weiss constant*. The internal field accounted for the ability of a ferromagnetic material to sustain flux in the absence of external field. Incorporating the effects of temperature upon the ability of the atomic magnetic dipoles to retain a preferred orientation, a model solution provides a value for M in the absence of an external field – but only below a certain temperature. Above this critical temperature, the *Curie* temperature, the ability to sustain the independent magnetisation disappears and the material reverts to the paramagnetic state.

Good agreement exists between predicted and measured values of temperature at which the *Curie effect* is observed in common ferromagnetic materials. However, predicted values of the internal flux density have the order of 10^3 T – very much larger than the values of B measurable in laboratory samples of ferromagnetic material – and 'spontaneous magnetisation' is never demonstrated by fresh or previously de-magnetised samples *en masse*.

To explain the deficiencies apparent in the theory of spontaneous magnetisation below the Curie temperature, Weiss postulated the idea of *magnetic domains*, each domain containing many atoms or molecules, though of small physical size. Each crystal of ferromagnetic material would contain many domains but the shape, size and magnetic orientation of these domains would vary according to the level and direction of the external field. Every domain would exhibit spontaneous magnetisation and, in the unmagnetised sample be small in size, with magnetisation oriented at random.

Subsequent experimentation has confirmed the existence of these domains with sizes ranging from 10^{-3} mm^3 to 1 mm^3, or greater. Single crystals of materials such as iron and cobalt also possess the property of *anisotropy* – that is, the ease with which the material can be magnetised is dependent upon the orientation of the external inducing magnetic field with respect to the crystal axis.

In an unmagnetised single crystal the domains are oriented in such a manner as to minimise the external magnetic field since the creation of an external field would require an energy transfer. For this reason adjacent 'principal' domains have magnetisation

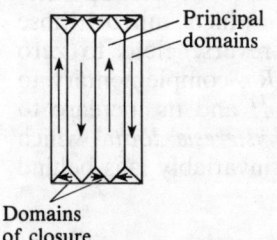

Figure 3.7 Idealised domain structure in an unmagnetised single crystal of iron.

directed oppositely. The external field is reduced further if small 'domains of closure' exist at the surfaces near the poles of the principal domains, as shown in the possible arrangement of Fig. 3.7. The size of the domains in an unmagnetised sample is determined by a compromise between two energy considerations.

A large number of small domains requires extensive *boundaries* between adjacent domains – known as *Bloch walls* – across which the electron spins reverse direction in a series of stages or increments corresponding to the number of atoms in the wall. Energy is required to maintain these boundaries. Conversely, the incorporation of larger domain sizes within the sample means that the domains of closure occupy a relatively larger proportion of the total crystal volume and, as these correspond to magnetisation of the sample in a 'hard' direction, the anisotropy energy requirement will be increased.

Starting with an unmagnetised sample, a zero-valued H corresponds to a zero-valued B, with all the domains in the individual crystals effectively oriented at random but corresponding principally to 'easy' directions of magnetism in each individual crystal. On application of a *small* external field the polycrystalline sample responds in such a way that domains whose magnetisation effectively parallels the applied field increase in size by movement of the Bloch wall separating such a domain from an adjacent one whose magnetisation is parallel, but oppositely directed. As the applied field is small the wall movement is small and reversible.

Over such small excursions of external field the permeability of the material is essentially high and constant. If the applied field is increased further, however, the boundary wall movement between domains is much greater and its displacement is achieved through a series of jerks. This involves the expenditure of energy which is not recovered when the applied field is reduced or removed – so accounting for the phenomenon of hysteresis indicated in Fig. 3.6.

The range of magnetisation corresponding to the 'irreversible' domain boundary wall movements is characterised by reducing permeability as H is increased. At higher values of H – beyond Z on the curve for ferromagnetic material in Fig. 3.6 – further increases in H result in small, but linear, increases in B. Increased magnetisation of the sample is achieved by *rotation* of the direction of magnetisation of whole domains – a process which is difficult but reversible.

3.4 The Hysteresis Loop

In the foregoing we have accounted for the behaviour of polycrystalline ferromagnetic materials when, starting from an unmagnetised state, the magnetic field intensity is increased to some maximum value and is then reduced to zero. If the excursion is more than minimal, *residual flux density* or *remanence*, B_r, remains (OR in Fig. 3.8) due to the non-elastic displacements of the boundary walls between magnetic domains. This residual flux density may be reduced to zero if the magnetising force is applied

62 Electrical machines and drive systems

in the reverse direction and is increased from zero to the value OC, restoring the positions of the domain boundaries to their initial locations. The value of H equivalent to OC is called the *coercive force* or *coercivity*, H_c. Further increase of H in the opposite direction leads to saturated conditions Z' which mirror those previously encountered. Reduction of the reverse field to zero results in a value of residual flux density OR', complementary to OR. Restoration of the original polarity to H and its increase to saturation completes the traverse of the *hysteresis loop*, which demonstrates that the change in flux density invariably lags behind a change in field intensity.

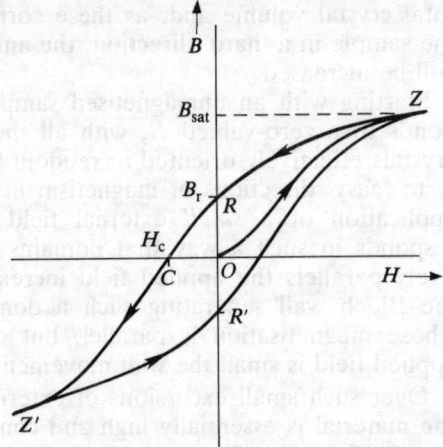

Figure 3.8 Complete hysteresis loop of ferromagnetic material.

Depending upon the application intended for the magnetic material the hysteresis loop should tend towards different optimal shapes. A good permanent magnet material should obviously have a large value of residual flux density but should also be able to resist the demagnetising influence of a reverse magnetic field. Its coercive force should therefore be large but, more significantly, the *slope* of the hysteresis loop where it crosses the B axis should be low, i.e. a 'flat' hysteresis loop is required. For an application in which the direction of the magnetic field and flux reverse frequently the opposite is true: low values of residual flux density and coercive force together with high permeability are required, corresponding to a hysteresis loop which is tall and thin. Such a characteristic, included in Fig. 3.9, is provided by a 'soft' magnetic material whose elastic domain boundaries are able to move reversibly with ease. This requires a crystal structure free from the impurities or irregularities whose effect is to inhibit the free movement of the boundary walls. Practical soft magnetic steels are usually prepared by a heat treatment which reduces strain within the crystal lattice and subsequent slow cooling. Contrasting qualities are required of a permanent magnet material within which lattice imperfections are deliberately introduced.

The area enclosed by the hysteresis loop of a 'soft' material is ideally small in order to minimise energy losses due to the hysteresis effects. Equation [3.7] has demonstrated that the energy supplied by the source when a magnetic material is subjected to an increase of H from zero to some peak value is equal to the product of the volume of material and the area between the B axis and appropriate limits on the B/H curve for increasing H. On reducing H to zero an amount of energy is recovered by the source proportional to the area defined by the same axis and the *reducing* B/H relationship. The difference between the areas is half the area enclosed by the complete hysteresis loop. If the material is incorporated in a 50 Hz transformer, for example, the hysteresis loss in watts is 50 times the product of the volume of the core (m^3) and the area of the hysteresis loop traversed (Wb A m^{-3}).

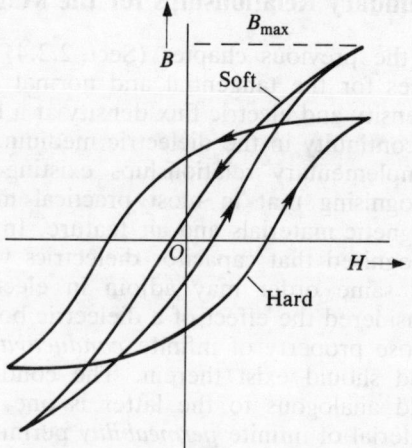

Figure 3.9 Hysteresis loops for hard and soft magnetic materials.

3.5 Eddy Current Loss

Hysteresis is not the only cause of energy loss which is manifested as heat developed in a magnetic circuit subjected to alternating excitation. The changing flux induces e.m.f. and current in closed elementary paths within the iron, in much the same manner as the coupled flux induces e.m.f. in encircling copper windings. Such *eddy currents* are minimised by using magnetic materials of high resistivity and employing core packs of electrically insulated sheets or *laminations*, built up with layers parallel to the direction of the magnetic flux. This artifice limits the size of eddy-current circuits to that having a width equal to the lamination thickness, keeping low the level of flux enclosed by a typical eddy-current path and limiting the values of e.m.f. and current. Section 4.6 considers the dependence of eddy-current loss on such parameters as supply frequency, peak flux density, resistivity and thickness of laminations in the context of a transformer magnetic core.

In practice, the effect of eddy currents is to modify the shape of the hysteresis loop, making it fatter and thus increasing its area as the loop is transversed more rapidly. Such a *dynamic hysteresis loop* relates, of course, to a particular lamination thickness, frequency and amplitude of flux-density swing.

The qualitative treatment of ferromagnetism given above has not made detailed reference to the specifications of ferromagnetic materials available to the designer today. The characteristics of materials appropriate to a particular application are frequently distinctive and will be further discussed in the context provided by the subject areas of later chapters. Suffice it to say for the moment that the availability of suitable magnetic materials is of great importance in magnetic systems, and the continual improvement in the capability and efficiency of many items of electrical plant is critically dependent upon the further development of magnetic materials.

3.6 Boundary Relationships for the Magnetic Field

In the previous chapter (Sect. 2.2.4) we examined the consequences for the tangential and normal components of electric field intensity and electric flux density at a boundary which constitutes a discontinuity in the dielectric medium. We shall now establish the complementary relationships existing in magnetic field systems, recognising that in most practical magnetic circuits both ferromagnetic materials and air feature. In the studies of Chapter 2 we recognised that capacitor dielectrics with relative permittivities of the same order may adjoin in electric field systems, and also considered the effect of a dielectric bounded by an ideal conductor whose property of infinite *conductivity* required that zero electric field should exist therein. The condition in the electromagnetic field analogous to the latter is one in which the presence of a material of infinite *permeability* permits any value of magnetic flux density to exist within it coincidentally with zero magnetic field intensity. Whilst magnetic materials with infinite permeability are not available, ferromagnetic materials in conjunction with air tend towards this idealised composite so that in many applications one might ignore the effect of the air, or the iron, as appropriate.

In establishing the general conditions at a boundary between dissimilar magnetic materials we consider the configuration of Fig. 3.10, in which magnetic flux is incident at an angle θ_1 to the normal to a plane surface which separates media of permeabilities μ_1 and μ_2. At that surface the facility for current flow with uniform density, directed at right-angles to the incident flux, also exists. We seek to determine the relationship between the normal and tangential components of magnetic flux density and field intensity existing on both sides of the boundary.

Consider an elementary volume defined by plane surfaces parallel with, and either side of, the boundary. The defining surfaces are of any shape, but identical, separated by infinitesimally small distance Δy. Due to the solenoidal property of B (eqn [2.29]), as $\Delta y \to 0$ the flux crossing the upper defining surface must equal that crossing the lower surface. It follows then that the *normal* component B_n must be identical on both sides of the boundary, i.e.

$$B_{n_1} = B_{n_2} \qquad [3.9]$$

or

$$B_1 \cos \theta_1 = B_2 \cos \theta_2.$$

Figure 3.10 Field conditions at a boundary between media of differing permeability.

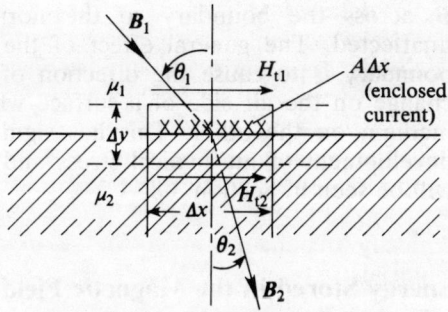

Considering now the elementary path shown in Fig. 3.10 to enclose the current $A\Delta x$ flowing at right-angles to the plane of the rectangular path with sides Δx and Δy. Integrating the effects of the magnetic field components over this path we get, from Ampère's law, eqn [2.27]

$$H_{t1}\Delta x - H_{t2}\Delta x = A\Delta x, \text{ if } \Delta y \to 0.$$

Thus, at the boundary

$$H_{t1} - H_{t2} = A. \quad [3.10]$$

In particular, if no current exists at the boundary between media of different permeabilities

$$H_{t1} = H_{t2} \quad [3.11]$$

or

$$H_1 \sin \theta_1 = H_2 \sin \theta_2.$$

Taking the quotient of both sides of the alternative forms of eqns [3.9] and [3.11], we get

$$\frac{B_1 \cos \theta_1}{H_1 \sin \theta_1} = \frac{B_2 \cos \theta_2}{H_2 \sin \theta_2}$$

giving

$$\frac{\tan \theta_2}{\tan \theta_1} = \frac{\mu_2}{\mu_1}. \quad [3.12]$$

The angle θ is shown by eqn [3.12] to be smaller on the side of the boundary having the lower permeability. Hence, if $\mu_2 = 1000\mu_1$, say, as corresponds to a ferromagnetic material boundary with air, *in the absence of current flow at the boundary the direction of the field (and flow) lines on the low permeability side can never depart much from the normal to the boundary*. The practical consequence of this is that *when flux lines enter or emerge from a block of ferromagnetic material in air, they do so effectively at right-angles to the surface.*

The simple relationship of eqn [3.12] does not apply when current exists at the boundary. Now the tangential components of H on the two sides of the boundary differ by the *linear current density A*, current per unit length, directed at right-angles to both the plane of the elementary path shown in Fig. 3.10 and to the direction of the field and flux lines. The existence of current at the surface has no effect on the continuity of the *normal* component of

B across the boundary so the normal components of *H* are unaffected. The general effect of the presence of current at the boundary is to cause the direction of the magnetic field lines to change on the air side of a surface which has a high permeability medium on the other. This has significant consequences for the development of mechanical stress and force at the boundary – as will be seen in Section 3.8.

3.7 Energy Stored in the Magnetic Field, Fringing

Prior to evaluating the mechanical stress developed at bounding surfaces between air and ferromagnetic materials we need to establish that a magnetic field is capable of storing energy. That this is so is clearly demonstrated by the common observation that a permanent magnet is able to lift an iron object, and the recognition from Section 3.3 that energy must necessarily be expended in producing a permanent magnet. The energy released when a permanent magnet pulls an object towards its pole faces must previously have resided in the field of the magnet. The energy stored per unit volume of space in which the field is known may readily be calculated by considering an idealised electromagnetic circuit for which the changes in the energy balance equation due to an incremental change in the field volume are carefully evaluated.

Figure 3.11 shows such a circuit which magnetically comprises a ferromagnetic material in series with an airgap which is capable of variation in length. The magnetic circuit is linked by a coil of *N* turns which may be connected to a battery. We expect that energisation of the coil with current will create forces on the fixed and movable pieces of iron, acting to reduce the airgap length. Even if no change in the airgap dimensions is permitted we would anticipate that energy will be absorbed by the magnetic circuit, since the coil will have received energy from the electrical source.

In order to simplify the analysis we will make the following assumptions.

(i) The ferromagnetic material is of infinite permeability. This means that the field is concentrated in the airgaps.

(ii) The only significant airgap is the one of length *x* which has flux density *B* between its plane surfaces defined by the ferromagnetic material, both uniform and directed normally to those surfaces of area *a*.

(iii) Each turn of the *N*-turn coil has zero resistance, is circular of radius *r* and links the same flux $\Phi = Ba$.

Consider one turn of the coil; due to symmetry its magnetic vector potential *A* will be everywhere uniform and of value

$$|A| = \frac{\Phi}{2\pi r}$$

from eqn [2.33]. From eqn [2.39]

$$|E| = -\frac{\partial A}{\partial t} = -\frac{1}{2\pi r}\frac{d\Phi}{dt}$$

Force and torque development in magnetic circuits. Properties of magnetic materials 67

The e.m.f. per turn

$$\int \mathbf{E} \cdot d\mathbf{l} = -\frac{d\Phi}{dt}.$$

Hence, for the N-turn coil, the total e.m.f.

$$E = -N\frac{d\Phi}{dt} \qquad [3.13]$$

and the terminal voltage, instantaneously,

$$v_L = -E = N\frac{d\Phi}{dt}. \qquad [3.14]$$

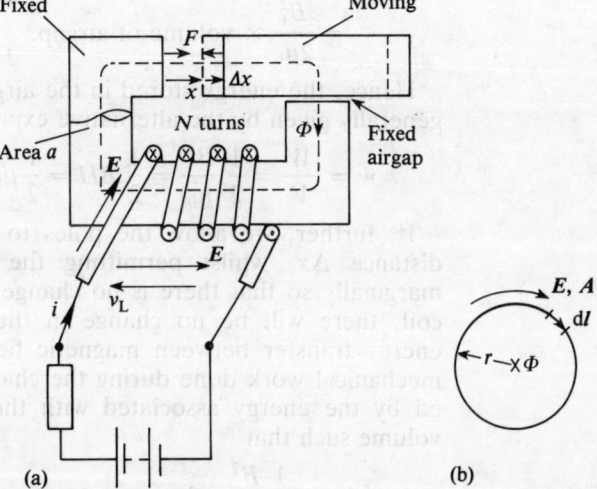

Figure 3.11 A composite magnetic circuit with an exciting coil and an electrical energy source.

Let the current flowing in each turn of the coil be instantaneously i. The total current linked by the flux path shown in Fig. 3.11 is equal to Ni which, by Ampère's law, is equal to the line integral of $\mathbf{H}$ over the flux path shown. Since we are ignoring the reluctance of the iron and the smaller airgap, the magnetic field intensity within the main gap of length x will be uniform and of value

$$H_g = \frac{Ni}{x}.$$

As we are also ignoring 'fringing' by assuming that the gap flux density is uniform over area a, we get for the flux density everywhere in the gap

$$B = \mu_0 \frac{Ni}{x}$$

and a value of flux Φ common to the magnetic circuit and each turn of the coil given by

$$\Phi = Ba = \mu_0 Ni \frac{a}{x}.$$

(*Note:* For this elementary magnetic circuit Ni is the m.m.f. of the coil and

$$\mu_0 \frac{a}{x} \quad \left[= \text{permeability} \times \frac{\text{area}}{\text{length}} \right]$$

is the *permeance P* of the magnetic circuit, relating flux to m.m.f.)

If the current flowing in the coil undergoes a change from $i = 0$ to $i = I$ in time T as the flux is increased from zero to a final value Φ_f, the energy W supplied by the source of electrical energy to the loss-free coil is given by

$$W = \int_0^T v_L \, i \, dt = \int_0^T N \frac{d\Phi}{dt} i \, dt = \int_0^{\Phi_f} N i \, d\Phi$$

$$= \int_0^{\Phi_f} \frac{x}{\mu_0 a} \Phi \, d\Phi = \frac{xa}{2\mu_0} \left(\frac{\Phi_f}{a}\right)^2$$

$$= \frac{B_f^2}{2\mu_0} \times \text{volume of airgap}.$$

Hence, the energy stored in the airgap per unit volume of gap is generally given by the alternative expressions

$$w = \frac{W}{V} = \frac{1}{2} \frac{B^2}{\mu_0} = \frac{1}{2} BH = \frac{1}{2} \mu_0 H^2 \, \text{J m}^{-3}. \quad [3.15]$$

If, further, we allow the poles to increase their separation by distance Δx, whilst permitting the source current to increase marginally so that there is no change in the flux embraced by the coil, there will be no change in the airgap flux density or any energy transfer between magnetic field and electric battery. The mechanical work done during the change must be precisely matched by the energy associated with the incremental change in gap volume such that

$$F \Delta x = \frac{1}{2} \frac{B^2}{\mu_0} a \Delta x,$$

giving

$$\frac{F}{a} = \frac{1}{2} \frac{B^2}{\mu_0} = \frac{1}{2} BH = \frac{1}{2} \mu_0 H^2. \quad [3.16]$$

In eqn [3.16] F/a is the mechanical force per unit area developed normally at the plane surfaces defining the airgap. It can be identified (Sect. 3.8.1) with a tensile stress t_m directed along the lines of B or H at the iron/airgap boundary, such as to oppose an increase the length of the gap. Mechanical energy would need to be supplied to the system if the gap length were to be increased whilst maintaining the same energy density within the gap.

3.7.1 Fringing

The validity of the method used above in evaluating the relationships for energy stored per unit volume of airgap, eqn [3.15], is strictly dependent upon the assumption of a sharp discontinuity of flux density in the region of the gap defined by the projection of the 'iron' pole pieces. Such an abrupt change does not occur in practice: the flux diverges as it leaves one pole piece to converge again on re-entry into the high-permeability iron. The extent to which fringing of the flux takes place is greater as the length of the

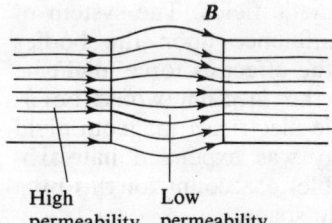

Figure 3.12 Effect of fringing on field and flux lines in composite magnetic circuits.

airgap is increased. We can explain and account for the fringing evident in the same way as proved appropriate in the case of electric flux, Section 2.3.2, by assigning to magnetic flux lines the capability of exerting a lateral 'bursting' stress in addition to the tensile longitudinal stress which accounts for the force exerted on the pole pieces given in eqn [3.16].

The effect of the lateral stress in endowing magnetic flux lines with the ability 'to elbow each other apart' is demonstrated in Fig. 3.12.

3.8 Maxwell's Stresses in the Magnetic Field

In Section 2.4.10 of the previous chapter we speculated that the system of forces observed to act on susceptible objects placed in the magnetic field resulting from the flow of electric current might well be accounted for through a system of mechanical stresses existing at all points in the field, with components along and at right-angles to the direction of the field. We postulated values for the magnitudes of these stresses and gave the proposal some credibility by calculating the force exerted over an *infinite* plane surface located on the perpendicular bisector of the plane joining parallel filaments of current. For a spacing of 1 m, the calculated force of 2×10^{-7} N per metre length agreed with the definitive force appropriate to unit current of 1 A flowing in each conductor and capable of direct measurement *on the conductors themselves*.

André Marie Ampère is regarded as being the first to quantify the relationship, experimentally verifiable, between the magnitude of the force developed on parallel conductors of length l, the value of the current I, the separation distance d and a property of the medium – permeability μ. The relationship, derived as eqn [2.38], is restated thus

$$\frac{F}{l} = \frac{\mu}{2d} \frac{I^2}{\pi}.$$

Now, the early experimenters of Ampère's day were not unfamiliar with the fact that mechanical forces were developed on bodies associated with electrostatic and magnetic effects. Forces were commonly experienced by electrostatically charged rods and by magnetic materials suspended in the earth's magnetic field. [The Chinese have been credited with use of the compass around 2 500 BC and the fact that iron was attracted by lodestone was reported by Thales of Miletus in the sixth century BC]. Ampère's conception of the means by which these forces were achieved, however, in common with other investigators such as Coulomb, was simply one of 'action at a distance', with no real attempt being made to explain the mechanism by which such influences were brought to bear. Michael Faraday later argued that the effects must be brought about by 'continuous action between contiguous portions of an intervening medium' and postulated the *aether* as a medium subjected to a state of mechanical stress which corresponded to the presence of appropriate electric or magnetic fields. The system of forces was detectable only by their influence upon the bodies concerned. Whilst the manifestation of the effect as force might be localised, the distribution of stress in the medium would be as diffuse and widespread as the responsible electric or magnetic field system. As it was apparent that energy was expended in establishing these fields, it would be reasonable to account for this by a distribution of *strain energy* over the field space.

It was James Clerk Maxwell who formulated mathematically the clear perceptions of Faraday's reasoning, for both electrostatic and magnetic fields. In Chapter XI of his *Treatise on Electricity and Magnetism*, Maxwell examines mathematically the system of forces acting upon a current-carrying element placed within a magnetic field and determines the nature of the mechanical stress system in the surrounding medium, satisfying simultaneously the requirements that:

(i) every element of the field space is in mechanical equilibrium; and
(ii) the resulting force on the current element agrees with Ampère's formulation.

The force exerted on the current element is then recognised as due to the boundary effects of mechanical stress distributed throughout the field space.

Maxwell's analysis for the magnetic field was conducted in very general terms, considering a medium in which B and H did not necessarily act in the same direction and for which a very complicated system of stresses results. For a medium in which B and H are co-directed the stress system reduces to one in which two distinguishable stresses exist – a simple *compressive* stress perpendicular to the lines of force and a simple *tensile* stress along the lines of force. Further simplification results for a medium in which B and H are linearly related, as through permeability μ_0 for air or free space, the lateral and longitudinal stresses are then identical and expressible variously as

$$t_m = p_m = \frac{1}{2}\mu_0^2 H, \frac{1}{2}BH \text{ or } \frac{1}{2}\frac{B^2}{\mu_0}.$$

In a medium such as air, therefore, knowledge of the distribution of B and H gives a clear picture of the state of stress within the field. The stresses are larger as the square of B or H. In general terms it may be said that the stress system is comparable to one in which the lines of B or H are represented as stretched elastic strings – tending to shorten in length and to dilate, 'elbowing' each other apart.

In practical experience, the tensile stress system is easy to recognise, the compressive stresses perpendicular to the field lines less so. The calculation of magnetic forces by this method is easy in cases where only one system of stress is effective due, for example, to considerations of symmetry. By this expedient simple idealised conditions may be set up to confirm the expressions for the components of stress deduced, more generally, by Maxwell.

In the analyses which follow we shall consider incremental

Force and torque development in magnetic circuits. Properties of magnetic materials 71

changes in the configuration of particular composite magnetic circuits comprising air and an infinitely permeable material (iron). The adjustments will be such as to render unchanged the flux linkage of the electric circuit responsible for setting up the initial field conditions. According to eqn [3.15] the initial field conditions will correspond to a distribution of energy throughout the air space of local density

$$w = \frac{1}{2}\frac{B^2}{\mu_0}.$$

Any consequential redistribution of energy will involve transfers between mechanical and field formats alone, the electrical power flow being zero.

3.8.1 Mechanical stress along a line of force

Consider a region of air space in which the field and flux densities are uniform, impinging normally upon a ferromagnetic surface of area a as shown in Fig. 3.13. Let the surface move through distance Δx in the direction of the lines of force, so increasing the volume of the field space in air by $a\Delta x$; at the same time adjusting the exciting coil current to maintain constant flux density and hence constant total flux.

The *increase* in stored magnetic field energy

$$\Delta W = \frac{1}{2}\frac{B^2}{\mu_0}a\Delta x.$$

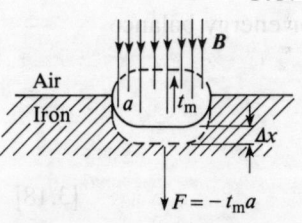

Figure 3.13 Pertaining to the calculation of mechanical stress due to magnetic flux normal to a plane surface.

The mechanical work done in the direction of the displacement

$$\Delta W_{\text{mech}} = F\Delta x.$$

Here F is the force on the surface directed as shown in the figure, equal to $-t_m a\Delta x$, where t_m is the *tensile stress* at the surface.

For energy balance

$$\frac{1}{2}\frac{B^2}{\mu_0}a\Delta x - t_m a\Delta x = 0.$$

Hence

$$t_m = \frac{B^2}{2\mu_0}, \qquad [3.17]$$

i.e. the stress normal to the lines of force is tensile (tending to shorten) and is of value $B^2/2\mu_0$. The result is, not surprisingly, identical with eqn [3.16], as identical procedures have been followed in the derivations.

3.8.2 Mechanical stress at right-angles to a line of force

Now consider a tube of flux Φ, conveniently of circular cross-section, enclosed in air within a cylindrical volume such that the flux lines are parallel to the sides, normal to the plane upper and lower surfaces and of uniform flux density B.

Figure 3.14 shows the flux to be confined laterally within the described volume by a surrounding medium of zero permeability,

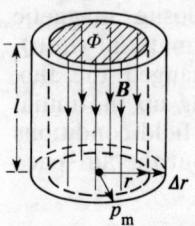

Figure 3.14 Pertaining to the calculation of mechanical stress due to magnetic flux parallel to a plane surface.

so that an abrupt discontinuity of flux occurs at the bounding surface.

Let the radius of the confining surface be increased by Δr, so that the enclosed volume increases from $\pi r^2 l$ to $\pi (r + \Delta r)^2 l$, at the same time adjusting the magnetic field intensity to maintain constant flux Φ with a consequential reduction in the flux density. The consequential *increase* in stored magnetic field energy

$$\Delta W = \frac{\Phi^2 l}{2\mu_0} \left[\frac{1}{\pi (r + \Delta r)^2} - \frac{1}{\pi r^2} \right]$$

$$= \frac{\Phi^2 l}{2\pi\mu_0} \left[-\frac{2r\Delta r}{r^4} \right] \text{ since } \Delta r \ll r$$

The mechanical work done in the direction of the displacement

$$\Delta W_{\text{mech}} = 2\pi r l p_m \Delta r,$$

where p_m is the radially directed stress. For energy balance

$$\frac{\Phi^2 l}{2\pi\mu_0} \left[-\frac{2r\Delta r}{r^4} \right] + 2\pi r l p_m \Delta r = 0.$$

Therefore

$$p_m = \frac{2}{\mu_0} \left[\frac{\Phi}{2\pi r^2} \right]^2 = \frac{B^2}{2\mu_0}. \qquad [3.18]$$

Equation [3.18] declares that the lateral stress at the boundary acts away from the region where the flux lines are most dense and also has value $B^2/2\mu_0$.

3.8.3 *Normal and tangential mechanical stresses at iron/air boundaries*

The components of stress t_m (along) and p_m (at right-angles to) the lines of force are called the *principal stresses* at a point. In practical applications, where it is desired to evaluate the stresses giving rise to mechanical force on ferromagnetic materials bounding an air space in which magnetic field energy exists, it is of greater value to express the developed stresses in terms of components *normal* and *tangential* to the boundary. Normal stresses are significant in the operation of electromechanical linear-motion actuators, such as relays, solenoids and contactors, whereas tangential stresses are more relevant to the operation of rotary actuators and rotating machinery. Generally speaking, both tangential and normal stresses will be developed, but symmetry of the magnetic structure will usually lead to the effective cancellation of one system of stresses.

It should be apparent that where magnetic field and flux lines cross normally an iron/air boundary the resultant stress on the surface is entirely tensile – the lateral stresses cancel. Similarly, in a theoretical situation in which the field lines in air are parallel to a plane ferromagnetic surface the force on the surface is entirely compressive.

In practical magnetic circuits having an airgap in series with highly permeable iron the flux crosses the boundary normally. For the flux to cross at an angle other than 90° to the bounding surface it is necessary to provide m.m.f. (current) on that surface, producing a *tangential* field component in the airgap (Sect. 3.6). Figure 3.15 shows limiting and general cases of magnetic flux traversing plane boundaries between air and a highly permeable ferromagnetic material.

Figure 3.15 Limiting and general cases of magnetic flux crossing iron/air boundaries.

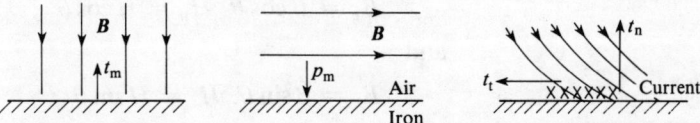

Referring to Fig. 3.16, let t_m and p_m be the principal stresses, respectively tensile and compressive (bursting), along and at right-angles to the field lines in air which are incident upon an iron/air boundary. Let t_n and t_t be, respectively, the normal and tangential components of stress developed over an elemental surface Δs at the iron/air boundary and let θ be the inclination of the field lines to the normal to the elemental surface over which the field is assumed to be uniform.

Figure 3.16 Relationship between principal stresses and normal and tangential components at an iron/air boundary.

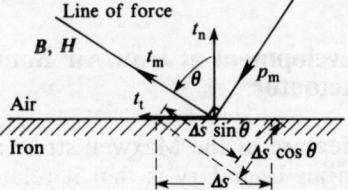

Now the *force* associated with a stress is the product of the stress and the projection of the area at right-angles to its line of action. Hence, resolving forces associated with the principal stresses acting on the elemental area we get

$$\text{resultant normal force} = t_m(\Delta s \cos\theta)\cos\theta - p_m(\Delta s \sin\theta)\sin\theta$$

and

$$\text{resultant tangential force} = t_m(\Delta s \cos\theta)\sin\theta + p_m(\Delta s \sin\theta)\cos\theta.$$

Hence

$$\text{resultant normal } stress = t_m \cos^2\theta - p_m \sin^2\theta$$

and

$$\text{resultant tangential } stress = (t_m + p_m)\cos\theta\sin\theta.$$

Since

$$t_m = p_m = \frac{B^2}{2\mu_0} = \frac{1}{2}BH$$

$$t_n = \tfrac{1}{2}BH(\cos^2\theta - \sin^2\theta) \qquad [3.19]$$

$$t_t = \tfrac{1}{2}BH\sin^2\theta. \qquad [3.20]$$

Thus, for given B (and H), t_n is a maximum when θ is zero (as expected) but t_t is a maximum when θ is 45°. Further algebraic manipulation of the equations shows that the resultant stress is always inclined at 2θ to the normal to the boundary.

It is particularly convenient to express t_n and t_t in terms of the normal and tangential components of $\boldsymbol{B}$ and $\boldsymbol{H}$. Thus, if

$$B_n = B\cos\theta, \quad H_n = H\cos\theta$$

and

$$B_t = B\sin\theta, \quad H_t = H\sin\theta$$

then

$$t_n = \tfrac{1}{2}(B_n H_n - B_t H_t) \qquad [3.21]$$
$$t_t = B_n H_t \qquad [3.22]$$

The simple form of eqn [3.22] is particularly noteworthy, as also are the observations that, provided θ is small, i.e. $< 45°$, t_n and t_t are directed such that their components along the field line are tensile. For $\theta > 45°$ the component of t_t along the field line is again tensile, that for t_n is compressive.

3.9 Force Development at Iron/Air Boundaries Appropriate to a Linear Actuator

The key feature of the Maxwell stress approach to force evaluation at an iron/air boundary is that it relates the force density directly to the magnetic condition of the surface, that is the flux density magnitude and direction on the air side of the boundary – nothing else. Other expressions for force often employed incorporate magnetic or electric *circuit* concepts such as reluctance R, current I and self inductance L. Thus we may develop, from the same virtual work considerations employed in Section 3.8.1., the alternative expressions given in eqn [3.23] for the force acting normal to a plane surface across which the flux Φ is directed normally and uniformly with density B

$$F = -\frac{1}{2}\Phi^2 \frac{dR}{dx}, \quad F = \frac{1}{2} I^2 \frac{dL}{dx}. \qquad [3.23]$$

The sign is significant and the derivative form serves to identify the reluctance or inductance term with the variable airgap length x. Constant components of R and L due say, to, auxiliary airgaps or leakage flux are thus excluded. Linear relationships are assumed between Φ and I for given x but not between R (or L) and x.

The Maxwell stress equation equivalent to eqns [3.23] is

$$F = t_n a = \frac{1}{2}\frac{B^2}{\mu_0} a, \qquad [3.24]$$

where a is the surface area.

Equations [3.23] may be deduced from the more fundamental formulation of eqn [3.24] for simple specific situations like that shown in Fig. 3.17, in which the magnetic circuit has reluctance defined solely by the variable airgap of length x and constant area a. Thus

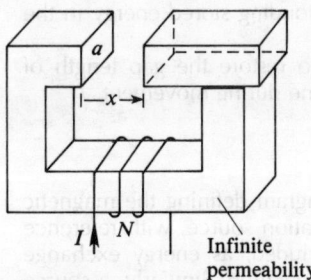

Figure 3.17 Simple magnetic circuit with variable airgap length.

$$\left.\begin{aligned}
F &= \frac{1}{2}\frac{B^2}{\mu_0}a = \frac{1}{2}Ba\frac{B}{\mu_0} = \frac{1}{2}\Phi H \\
&= \frac{1}{2}\Phi\frac{(NI)}{x} = \frac{1}{2}\Phi^2\frac{R}{x} \\
&= \frac{1}{2}\frac{N\Phi}{I}\frac{I^2}{x} = \frac{1}{2}I^2\frac{L}{x}.
\end{aligned}\right\} \quad [3.25]$$

The differential forms of eqns [3.25] given in eqns [3.23] have merit in particular situations, especially the one involving circuit current and self-inductance dependence on airgap length. In many practical circumstances it is not feasible to determine accurately the flux conditions at the pole surfaces, and the *distribution* of the mechanical stress at the surfaces is of no account. Furthermore, in linear motion transducers where the distance travelled by the moving element may be extensive, the experimental derivation of circuit inductance variation with gap length or armature displacement is much easier to establish than is an adequate knowledge of the flux conditions within the magnetic circuit.

Electromagnetic linear motion actuators or transducers find many common uses today. Included within this category are the familiar hinged armature relay and the heavier duty solenoid. In some applications the movements may be restricted to short distances, so that changes in airgap length during operation are small and the developed forces are largely independent of position. Mechanical linkages and levers may be used to provide increased displacements, as in robotic systems, with positional feedback enhancing the degree of control. In many such applications a complex system study is called for, taking into consideration the inertial time constants of the interrelated electrical and mechanical systems, losses manifested as friction, iron and copper loss, and the characteristics of any feedback incorporated. Such details are beyond the scope of the present book but we shall now examine quantitatively the electrical and magnetic energy considerations relevant to simple electro-magnetic circuits subjected to changes in airgap length.

Worked example 3.1 Two pole surfaces of highly permeable iron define the airgap shown in Fig. 3.18 with cross-sectional area 6 cm², gap length g and gap flux 120 μWb.

Figure 3.18 Variable airgap defined by adjustable pole pieces.

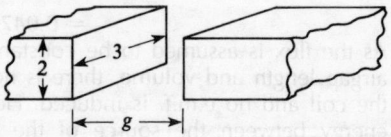

(i) Calculate the force exerted on the pole surfaces when $g = 5$ mm and the corresponding energy stored in the magnetic field.
(ii) If the gap length is reduced to 2 mm by rapid pole movement with constant flux, calculate the mechanical work done during the motion.
(iii) Assuming that the coil current recovers to its initial value, calculate the gap flux when steady conditions appropriate to the reduced gap

length are attained. Calculate the corresponding stored energy in the airgap.

(iv) If the poles are now separated slowly to restore the gap length of 5 mm, calculate the mechanical work done during movement.

Solution Here it is strictly necessary to complete the diagram defining the magnetic circuit to include the exciting coil and its excitation source, with reference directions of coil current and flux linkage included, as energy exchange with the source occurs in parts of the cycle of operation. Similarly, a source resistor R should be incorporated as in Fig. 3.19 (assuming the source is one of nominally constant voltage E) to limit the current in the steady state. Since all the iron is assumed to be of infinite permeability, the magnetic field energy will be located in the adjustable airgap.

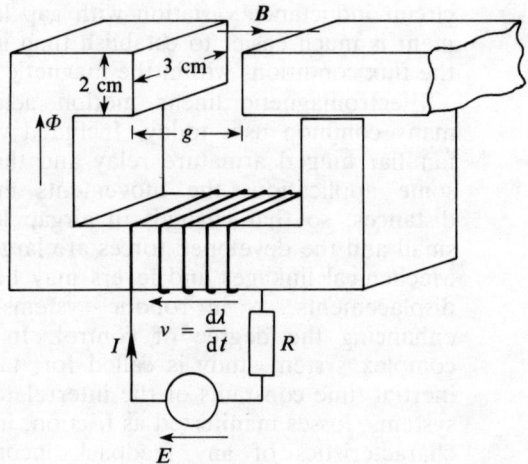

Figure 3.19 Completed electric and magnetic circuits for Worked Example 3.1.

(i) The attractive force between the pole surfaces = $\frac{1}{2}B^2/\mu_0$ × pole area

$$= \left(\frac{120 \times 10^{-6}}{6 \times 10^{-4}}\right)^2 \frac{6 \times 10^{-4}}{2 \times 4\pi \times 10^{-7}}$$

$$= 9.55 \text{ N.}$$

Energy stored in airgap = $\frac{1}{2}B^2/\mu_0$ × airgap volume

$$= \left(\frac{120 \times 10^{-6}}{6 \times 10^{-4}}\right)^2 \frac{6 \times 5 \times 10^{-7}}{2 \times 4\pi \times 10^{-7}}$$

$$= 0.047\,75 \text{ J.}$$

(ii) As the flux is assumed to be constant during the rapid reduction in airgap length and volume, there is no change in the flux linkage of the coil and no e.m.f. is induced. Hence there is no interchange of energy between the source of the exciting current and the field system. The mechanical work done during a change in gap length therefore equals the change in gap energy.

$$\text{Mechanical work done} = 0.048 \times \frac{(5-2)}{5} = 0.029 \text{ J.}$$

(iii) 'Assuming that the coil current recovers to its initial value, calculate the gap flux when steady conditions appropriate to the reduced gap length are attained'

Force and torque development in magnetic circuits. Properties of magnetic materials

This part of the question acknowledges that a transient period exists during and after gap adjustment. Whilst gap length is changing the m.m.f. cannot remain constant unless the excitation source is one of constant current. As a constant voltage source is assumed, *steady-state* exciting m.m.f. must correspond to current E/R. The associated steady-state *flux* is a function of gap length g. When rapid changes of g occur, the *change* in coil flux linkage $\Delta\lambda$ may be small compared with λ but its rate of change with time will be large, giving rise to a reduced exciting current I if the change $\Delta\lambda$ is positive, as with reduction in gap length. During periods of flux change, energy transfers between electrical circuit and magnetic field systems occur. With the notation of Fig. 3.19, positive values for $d\lambda/dt$ and I correspond to electrical power absorption by the coil.

The new value of the gap flux in the steady state at constant m.m.f.

= original value × new permeance/old permeance

$= 1.2 \times 10^{-4} \times 5/2 = 300 \; \mu\text{Wb}$.

Stored energy in the gap $= \frac{1}{2} B_{\text{new}}^2 / \mu_0 \times \text{vol}_{\text{new}}$

$\equiv 0.047\,75 \times 5/2$

$= 0.119\,38 \; \text{J}$.

(iv) '*If the poles are now separated slowly to restore the gap length of 5 mm calculate the mechanical work done during movement.*'

A low rate of change of gap length corresponds to a small induced coil e.m.f., hence the coil current is determined solely by source e.m.f. and circuit resistance. Flux must *reduce* during gap extension as reluctance is increasing and initial and eventual currents are the same. During this change $d\lambda/dt$ is negative and the coil returns electrical energy to source.

In evaluating the mechanical work done by integrating a variable force on a pole piece over the distance moved, an expression for the force must be used in which gap length appears as a variable. Thus

$$W_m = \int_{g=0.002}^{g=0.005} F(g) dg$$

where

$$F(g) = \frac{1}{2} B^2(g)/\mu_0 \times \text{area}$$

and the change in B occurs at constant current for which $B \propto g^{-1}$. Putting $B(g) = Kg^{-1}$ gives $F(g) = K'g^{-2}$. From the answer to part (i), at $g = 5$ mm force $F = 9.55$ N. Hence $K' = 9.55 \times 25 \times 10^{-6} = 238.75 \times 10^{-6}$ N m². Thus

$$W_m = \int_{g=0.002}^{g=0.005} \frac{238.75 \times 10^{-6}}{g^2}$$

$$= -238.75 \times 10^{-6} \left| \frac{1}{g} \right|_{0.002}^{0.005} = 0.0716 \; \text{J}.$$

During gap extension the exciting coil *absorbs* energy W_e of value $\int I d\lambda = NI \, \Delta\Phi$. NI is the coil m.m.f., held constant at a value of flux/permeance given, for example, by the *initial* conditions. $\Delta\Phi$ is the ($-$ve) increase in flux from 300 μWb to 120 μWb.

Thus

$$W_e = -\frac{120 \times 10^{-6} \times 5 \times 10^{-3}}{\mu_0 \times 6 \times 10^{-4}} \cdot 180 \times 10^{-6}$$

$$= -0.1432 \; \text{J}.$$

The *reduction* in stored magnetic field energy as the gap increases is given by the solutions to parts (iii) and (i) as $W_f = 0.11938 - 0.04775 = 0.0716$ J. Hence the *input* mechanical energy required, which is equal to the difference between the energy returned to the source and the reduction in stored field energy, is $W_m = 0.1432 - 0.0716 = 0.0716$ J.

This result confirms that arrived at more directly above. It is no coincidence that the mechanical work and the change in stored magnetic field energy are equal to each other at half the value of energy transfer associated with the electrical source. This '50 : 50 rule' for the division of electrical energy between mechanical energy and the change in stored field energy is characteristic of airgap changes brought about in a linear magnetic circuit at constant current. As noted earlier, if the change is effected at constant *flux* the mechanical energy is entirely accounted for by change in stored magnetic field energy.

Worked example 3.2 A moving-iron ammeter provided on test the following relationship between coil self inductance and pointer deflection.

Deflection (degrees)	0	20	40	60	80	100	120	140
Inductance (mH)	2.50	2.63	2.75	2.88	3.00	3.10	3.18	3.25

(i) Estimate the spring restraint torque required if full-scale deflection of 120° is to be achieved at a current of 1 A.
(ii) Assuming that the restraining spring has a linear rate (i.e. the torque is proportional to the deflection) estimate the current values required for deflections of (a) one-half and (b) one-quarter full scale.

Solution The way that the data has been expressed provides the clue that an expression for force (or torque) is needed in which inductance appears explicitly. The equation

$$F = \frac{1}{2} I^2 \frac{L}{x}$$

was deduced as eqn [3.25] for the particular configuration of Fig. 3.17 in which the m.m.f. was proportional to current and L inversely proportional to x. In a moving-iron instrument, however, the construction is usually such (Fig. 3.20) that a curved ferromagnetic rod is drawn into a curved solenoidal coil, so that the flux path and the effective turns are dependent upon the deflection θ. An idealised model may be used to demonstrate that inductance L is proportional to θ. If one incorporates a constant 'leakage' flux in the idealised model and assumes the inductance to made up of two components – one strictly proportional to θ and the other constant – then use of the differential form of the torque expression analogous to eqn [3.23], i.e.

$$T = \frac{1}{2} I^2 \frac{dL}{d\theta} \qquad [3.26]$$

provides for torque proportional to current squared (assuming no saturation effects) and independent of θ.

Force and torque development in magnetic circuits. Properties of magnetic materials

Figure 3.20 Moving iron instrument movement.

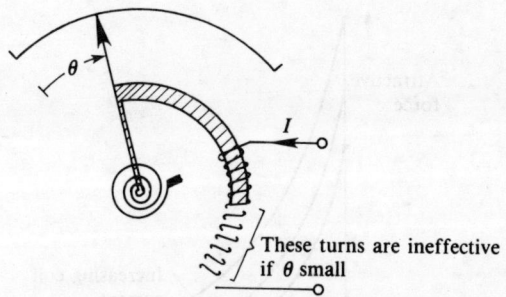

The problem under review involves a system for which the slope of the L/θ relationship is not constant, indicating a flux distribution altering in a complex manner as θ changes.

(i) At $\theta = 120°$

$$\frac{dL}{d\theta} \simeq \frac{3.25 - 3.10}{140 - 100} \cdot \frac{180}{\pi} = 0.215 \text{ mH/rad}.$$

Hence

$$T_{120°} = \tfrac{1}{2}(1.0)^2 \times 0.215 = 0.1075 \text{ mN m}$$

for a current of 1.0 A to correspond to a deflection of 120°.

(ii) (a) At $\theta = 60°$

$$\frac{dL}{d\theta} = \frac{3.00 - 2.75}{80 - 40} \cdot \frac{180}{\pi} = 0.358 \text{ mH/rad}.$$

For a spring-restraint torque equal to half that corresponding to $\theta = 120°$

$$T_{60°} = \tfrac{1}{2}I^2 \times 0.358 = 0.0537$$

giving $I_{60°} = 0.548$ A.

(iii) At $\theta = 30°$

$$\frac{dL}{d\theta} = \frac{2.75 - 2.63}{40 - 20} \cdot \frac{180}{\pi} = 0.344 \text{ mH/rad},$$

$$T_{30°} = \tfrac{1}{2}I^2 \times 0.344 = 0.0269,$$

giving $I_{30°} = 0.395$ A.

The range of applications for which attraction-type reluctance actuators are employed is limited in view of the highly non-linear, inverse relationship between force and distance of travel typified by Fig. 3.21. As the force is independent of the direction of current flow such devices respond to alternating as well as to direct current, although modification may be needed to average the time dependency of the force. Most applications are of a two-state, 'on–off' nature incorporating a return spring, otherwise two complementary units may be employed.

3.10 Rotary Motion Actuators, Torque Development

The examples considered so far have involved essentially linear displacements and effective force development due to that component of stress along the magnetic field lines. Mechanical stress

Figure 3.21 Force/travel distance relationship for a linear actuator, (attraction type).

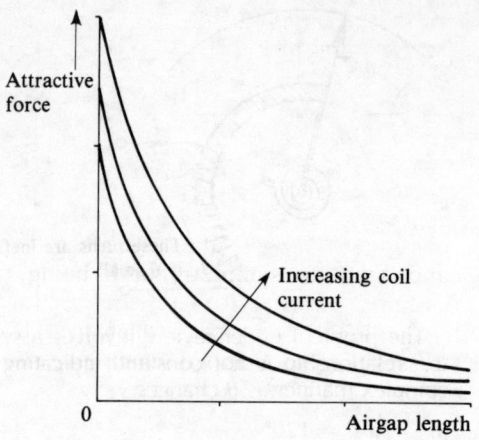

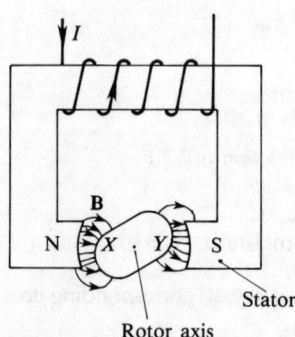

Figure 3.22 Alignment-type rotary actuator.

evaluation has been confined to iron/air boundary surfaces at the air sides of which the field lines enter or emerge normally with uniform flux densities. An illustration in which a consideration of the system of lateral mechanical stresses in the air space may prove advantageous is provided by the asymmetrical rotor system of Fig. 3.22.

Here, if we assume infinitely permeable iron, the flux will cross the iron/airgap boundaries normally with the mechanical stress on the rotor surface, for example, everywhere normal to the surface. The net torque developed is due to the higher flux densities prevailing in the regions X and Y when the rotor is displaced asymmetrically with respect to the stator as shown. Clearly, in practice, the consequential stress and torque development is critically dependent upon the actual geometry. If simplifying assumptions are made regarding the field pattern in the airgap, approximate expressions for the torque may be deduced. Further, for any assumed flux distribution in the airgap, all valid alternative methods of torque calculation should yield appropriate, consistent results.

Rotor torque, for example, results from the effective sum of those components of mechanical stress acting at right-angles to a radius, the integration being carried out over the whole of the closed rotor iron surface bounded by the airgap. This value for torque must be matched by integration of the effects of stresses over *any* closed surface which encloses the rotor including, in the limit, the stator iron/airgap boundary. This is the case as the torques on stator and rotor must be in equilibrium with each other, as also must be the stress developed in the intervening space.

The above property may be utilised in estimating the torque developed by the salient-pole device illustrated in Figs 3.22 and 3.23 for an assumed flux distribution which crosses the iron/air boundaries normally and is non-existent away from the overlapping pole surfaces. Rotor and stator torque may be calculated by considering the stresses developed over the closed surface shown surrounding the rotor iron and completely within the airgap such that sections GK and MH are very close to the stator poles and sections JE and LF are very close to the rotor poles. In these

Figure 3.23 Closed surface embracing the salient-pole rotor of Fig. 3.22.

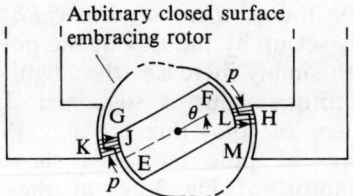

regions the *tensile* stresses across the surface (where they exist) cancel due to symmetry and, being radially directed, contribute nothing to torque. Over the sections of surface KJ and LH, however, of area lg parallel to the assumed flux path, *lateral* stresses p are developed which give rise to a torque of value

$$T = 2rlg\left(\frac{B^2}{2\mu_0}\right) = rlg\frac{B^2}{\mu_0}, \qquad [3.27]$$

where: r is the mean distance from the axis to the airgap,
g is the gap length, and
l is the axial length.

Equation [3.27] appears to suggest that the torque is independent of the relative positions of the stator and rotor poles. Over a large range of values of deflection this is true, but a practical device will yield torque/deflection characteristics of the type shown in Fig. 3.24, the gradient and extent of the non-constant portion being dependent upon the shape of the poles. In particular, an actual device exhibits stable equilibrium at $\theta = 0°$, when the stator and rotor poles are co-aligned. At any general displacement θ the torque acts in such a direction as to tend to restore the movable rotor to this position of equilibrium. The apparent error in the torque calculation is not due to any defect in the method employed but is to the simplifying conditions assumed, which ignored fringing flux in considering only *radial* flux in those areas where stator and rotor poles overlap.

Figure 3.24 Torque/deflection relationship for a rotary actuator, (alignment type).

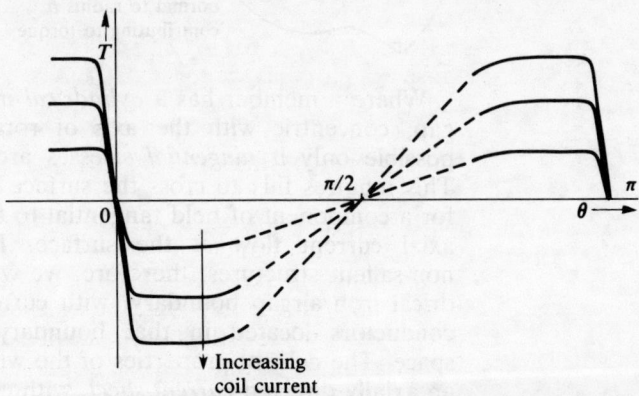

Alternative methods of torque calculation involving incremental changes in inductance (or stored energy) which make the same assumptions regarding the flux distribution lead to the same unreliable torque expression. Over the range of values of θ for which the torque is substantially constant the value of the torque is seen to be independent of the *width* of the poles. The torque is effectively set up by the flux at the pole edges and extension of the pole width simply increases the angular movement over which the constant torque value is sustained. Larger values of torque for given values of total flux crossing the gap may be achieved by castellating the pole faces, as shown in the restricted motion 'torque motor' of Fig. 3.27, at the expense of reducing proportionately the displacement angle over which the torque is maintained substantially constant.

The tendency of salient poles to align themselves with the axis of an excited field system is utilised in the continuous rotation reluctance and stepping motors considered in Chapter 7.

3.11 Torque Development in Rotating Machines with Cylindrical or Salient Iron/Airgap Surfaces

Torque on the stator or rotor members bounding an airgap is produced as a net result of the summation of mechanical stresses distributed over the appropriate surface bounding the airgap. If the member is unexcited, having no current to influence the field in the airgap, the stresses developed on the iron surface ideally are normal to that surface. If that surface is non-cylindrical, as with the salient-pole structure shown in Fig. 3.25, these stresses will have a component normal to a radius drawn from the axis of rotation, so providing a component of torque.

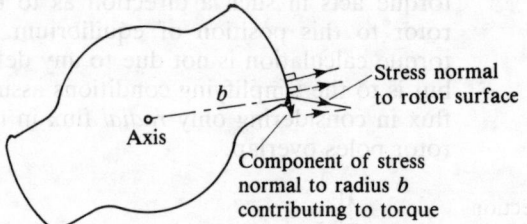

Figure 3.25 Mechanical stress development at the surface of salient-pole rotor.

Where a member has a *cylindrical* iron surface bounding the air gap, concentric with the axis of rotation, torque production is possible only if *tangential* stresses are developed at the surface. This requires flux to cross the surface other than normally, calling for a component of field tangential to the surface and necessitating axial current flow at the surface. In our idealised model of non-salient structures, therefore, we will consider a smooth, cylindrical iron/airgap boundary, with current-carrying axially directed conductors located in that boundary and occupying negligible space. The current properties of the winding will be represented as an axially directed *current sheet*, with current distribution appropriate to the types of winding which are discussed in Chapter 5 and subsequently. As demonstrated in that chapter, current location at an iron/airgap boundary produces in the airgap both tangential and normal components of field. The boundary for a cylindrical surface is everywhere at a constant radius drawn from the axis of rotation, hence summation of the tangential stresses leads directly to torque

Force and torque development in magnetic circuits. Properties of magnetic materials 83

evaluation. Excitation of the winding leads inevitably to the development of distributed tangential stresses at the boundary with the airgap. Whether or not *resultant torque* is developed depends upon the magnitude and space relationships between normal and tangential components of field.

It is helpful to recall eqn [3.22] at this juncture, i.e.

$$t_t = B_n H_t,$$

applying it at an arbitrary point on the airgap/iron surface and noting that the *radial* flux density B_n is dependent upon current at both sides of the airgap, whereas H_t is dependent upon current at the surface concerned alone. These relationships are confirmed in Section 5.4.

3.12 Effects of Slotting

In most practical machine windings the current-carrying conductors are not placed on the surface bounding an airgap but are contained within *slots* cut into the ferromagnetic material and open to the airgap. Such an arrangement gives good mechanical support to the conductors but obviously creates significant local irregularities in the airgap/iron boundary and the distribution of current at that surface. It is arguable, therefore, that the representation of current as a sheet with a simple distribution in space might be an unrealistic alternative to its typical location in slots. Such criticism is valid in that the stress distribution at the surface of the slotted core is locally radically changed (Fig. 3.26), being everywhere at right-angles to the iron surface and generally different *at each side* of the slot. This difference which can account for net torque is due, however, to the tangential field effects of the current contained in the slot alone and the detailed elements of practical machine construction need not invalidate the inherent simplicity of the model.

Apparent from Fig. 3.26 is the fact that in slotted iron structures the bulk of the mechanical stress is developed radially on the teeth which separate the slots, contributing nothing to torque. The conductors within the slots are subjected to force according to eqn [3.3] but, being located in regions of relatively low flux density, the conductors are responsible for only a small fraction of the total torque developed. The major part of the torque arises from stresses developed at the sides of the teeth.

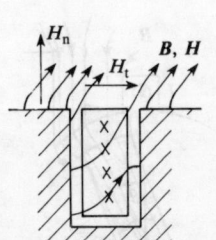

Figure 3.26 Magnetic field and flux at the surface of slotted iron core.

3.13 Rotary Actuator with Permanent-magnet Polarisation

The *Laws relay* is an example of a reluctance-type rotary actuator which employs permanent magnets located on the same member as a 'control' winding to produce a *polarising* field. This field is responsible for the majority of the flux which accounts for the development of torque. The control field effectively 'steers' the flux produced by the polarising magnets in an appropriate direction and the salient, soft-iron rotor follows in accordance with the alignment principle.

The relay is illustrated in Fig. 3.27 which shows a section at right-angles to the axis of rotation and the plane of the coil. In the absence of current excitation the rotor lines up with the axis AA' about which the flux due to the polarising magnets distributes itself symmetrically. The rotor is then held in a position of stable equilibrium – to which position it will tend to return if given a displacement in either direction. When, however, the coil is excited with current of the polarity indicated, the effect of its field is a redistribution of flux such that the portion carried by the right-hand side of the upper stator pole increases at the expense of the left-hand portion, with complementary effects at the lower pole. Since the permeability of the iron is very much greater than that of air, all flux enters or emerges from the iron at right-angles to its boundary with air. Torque on stator and rotor members is due to components of mechanical stress at these surfaces developed at right-angles to a radius drawn from the axis of rotation. The tensile stresses directed along the flux lines normal to the surface at the appropriate sides of the poles and the stator slots are seen to be effective in producing net clockwise rotor torque.

Clearly, the relationship between torque and angular displacement is very much dependent upon the geometric configuration, but suitable design can achieve a high torque/angular displacement rotor with low inertia. If the load torque is small the rotor position is consequently able to follow closely changes in the control

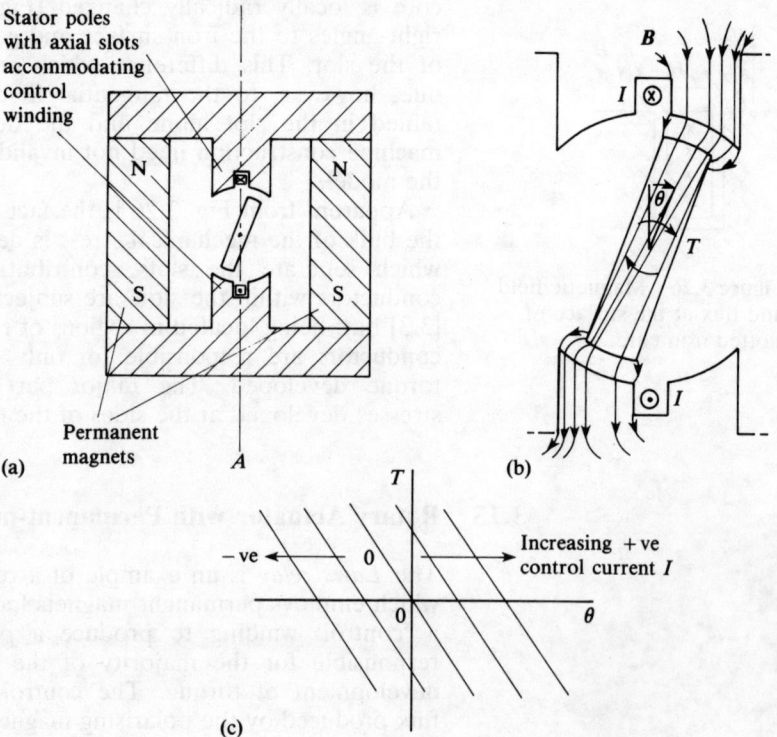

Figure 3.27 Rotary actuator with permanent magnet polarisation: (a) general arrangement; (b) flux detail, (c) torque/displacement characteristic, dependence on control winding current.

winding current. The device finds use in valve control on hydraulic servo systems.

3.14 Direct Current Moving-coil Actuator

This device is well known for its application in the permanent-magnet moving-coil current meter. A multiturn rectangular coil is suspended such that its active coil sides lie in the field of a permanent magnet. The reluctance of the magnetic circuit is kept low by the incorporation of a cylindrical soft-iron core concentric with the axis of coil rotation. Figure 3.28(a) shows the stationary poles shaped such that the permanent-magnet flux density is uniform and radial across the gap. When excited with current the suspended coil experiences a torque due to the force on the active conductor positioned in the airgap, given by eqn [3.3], i.e. $d\mathbf{F} = I(d\mathbf{l} \times \mathbf{B})$ acting at radius r. The torque dependence on position and coil current is shown in Fig. 3.28(b).

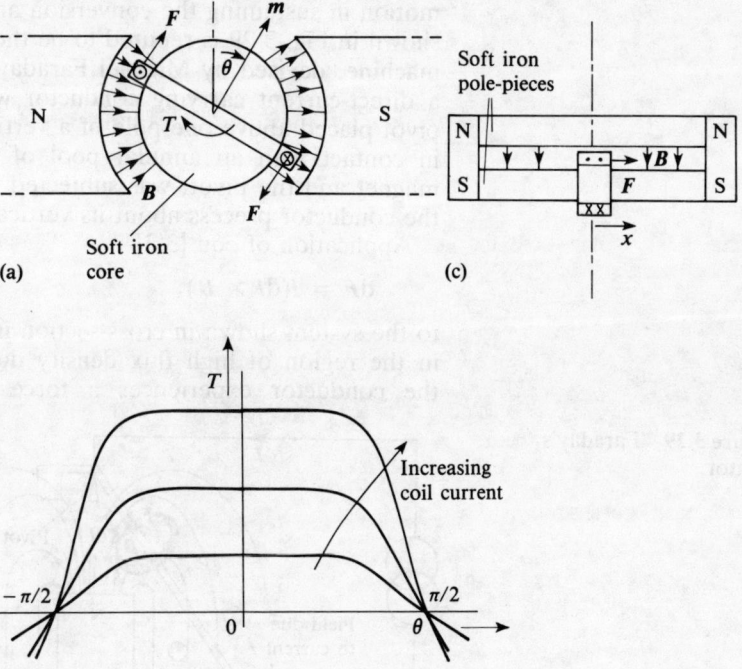

Figure 3.28 Direct-current moving-coil rotary actuator: (a) general arrangement, (b) torque/displacement relationship, (c) linear motion equivalent.

Usually the rotor coil motion is restricted to less than 180° and is opposed by a torsional spring with a linear *rate*, that is the restraining torque is proportional to the angular displacement. This provides for a position of equilibrium in which the deflection is proportional to the exciting coil current, on the asumption that friction or load torques are negligible. Reversal of coil current reverses the direction of the torque.

A variant providing for linear motion is shown in Fig. 3.28(c). This employs two identical permanent magnets and soft-iron pole pieces which provide a uniform field in the rectangular space through which the coil can move. The low-inertia coil straddles the lower bar and has a limited active conductor length. Its self inductance is not insignificant and its field tends to modify the flux distribution in the gap. In consequence, the force capability of the device is limited but applications are found in instrumentation, typified by the pen drive of a chart recorder.

3.15 Faraday's Rotator

All the devices we have so far considered are capable of limited motion only, whether rotational or linear. We shall now look at two elementary devices which are capable of continuous rotation and hence are perhaps better classified as *motors* (and/or *generators*, as the energy conversion process is essentially reversible). This being the case, we shall not simply be content with establishing the mechanism by which the devices produce *torque*, but proceed a step further to show the consequences of continuous motion in sustaining the conversion and flow of energy. The device shown in Fig. 3.29 is reputed to be the world's first rotating electric machine, devised by Michael Faraday in 1821. He discovered that a direct-current carrying conductor with one end supported by a pivot placed above one pole of a vertical bar magnet and the other in contact with an annular pool of mercury concentric with the magnet and the pivot, was subjected to a force which would make the conductor precess about its vertical support axis.

Application of eqn [3.3]

$$d\mathbf{F} = I(d\mathbf{l} \times \mathbf{B})$$

to the system shown in cross-section in Fig. 3.29 demonstrates that, in the region of high flux density due to the permanent magnet, the conductor experiences a force tending to rotate it in an

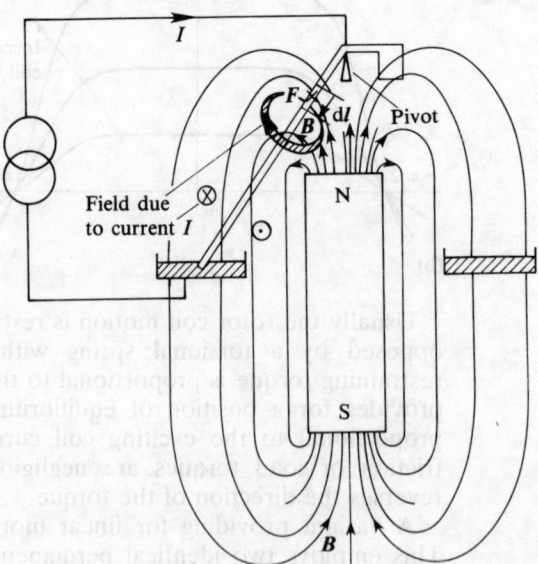

Figure 3.29 Faraday's rotator.

anticlockwise direction as viewed from below. The magnetic field due to the conductor current tends to increase the magnetic flux density behind the conductor as it moves, and to reduce the flux density ahead, in sympathy with the view that the lateral stresses in the air medium are responsible for the torque. If the development of this force results in motion, the velocity of each element dl of the conductor will be at right-angles to the axis of the conductor and orthogonal also to some component of external flux density B. Application of eqn [2.39]

$$E = v \times B,$$

shows that the electromagnetically induced electric field within the conductor is predominantly in such a direction as to oppose the flow of current. The conductor thus absorbs electrical energy from the source, as we would expect.

Such a device, then, is an elementary electric motor. It is called a *homopolar* device because its active conductor experiences a magnetic field of unchanging polarity as it moves.

3.16 Faraday's Disc

For reasons which should be apparent, the rotator described above is quite ineffective as an energy converter. A more useful device, also due to Faraday, replaces the pivoted conductor by a pivoted, circular conducting disc across which current flows radially via rubbing contacts with the disc centre and rim. Clearly, the current density will reduce as the radius of an elementary annular ring increases. Torque is produced by the interaction of the field produced by the disc current and a uniform external field applied across the disc. Application of eqn [3.3]

$$d\boldsymbol{F} = I(d\boldsymbol{l} \times \boldsymbol{B})$$

to the element of conductor dl shown in the cross-sectional view of Fig. 3.20(a) demonstrates anticlockwise torque development on the disc when viewed from above, with current and external flux density directed as shown. Again, eqn [2.39]

$$\boldsymbol{E} = \boldsymbol{v} \times \boldsymbol{B}$$

confirms that, if the disc is allowed to rotate under the influence of the torque, the electromagnetically induced electric field, radially directed, opposes the flow of current derived from an appropriate source, thus putting the source on load. If the disc were *driven* in the opposite direction the energy flows would reverse, and the device would function as a generator.

The relationship between torque, current, uniform external flux density and the outer and inner radii of the disc is given by eqn [3.28].

$$T = \tfrac{1}{2}BI(r_0^2 - r_i^2). \tag{3.28}$$

Proof With reference to Fig. 3.30(c), consider an elementary section of an annular

88 Electrical machines and drive systems

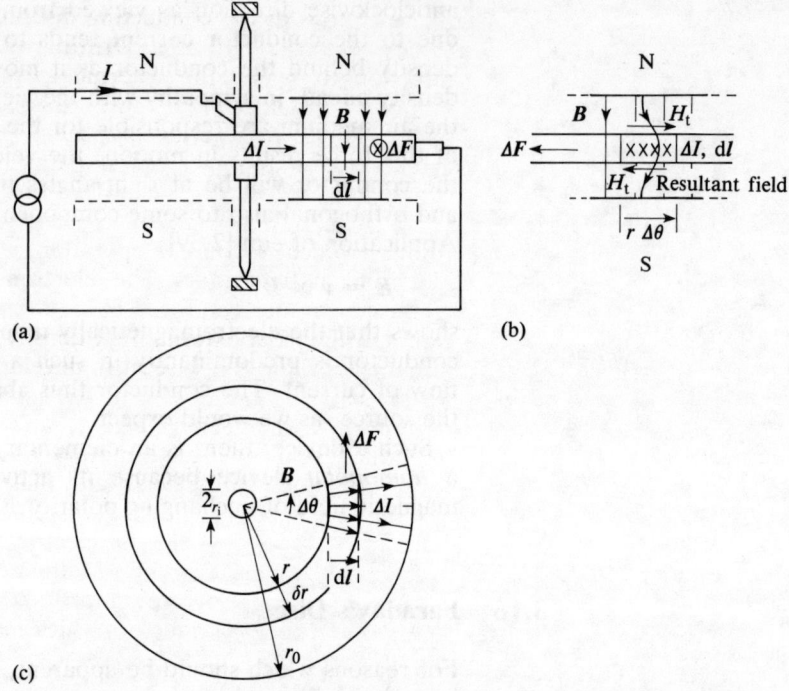

Figure 3.30 Faraday's disc: (a) elevation, (b) slice across part of the disc in the plane of an axial external magnetic field, (c) view from above the disc.

ring at radius r, subtending angle $\Delta\theta$ at the centre of the disc and carrying current ΔI in the radial direction. The current flow is at right-angles to the axially directed external flux density B and elemental force ΔF is developed tangentially, of value

$$\Delta F = \Delta I (dl \times B)$$

$$\Delta F = I \frac{\Delta\theta}{2\pi} \delta r B. \qquad [3.29]$$

The same expression for ΔF is obtained by integrating the tangential *stress* over the upper and lower surfaces of the elementary section having total area $2r\Delta\theta\delta r$. Over this surface, bounding the elementary section viewed in Fig. 3.30(c), the *tangential* field due to the current is given by

$$H_t = \frac{\Delta I}{2r\Delta\theta} = \frac{I}{4\pi r}.$$

Since the *normal* flux density at each surface is B, the tangential stress on both surfaces, given by eqn [3.22] is

$$t_t = B_n H_t = \frac{BI}{4\pi r}.$$

Stresses on both surfaces are directed to the left in Fig. 3.30(b) and give rise to elemental force, in agreement with eqn [3.29].

The torque contribution due to all such elements of the annular ring is obtained by integrating eqn [3.29] with respect to θ and evaluating between limits 2π radians apart, then multiplying by the radius r at which the force due to the annulus acts. Total torque is then evaluated as

$$T = \int_{r_i}^{r_0} BIr dr = \frac{1}{2} BI(r_0^2 - r_i^2). \qquad [Q.E.D.]$$

In deriving eqn [3.28] we have assumed radial current flow. The free electrons in the conducting disc, however, are not constrained to flow in particular paths. Drifting with a component in the radial direction to satisfy the external current requirement their radial motion is at right-angles to the axial external field and consequently the electrons experience a *tangential* force which makes their actual motion follow a spiral path. For a disc material of high conductivity, however, the current flow is predominantly radial.

Up to this point we have not considered the effect of disc rotation which, for *motor* operation, will be in the direction of the electromagnetic torque. The electrons in the disc experience in consequence another force at right-angles to their *tangential* motion and the axial external flux density. This force acts radially and accounts for the 'back' e.m.f. developed between the brush contacts, opposing current flow for motor action and assisting the current flow when generating.

Permanent-magnet disc motors have the advantage of a high ratio of torque to inertia and find application where a quick response is necessary, with ratings up to about 5 kW. The basic homopolar form develops its electrical rating as a product of low voltage and high current. Multipole, *heteropolar* variants are preferred with the rotor current constrained to flow in paths defined by copper conductors which may be 'printed' directly on to the insulating disc.

3.17 Permanent Magnetism

In Section 3.4 it was noted that a ferromagnetic material subjected to an external magnetic field which is subsequently reduced to zero possesses residual or remanent flux at zero H due to hysteresis. Ideal permanent magnet material will have a high value of *remanence*, or residual flux density B_r, and an ability to maintain this high flux density when the applied external field is reversed in polarity. Permanent-magnet field systems offer several desirable features in application to electrical machines where torque is developed by the interaction of a nominally constant magnetic field with one derived from a variable armature current, If such a polarising field is derived from a permanent magnet, the exciting winding and current source of the electromagnetic alternative are dispensed with. Costs may be saved in manufacturing due to simpler assembly and more efficient running secured through the absence of field circuit copper loss. Furthermore, temperature changes with load may have a lower influence on permanent magnet flux than that derived from an electromagnet whose coil resistance is temperature dependent. Disadvantages also accrue to permanent-magnet field systems, of which the most important are limited controllability and the possibility of loss of permanent magnetism resulting from abnormal loading or dismantling of the magnetic circuit during repair.

Not all permanent-magnet materials suffer this disadvantage, however, and the attributes of recent developments have given rise to a rapid increase in the demand for permanent magnet machines. Such growth is set to continue as properties improve and costs fall.

3.17.1 Magnetic circuit modelling of permanent-magnet systems

Permanent-magnet applications to electrical machines are characterised by having a magnetic circuit which includes an airgap and links externally sourced armature current. Whilst the effect of the airgap is invariably to tend to demagnetise the 'permanent' magnet, that of the armature current may assist or oppose the permanent-magnet field in the different regions of the magnetic material. Since the B/H relationship for the latter is essentially non-linear, and the distributed nature of the armature current renders strictly inappropriate the designation of simple leakage flux paths, the precise evaluation of permanent-magnet motor field conditions under load requires the use of modern computer methods of field analysis. Nevertheless, the principles involved may be established by considering the behaviour of a simplified model magnetic circuit comprising a permanent magnet in series with an airgap, joined by soft-iron pole pieces of infinite permeability. Where leakage flux paths are considered, these are regarded as being in parallel with the airgap. The nature and influence of armature current on the resultant flux is essentially different for a.c. and d.c. machines and the distinctive effects will be discussed in later chapters devoted to the study of these particular machine types.

3.17.2 The demagnetisation characteristics of permanent-magnet materials

Suppose that a sample of iron alloyed with aluminium, nickel and cobalt (Alnico) in the form of a toroid is subjected to a cycle of magnetisation in the manner by which the hysteresis loop of Fig. 3.8 was evaluated. The first quadrant of the B/H plot demonstrates that the material has been driven well into saturation. In the application of permanent-magnet materials it is essentially the *second* quadrant of the B/H plot which is the main region of interest, where the demagnetisation field acts to reduce the flux density below the remanence value B_r, at first gradually but then with increasing slope beyond a 'knee point'. For a small value of $-H$, its subsequent removal may result in recovery of the flux density to the remanent value B_r. If, however, the value of the demagnetising field is larger (H_X in Fig. 3.31) prior to its reduction towards zero, the subsequent increase in B is less than that corresponding to the earlier reduction in B.

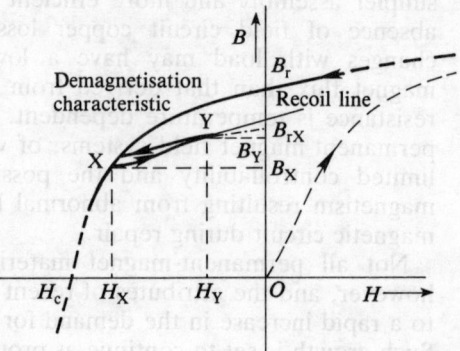

Figure 3.31 Demagnetisation characteristic of permanent-magnet material.

Force and torque development in magnetic circuits. Properties of magnetic materials 91

If the reduction in demagnetising field takes H to zero, the new value of residual flux density $B_{rX} < B_r$. If the reduction is less (to H_Y in Fig. 3.31), subsequent increase to H_X may restore the flux density to very nearly B_X. During these changes in H the changes in B follow a *minor hysteresis loop*, whose average slope $(B_Y - B_X)/(H_Y - H_X)$ is somewhat less than the average slope of the major hysteresis loop in this region $(B_r - B_X)/(O - H_X)$. With many materials the restoration of the flux density to B_X when H is restored to H_X from H_Y is not quite accomplished, but repetition of the excitation change procedure $H_X \to H_Y \to H_X$ gives rise to a reducing discrepancy between the start and finish values for B at H_X. Such a procedure is repeated several times during *stabilisation* prior to service in applications where similar changes in magnetisation will be encountered, so ensuring consistent performance – provided H_X is not exceeded.

The line XY in Fig. 3.31 is called the *recoil line*, having, in good permanent-magnet materials, a slope which almost equates to that of the major hysteresis loop in the region and tends to the value μ_0. The integrity of the recoil line should also extend over a large range of values of H. The more-recently developed permanent-magnet materials exhibit these features, combining them with high values of residual flux density B_r. Traditional nickel compounds of iron offer high B_r but pronounced curvature in the second quadrant of B/H and a correspondingly low value of coercive force (*coercivity*). The practical effect of this is to limit the value of demagnetising field which the magnet might encounter in service without significant loss of remanence B_r when the demagnetising influence is removed. Freely available ceramic compounds of iron, e.g. *strontium ferrite* may be made to have a high coercive force, and recoil lines with low *recoil permeability* approximating to that of the major hysteresis loop over the whole of the second quadrant, but with a low remanence. The modern *rare-earth* compounds with *cobalt* only have the disadvantage of high cost.

Figure 3.32 shows a portion of the major B/H loop for the three

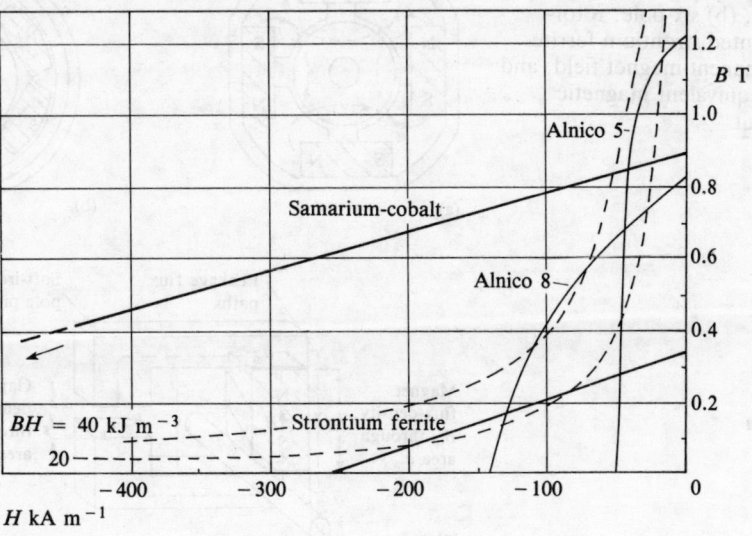

Figure 3.32 Second-quadrant B/H characteristic for permanent-magnet materials: Alnico 5, Alnico 8, strontium ferrite, samarium-cobalt.

main types of permanent-magnet material with, for comparison purposes, that of a low hysteresis-loss transformer steel. For samarium-cobalt and strontium ferrite only, the B/H characteristic shown coincides with the recoil line throughout the entire second quadrant.

3.17.3 Demagnetisation by incorporation of an airgap

The usual purpose of a permanent magnet is to produce magnetic flux density in an airgap. Many ingenious practical arrangements are possible and those appropriate to an electrical machine application normally involve a composite of magnetic material, and soft-iron pole pieces (pole shoes) to define the airgap, with the circuit completed via a laminated soft-iron structure supporting the armature winding.

Two such configurations for d.c. machines are shown in Fig. 3.33(a) and (b), for which the essential features are illustrated in the simplified arrangement of Fig. 3.33(c), including current linked by the magnetic circuit to represent the effect of armature current. For the present, however, such current will be disregarded and the effect of the airgap alone considered. If the effects of flux leakage and fringing are ignored and the airgap is assumed to be of uniform cross-sectional area a_g and length l_g, with the permanent-magnet dimensions similarly a_m and l_m, application of Ampère's law, eqn [2.27], with the magnetic circuit closed through pole pieces of infinite permeability gives

$$\oint \mathbf{H} \cdot d\mathbf{l} = H_m l_m - H_g l_g = 0$$

thus

$$H_m = -H_g \frac{l_g}{l_m}. \qquad [3.30]$$

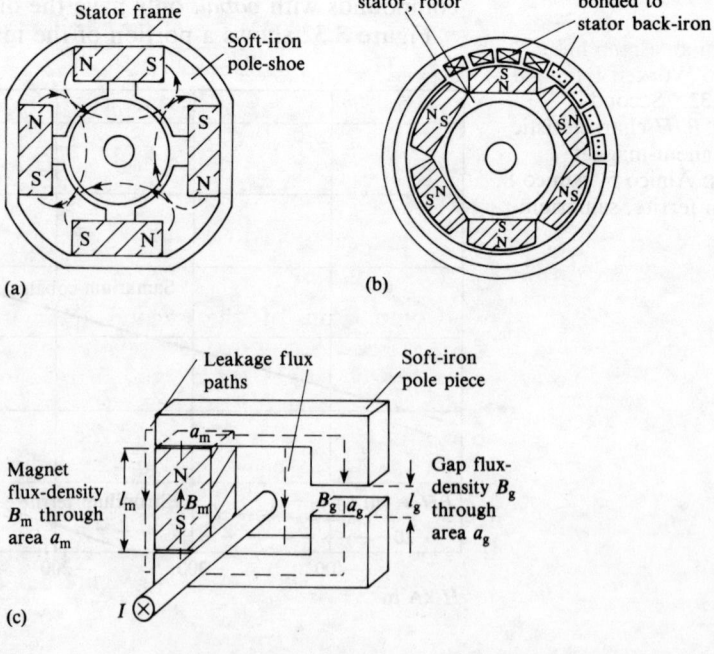

Figure 3.33 Permanent magnet d.c. machines: (a) four-pole, stator-mounted Alnico permanent-magnet field, (b) six-pole, rotor-mounted strontium ferrite permanent-magnet field, and (c) equivalent magnetic circuit.

Since $B_g = \mu_0 H_g$ and a common value flux traverses the circuit

$$H_m = -\frac{B_g}{\mu_0}\frac{l_g}{l_m}$$

and

$$B_m = B_g \frac{a_g}{a_m}. \quad [3.31]$$

Thus the magnet flux density and field intensity are related through

$$\frac{B_m}{H_m} = -\mu_0 \cdot \frac{a_g}{a_m}\frac{l_m}{l_g}. \quad [3.32]$$

Equation [3.32] defines the equation of a *load line* plotted with B and H co-ordinates, which requires the operating point of the airgapped magnet to lie at its intersection with the demagnetisation characteristic defined by an appropriate recoil line. See Fig. 3.34(b). The minus sign confirms that the load line lies in the second (and fourth) quadrant. Increasing the value of l_g reduces the slope of the load line – called the *airgap line* – and increases the degree of demagnetisation of the magnet. The airgap line has the effect of referring the airgap reluctance to the dimensions of the magnet. Up to a certain limit, increasing the airgap length increases the magnetic field energy stored therein, which is given by $\tfrac{1}{2}B_g H_g a_g l_g$. Substitution for B_g and H_g using eqns [3.31] and [3.30] shows that the gap-stored energy equates to $\tfrac{1}{2}B_m H_m a_m l_m$, and hence is a maximum for a given magnet volume when the operating condition is such that the product $B_m H_m$ peaks.

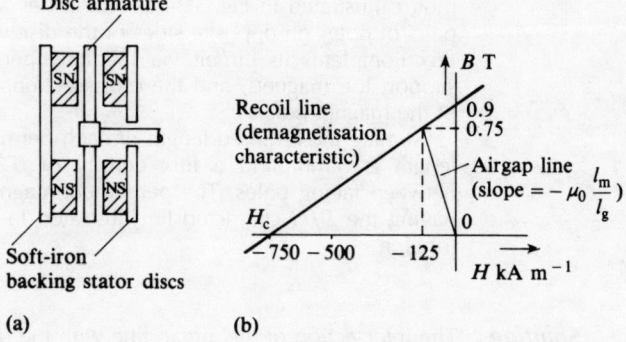

Figure 3.34 (a) Disc armature d.c. motor with ferrite permanent-magnet field. (b) Demagnetisation characteristic for samarium-cobalt with an airgap line relevant to Worked Example 3.3.

Comparison of the second quadrant B/H characteristics of Alnico and strontium ferrite presented in Fig. 3.32 suggests that the higher values of magnet *energy product* $B_m H_m$ are achieved with a combination of large B_m and small H_m for Alnico, and small B_m and large H_m for ferrite. For given airgap dimensions a_g and l_g, the high-slope airgap line indicated for Alnico is achieved by proportioning the magnet dimensions such that l_m/a_m is large, i.e. the magnet is long and thin. Conversely, for optimal use of ferrite material – and the rare-earth compounds – the magnet is proportioned short and fat. Figure 3.33(a) and (b) illustrate these points.

3.17.3.1 Leakage

The foregoing considerations have disregarded leakage flux in paths by passing the airgap of interest. Leakage flux is driven by the difference in magnetic (scalar) potential existing between the pole surfaces of the permanent magnet, which ideally equates to that between the pole pieces which bound the airgap in Fig. 3.33(c). For a required gap flux density the leakage flux becomes a larger proportion of the total magnet flux as the airgap length is increased. The *leakage factor* k_1 is used to define the ratio of total magnet flux to useful airgap flux, so that eqn [3.31] becomes

$$B_m = k_1 B_g \frac{a_g}{a_m}$$

and the airgap line is redefined by the identity

$$\frac{B_m}{H_m} = - k_1 \mu_0 \frac{a_g}{a_m} \frac{l_m}{l_g}. \qquad [3.33]$$

The value of k_1 depends on the magnetic circuit configuration and is minimal where the airgap is short and in close proximity to the magnet. Good design requires the accurate prediction and minimisation of leakage flux in order to minimise magnet volume and hence cost. Modern methods of field analysis enable calculations to be carried out such that the effects of incorporating saturable iron and the magnetic and thermal history of the materials involved may be established.

Worked example 3.3 The permanent-magnet field system of the double-sided (heteropolar) disc motor illustrated in Fig. 3.34(a) establishes an axially directed flux between pairs of poles on opposite sides of the disc winding. The permanent-magnet flux completes its circuit via infinitely permeable mild steel rings which support the magnets, and the cross-sectional area of the airgap equals that of the magnet poles.

Estimate the required length of each permanent magnet if the total airgap length is 5 mm and a flux density of 0.75 T is required in the airgap between facing poles. The permanent-magnet material is samarium-cobalt, having the *B/H* characteristic illustrated in Fig. 3.32. Neglect leakage and fringing.

Solution The intersection of the airgap line with the demagnetisation characteristic of samarium-cobalt is required at $B_m = B_g = 0.75$ T. From eqn [3.32].

$$\text{Slope of the airgap line} = -\mu_0 \frac{a_g}{a_m} \frac{l_m}{l_g} = -\mu_0 \frac{l_m}{l_g}$$

$$= -4\pi \times 10^{-7} \times 0.2 \times 10^3 \, l_m.$$

Intersection of the airgap line and the demagnetisation characteristic is shown in Fig. 3.34(b) to occur at $B_m = 0.75$ T, $H_m = -750 \times 10^3 \times 0.15/0.9 = -125 \times 10^3$ A m^{-1}. Equating B_m/H_m to the slope of the airgap line gives $l_m = 0.75/(125 \times 10^3 \times 4\pi \times 10^{-7} \times 0.2 \times 10^3) = 0.024$ m. As a pair of magnets is concerned with each gap crossing, the length of each magnet is 12 mm.

3.17.4 Effect of linked current on permanent-magnet circuits

As noted in the previous section, the effect of creating an airgap within a magnet circuit incorporating a permanent magnet is to exert a demagnetising influence on the magnet. If the magnetic circuit embraces current, the sum of the $H \cdot l$ products around the loop is no longer zero but equates to the enclosed current. Depending on the polarity of the current, its influence may be to increase or reduce the flux density. If the effect is demagnetising and an Alnico-type material is in use, care must be taken to ensure that the operating point is not displaced from the recoil line on to the demagnetising characteristic if it is required that the original magnet flux should be recovered when the external field is removed.

The effect of superimposed electromagnetic excitation on the operating point of a permanent-magnet circuit incorporating an airgap may be accommodated by considering the base of the airgap line to be displaced along the H axis by an amount H_e equivalent to the linked circuit current divided by the length l_m of the permanent magnet. If the effect is demagnetising the airgap line is moved to the left, as illustrated in Fig. 3.35.

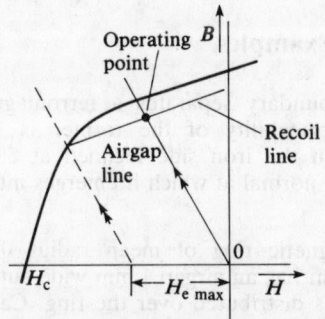

Figure 3.35 Demagnetisation characteristic for Alnico permanent-magnet material illustrating effect of demagnetising linked current on airgap line location.

3.17.5 Review of permanent-magnet materials

The metallic alloys of aluminium, nickel and cobalt in varying proportions with iron are characterised by high remanence (up to 1.2 T), low coercivity (up to 120 kA m^{-1}), and a maximum energy product $(B_m H_m)_{max}$ of around 100 kJ m^{-3}. The slope of a typical recoil line, the *recoil permeability*, is not less than $2\mu_0$ in a restricted region of linear response to changes in excitation. Being mechanically hard materials, such alloys present machining problems and the curvature of the B/H characteristic necessitates magnetisation after final assembly when already incorporated in electrical machines. One inherent advantage is their good stability of magnetic parameters against changes in operating temperature. Ceramic mixtures of powdered oxides of iron, such as strontium ferrite, have a linear demagnetisation characteristic with a high coercivity of up to 250 kA m^{-1}, a recoil permeability of typically, 1.1 μ_0 and a corresponding low remanence of about 0.4 T. Although their maximum energy product is much less than that attainable with Alnico materials and the effect of temperature is significant, strontium ferrites are much used in d.c. motors. The desired permanent-magnet field pattern, which may be complex, is established by forming the material into the required final shape in a magnetic field before firing.

The rare-earth materials such as samarium-cobalt offer high values of remanence and coercivity at, typically, 0.9 T and 750 kA m^{-1}, with a linear demagnetisation characteristic of slope $\sim \mu_0$ and low temperature coefficients. Costly, but with high energy densities, they are used in aerospace applications and for computer memory disk drives. Much current development is based on the use of alloys incorporating neodymium-iron-boron. Maximum energy densities in excess of 400 kJ m^{-3} are achievable in anisotropic form, but an isotropic variant suitable for injection moulding offers energy densities at around 40 kJ m^{-3}, greater than that of the popular ferrites. Increasing use of this material is expected as the price falls in response to the developing market for permanent-magnet machines, the supply of samarium being limited to a much greater extent than that of neodymium. The principal disadvantages of NdFeB alloys at the present time are a relatively low Curie temperature and high temperature coefficient of coercivity.

3.18 Tutorial Examples

3.1 A plane boundary separates a ferromagnetic medium from air. If the relative permeability of the former is 1000 and flux approaches the boundary on the iron side inclined at 88° to the normal, calculate the angle to the normal at which it emerges into air.

[1.64°]

3.2 A ferromagnetic ring of mean radius 0.3 m and circular cross-section radius 0.08 m has an airgap 1 mm wide cut in a radial plane. A winding of 1000 turns is distributed over the ring. Calculate the current necessary to develop a flux density of 0.5 T in the series magnetic circuit. Neglect fringing and assume that $B = 0.5$ T, $H = 1600$ A T m^{-1} is a point on the (non-linear) magnetisation characteristic for the iron.

[3.41 A]

3.3 If the current through the winding of **3.2** is increased to 4.5 A, calculate the force developed on the plane iron/air surfaces defining the airgap. Assume a value of 1000 for the relative permeability of the ferromagnetic material at the appropriate operating point.

[30.7 kN]

3.4 A cylindrical plunger solenoid is shown in cross-section in Fig. 3.36 and has a plunger of cross-sectional area 15 cm^2 free to move through a circular hole in the surrounding magnetic material. The coil has 3000 turns, 8 Ω resistance and is connected to a 12 V d.c. supply. The magnetic material may be considered to be infinitely permeable up to its saturation flux density of 1.6 T and the airgap between plunger and outer shell can be considered to be negligible.

(i) Determine a general expression for the static force F on the plunger as a function of the working airgap length g and the coil current I.

(ii) Over what range of gap length g is the force of the plunger essentially constant because saturation flux density has been reached, and what is the magnitude of this force?

Figure 3.36 Plunger solenoid.

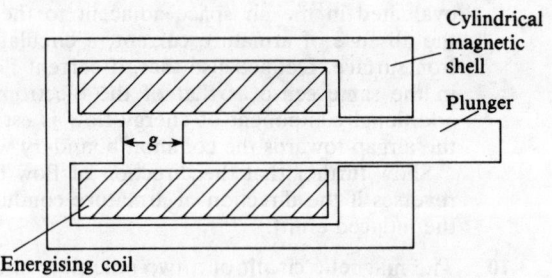

(iii) Suppose the plunger is constrained to move slowly from a gap of 1 cm to the fully closed position, what will be the mechanical work done?

(iv) If the plunger is allowed to close so quickly from an initial gap of 1 cm that the flux linkage of the coil does not change during motion, what is then work done?

U. of B. [$F = \frac{1}{2}(NI)^2 \mu_0 a g^{-2}$; $g < 3.534$ mm, 1 530 N; 8.91 J; 1.19 J]

3.5 Use an energy-method implied equation [3.23] to *estimate* the force developed on the conductor identified in Fig. 3.1 if it is held stationary but in contact with long, parallel, fixed rails whose remote ends are connected, not by resistor R, but by a 100 A current source. The rails are 0.2 m apart and of radius 5 mm. The inductance per loop metre of (long) parallel conductors (excluding internal flux) is given by eqn [2.50]. Is the force uniformly distributed over the conductor length?

[7.4×10^{-3} N; *No*]

3.6 An external field system applied to a Faraday's disc motor establishes an axial flux density of 0.7 T over the entire disc surface. If the inner and outer radii of the rotor are 20 mm and 100 mm, respectively, evaluate the common factor that relates radial induced e.m.f. to angular velocity and electromagnetic torque to current supplied.

[3.36×10^{-3} V s rad^{-1}, N m A^{-1}]

3.7 If the rotating disc of **3.6** is 2 mm thick and made of copper having conductivity 5.7×10^7 S m^{-1}, calculate the d.c. resistance presented by the disc to radial current flow.

[2.25×10^{-6} Ω]

3.8 Show that a local application of external magnetic flux normal to a rotating disc of finite conductivity exerts a braking torque approximately proportional to the speed.

Show how the (homopolar) disc motor may be reconfigured with the solid disc replaced by radial conductors resembling the spokes of a wheel, connected at rim and axle. Suggest also how the radial conductors lying under adjacent poles might be interconnected to effect the summation of motionally induced e.m.f.s in a *heteropolar* variant for which the active conductors experience an external field of alternating polarity as the rotor rotates. Assume a single-circuit (armature) winding with one conductor per pole connected to the external circuit via slip rings at the inner radius. The device is an elementary a.c. generator.

3.9 In an unloaded d.c. machine, parallel, axially directed armature conductors on the surface of a non-conducting ferromagnetic cylinder sweep past a stationary, radial field system derived from a d.c. electromagnet or permanent magnet, such that the conductors lie at right-angles to both the direction of motion and the field flux. Show that the Poynting vector evaluated in the air space adjacent to the rotor conductors indicates, in the absence of armature current, a circulation of energy tangential to the iron surface. Deduce also that, if current flows in the armature conductors in the same sense as that of the electromagnetically induced e.m.f., an additional component of energy flow is established which is directed from the airgap towards the common boundary with iron and conductor.

Show further that the direction of flow of the latter energy component reverses if the direction of armature conductor current changes to oppose the induced e.m.f.

3.10 The magnetic circuit of a two-pole permanent-magnet d.c. motor is shown in cross-section in Fig. 3.37. The rotor radius is 22 mm and the airgap length is 2.5 mm. The stator magnets are to be magnetised radially and made of strontium ferrite having a demagnetisation characteristic such that both the second quadrant B/H relationship and the recoil line are given by the equation

$$B(\text{T}) = 0.34 + 1.308 \times 10^{-6} H(\text{A m}^{-1})$$

Determine the radial thickness of each magnet sector required to produce a radial flux density of 0.25 T in the airgap. Assume a leakage coefficient of 1.1, neglect fringing and the reluctance of the soft-iron stator yoke and rotor core. Determine also the radial flux/pole if the pole arc/pole pitch ratio is 0.7 and the axial length of stator pole and rotor core is 50 mm.
[Note: The pole arc/pole pitch ratio defines the effective proportion of rotor circumference experiencing radial flux due to the stator poles.]

[6.6 mm, 0.605 mWb]

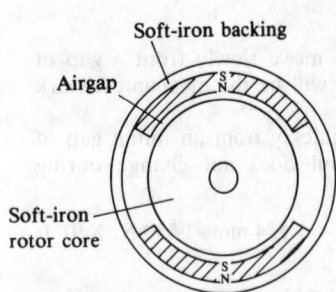

Figure 3.37 Magnetic circuit of a two-pole permanent-magnet d.c. motor.

3.11 A permanent magnet is required to develop a flux density of 0.6 T in an airgap of length 3 mm and square cross-sectional area 60×60 mm^2. The airgap is defined by soft-iron pole pieces of infinite permeability with the general configuration of Fig. 3.33(c). Estimate the length and cross-sectional area of a rectangular permanent magnet having a minimal volume of (i) Alnico 5, and (ii) strontium ferrite with the characteristics given in Fig. 3.32. Assume a leakage factor of 1.5 for Alnico and 2.0 for the ferrite.

[35.8 mm, 3249 mm^2; 13 mm; 21 600 mm^2]

3.12 The permanent-magnet circuit of **3.11** is configured using strontium ferrite of length 13 mm and cross-sectional area 21 600 mm^2. Calculate the field energy stored in the airgap and compare with one-half the energy product times volume of magnet material. Calculate also the new values of gap flux density and required energies on reduction of the airgap length from 3 mm to 2 mm if the leakage factor falls from 2.0 to 1.5.

[1.547 J, 3.09 J; 0.82 T, 1.93 J, 2.91 J]

3.13 A permanent-magnet system incorporating Alnico 8 material is stabilised along a recoil line which intersects the demagnetisation characteristic given in Fig. 3.32 at a magnet flux density of 0.35 T and provides 0.7 T at $H = 0$. The magnet is dimensioned so as to develop flux in a specified airgap defined by infinitely permeable soft iron, such that the magnet flux density is 0.5 T. If the length of the magnet is 50 mm, calculate the maximum demagnetising ampere-turns permitted to link the circuit without incurring any reduction in permanent magnetisation.

[3550 A]

3.14 Account qualitatively for the change in stored magnetic energy distributed between the main airgap field, leakage field and permanent magnet on removal of the soft-iron rotor of a d.c. machine with permanent-magnet stator poles. An input of mechanical energy is necessary to achieve this.

3.15 A homopolar d.c. machine has a permanent-magnet field directed radially across the airgap. The uniformly distributed and axially directed armature current I is carried by a thin conducting cylinder at radius r_0, concentric with cylindrical N and S pole surfaces. Calculate the magnetic field distribution within the airgap due to the armature current.

Hence show that integration of the tangential mechanical stress over the inner surface, radius r_i, of the outer pole gives rise to stator torque of value $T = B_n I l r_i$, where B_n is the radial flux density at the inner surface of the outer pole and l is the axial length of stator and rotor.

Show also that the same value of torque is developed on the conducting cylinder which constitutes the rotor and sketch the resultant field distribution over both parts of the airgap.

4 The transformer

4.1 Introduction

In this chapter we look closely at the particular electrical 'machine' which is mainly responsible for the provision of commercial electrical power supplies at virtually every industrial conurbation or isolated hamlet within all developed economies. The standard domestic service facility provided by a local electricity board in the United Kingdom makes available, on demand, a reliable power supply of the order of 10^4 W. The quality of the supply is indicated by its availability and voltage waveshape (ideally a pure sinusoid) and the stability of its voltage amplitude and frequency.

A clue to the identity of the device in question is incorporated in the above statement implying that the supply is of alternating not direct current. Many useful applications of electrical power require a d.c. supply so that a.c. is not immediately appropriate – *rectification* is required before use. The probability is, however, that some change in a.c. voltage level will be required before conversion to the correct d.c. level and it is the *electrical transformer* which transforms electrical power at one level of voltage to, ideally, the same power at another voltage level. The power transformer is an essential element of an a.c. power transmission and distribution network, as electrical energy is most economically transmitted over long distances at high voltage and low current. Such high voltages around 275 kV or 400 kV are not appropriate to power generation or distribution at around 33 kV, with further reduction to 400/230 V for domestic three-phase/single-phase supplies.

Electrical transformers vary vastly in power rating – from several hundred MW in a power transmission network to 1 MW or less in electronic equipment. All have the feature of a magnetic core linked by more than one electric circuit. The principal electromagnetically induced e.m.f. in each winding is then proportional to the number of turns linked by the common or mutual flux.

In Section 2.4.11 we recognised that *mutual inductance* exists between electrical circuits which share a common magnetic circuit. A change of current in one circuit affects the flux linkage of the other. Equations [2.42] and [2.43] relate the e.m.f.s induced in one circuit to the rate of change of current in the other through mutual inductance coefficients. It may readily be shown that, in the case of a linear magnetic circuit with $B \propto H$, the mutual inductance coefficients M_{12} and M_{21} applicable to circuits 1 and 2 which share a common flux Φ_m are equal.

4.2 Voltage and Current Relationships in an Ideal Two-winding Transformer

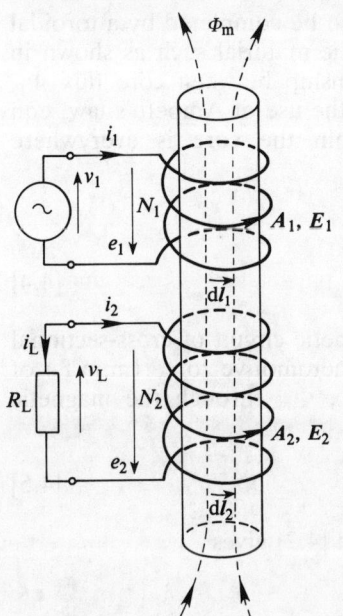

Figure 4.1 Two electrical circuits linked by common flux Φ_m.

The fundamental concepts relating to electromagnetic induction introduced in Section 2.5 may be recalled in the static coupled circuits of Fig. 4.1 in which a magnetic core of circular cross-section carries a mutual flux Φ_m which is assumed to be the only source of flux linkage for each circuit. Around each circuit the induced e.m.f. is derived as the line integral of the electromagnetically induced electric field E, which is equal and opposite to the time rate of change of magnetic vector potential A, which is itself related through its line integral to the enclosed flux. These relationships introduced in Chapter 2 are summarised thus:

$$\left. \begin{array}{l} e = \int E \cdot dl \\ E = -\dfrac{\partial A}{\partial t} \\ \oint A \cdot dl = \Phi. \end{array} \right\}$$

If we assume that E, A and dl are all directed along a plane at right-angles to the flux Φ_m we can calculate the induced e.m.f. per turn as

$$\begin{aligned} e_{\text{turn}} &= \oint -\frac{d}{dt}(A \cdot dl) \\ &= -\frac{d}{dt}\Phi_m \end{aligned}$$

c.f. eqn [2.40]. Hence, if winding 1 has N_1 turns, closely wound together and linking the same flux Φ_m, the winding e.m.f. is given by

$$e_1 = -N_1 \frac{d\Phi_m}{dt}. \qquad [4.1]$$

Further, if winding 2 has N_2 turns, also closely wound with each turn linking the same flux Φ_m, this winding has e.m.f.

$$e_2 = -N_2 \frac{d\Phi_m}{dt}. \qquad [4.2]$$

It immediately follows from eqns [4.1] and [4.2] that the ratio of the induced e.m.f.s in the two mutually coupled circuits is given by the ratio of the winding turns, i.e.

$$\frac{e_1}{e_2} = \frac{N_1}{N_2}. \qquad [4.3]$$

This very simple relationship lies at the heart of transformer operation, but it is critically dependent upon the uniformity of the time rate of change of magnetic flux linking each turn in each winding. In practice, this is difficult to achieve if Φ_m is to be regarded as the only magnetic flux associated with each winding. Thus far we have not considered how Φ_m might be created, or the nature of the magnetic circuit within which it is supposed to flow. These details are strictly irrelevant to the validity of eqn [4.3]. Figure 4.1, however, has incorporated a *source* associated with winding 1, therefore one might presume that the flux Φ_m is established by the magnetic field created by the current flowing in the designated *primary* winding 1.

If we consider the magnetic circuit to be completed by a toroidal core of high-permeability ferromagnetic material such as shown in Fig. 4.2, we can establish the relationship between core flux Φ_m and primary winding current through the use of Ampère's law, eqn [2.27], assuming that the field within the core is everywhere uniform at value $H = B/\mu$. Thus

$$\oint H \cdot dl = N_1 i_1$$

$$\Phi_m = Ba = \mu \frac{a}{l} N_1 i_1 = P N_1 i_1, \qquad [4.4]$$

where P is the *permeance* of the magnetic circuit of cross-sectional area a and length l. $N_1 i_1$ is the magnetomotive force (m.m.f.) of winding 1 responsible for driving flux Φ_m through the magnetic circuit, whose permeance is given by

$$P = \mu \frac{a}{l}. \qquad [4.5]$$

Substituting eqn [4.4] in eqns [4.1] and [4.2] gives

$$e_1 = -N_1^2 P \frac{di_1}{dt}$$

$$e_2 = -N_2 N_1 P \frac{di_1}{dt}.$$

By analogy with eqns [2.41] and [2.43], $N_1^2 P$ is the *self inductance* L_1 of winding 1 and $N_2 N_1 P$ is the *mutual inductance* M_{21} relating the e.m.f. induced in winding 2 to the rate of change of current in winding 1.

If we interchange the roles of windings 1 and 2 so that winding 2 constitutes the exciting source of m.m.f. analogous expressions are developed

$$\left.\begin{array}{l} e_2 = -N_2^2 P \dfrac{di_2}{dt} \\[6pt] e_1 = -N_1 N_2 P \dfrac{di_2}{dt} = -M_{12} \dfrac{di_2}{dt}. \end{array}\right\}$$

Thus $N_2^2 P$ is the self inductance of winding 2 and M_{12}, relating e.m.f. in winding 1 to the rate of change of current in winding 2, has the same value as M_{21}.

A practical mutual inductor or transformer does not conform to the ideal relationships implied above, on account of the impossibility of arranging for the flux linked by each turn on each winding to be identical. *Leakage flux*, considered in detail in Section 4.4, accounts for the discrepancy, linking the turns of each winding differently. The effects of leakage flux may be minimised by arranging the coils to be in close proximity to each other and by use of high-permeability core materials. This latter property enables the necessary core flux to be produced with minimal *magnetising current* i_1 required to flow in the primary winding of N_1 turns.

Figure 4.2 Magnetic circuit of Fig. 4.1 completed via toroidal core of permeability μ.

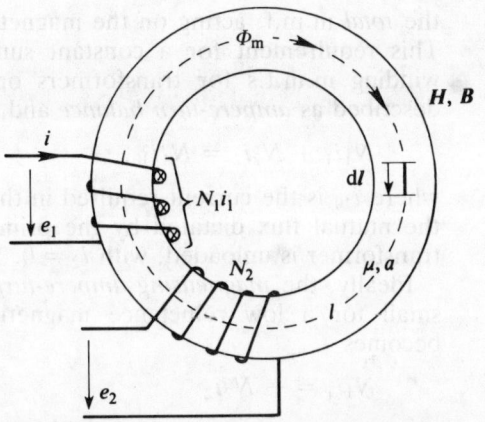

There is essentially no difference between the primary winding and that of N_2 turns designated the *secondary* winding. The primary winding is perhaps best regarded as the one connected to a source of supply, and towards which the average (real) power normally flows. Suppose now that a load resistor is connected across the secondary winding terminals as shown dashed in Fig. 4.1. The electromagnetically induced e.m.f. $e_2 = \oint \boldsymbol{E}_2 \cdot d\boldsymbol{l}$ in the circuit will endeavour to establish current i_2 in the same circuit, such that $e_2/i_2 = v_L/i_L = R_L$ at each instant in time. The waveform of load current i_L will be required by Ohm's law to match the waveform of voltage across the load resistance, and this appears to impose constraints on the time variation of mutual flux Φ_m through eqn [4.2]. The mutual flux is related via eqn [4.1] to the induced e.m.f. e_1 of winding 1, which, disregarding the effect of the leakage flux of the primary winding, is equal and opposite to the primary terminal voltage v_1. If, therefore, the source of supply to the primary winding is one of well-defined voltage, the core flux magnitude and waveshape will be as required through eqn [4.1], with consequences for the waveshape of the secondary terminal voltage and current. Most power transformers and voltage transformers are operated in this fashion, whereby the mutual flux is determined by the primary supply voltage. The secondary e.m.f. responds via eqn [4.2] and the secondary current results from the impression of this e.m.f. across the load impedance.

Once secondary winding current begins to flow, its magnetic field acts to modify the mutual flux in the transformer core. By Lenz's law, Section 2.5.2, the magnetic effect of the induced current is inherently to oppose the specific change in flux which brought about the e.m.f. responsible for the current. Thus, at a time when the flux is *increasing*, the current developed in a secondary winding supplying any load will seek to develop an additional m.m.f. which acts on the magnetic core to oppose that of the primary winding, which might be considered to be originally responsible for the mutual flux. If the mutual flux is required to maintain a particular magnitude and waveshape, due to the connection of the primary winding to a defined voltage source, the attempts of the secondary winding m.m.f. to effect any change are opposed by the modification of the primary winding current so that the *total* m.m.f. acting on the magnetic circuit remains unchanged. This requirement for a constant sum of primary and secondary winding m.m.f.s for transformers operating at 'constant flux' is described as *ampere-turn balance* and may be stated thus

$$N_1 i_1 + N_2 i_2 = N_1 i_{10}, \qquad [4.6]$$

where i_{10} is the current required in the primary winding to provide the mutual flux dictated by the primary voltage source when the transformer is unloaded, with $i_2 = 0$.

Ideally, the *magnetising ampere-turns* required, $N_1 i_{10}$, should be small for a low reluctance magnetic circuit, so that eqn [4.6] becomes

$$N_1 i_1 = - N_2 i_2$$

or

$$\frac{i_2}{i_1} = - \frac{N_2}{N_2}. \qquad [4.7]$$

This condition is practically achieved in iron-cored transformers operating at full load current.

4.3 Effect of Core Saturation

In the derivation of eqns [4.1]–[4.7], no restriction has been placed on the time variation of current, voltage, e.m.f. and mutual flux. In practice, however, there is a limit to the flux density attainable in the iron core of a transformer. Above saturation levels the magnetising current requirement increases disproportionately to increases in flux. If the primary winding is connected to a source of constant d.c. voltage, eqn [4.1] requires that the mutual flux increases indefinitely at a constant rate. If the source is able to provide the ever-increasing magnetising current, the constant rate of change of flux requirement will be met and a constant e.m.f. will be induced in the secondary circuit. In practice, however, source impedance will limit the magnetising current flowing in the primary winding, the core flux will reduce its rate of increase and the primary and secondary e.m.f.s will fall below their initial values. The problem of an ever-increasing flux requirement will be avoided, however, if the polarity of the primary source voltage is reversed periodically, e.g. if a square-wave alternating voltage is applied. Primary and secondary e.m.f.s should then have correspondingly square waveforms – provided the frequency is not too low – but the magnetising current waveshape will ideally be *triangular*, unlike the square-wave load current appropriate to a resistive load connected across the secondary winding terminals.

Power transformers are generally operated from constant-voltage, constant-frequency *sinusoidal* sources. The core flux, winding-induced e.m.f.s, load and magnetising currents are correspondingly all sinusoidally time variant with appropriate phase differences. An exception to this rule strictly applies in the case of a magnetising current which contains time harmonics due to the non-linear nature of the B/H relationship of the core material.

Provided that the source does not present a high impedance to the flow of the harmonic current content of the magnetising current, the sinusoidal waveshape of the secondary e.m.f. (and the load current for a constant-impedance load) should be preserved.

The transformer which we have considered to date is closely akin to an *ideal* transformer which has the properties of no leakage flux and a magnetic circuit of infinite permeance, so that the flux requirement is met without the need for magnetising current flow in the primary winding. Such a transformer satisfies eqns [4.3] and [4.7], which, written in terms of r.m.s. values of phasor quantities, become

$$\frac{E_1}{E_2} = \frac{N_1}{N_2}, \quad \frac{I_2}{I_1} = -\frac{N_1}{N_2}. \qquad [4.8]$$

The first degree of imperfection admits a magnetic circuit of finite but constant permeance over the range of flux density required. Equation [4.7] is then no longer generally valid, but eqn [4.6] is expressed thus in terms of r.m.s. phasor quantities

$$N_1 I_1 + N_2 I_2 = N_1 I_{10}. \qquad [4.9]$$

4.4 Leakage Flux

Equation [4.9] above indicates that the practical requirement for magnetising current creates a discrepancy in the ampere-turn balance between windings. Leakage flux provides a complementary feature in accounting for a departure of the terminal voltage ratio of a transformer from the turns ratio. Just as the error in ampere-turn balance is identified with magnetising current – which is roughly proportional to mutual flux and supply voltage – so the error in voltage ratio due to leakage flux may be shown to be dependent upon load current, the precise relationship being affected by other contributory influences such as power factor.

As mentioned in Section 2.4.11, leakage flux is that flux which links one winding only and does not flow in the low-reluctance path of the mutual flux. Leakage flux associated with one winding of a two-winding transformer is shown in Fig. 4.3 to be characterised by its flow largely in air paths of high (and constant) reluctance and the creation of *partial flux linkages*, with the winding responsible through its m.m.f.

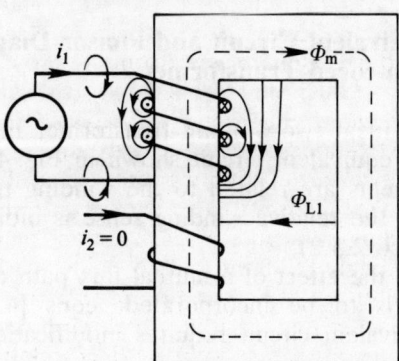

Figure 4.3 Leakage flux paths associated with winding 1 of two-winding transformer.

The precise calculation of leakage flux is a difficult task due to its distributed nature. Its effect is similar to the mutual flux, in being responsible for a component of the total electromagnetically induced e.m.f. in the linked winding but it makes no contribution to the e.m.f. of any other winding. Since leakage flux paths are predominantly in air with a much higher reluctance than that of the mutual flux, only a small proportion of the total flux linking a power transformer winding is accounted for by leakage flux. Saturation of iron in the mutual flux path has virtually no effect on the reluctance of the leakage flux paths, hence the induced e.m.f. due to leakage flux set up by a sinusoidal current is also sinusoidal, with proportionate amplitude and 90° phase shift. Whereas it is strictly the integrated flux changes experienced by each *part* of the winding which accounts for the total leakage e.m.f., an effective value of leakage flux associated with the total number of turns N_1, say, is obtained through dividing the aggregated partial flux linkages by N_1. Relatively simple experimental tests enable the leakage inductances of an assembled transformer to be measured, particularly if, as is often the case, there is no need to distinguish between the leakage reactances of each winding under sinusoidal conditions.

Constructional details greatly influence the relationship between leakage flux and mutual flux. Large transformer cores are built up from sheets of laminated high-silicon steel, cold rolled in the direction of eventual magnetisation to improve their magnetic properties (see Sect. 3.5). *Mitred joints* (Fig. 4.4) are employed at the corners to avoid the higher iron loss and reluctance associated with flux paths at right-angles to the direction of rolling, the sheets of steel being butt jointed and interleaved as the core is built up layer by layer.

Bearing in mind the need for adequate insulation between separate windings, and between windings and core, the proximity of windings to each other may be adjusted by the designer in order to control the leakage flux and, incidentally, the winding capacitance distribution. Whilst in most transformer applications a low leakage reactance is desirable, this is not always so.

When short circuits on the secondary side of a transformer connected to a constant-voltage primary supply are a regular occurrence, the consequential current may be limited to a safe level by a suitably high value of leakage reactance.

4.5 Equivalent Circuit and Phasor Diagram of a Two-winding Iron-cored Transformer

An ideal two-winding transformer may be simply represented by the equivalent circuit shown in Fig. 4.5, in which the e.m.f.s and currents are related to the winding turns by eqns [4.3] and [4.7], and the relative winding sense is indicated by the appended dots, (Sect. 2.5.2).

If the effect of a mutual flux path of finite but linear permeability is to be incorporated, eqns [4.3] and [4.6] hold and the equivalent circuit requires modification by the incorporation of a

Figure 4.4 (a) Assembly of a 500 kVA, oil cooled, three-phase transformer core. (Photograph by courtesy of Brush Transformers Ltd Loughborough, UK.)
(b) Small transformer core assembled from interleaved 'U' and 'T' shaped stampings. Appropriate for 'shell-type', single-phase transformer with windings encircling the central limb or three-phase, 'core-type' transformer with phase windings on each limb.

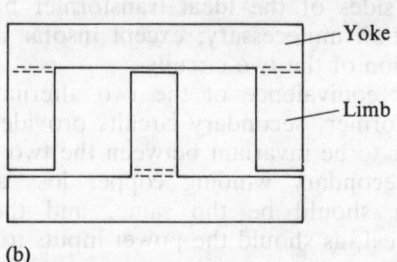

(b)

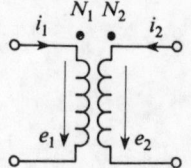

Figure 4.5 Equivalent circuit of an ideal transformer.

pure inductance carrying the magnetising current i_0 and associated with the e.m.f. of the particular winding with respect to which the magnetising current has been declared. In the case depicted in Fig. 4.6, this is the *primary* winding. The effect of leakage flux appropriate to each winding may now be recognised by the incorporation of a corresponding series inductance carrying winding current and the effect of winding resistance similarly treated. The distributed nature of the circuit parameters is not acknowledged, neither is the effect of winding *capacitance* in this model which is appropriate only to low-frequency operation.

If one represents losses in the iron core carrying mutual flux Φ_m by a resistor in parallel with the magnetising inductance (refer to Sect. 3.4, 3.5 and 4.7), the complete equivalent circuit is as shown in Fig. 4.6, with the values of the shunt elements constant over a limited range of mutual flux Φ_m. In this figure the reference direction of secondary current has been reversed in order that i_2 will have generally the same polarity as i_1. For similar reasons, the winding e.m.f.s are depicted as $-e_1$ and $-e_2$, as these more nearly identify with the terminal voltages v_1 and v_2. The advantages of the change in notation will be apparent as we move towards the

Figure 4.6 Equivalent circuit of a real transformer.

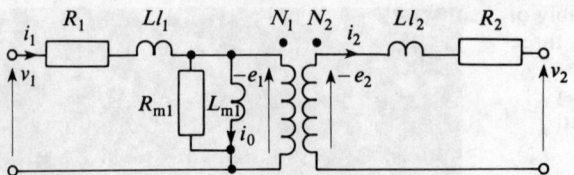

development of the phasor diagram of the transformer. Most transformers are designed for sinusoidal operation, often at constant frequency, and it is usual to model these in terms of *reactance* rather than inductance. *Further discussion of transformers will, except when stated to the contrary, assume sinusoidal voltages and currents, with winding inductances modelled as reactances.*

4.5.1 Referred parameters

Simplification in transformer representation may be afforded by eliminating the ideal transformer which, apparently, is redundant only when winding turns N_1 and N_2 are equal. If, however, the values of all the circuit parameters at one side of the ideal transformer are modified by appropriate constant factors which relate to the winding/turns ratio, then the e.m.f.s and currents at both sides of the ideal transformer become equal, rendering its retention unnecessary, except insofar as it indicates the electrical isolation of the two circuits.

For equivalence of the two alternative representations of the transformer, secondary circuits provided by Fig. 4.7, we need the power to be invariant between the two circuits in all measures, i.e. the secondary winding copper loss and the leakage reactance power should be the same, and the load powers should be identical, as should the power inputs to the secondary circuits also, i.e.

$$(I'_2)^2 X'_{l2} = I_2^2 X_{l2}$$
$$(I'_2)^2 R'_2 = I_2^2 R_2$$
$$V'_2 I'_2 = V_2 I_2$$

with

$$E'_2 I'_2 = E_2 I_2.$$

Figure 4.7 Equivalent representations of a transformer secondary circuit.
(a) referred to primary turns.
(b) natural format.

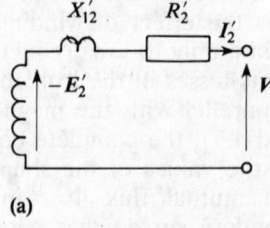

(a)

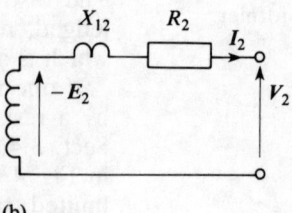

(b)

Now, if we propose that

$$E'_2 = \frac{N_1}{N_2} E_2$$

and

$$I_2' = \frac{N_2}{N_1} I_2,$$

which will enable E_2' to be equated with E_1 and I_2' with I_1 to achieve elimination of the ideal transformer, simultaneous satisfaction of the above four equations is achieved, provided

$$X_{12}' = \left(\frac{N_1}{N_2}\right)^2 X_{12}$$
$$R_2' = \left(\frac{N_1}{N_2}\right)^2 R_2$$
$$V_2' = \left(\frac{N_1}{N_2}\right) V_2 \qquad [4.10]$$

with

$$I_2' = \left(\frac{N_2}{N_1}\right) I_2.$$

The above process is described as 'referring the secondary circuit to the primary winding number of turns'. The referring factor for all voltages is turns ratio, for current inverse turns ratio, and for all impedances it is (turns ratio)^2. Thus all impedances in the secondary circuit, including the load impedance, are multiplied by this factor. The procedure is essentially reciprocal and all primary circuit parameters may equally be referred to the secondary winding number of turns, including the magnetising reactance and core-loss resistance, using the inverse turns ratio N_2/N_1 for voltage, turns ratio N_1/N_2 for current and (inverse turns-ratio)^2 for impedance.

The complete equivalent circuit of a transformer under steady-state a.c. conditions is therefore modelled as in Fig. 4.8. where the secondary winding parameters have been referred to the primary winding.

Figure 4.8 Equivalent circuit of transformer referred to the primary winding.

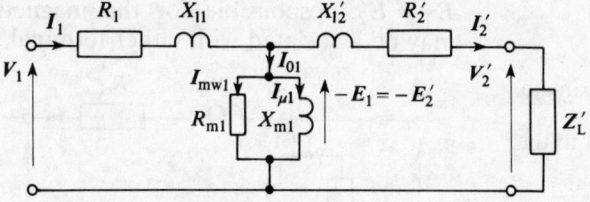

In presenting the phasor diagram of a transformer it is usual to acknowledge that the mutual flux in the core is most appropriate as the reference quantity, by endeavouring to arrange that this quantity – together with the true magnetising current component I_μ responsible for its production – is directed horizontally from left to right.

The phasor diagram for a transformer supplying an inductive load is shown in Fig. 4.9. Here the leakage reactance and resistance voltages, with the magnetising current, have been exaggerated for clarity.

Figure 4.9 Transformer phasor diagram for lagging power factor load.

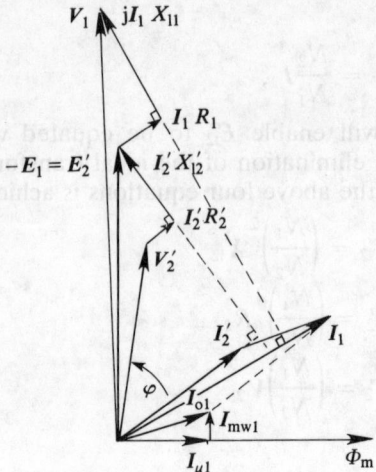

Since the mutual flux magnetic circuit of an iron-cored transformer is of high permeance and low loss, the magnetising current I_{o1} is typically less than 5% of the rated primary current, and little error is caused by moving the magnetising branch in the equivalent circuit representation of the loaded transformer to the input terminals. This makes further simplification possible in that the resistance and leakage of the primary and secondary winding *referred to the primary number of turns* may be combined to yield *equivalent* leakage impedance components referred to the primary turns.

Thus in the approximate equivalent circuit of Fig. 4.10

$$R_{e1} = R_1 + R_2'; \quad X_{e1} = X_{l1} + X_{l2}'. \qquad [4.11]$$

One consequence of this simplification is that the separate identities of the winding e.m.f.s induced by the mutual flux are lost. A well-designed transformer operating near its rated voltage over its normal range of load current will, however, demonstrate little distinction between V_2' and V_1, suggesting that the e.m.f. $E_1 = E_2'$, responsible for the magnetising current and core loss may be associated with either terminal voltage, as most expedient.

Figure 4.10 Approximate equivalent circuit of a transformer referred to the primary winding.

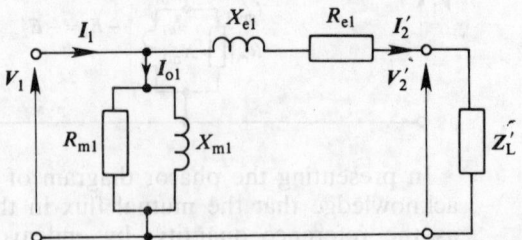

4.6 Voltage Regulation of a Transformer

It is readily apparent that the difference, or error, between V_2' and V_1 is given by the phasor sum of the voltage drops developed across the leakage impedances of primary and secondary (referred) windings. Clearly, this voltage sum is dependent upon load current magnitude and phase relative to terminal voltage. The voltage regulation ε of a transformer is defined *as the rise in secondary terminal voltage which occurs when the load current is reduced from rated, or full load, value to zero*. It is usually expressed per unit (p.u.), i.e. as the difference in voltage magnitude divided by rated voltage.

Thus

$$\varepsilon_{\text{p.u.}} = \frac{|V_1| - |V_2'|}{|V_2'|},$$

evaluated for rated V_2' and I_2'. Magnetising current is not significant in influencing the voltage regulation of a transformer at full load current. The phasor diagram of voltage and current which is appropriate to the approximate equivalent circuit of Fig. 4.10 is shown in Fig. 4.11 for an inductive load of power factor $\cos\phi$, taking V_2' as reference. Voltage phasor addition gives

$$V_1 = V_2' + I_2'R_{\text{e1}}\cos\phi - jI_2'R_{\text{e1}}\sin\phi + I_2'X_{\text{e1}}\sin\phi$$
$$+ jI_2'X_{\text{e1}}\cos\phi.$$

By Pythagoras

$$V_1^2 = [V_2' + I_2'(R_{\text{e1}}\cos\phi + X_{\text{e1}}\sin\phi)]^2 + [I_2'(X_{\text{e1}}\cos\phi - R_{\text{e1}}\sin\phi)]^2$$
$$= (V_2')^2 + 2V_2'I_2'(R_{\text{e1}}\cos\phi + X_{\text{e1}}\sin\phi) + \text{smaller terms}.$$

Therefore

$$V_1 \simeq V_2'\left[1 + \frac{2I_2'}{V_2'}(R_{\text{e1}}\cos\phi + X_{\text{e1}}\sin\phi)\right]^{1/2}$$
$$\simeq V_2'\left[1 + \frac{I_2'}{V_2'}(R_{\text{e1}}\cos\phi + X_{\text{e1}}\sin\phi)\right]$$

$$\varepsilon_{\text{p.u.}} = \frac{V_2 - V_2'}{V_2'} \simeq \frac{I_2'}{V_2'}(R_{\text{e1}}\cos\phi + X_{\text{e1}}\sin\phi). \qquad [4.12]$$

Now

$$\frac{I_2'R_{\text{e1}}}{V_2'}$$

is a dimensionless quantity which, if I_2' and V_2' are the rated values of secondary terminal current and voltages referred to the primary winding, is called the *per-unit equivalent resistance of the transformer*. $R_{\text{e p.u.}}$ is equal to the voltage developed across the equivalent resistance when rated current flows, divided by rated voltage, with all quantities consistently referred to the primary winding. A little manipulation will show that the same value for $R_{\text{e p.u.}}$ applies if all quantities involved in the definition are referred to the *secondary* winding, i.e. I_2, V_2 and $R_{\text{e2}} = R_{\text{e1}}(N_2/N_1)^2$. This property accounts in part for the popular practice of expressing transformer impedances in per unit. Another important reason is that, over a wide range of ratings, transformers of similar construction give rise to comparable values of impedance when expressed p.u. Equivalent resistance p.u. may be expressed also as the copper loss at rated current divided by rated VA.

Figure 4.11 Phasor diagram of a transformer supplying are inductive load, with magnetising current neglected.

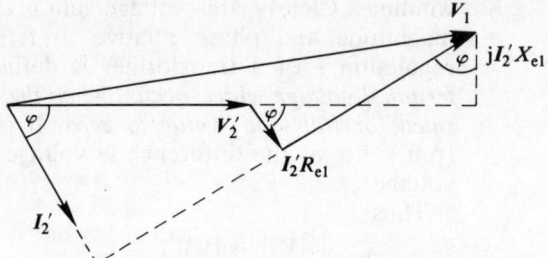

Complementary remarks apply to the per unit equivalent (leakage) reactance of the transformer. Thus

$$X_{\text{ep.u.}} = \frac{I'_2 X_{e1}}{V'_2} = \frac{I_2 X_{e2}}{V_2} = \frac{I_2^2 X_{e2}}{V_2 I_2} = \frac{I_1^2 X_{e1}}{V_1 I_1}.$$

In these relationships no distinction is made between I'_2 and I_1, or I'_1 and I_2, since they relate to rated load conditions for which the effect of magnetising current in this context may be ignored.

In terms of p.u. equivalent resistance and leakage reactance, eqn [4.12] may be re-written to express voltage regulation as follows

$$\varepsilon_{\text{p.u.}} = R_{\text{ep.u.}} \cos \phi + X_{\text{ep.u.}} \sin \phi. \qquad [4.13]$$

Normally, with power transformers, $X_{\text{ep.u.}} \gg R_{\text{ep.u.}}$, and eqn [4.13] demonstrates that transformer voltage regulation may be kept low by maintaining the power factor of the load at a high value. Capacitive loads are responsible for negative voltage regulation, i.e. a secondary voltage increase on load.

4.7 Magnetising Current Requirements and Energy Losses in Transformer Cores

As noted in Sections 3.3 and 3.4, high-permeability ferromagnetic materials exhibit saturation and hysteresis. That is, when the magnetic field intensity is changing the flux density change depends not simply on the instantaneous value of the field, but also on the recent history of the material. In a transformer which is connected to a sinusoidal constant voltage supply, the core flux maintains an essentially sinusoidal variation of constant amplitude. This is such as to induce within the primary winding the e.m.f. $-E_1$ which differs from the supply voltage V_1 by the small primary leakage impedance voltage drop. The *transformer e.m.f. equation* relates the amplitude of the mutual flux $\Phi_m = \hat{\Phi}_m \sin \omega t$, say, to the r.m.s. value of winding e.m.f. via the appropriate winding turns and supply frequency. Thus, from eqn [4.1] for the primary winding,

$$e_1 = -N_1 \frac{d\Phi_m}{dt} = -N_1 \frac{d}{dt}(\hat{\Phi}_m \sin \omega t)$$
$$= -2\pi f N_1 \hat{\Phi}_m \cos 2\pi f t,$$

giving

$$E_1 \text{ (r.m.s.)} = \sqrt{2} \pi f N_1 \hat{\Phi}_m,$$

i.e.

$$E_1 = 4.44\, fN_1\hat{\Phi}_m. \qquad [4.14]$$

The hysteresis and non-linearity demonstrated in the B/H relationship for iron-core materials means that a sinusoidal flux density variation must correspond to a non-sinusoidal magnetising current waveshape.

Figure 4.12 shows that the waveform of the magnetising current is characteristically 'peaky', containing odd harmonics and a fundamental which lags the sinusoidal flux density wave by a small angle. The component of the fundamental current in phase with B is modelled in the equivalent circuit of Fig. 4.8 as $I_{\mu 1}$, the true magnetising current. That component which lags B by 90° is modelled as a part of I_{mw1}, the 'wattful' component of the 'no-load' current which accounts for the *hysteresis and eddy-current losses* in the core.

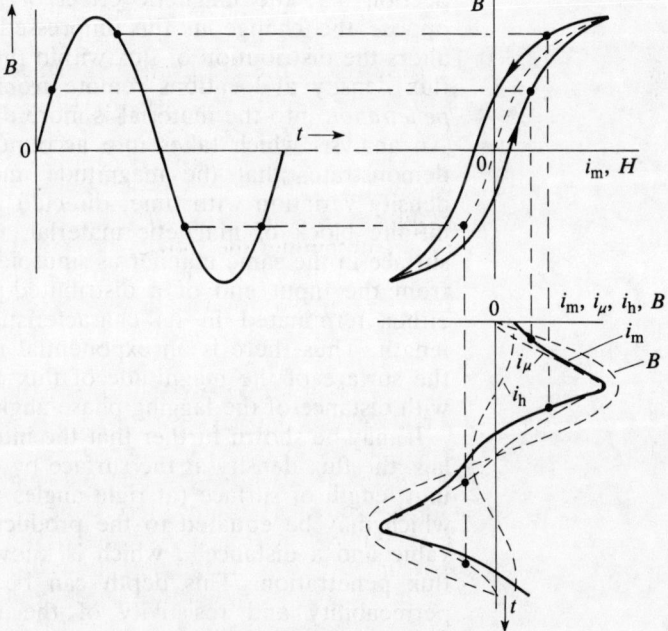

Figure 4.12 Magnetising current waveform for a transformer with a sinusoidal flux density. i_μ and i_h are, respectively, fundamental true magnetising and hysteresis loss components of magnetising current i_m.

An a.c. equivalent circuit such as Fig. 4.8 is strictly valid at one frequency only and therefore the harmonic components of the magnetising current are unrepresented. Harmonic currents in conjunction with a sinusoidal voltage source at fundamental frequency cannot account for transfers of average power between source and magnetising circuit, hence the use of a linear resistor in a fundamental frequency equivalent circuit to account for losses in the magnetic circuit is valid, although the value of the modelling resistor strictly requires amendment in accordance with the amplitude of the flux swing experienced. As shown in Section 3.4 the hysteresis energy loss in W m^{-3} per cycle of magnetisation is equal to the area embraced by the hysteresis loop. Steinmetz demonstrated that this energy loss in a wide range of typical transformer steels was roughly proportional to the amplitude of the flux swing raised to the power 1.6.

The use of a fixed value resistor for R_{ml} in the magnetising branch of Fig. 4.10 corresponds to a core loss dependency on flux density raised to the power 2, since the average power dissipated in the resistor is proportional to E_1^2. Such a representation is valid for the eddy-current losses developed in the core material, which will be laminated in order to minimise the *area* embraced by the closed eddy-current paths.

Whereas hysteresis power loss is proportional to frequency this is not the case for eddy-current loss. A fairly simple analysis demonstrates that eddy-current loss should be proportional to (frequency)^2, as the eddy e.m.f.s which will be responsible for driving the eddy currents are proportional to the time derivative of the sinusoidal flux. However, depending upon the frequency and the thickness of the laminations it may not be valid to assume a uniform flux density throughout the iron. As we have noted in Section 3.5, the magnetic effect of induced eddy currents is to oppose the change in flux impressed externally. Their influence alters the distribution of flux within the core material, reducing the flux density at locations remote from the core boundaries. *Flux penetration* into the material is more difficult at higher frequencies. An analysis which takes into account permeability and resistivity demonstrates that the magnitude and phase of sinusoidal flux-density variation with time, directed parallel to the surface of an infinite block of magnetic material, vary with distance from the surface in the same manner as sinusoidal voltages relate to distance from the input end of a distributed-parameter transmission line, either terminated in its characteristic impedance or of infinite length. Thus there is an exponential reduction with distance from the surface of the magnitude of flux density and a linear increase with distance of the lagging phase-angle.

It may be shown further that the integrated flux within the block lags the flux density at the surface by 45° and has a magnitude per unit length of surface (at right-angles to the direction of the field) which may be equated to the product of the surface flux density value and a distance d which is known as the effective depth of flux penetration. This depth can be evaluated in terms of the permeability and resistivity of the medium and the operating frequency as

$$d = \sqrt{\frac{\rho}{\mu\omega}}.$$

Magnetic cores are built up from insulated laminations having this order of thickness.

Having noted that the magnetising impedance of an iron-cored transformer is non-linear – such that a sinusoidal flux-change requires a non-sinusoidal magnetising current – it is not immediately obvious as to the source of the harmonic current required. These harmonic components of magnetising current are not insignificant. To reduce the volume and weight of core material, power transformers are operated well into the saturated region, and the dominant third-harmonic component of magnetising current may be 40% of the fundamental. If we consider the transformer to be supplied from a constant voltage source of sinusoidal waveshape then, by definition, this source cannot supply harmonic e.m.f.s or current but will ideally present zero impedance to impressed harmonic current flow. If harmonic current components are absent from the magnetising current the transformer flux waveform is distorted from the sinusoidal, so introducing harmonic e.m.f.s in the primary (and secondary) windings. The primary harmonic e.m.f.s are responsible for the corresponding harmonic currents which are able to circulate in the primary winding and source, restoring the flux waveform to a near-sinusoidal shape. Thus the harmonic magnetising current requirements are accommodated by a *slight* distortion of the mutual flux waveshape. It is essential, however, that the primary source should present a low impedance to harmonic current flow in order that this distortion, reflected in the secondary voltage waveshape, is negligible.

In *three-phase transformer* applications where the windings may be connected in *star*, the use of isolated star points is undesirable as it prohibits the flow of *triplen* harmonic components of magnetising current in at least one winding. Triplen harmonics are odd multiples of three, including the third. Significant distortion of the phase voltage waveforms will thus be caused, although the distortion will not be apparent in the line-to-line voltage waveforms which constitute the *difference* of phase to neutral voltages. If just one winding is connected in delta, or alternatively in star, with the star point connected to the supply neutral (or to the star point of the load), a path will be provided wherein the triplen harmonic current can flow so as to minimise the distortion of the flux waveform.

4.7.1 *Magnetising current inrush*

As noted above, the steady-state magnetising current of an iron-cored transformer contains harmonics. Since the waveform is normally symmetrical about the time axis, even harmonics will be absent. The reason for this is that the flux excursion is symmetrical about zero, the B–H loop traversed is centred about $B = 0$, $H = 0$.

When a transformer is de-energised, however, the primary m.m.f. may be reduced rapidly to zero. The secondary m.m.f. will immediately adjust in an endeavour to maintain momentarily the level of flux linked by the still-closed secondary winding, but will then decay at a rate influenced by the nature of the connected load. When the secondary current has fallen to zero the transformer iron may be left with a fairly high value of residual flux-density, particularly if the air spaces in the magnetic core are negligible. On connection of the transformer primary winding to an a.c. voltage source the initial rate of change of flux must match the initial instantaneous voltage impressed. Ideally, a transformer with no residual flux should be connected to a sinusoidal voltage source at the instant of voltage maximum – which corresponds to a core flux requirement of zero but maximum rate of change. The core flux will then be immediately established as a pure sinusoid.

If, however, the transformer is connected at an instant of voltage zero, this would correspond in the prospective steady-state to a core flux of maximum value and about to reduce. If the initial flux is in fact zero, the effect of the supply voltage will be to drive the flux unidirectionally for the first half-cycle towards a level of flux which is twice the eventual steady-state amplitude. Thus the transformer is driven well into saturation and the corresponding magnetising current has extremely high peak values which may exceed appreciably the rated value, and even the short-circuit value, subjecting the windings to large mechanical forces. Due to energy losses in the winding resistance and the core material itself, the high levels of flux in the core are not sustained and the d.c. component of flux falls to zero in due course, as shown in Fig. 4.13.

The magnetising current during this period of inrush is seen in Fig. 4.13 to possess a large d.c. component and even harmonics which are absent in the steady state. This feature may be employed to enhance the stability of schemes of transformer *current differential protection* which operate on the basis of fundamental frequency ampere-turn balance of primary and secondary windings under normal operation. Unless the special nature of magnetising inrush current is taken into account, the protection may not be able to distinguish between the switch-in condition and that of an internal fault, since both feature a primary/secondary current ratio which is not in accordance with that predicted for ampere-turn balance via the normal turns ratio.

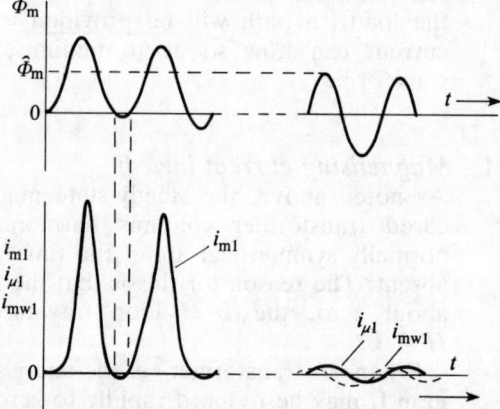

Figure 4.13 Core flux and magnetising current of a transformer with zero residual flux energised at the instant of voltage zero.

If significant residual flux is present in the transformer core the switching transient current situation is improved or worsened, depending upon the polarity of the residual flux and that of the initial change. If magnetising inrush currents are likely to pose problems on systems, synchronised switching-in may be arranged to close the primary circuit at voltage maximum. In some systems magnetising inrush current magnitude is limited by the impedance of the supply.

4.8 Transformer Efficiency

The power efficiency of any system is the ratio of output power to input power. For a transformer the energy is averaged over half the supply period. Alternative formulations follow from the recognition that the output power must equal the input power less the internal power losses, e.g.

$$\text{efficiency } \eta = \frac{\text{output}}{\text{output} + \text{copper losses} + \text{iron losses}}. \quad [4.15]$$

For a transformer operating from a constant voltage, constant frequency supply, the mutual flux in the core is essentially sinusoidal and of constant amplitude. Hence the iron losses are virtually independent of load current. The copper losses, on the other hand, are proportional to (load current)^2, i.e.

$$P_{Cu} = (I'_2)^2 R_{e1}$$

The output power of the single-phase transformer is given by $V'_2 I'_2 \cos\phi$, where $\cos\phi$ is the load power factor. The losses are independent of power factor so it is apparent that the efficiency, like voltage regulation, is dependent upon power factor. The relationship is shown in Fig. 4.14. Efficiency is maximised at unity power factor load. Consideration of load current as the sole variable in eqn [4.15] leads to the observation that maximum efficiency occurs at the particular value of load current at which the copper losses and the iron losses are equal.

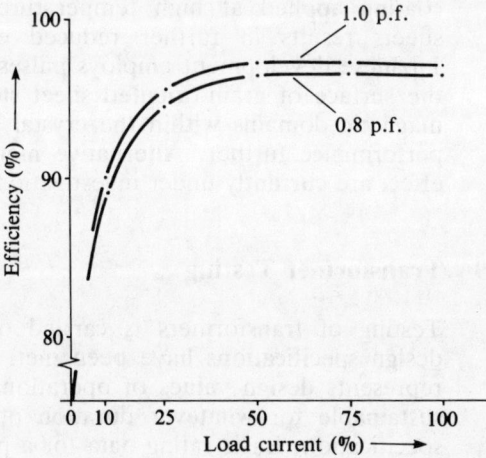

Figure 4.14 Dependence of transformer efficiency on load power factor.

Power transformers have quite considerable iron loss, since the operating flux density is high to minimise iron volume, but they are usually operated at higher levels of load than correspond to maximum efficiency to justify the initial capital cost. Transformers are designed with thermal limitations in mind, that is, the *heating* effects of the losses over the duty cycle envisaged. Such considerations are relevant to all types of electrical machines.

Although the efficiencies of power transformers are very high, development work carried out to reduce losses further is worthwhile. A transformer operating at its rated load condition of 220 MVA at 0.9 p.f. with an efficiency of 99% will develop 2 MW of loss manifested as heat. This has to be dissipated in the vicinity of the transformer, entailing the use of ancillary equipment including heat exchangers for circulating oil coolant. At times of expensive energy the direct cost of the losses is also significant. Such losses incurred over the service life of the equipment may be converted into an equivalent lump sum which would generate sufficient income to fund the continuing losses. Such a capital sum is dependent upon current and prospective energy costs and rates of interest on money. At the present time (1988) power transformer no-load or iron loss is capitalised in the UK at around £5000 per kW of loss. This is higher than the comparable figure for load or copper loss which is dependent upon the *load factor* of the equipment, i.e. the mean percentage of rated current to which it is subject. In power networks generator transformers associated with base load generating stations are operated at the highest load factors, with capitalised load losses of the order of £3000 per kW.

The fact that, in modern transformers, the copper loss on full load may exceed the no-load loss by a factor approaching 10 is due to developments in transformer steel and methods of construction. Over recent times each twenty-year period has witnessed a halving of the iron losses attainable. Improved grain orientation procedures produce fewer but larger grains on recrystallisation after annealing of the cold-rolled steel sheet and provide a reduction in hysteresis loss. A further decrease in no-load loss may be achieved by subjecting the laminations to tensile mechanical stress resulting from the different expansion coefficients of the steel and a glass coating applied at high temperature. The use of thinner steel sheets results in further reduced eddy-current loss. A recent Japanese development employs pulses from a laser beam to scribe the surface of grain-oriented sheet steel, reducing the size of the magnetic domains within the crystal structure and improving loss performance further. Alternative methods of producing the same effect are currently under investigation.[14]

4.9 Transformer Testing

Testing of transformers is carried out primarily to ensure that design specifications have been met. The *rating* of a transformer represents design values of operational variables which should be sustainable for whatever duration of time is appropriate to the specification. Basic rating data for a power transformer is in terms of the r.m.s. volt-ampere product of either primary or secondary winding and the r.m.s. voltage of each winding. A power transformer operating at rated voltage and rated VA will develop iron and copper losses appropriate to full-load conditions: the heat generated raises the temperature of the materials, ultimately to a level at which the heat removed by cooling, due either to natural processes (such as convection and radiation) or to the forced circulation of coolant, matches the rate of heat production. According to the duty cycle imposed on the transformer, limits of permissible temperature rise must not be exceeded or failure of insulation may result. The insulation must be capable of withstanding the most severe electrical stresses which the transformer might encounter in service. In small and very large transformers this is likely to be related to the peak service voltage. In medium-voltage distribution transformers this is likely to be associated with lightning strikes on the power network. The steep wavefront characteristic of lightning and switching surges accounts for the fact that the distribution of electrical stress in the dielectric is affected by the capacitance distribution between turns, with the turns near the ends of the windings being most at risk unless special precautions are taken.

The *direct* load testing of transformers may not always be feasible due to the power demands of highly rated plant, unless two similar transformers are available for a 'back-to-back' test (see Tutorial Example 4.10). Where possible, low-power tests are carried out, a comprehensive programme of which should enable full-load performance to be predicted accurately. For example, a *no-load* test carried out at rated voltage and frequency will excite the magnetic circuit normally and enable the iron losses to be established and the parameters of the magnetising branch of the equivalent circuit to be deduced. A *short-circuit* test carried out at rated current will require a very low applied voltage and core flux, with correspondingly negligible magnetising current and core loss. Measurement of input power (interpreted as copper loss), voltage and current will enable the equivalent *leakage impedance* of the transformer to be established. It is usual to apportion this equally between primary and secondary windings (with one winding necessarily referred to the other), as justified by the reasoning given in Section 4.10. A production transformer may exceptionally be tested under short-circuit conditions with full rated voltage applied in order to check the mechanical integrity of the winding supports. The windings are subjected to significant short-circuit forces which act to draw together adjacent turns of the same winding and to push apart adjacent turns of different windings.

4.10 Equivalence of Winding Resistance and Leakage Reactance Values Referred to the Same Number of Turns

Figure 4.15 shows a section through a single-phase shell-type transformer. Window area A is available to accommodate both primary and secondary windings. Logically, the winding space should be equally shared between the windings so that, with a turns ratio of N_1/N_2, the space available for each turn and its insulation will be $A/2N_1$ and $A/2N_2$, as appropriate. If the load current *density* in each winding is to be equal, and the same proportion of the winding area is occupied by insulation, the rate of heat production due to copper loss in each winding will be the same, assuming the same *length of mean turn* for each winding. On this basis, the length of each winding will be proportional to its

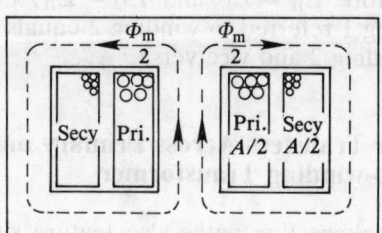

Figure 4.15 Single-phase shell-type transformer with window area A.

turns and the cross-sectional area inversely so. It follows that the resistance of winding 1 will be related to that of winding 2 by the (turns ratio)^2, i.e. $(N_1/N_2)^2$. Thus the resistance of the primary winding referred to the secondary winding will equal the actual resistance of the secondary winding, and vice versa.

A similar argument applies to the treatment of leakage reactance. In Fig. 4.16 the leakage flux is somewhat unrealistically represented, in that partial leakage flux linkages are not shown and the leakage flux is concentrated in the space between the windings. Figures 4.16(a) and (b) correspond to the same instant in time, with the leakage flux component due to the primary m.m.f. in (a) and due to the secondary m.m.f. in (b). The winding currents are typically oppositely directed, as shown, since their algebraic sum constitutes the m.m.f. responsible for the mutual flux Φ_m flowing in each half of the iron core. The net m.m.f. driving the mutual flux is much less than that responsible for the leakage flux of each winding. It is important to recognise that the leakage flux components act together in the air space between the windings and are individually proportional to the load current values i_1 and i_2, whereas the mutual flux flowing in the core is largely independent of the values of i_1 and i_2, if the transformer is connected to a constant voltage source. It is in order to add directly the leakage flux components since the large air-path element gives the appropriate magnetic circuits a linear flux/m.m.f. relationship.

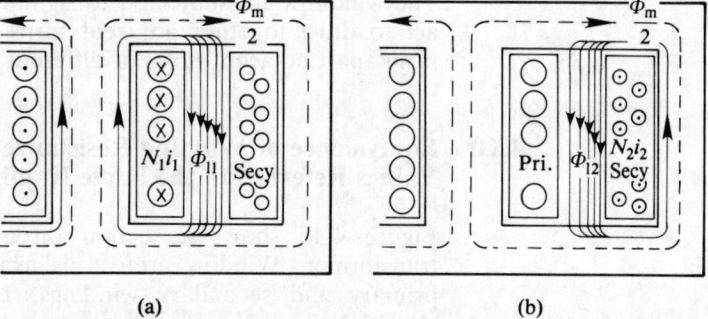

Figure 4.16 Leakage flux components due to primary and secondary transformer windings at the same instant in time: (a) Φ_{11} largely in the air space between windings due to m.m.f. $N_1 i_1$; (b) Φ_{12} largely in the air space between windings due to m.m.f. $N_2 i_2$; Φ_m = mutual flux in the right-hand half of the core due in each case to m.m.f. $(N_1 i_1 - N_2 i_2)$.

Since the leakage flux paths have reluctance defined largely by the common air path, the ratio of the leakage fluxes Φ_{11}/Φ_{12} is equal to the ratio of the m.m.f.s $N_1 i_1/N_2 i_2$. Hence the leakage inductances are related thus

$$\frac{L_{11}}{L_{12}} = \frac{N_1 \Phi_{11}}{i_1} \frac{i_2}{N_2 \Phi_{12}} = \left(\frac{N_1}{N_2}\right)^2.$$

Therefore $L'_{11} = L_{12}$ and $L'_{12} = L_{11}$, i.e. the leakage reactance of winding 1 referred to winding 2 equals the actual leakage reactance of winding 2 and vice versa.

4.11 Power Transfers Across Primary and Secondary Windings of a Two-winding Transformer

The leakage flux paths also feature significantly when considering the Poynting vector mechanism by which power is transferred from one winding to the other. The interchange of energy takes place across the leakage flux paths. If one considers an instant in time when the primary current flow is directed as shown in Fig. 4.16(a) and the mutual flux Φ_m is *decreasing*, eqn [2.40] shows that the direction of the primary-winding electromagnetically induced electric field E_1 due to the mutual flux is in the same sense as that shown for primary current i_1. If the primary current i_1 is also flowing in its reference direction, the H_1 field due to the N_1 turns will be directed as shown in the figure for the leakage flux. The direction of the power flow vector $S_1 = E_1 \times H_1$ is correspondingly into the primary winding from the surrounding air space which accommodates the leakage flux. Simultaneously, the secondary winding electric field E_2 induced by the mutual flux is directed in the same sense as for the primary winding and oppositely to that shown for i_2 in Fig. 4.16(b). E_2 interacts with the H_2 field due to the N_2 secondary turns, directed as shown for the leakage flux, to produce a Poynting vector $S_2 = E_2 \times H_2$ directed out of the secondary winding into the leakage flux space separating the two windings. It is readily shown that the power density vectors S_1 and S_2 associated with each winding surface have almost identical values. The iron core plays no part in the transfer of electrical power through the medium of the electric and magnetic fields existing in the leakage field region between the windings.

Although the above description relates to concentric, cylindrical windings, the same general conclusions apply to more complex winding arrangements. The calculation of actual electric and magnetic fields and corresponding power densities in the general regions of space remote from the conductor surfaces is very complex.

4.12 Polyphase (Three-phase) Transformer Connections and Core Arrangements

Polyphase power systems are designed with balanced operation in mind. That is, normal system operation is such that the phase currents and voltages are equal in magnitude and differ in time phase by $2\pi/m$ radians, where m is the number of phases. Connections between phases are such that the number of necessary external connections is minimised; the most popular are star (wye) and mesh (delta) connections. Star connections are more appropriate for transformer windings connected to distribution circuits, as the star point is available for single-phase load connections. If the load is unbalanced the phase voltages will be affected in a manner dependent upon the connections of both primary and secondary windings and the system beyond.

The configuration of the transformer magnetic circuit is also relevant. Economies in construction costs may be realised in the *core-type* arrangement shown in Fig. 4.17(a), where each of the three limbs are embraced by primary and secondary windings of one of the three phases. For balanced voltages the phase fluxes differ in phase by 120° and hence sum to zero at all instants. Thus the flux linking phase *a* completes its circuit via the limbs carrying phases *b* and *c*. The effective length of the magnetic circuit/phase is reduced, with correspondingly lower magnetising current and iron loss than would be the case if the transformer were comprised of three similarly rated single-phase units with independent magnetic circuits. The magnetic core configuration employed influences also the behaviour under unbalanced conditions which arise from abnormal loading or system faults. The three-limb core-type construction tends to inhibit the flow of triplen harmonic fluxes (odd multiples of the third harmonic) and those components of flux which would be created by components of current in the three-phase windings which are in phase with each other. To complete a closed path, such fluxes are required to traverse a high-reluctance region outside the iron core. The effect of discouraging triplen harmonic fluxes is to render triplen harmonic e.m.f.s in the phase voltages unlikely. The effect of discouraging flux due to co-phasal current components is to reduce the magnetising impedance of the transformer as presented to *zero phase-sequence* current which flows in the system when subjected to unbalanced phase loading or faults involving earth.

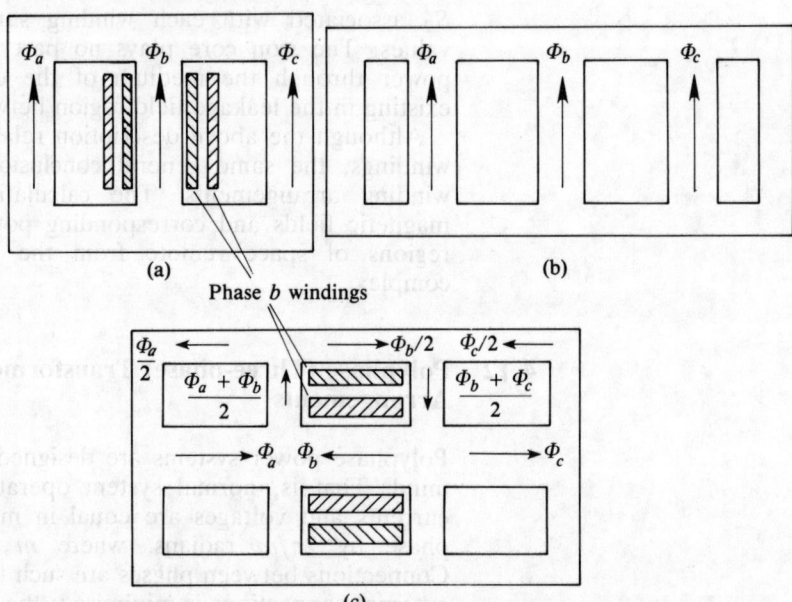

Figure 4.17 Magnetic circuits for three-phase transformer units: (a) three-limb core-type; (b) five-limb core-type, and (c) shell-type. (Core flux paths for one phase only are shown.)

In our discussion of three-phase transformers we shall make our assessment initially on the basis of an assumed balanced three-phase operation, recognising that starvation of the triplen harmonics in the magnetising current will introduce triplen harmonic distortion in the flux waveform – if a low-reluctance path for triplen harmonic flux exists. Also, we shall assume that each limb of the core satisfies the requirement for ampere-turn balance of the load current element of each winding.

(i) *Star/star connection (Fig. 4.18(a))*. Here balanced load currents in the secondary windings are reflected as balanced load currents in the primary windings. If both star points are isolated, triplen harmonic components of magnetising current are unable to flow, with corresponding distortion of the flux waveshape – assuming a low-reluctance path for triplen harmonic fluxes. The consequent harmonic e.m.f.s distort the phase e.m.f.s, but the line e.m.f.s (i.e. between line terminals) remain sinusoidal. The distortion gives rise to 'neutral-point oscillation', which refers to the triplen harmonic voltages observable between the star point of a transformer winding and the neutral point of the supply or load. Its effect is minimised if three-phase core-type construction of the magnetic circuit is employed.

If at least one of the winding star points is connected to an external circuit which permits triplen harmonic currents to circulate, the flux waveform is distorted only by the small amount necessary to generate the triplen harmonic e.m.f.s which correspond to the triplen harmonic requirements of the magnetising current.

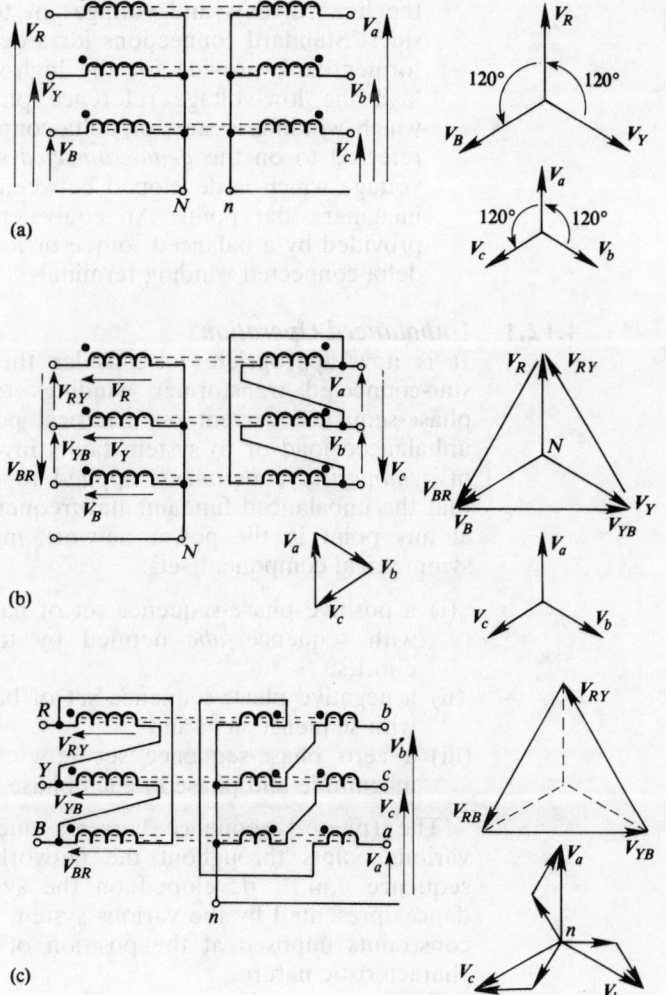

Figure 4.18 Three-phase transformer connections: (a) star/star, (b) star/delta, and (c) delta/zig-zag.

(ii) *Delta/star and star/delta connections (Fig. 4.18(b))*. If the star-connected winding has its neutral point isolated, triplen harmonic currents are unable to flow in the line connections to either winding but may circulate within the delta winding. Triplen harmonic currents are *co-phasal* (i.e. in phase with each other) in each phase of a three-phase winding. A slight distortion of the flux waveform will account for the triplen harmonic e.m.f.s necessary to generate the harmonic requirements of the magnetising current for the near-sinusoidal flux.

Star/delta configurations are used when a large voltage transformation ratio is required as the $\sqrt{3}$ relationship between line (to line) and phase (to neutral) voltages is incorporated in the relationship between primary and secondary line (to line) voltages. The line voltages on the two sides of the transformer differ in phase by $\pm 30°$ and thence in steps of $60°$, according to the precise interconnection of the phase windings. Care must be taken in paralleling circuits which contain transformers with significant phase shifts introduced between the line currents and voltages of the primary and secondary sides. Standard connections for a delta/star or star/delta transformer bank provide for the high-voltage reference phase to lead the low-voltage reference phase by $30°$, regardless of which winding is star- or delta-connected. The 'phase-voltage' referred to on the *delta-connected* side is the 'equivalent-star' voltage which is developed between the line terminals and an imaginary star point. An equivalent neutral point would be provided by a balanced source or load connected in star to the delta-connected winding terminals.

4.12.1 *Unbalanced Operation*

It is now appropriate to consider the behaviour of delta- and star-connected transformer windings towards the flow of zero phase-sequence current on electrical power systems subjected to unbalanced load or to system faults involving ground. The theory of *symmetrical components* applied to three-phase systems states that the unbalanced fundamental frequency currents (and voltages) at any point in the power network may be resolved into three symmetrical component sets:

(i) a positive phase-sequence set of balanced currents (voltages), with sequence *abc* defined by the synchronous generator e.m.f.s;
(ii) a negative phase-sequence set of balanced currents (voltages) with sequence *acb*; and
(iii) a zero phase-sequence set of currents (voltages), equal in magnitude and phase in each phase of the system.

The (phase-) sequence currents and voltages existing at the various points throughout the network depend on the positive sequence e.m.f.s developed on the system, the sequence impedances presented by the various system components and particular constraints imposed at the position of unbalance or fault by its characteristic nature.

Being static devices, transformers respond identically to positive and negative sequence currents and voltages. Their response to zero-sequence quantities is fundamentally different, however, and depends critically on the winding connections and the magnetic circuit configurations. The phase interconnection determines whether zero-sequence currents may flow in a winding and whether they can be admitted from the connected system or can necessarily be induced by transformer action. Thus definitive rules may be stated as follows.

(i) Zero-sequence current can flow in a *star-connected* winding only if the star point is earthed. The phase and line currents are identical.

(ii) Zero-sequence current can flow in a *delta-connected* winding only if electromagnetically induced via a mutually-coupled winding which is connected earthed star and supplied with zero-sequence current.

Common three-phase transformer connections therefore respond as follows when exposed to sources of zero-sequence current.

(i) *Star/star with isolated neutrals.* No zero-sequence current flow possible in either circuit. Primary and secondary windings both present infinite impedance.

(ii) *Star/star with both neutrals earthed.* Zero-sequence current flows in both primary and secondary windings, limited by the low-leakage impedances of the windings and zero-sequence impedances elsewhere on the system. With this connection, the effective leakage impedance per phase is virtually identical for positive-, negative- and zero-sequence currents.

(iii) *Star/star with earthed-star secondary only.* Zero-sequence current flow possible in the secondary circuit only. Hence transformation of zero-sequence current is not permitted. The secondary winding presents a high magnetising impedance to an external source of zero-sequence current; this impedance may be reduced to a low value if a three-phase core-type magnetic circuit is employed. The primary winding presents infinite impedance to zero-sequence current and acts as an effective open circuit.

(iv) *Star with isolated neutral/delta.* No zero-sequence current flow possible in line conductors of either winding. Both windings present infinite zero-sequence impedance.

(v) *Earthed-star/delta.* Zero-sequence current may flow in the phase windings of the star-connected winding via the line terminals and neutral point, thus inducing a circulating current in the delta-connected secondary. The difference between secondary phase currents is zero, hence the line currents on the delta side are zero also, and this winding presents infinite impedance to external zero-sequence current flow. The primary winding presents only a low leakage impedance to zero-sequence line currents flowing to ground via the earthed star point.

High-voltage transmission systems frequently employ earthed-star/earthed-star transformer connections in order that the $\sqrt{3}$ factor relating line and phase voltage may be used to advantage in reducing the insulation requirements of the windings. Such transformers are usually provided with a delta-connected tertiary winding for the purposes of flux waveshape improvement by allowing triplen harmonic currents to circulate therein and by the provision of a low zero-sequence impedance to earth for both primary and secondary circuits.

A special type of winding, known as an *interconnected-star* or *zig-zag* winding, is frequently employed in circumstances where balanced three-phase voltages may be developed between, or impressed across, the line terminals, whilst components of line current which are similarly directed at all instants are required to flow through the winding unimpeded. Such circumstances include:

(i) a.c. triplen harmonic currents; and
(ii) zero-sequence currents associated with unbalanced loading on faulted systems involving connection to earth.

From Fig. 4.18(c) it may be deduced that identical currents flowing into the interconnected-star winding at terminals *a*, *b* and *c*, and out at terminal *n*, produce *no net m.m.f.* on each limb of the magnetic core and hence encounter no inductive opposition to their flow (apart from leakage inductance effects), if the currents alternate. On the other hand, balanced three-phase currents of positive or negative phase sequence develop on each limb of the core an effective winding m.m.f. which is given by the difference of two identical components which are 120° out of phase with each other, and which are equivalent to the sum of two identical components having a phase difference of 60°. If no additional winding is present to carry a complementary current, the zig-zag winding current will be limited to that required to excite the core to the level of magnetic flux density necessary to support the applied terminal voltage. In this mode, where no secondary winding is present, the interconnected-star winding presents a high magnetising impedance to the flow of balanced, positive- or negative-sequence current, and a much lower leakage impedance to neutral for triplen harmonic- or zero-sequence current flow. The interconnected-star *reactor* with neutral point earthed is installed on power systems to provide an effective short circuit to earth for zero-sequence current, whilst having negligible influence on the flow of positive- or negative-sequence current.

4.13 Autotransformers

In contrast with double-wound transformers, autotransformers employ only one winding per phase. Electrical isolation of the primary and secondary circuits is consequently lost, but there are significant economic advantages to be gained if the voltage ratio required is near to unity. A 1 : 1 ratio autotransformer is electrically equivalent to just the magnetising characteristics of a double-wound transformer, with zero voltage regulation and negligible copper loss. Such a device is, of course, a simple shunt reactor but, with a turns ratio not greater than 3 : 1, an autotransformer will have significantly smaller size, loss and leakage reactance than its

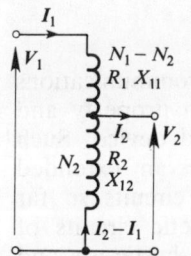

Figure 4.19 Single-phase autotransformer connections.

double-wound counterpart.

The essential connections of a single-phase autotransformer are depicted in Fig. 4.19 and it is readily shown that, for ampere-turn balance with magnetising current neglected, the current in the *common* part of the winding at $I_2 - I_1$ may alternatively be expressed as

$$\left(\frac{N_1 - N_2}{N_2}\right)I_1,$$

where N_1 and N_2 define the voltage ratio and, inversely, the current ratio of the transformer. This current flows in a portion of winding having resistance R_2 which, for the same current density as in that series portion carrying current I_1, has a value relative to that of the series winding given by

$$\frac{R_2}{R_1} = \left(\frac{\text{common turns}}{\text{series turns}}\right)\left(\frac{\text{series winding cross-sectional area}}{\text{common winding cross-sectional area}}\right)$$

$$= \frac{N_2}{N_1 - N_2}\frac{N_1 - N_2}{N_2} = 1.$$

With $R_2 = R_1 = R$, say, the total winding copper loss expressed in terms of an equivalent resistance R_{e1} referred to the total winding turns N_1 is, accordingly,

$$I_1^2 R_{e1} = I_1^2 R + (I_2 - I_1)^2 R$$

$$= I_1^2 R\left[1 + \left(\frac{N_1}{N_2} - 1\right)^2\right].$$

The loss clearly diminishes as the turns ratio approaches unity. With X_{11} and X_{12} respectively the leakage reactances of the series and common windings, the equivalent leakage reactance referred to the total 'primary' winding turns N_1 involves the same referring factor for 'secondary' leakage reactance X_{12}, i.e.

$$X_{e1} = X_{11} + X_{11}\left(\frac{N_1}{N_2} - 1\right)^2 X_{12}.$$

This relationship is readily proved by considering a short circuit at the 'secondary' terminals and evaluating the reactance presented to the 'primary' terminals, recognising that the secondary e.m.f. is absorbed by the leakage reactance of the common winding.

Short-time rated polyphase autotransformers find use in providing a reduced a.c. voltage whilst starting large induction motors in order to limit the starting current (and torque). A popular application is the *variable-ratio autotransformer* (single-phase or polyphase) in which a graphite brush connected to the line output terminal makes sliding contact with the exposed turns encircling a magnetic core of, commonly, toroidal form. The line input terminal is connected to a fixed point on the winding remote from the common (neutral) termination. The output voltage is almost infinitely variable (with increments equal to the induced voltage per turn) between zero and about 110% of the supply voltage, if the line input tapping point is located at 90% of the total winding length. With such a device the output voltage is a linear function of brush position.

4.14 High-frequency Transformers

Transformers find applications in electronic and communications systems because of their impedance transformation property and the circuit isolation provided by the double-wound device. Such transformers are often required to operate over an extended frequency range for which the simple equivalent circuits so far developed are not always appropriate. The magnetic circuits of high-frequency transformers are susceptible to high hysteresis and eddy-current losses unless low-hysteretic, high-resistivity core materials are employed. *Ferrite*, a ceramic, is such a material. Equation 4.14 shows that, for a given voltage per turn, higher frequencies require a lower peak flux, so high flux densities are seldom necessary. Magnetising reactance is proportional to frequency, which means that it can frequently be ignored at medium and high frequencies – but not at low frequencies where, due to the increased flux requirement, saturation of the core is more likely. Saturation has the effect of reducing the magnetising reactance further and introducing non-linearities. At high frequencies the effect of *capacitance* distributed between turns and between windings is highly significant.

A lumped equivalent circuit effective at high frequencies is shown in Fig. 4.20. In electronic circuit applications the impedance of the signal source frequently cannot be ignored, and this is incorporated in Fig. 4.20 as R_s.

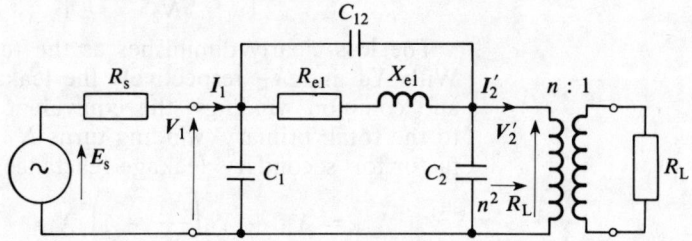

Figure 4.20 Equivalent circuit of a transformer at high frequency, with source resistance and resistive load.

4.15 Current Transformers

Electromagnetic *current transformers* (*c.t.s*) are theoretically identical with voltage transformers (*v.t.s*) and power transformers but have significant design and constructional differences. The purpose of a c.t. is to provide a secondary circuit with a replica of the current flowing in a primary circuit, maintaining integrity of waveshape, magnitude ratio and phase over a wide range of primary current values. The c.t. also, importantly, provides electrical isolation between high-voltage primary and low-voltage secondary circuits and, in high-power applications, a current reduction factor of the order of one hundred.

The secondary load on a c.t. is commonly known as its *burden* rather than as its load impedance. Whereas impedance is a voltage/current ratio, burden is a voltage-current product and is measured in VA. The current transformer is in many respects the *dual* of a voltage transformer. Minimum load (burden) for the c.t. corresponds to a short-circuit across the secondary winding – that for the v.t. has the secondary winding on open circuit. The *error* in a v.t. is due to load current flowing through the leakage impedance of primary and secondary windings, increasing with load current. That for a c.t. is due to magnetising current, increasing with core flux and hence with the voltage developed across the burden.

The equivalent current of a c.t. is identical with that of a voltage transformer or power transformer, but operational differences and design features mean that certain parameters may be ignored. The primary winding may be regarded as a *current feed* in the circumstances where the primary current is unaffected by changes in the secondary burden as, for example, with c.t.s employed in the monitoring of line currents flowing on electrical power networks. The current feed constitutes a high-impedance source, and thus enables the leakage impedance of the primary winding to be dispensed with. Further, in application to such systems the primary/secondary current ratio is likely to be high, of the order of 100 : 1, say, and therefore the number of secondary turns will be large and be comfortably provided by a winding in close proximity to the core which is usually toroidal in shape, as shown in Fig. 4.21(a). This arrangement provides for a low secondary leakage reactance which may often be neglected. The bulk of the insulation is provided on the primary winding which has few turns. In many applications, the minimum single turn is provided by a *bar-primary* – a single, straight conductor threaded through the centre of the toroid. The single turn is completed by the entire external circuit.

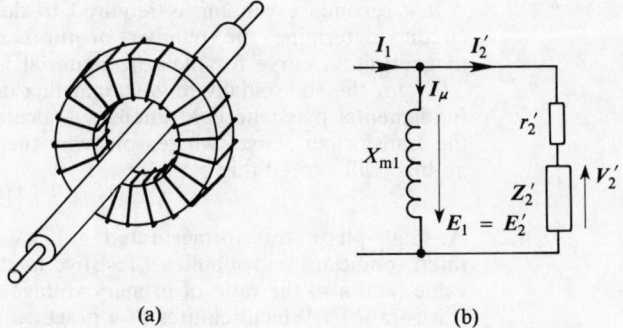

Figure 4.21 Construction and effective equivalent circuit of a current transformer.

Current transformer core materials are chosen to have low iron loss at normal flux densities. The core is frequently assembled in clock-spring fashion as tightly coiled strips of grain-oriented silicon steel, cold-rolled along the length of the strip so that high permeability is afforded to magnetisation around the eventual toroidal core form. Since the burden impedance is generally kept low to minimise the level of core flux required, the secondary winding *resistance* may be significant. Incorporating the magnetising reactance which appears connected in parallel with the series combination of secondary winding resistance and burden, the effective equivalent circuit of a c.t. is as shown in Fig. 4.21(b).

The size of a c.t. is largely determined by the size of its magnetic core which in turn relates to the burden impedance (enhanced by secondary winding resistance) and the secondary current, whose product establishes the core flux requirement via the transformer e.m.f. equation, eqn [4.14],

$$E_2 = 4.44 f N_2 \Phi_m.$$

The core cross-sectional area must be such as to prevent core saturation when the value of E_2 corresponds to maximum burden, otherwise the magnetising current I_{m1} will become excessive and I_2' will no longer approximate to I_1. Thus core saturation resulting from an excessive burden VA demand leads to unacceptable errors in the current ratio and secondary current waveshape.

Care must be taken in the use of c.t.s to ensure that there is no possibility of energising the primary circuit whilst the secondary winding is an open circuit. Under these circumstances the whole of the primary winding m.m.f. is available to magnetise the core which will rapidly saturate. The rapid rate of change of flux which occurs as the polarity of the saturated flux reverses on each zero-crossing of the primary current leads to large impulses of induced e.m.f. within the secondary winding and the likelihood of insulation failure.

4.17 Tutorial Examples

4.1 A 50 Hz single-phase transformer core has a cross-sectional area of 4 cm² and an effective length of 18 cm. Determine the minimum number of turns required on a 240 V (r.m.s.) primary winding if the peak flux density is not to exceed 1.4 T.

If a secondary winding is required to develop 110 V (r.m.s.) on open circuit, determine the number of turns on this winding. If the a.c. magnetisation curve for the core material is such that an r.m.s. value of 1.0 T for the sinusoidally time-variant flux density requires 140 A T m⁻¹ of fundamental magnetic field intensity, calculate the magnetising current of the transformer. Give two reasons why the measured primary current on no-load will exceed this.

[1930 turns; 885 turns, 13.1 mA]

4.2 A single-phase transformer rated at 5 kVA, 240:120 V operates under rated conditions and supplies a resistive load. Calculate the load resistance value, and also the ratio of primary voltage to primary current. Relate via (turns-ratio)². Which features of a practical transformer lead to discrepancy?

[2.88 Ω, 11.52 Ω]

4.3 A two-winding transformer with a linear, loss-free magnetic circuit has the following parameters relating to winding resistance, leakage inductance and magnetising inductance:

$R_1 = 0.432\ \Omega,\quad R_2 = 30\ \Omega;\quad L_{l1} = 2.88\ \text{mH},\quad L_{l2} = 0.20\ \text{H};$

$L_{m1} = 5\ \text{H},\quad L_{m2} = 347.22\ \text{H}.$

Determine the likely turns ratio, the self-inductance coefficients of each winding and the mutual-inductance coefficient.

[$N_1/N_2 = 0.12$, $L_1 = 5.00288$ H, $L_2 = 347.42$ H, $M = 41.67$ H]

4.4 The equivalent leakage impedance of a 2000/200 V, 10 kVA single-phase transformer is $8 + j150\ \Omega$ when referred to the primary winding. Calculate the change in voltage magnitude which will occur across the secondary terminals when the load current at 0.8 p.f. lag is reduced from the rated value to zero, the primary voltage being maintained.

[48.2 V]

4.5 Convert the transformer impedance data of **4.4** to p.u. and repeat the calculation.

$[Z_e = 0.02 + j0.375\ \text{p.u.}]$

4.6 The transformer of **4.4** dissipates 100 W as iron loss when excited at rated voltage and frequency. Determine its efficiency when supplying load current of (a) 10%, (b) 25%, (c) 50%, (d) 75%, (e) 100% rated current at (i) unity p.f., (ii) 0.8 p.f. lag.

[Solution presented as Fig. 4.14]

4.7 A 50 kVA, 2400/240 V, 60 Hz transformer gave the following test data:

Open-circuit test 240 V   5.43 A,   186.6 W

Short-circuit test 48.1 V   20.83 A,   616 W

Determine the parameters of the approximate equivalent circuit referred to the high-voltage side.

$[Z_{e1} = 1.42 + j1.82\ \Omega,\ Y_{o1} = (0.324 - j2.24) \times 10^{-4}\ \text{S}]$

4.8 Using the rating data and the solution to **4.7**, calculate, by means of a phasor diagram construction, the required primary voltage and its phase relative to the secondary terminal voltage for rated secondary voltage and current at (i) 0.8 p.f. lag, (ii) 0.8 p.f. lead.

Establish whether division of the equivalent impedance into equal parts and inclusion of the effect of magnetising current flowing in the primary winding has a significant effect on the full-load terminal voltage ratio.

Compare the calculated primary voltage on load with that predicted from use of eqn [4.12] for voltage regulation.

$[2446\ \angle\ 0.29°\ \text{V},\ 2401\ \angle\ 1.15°\ \text{V}]$

4.9 Open-circuit tests carried out on a 50 kVA, 11 kV/400 V, 50 Hz transformer gave rise to the following input power requirements:

Supply voltage	kV	11	13.20
Frequency	Hz	50	60
Total iron loss	W	700	912

Use a graph of total iron loss/frequency plotted against frequency to separate the hysteresis and eddy-current loss components and evaluate these components at (i) 50 Hz, (ii) 60 Hz with normal rated flux.

When the open-circuit test was repeated at 5.5 kV and 50 Hz, the total iron loss fell to 207 W. (iii) Calculate the Steinmetz index for hysteresis loss.

If the copper loss at rated current is 600 W, calculate the efficiency of the transformer when supplying rate voltage and current at unit p.f. at (iv) 50 Hz and (v) 60 Hz (reduced flux).

[(i) 400 W   300 W; (ii) 480 W   432 W; (iii) 1.6; (iv) 97.47%; (v) 97.54%]

4.10 Two identical single-phase transformers rated at 1 MVA, 33/11 kV have their low-voltage windings connected in parallel to an 11 kV supply and their high-voltage windings connected in series such that, normally, negligible current flows in the secondary circuit, whereas 8 A flows in each primary winding. The transformers are put on load for test purposes by incorporating a variable voltage source in the secondary circuit, and it is observed that a source voltage of 1920 V with an input power requirement of 10 kW is required to circulate rated current of 30.3 A. The 11 kV source supplies a total current of 16 A and 7.2 kW.

Draw the approximate equivalent circuit of the two transformers incorporating the two sources, employing equivalent leakage impedances referred to the 33 kV side and show that the source providing rated voltage supplies essentially the total iron loss and magnetising current, whereas that providing rated current supplies essentially the total copper loss and the total leakage impedance voltage of the two transformers.

Use the test data to estimate the leakage impedance of each transformer (in ohms referred to the high-voltage winding and in p.u.), the magnetising reactance in p.u. and the efficiency on full load at unity p.f.

[$5.45 + j31.2 \, \Omega$, $0.005 + j0.029$ p.u., 11.4 p.u., 99.1%]

4.11 A *pulse-transformer* is frequently used to couple the gate-cathode circuit of a thyristor to the electrically isolated collector circuit of a bipolar transistor with a view to injecting charge carriers into the gate of the thyristor at the instant the transistor is switched by base drive from cut-off to saturation. A current-limiting resistor R_s is connected in series with the pulse-transformer primary winding, between the transistor collector terminal and the d.c. voltage rail – maintained at V_{cc} above the common-emitter potential.

If the pulse-transformer load is considered resistive, of value R'_L when referred to the primary winding, the effective circuit may be represented by Fig. 4.22 in which L_{m1} represents the magnetising inductance referred to the primary winding, and the leakage inductance is considered negligible. R_s limits the magnetising current i_{m1} to prevent saturation of the transformer core.

Show that the voltage across the resistive load R'_L is given by

$$v'_2 = V_{cc} \frac{R'_L}{R'_L + R_s} \exp\left[\frac{-R'_L R_s t}{(R'_L + R_s) L_{m1}}\right].$$

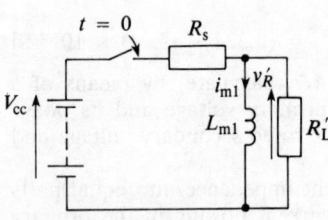

Figure 4.22 Effective circuit for Tutorial Example 4.11.

[*Note*: The parameters of the circuit are normally chosen such that the delay time-constant of the response is very much less than the duration of transistor saturation. This limits the energy input to the thyristor gate after gating has been achieved and is accomplished by incorporating airgaps in the transformer core.]

4.12 Two three-phase transformers are connected star/star and are paralleled on both primary and secondary sides. A balanced delta-connected load of impedance $2.4 + j1.8 \, \Omega$/phase is connected on the secondary side. The open-circuit secondary line voltage of one transformer is 400 V and that of the other is 403 V. Referred to the secondary winding, the equivalent leakage impedance of each transformer is $0.015 + j0.06 \, \Omega$. Calculate the current supplied by each transformer and the secondary terminal voltage.

[102.5 A, 125 A, 392 V]

4.13 The essential connections of a three-phase, three-pulse a.c./d.c. converter circuit are shown in Fig. 4.23. The smoothing inductance L ensures that the d.c. load current I_D is substantially constant, with a process of *natural commutation* enabling the load circuit to be completed via phases a, b and c in sequence; the instantaneous voltage v_d following the envelope of the phase-neutral voltages v_a, v_b, v_c, v_a, etc. The current in each phase, e.g. i_a, is inherently unidirectional, having value I_D for 1/3 of each cycle of supply, and is inherently incompatible with the requirements of an a.c. supply. Show that:

(i) the three phase currents have equal d.c. components $\frac{1}{3}I_D$;
(ii) the d.c. components of the converter current develop zero net d.c. m.m.f. on the magnetic core of a supply transformer whose secondary winding is connected in zig-zag:

(iii) if the converter supply transformer is provided with a delta primary and a star secondary, the primary line currents (equal to the difference of a pair of phase currents) have no d.c. component, with balance of primary and secondary ampere-turns possible on each limb of a core-type magnetic core construction (Fig. 4.17(a)); and

(iv) if the converter supply transformer is provided with a star primary and star secondary, balanced primary line currents with no d.c. component are achieved at the expense of *residual m.m.f.* $= \frac{1}{3}N_2 I_D$ on each limb of the core, where N_2 is the number of secondary turns/phase.

[*Note*: The existence of such a residual m.m.f., equal in magnitude on all three limbs of a *core-type* transformer is not intolerable. The flux path for such an m.m.f. is completed via the transformer tank and the air spaces which separate core and tank. The high reluctance of the circuit means that the level of d.c. flux in the core, about which the superimposed a.c. variations occur, is relatively low and the performance of the magnetic circuit is not grievously affected. If triplen harmonic ripple appears on the residual m.m.f., due to inadequate smoothing of the d.c. load current, co-phasal triplen harmonic e.m.f.s induced in the primary phase windings do not appear between the primary line terminals.]

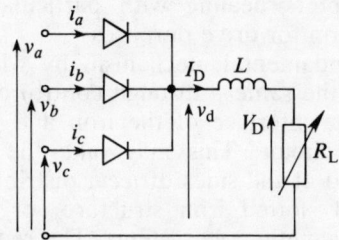

Figure 4.23 Essential connections of a three-phase three-pulse a.c./d.c. converter circuit.

4.14 A 50 Hz current transformer has a single-turn primary winding and a 200-turn secondary. The latter supplies a non-inductive burden of 1 Ω resistance with a normal current of 5 A. The corresponding core flux requires 80 A T. Draw a phasor diagram showing the currents, terminal voltage and flux referred to the secondary winding. Calculate (i) the maximum flux density for a core of 10 cm² net cross-section, (ii) the current ratio and phase angle. Neglect winding resistance, leakage reactance and core loss.

[(i) 0.113 T; (ii) 200.64, 4.57°]

4.15 The magnetising current drawn by a toroidal core c.t. of ratio 1000 : 5 A, when operating at rated current with a secondary burden of 1 Ω resistance, is 1.0 A at 0.4 p.f. lag. Calculate the ratio error, (nominal ratio − actual current ratio)/actual ratio, at 10% load. Assume that the Steinmetz coefficient is 1.6, that secondary resistance and eddy current loss are negligible and that the magnetising reactance is constant.

[−0.1%]

5 Windings for rotating machines

5.1 Introduction

Rotating electrical machines are essentially interlinked electrical and magnetic circuits, configured in such a way that mechanical torque, developed on both stationary and rotating members, is sustainable at a constant value in the steady state – even though the component parts of the electrical and magnetic circuits on stator and rotor are in relative motion. The relationship between torque and speed is an important operational feature of any machine, with significant differences being possible between contrasting machine types. Torque/speed characteristics are capable of adjustment through systems of control to be discussed further in later chapters dealing with particular machine types and their specification for drive purposes.

The fundamental mechanism by which each machine produces torque is the same – through control of the magnetic flux distribution at the interface of the iron and the airgap in the composite magnetic circuit. This may take the form of differential forces established at the sides of *teeth* distributed over the periphery of a cylindrical slotted iron structure, or at the sides of projecting, *salient*, magnetic pole systems. The redistribution of flux is brought about by the magnetic effect of current flowing through windings on one side, or both sides, of the airgap. Useful rotating machines are required to retain this particular distribution of flux with the rotor in continuous motion in order that the torque independence of rotor angular position is maintained. Hence, in the steady state, *there will be no relative motion between the magnetic fields of stator and rotor members*.

It is this feature which distinguishes the continuously rotating machine from the electromechanical transducers having angular movement discussed in Chapter 3 (Sect. 3.10 – 3.14). In these, the angular displacement of the rotor shaft is limited to less than one revolution. A device designed to develop a large torque over a restricted angular displacement is often described as a 'torque motor' to distinguish it from the motor having continuous rotation. Within the latter category fall the Faraday devices discussed in Sections 3.15 and 3.16.

5.1.1 The role of windings

If continuous electromechanical energy conversation is to occur without change in the value of the coupling magnetic field the electrical and mechanical power requirements must be complementary. This balance is accounted for by the development of e.m.f., within appropriate machine windings as a consequence of the motion, so as to promote or oppose the current flow responsible as a factor in torque development – in accordance with machine operation as generator or motor. The significance of the machine *windings* is thus seen to be two-fold. With the qualified inclusion of permanent-magnet machines, the magnetic field conditions at the airgap bounding surfaces responsible for torque development are established by winding *currents*. When the machine rotates, it is necessary that appropriate *electric field* conditions exist at the winding conductor locations to enable energy transfers to occur between the external electrical system and the machine.

As electric field conditions within electrical machines are a consequence of motion, and pre-existing *magnetic* fields, we shall defer consideration of winding e.m.f. until Section 5.8. Our initial concern will be with the magnetic field conditions established by winding currents.

The essential constructional features of d.c. machines and polyphase a.c. *synchronous* machines are depicted in the sectional view of Fig. 5.1, shown perpendicular to the axis of rotation. The rotor is essentially cylindrical with axially directed winding conductors located in slots (not shown) cut into the rotor iron periphery. The airgap between stator and rotor is non-uniform since the stator has salient poles, four in number. The machine is *heteropolar*, since the *armature* conductors on the rotor encounter stator poles of alternate N and S polarity as the rotor rotates. As saliency exists only on one side of the airgap, there is no change in the physical topology of the magnetic circuit with rotor rotation. Depending on the method of connecting the rotor conductors to the external circuit, the machine may be made capable of developing a steady torque when energised with either balanced, polyphase sinusoidal currents or steady direct current – provided that the stator *field* winding current is constant, and that the field of the rotor winding maintains constant value and position relative to the stationary stator field. Under these steady magnetic-field conditions it is apparent that no electromagnetically induced e.m.f. will be developed in the stator winding, eqn [2.39], and thus the stator field circuit constitutes simply, in the steady state, a *sink* for copper loss. The *rotor* conductors, on the other hand, have motion in the steady resultant magnetic field due to both stator and rotor currents and will, in general, develop electromagnetically induced e.m.f. The same rotor conductor interconnections which established rotor magnetic field conditions appropriate to the development of steady torque also provide for an effective harnessing of the rotor conductor e.m.f.s, such that the rotor winding as a whole may absorb, or supply, electrical power at a steady rate equivalent to the mechanical shaft power developed by the rotor, after accounting for losses.

In this context, the rotor winding is described as an 'armature' winding because it is associated with e.m.f. and electrical power flows which match the terminal voltage and power ratings. On the other hand, a 'field' winding, like the stator winding here, has no e.m.f. once steady flux conditions in the machine have been built up. Its *current* rating may well be of the same order as that of the armature winding, since its magnetic field is comparable with that of the armature winding, but its *power* rating simply equates to its copper loss.

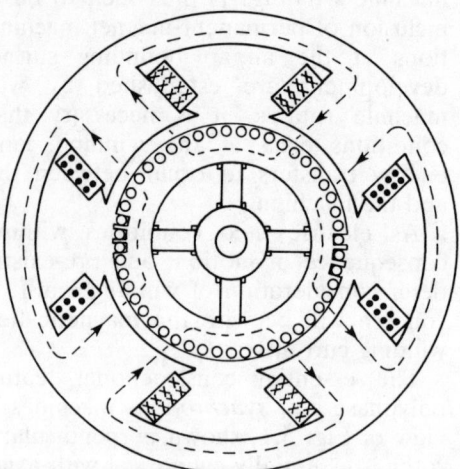

Figure 5.1 Section across the axis of rotation of a heteropolar four-pole machine with d.c. field excitation on stator salient poles.

The positions of armature and field windings on opposite sides of the airgap are essentially interchangeable. In d.c. machines and small polyphase a.c. synchronous machines it is usual to employ a stator field system and a rotating armature. In large a.c. synchronous machines (and induction machines) the high-current, high-voltage armature winding is more secure if mounted on the stator, with the d.c.-excited field pole system allowed to rotate.

5.2 The Magnetic Field of an Armature Winding

In a practical machine the armature winding is accommodated in slots cut into the laminated iron stampings making up the rotor or stator core, so that the conductors (insulated) are adjacent to the airgap and supported by the iron teeth. Due to the presence of the slots the airgap is irregular. In order to simplify the calculation of airgap fields it is common practice to ignore variations in the airgap permeance due to the presence of slots, and this we shall assume in our machine model. Simplifying assumptions are often also made regarding the distribution of winding current over the iron/airgap surface, but for the present we shall assume the current to be concentrated in conductors of finite width. Infinite permeability for the iron on either side of the airgap will be assumed so that the magnetic field intensity H within the iron is always zero. The airgap is assumed to be cylindrical in shape and with the gap length g (equal to the difference between the outer and inner radii) much smaller than these radii, thus obliging the flux density normal to the stator and rotor boundaries with the airgap at the same angular position to be uniform across the gap.

5.2.1 *Winding with one slot per pole*
First of all, we consider the field set up by four equally spaced

conductors on the rotor side of the airgap, carrying identical currents I parallel to the rotor axis but with adjacent currents of alternate polarity. Figure 5.2 shows the arrangement, Fig. 5.2(b) depicting a *developed* view with the airgap 'opened out'.

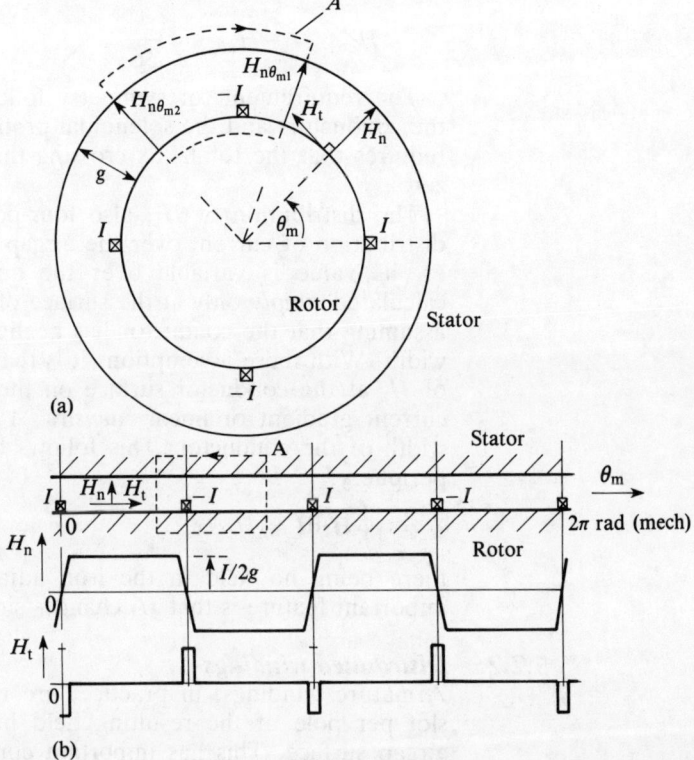

Figure 5.2 Radial and tangential magnetic field components established in a uniform airgap by an elementary four-pole winding with one slot/pole. H_n is essentially uniform across gap. H_t is evaluated at the energised rotor surface.

This figure shows the type of magnetic field intensity distribution produced by the four conductors in terms of its radial (normal) and tangential components. They are evaluated by the application of Ampère's law, eqn [2.27], which states that the line integral of magnetic field intensity H over any closed circuit is equal to the current enclosed, i.e.

$$\oint H \cdot dl = I.$$

In vector algebra parlance, the scalar (or dot) product of H and dl is involved in the integration. Thus, over the element dl of circuit, the appropriate value of H to use in the integration is the component in the direction of dl. The reference direction of H is right-handed about the direction of I, that is, a right-handed screw rotated in the direction of H would progress in the direction of I.

The forms of the distribution of the normal and tangential components H_n and H_t for the four-conductor system depicted in Fig. 5.2(b) are significantly different. That of H_n is approximately a square waveform, indicating a four-pole field, i.e. two pairs of adjacent north and south poles over the complete airgap surface. H_n is almost uniform across the gap and undergoes change only in the regions where current is present.

H_n is readily evaluated by considering a circuit such as A and applying eqn [2.27] to give

$$H_{n\theta_{m2}}g - H_{n\theta_{m1}}g = -I.$$

If, due to a requirement for symmetry $|H_{n\theta_{m1}}| = |H_{n\theta_{m2}}|$, then

$$H_{n\theta_{m1}} = -H_{n\theta_{m2}} = \frac{I}{2g}. \qquad [5.1]$$

The requirement for symmetry follows from the equispacing of the conductors and the solenoidal property of magnetic flux, which requires that the total flux crossing the airgap due to H_n must be zero.

The distribution of H_t, also four-pole, is seen to resemble the distribution of current over the airgap surface. Unlike H_n, however, its value is variable over the airgap length g. Its value is calculated *simply* only at the surface of the conductor, and then on assuming that the conductor has negligible thickness, though finite width. With these assumptions it is readily seen that the magnitude of H_t at the conductor surface on the airgap side is equal to the current gradient or linear '*density*' $A = I/\Delta x$, with Δx being the width of the conductor. This follows because, over the conductor periphery,

$$\oint H \cdot dl = H_t \Delta x,$$

there being no field in the iron adjacent to the conductor. An important feature is that H_t changes sign with current direction.

5.2.2. *Distributed windings*

Armature windings, in practice, are not concentrated in a single slot per pole of the resulting field but are spread out over the airgap surface. This has important consequences for the resultant field distribution, expressed in terms of normal and tangential components.

A design objective with *alternating-current* machines intended for connection to sinusoidal sources of supply is to produce near-sinusoidal flux distributions in the machine airgap under all conditions of load, requiring that any component of the resultant airgap field should be sinusoidally distributed also. Observation of the field waveforms of Fig. 5.2(b) shows both H_n and H_t to have significant space harmonics in addition to the fundamental four-pole field. Controlling the layout of the winding can reduce the significance of these harmonics and, indeed, effect selective elimination.

With *direct-current* machines the concern about the distribution of flux and field is that they should be as uniform as possible. This is to avoid magnetic saturation in regions of high flux density, such as at the armature teeth, and to obtain a uniform mechanical stressing of the armature iron. This effectively means distributing the armature winding over a large number of slots, so that the current and hence the H_t distribution approximates to a square waveform. As will be shown in Section 5.4.1 (Figure 5.10) the associated H_n distribution approximates to a *triangular* waveform with an apparent 90° space phase-difference with respect to H_t.

Windings for rotating machines 139

The desire to minimise *ripple* on the e.m.f. induced in the armature circuit reinforces the need to distribute the armature winding.

Owing to their requirement for connection to a *commutator*, the armature coils used for d.c. machines differ in detail from those used in a.c. machines. Typical coil shapes are shown in Figs. 5.3 and 5.4 for a.c. and d.c. machine windings, respectively. Coils consist of *active coil sides* located in the slots and responsible for the airgap field production, connected by sections whose magnetic field and electrical resistance effects impair the machine performance. Collectively, these inactive parts of the conductors are known as the *end-windings* of the machine.

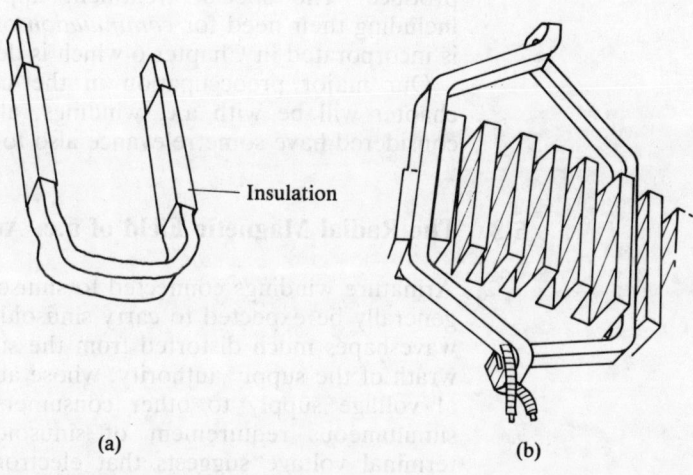

Figure 5.3 Alternating-current machine armature winding elements: (a) hairpin coil for 'semi-closed' slots, single-layer winding; and (b) 'pulled' multi-turn coil for double-layer winding with two coil sides/slot – open, with the winding being secured by wedges.

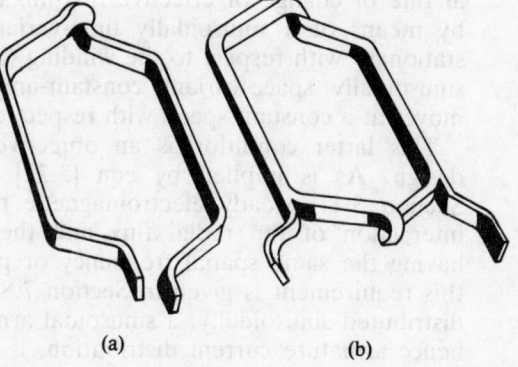

Figure 5.4 Direct-current machine armature coils: (a) single-turn (bar) coil, insulated, for double-layer 'lap' winding: and (b) multi-turn coil, insulated, for double-layer 'wave' winding.

Individual armature coils are connected with adjacent coils to form the armature winding. The resulting winding may be *continuous*, as for a d.c.-type winding with each coil connected to a *commutator segment* for periodic connection via a *brush* to an external circuit. Alternatively, coils may be connected in groups to form the phase windings of an a.c. machine, with the phase windings typically interconnected in star or delta. Coil interconnections may also be arranged in series or parallel configurations, as is primarily determined by the required relationships between the e.m.f.s induced in the active coil sides (by their motion at right-angles to the radial flux crossing the airgap) and the armature terminal voltage of the machine. When coils or groups of coils are connected in parallel, care must be taken to ensure that their electrical characteristics, including induced voltages, are identical so that their currents are identical. If, as is more common, the coils are connected in series, their currents are obliged to be equal. This equivalence of current within interconnected groups of coils is a simplifying influence in establishing the magnetic field of armature windings.

As noted above, d.c. armature windings have a rather different objective to that of a.c. windings in respect of the airgap field they produce. The special treatment appropriate to d.c. windings, including their need for *commutation* of the individual coil current, is incorporated in Chapter 6 which is devoted to d.c. machines.

Our major preoccupation in the next section of the present chapter will be with a.c. windings, although many of the issues considered have some relevance also to d.c. machine windings.

5.3 The Radial Magnetic Field of a.c. Armature Windings

Armature windings connected to sinusoidal a.c. power systems will generally be expected to carry sinusoidal currents. Indeed, current waveshapes much distorted from the sinusoidal may well incur the wrath of the supply authority, whose ability to maintain a sinusoidal voltage supply to other consumers would be impaired. The simultaneous requirement of sinusoidal current and sinusoidal terminal voltage suggests that electromagnetically induced armature winding e.m.f.s should be sinusoidal also, implying a sinusoidal rate of change of effective flux-linkage. This is achieved either by means of a sinusoidally time-variant pulsating flux, which is stationary with respect to the winding (as in a transformer) or by a sinusoidally space-variant constant-amplitude radial flux, which moves at a constant speed with respect to the winding.

This latter condition is an objective in rotating a.c. machine design. As is implied by eqn [3.22] and the considerations of Section 3.11, steady electromagnetic torque is produced by the interaction of the radial flux and the tangential armature field having the same spatial frequency or pole-pitch. Formal proof of this requirement is given in Section 7.8. Thus, if the radial flux is distributed sinusoidally, a sinusoidal armature tangential field, and hence armature current distribution, is indicated. In practice, it is not possible to arrange windings to give a pure sinusoidal current distribution, and thus the H_t field must contain *space* harmonics. These will have no effect on the torque, however, if the *radial* flux distribution is sinusoidal, although the *tangential* armature field has an indirect (leakage) effect on induced armature voltage. Hence we seek to minimise the harmonics present in the radial component of armature field which, in conjunction with the radial component of field produced by windings on the other side of the airgap, generates the radial flux wave.

For this reason we will examine in detail the radial field

Windings for rotating machines 141

produced by an armature coil, expressing this in terms of its fundamental and harmonic components using Fourier analysis. A more general configuration than that of Fig. 5.2 will be used, such that adjacent conductors are asymmetrically spaced. The *short-pitch* or *chorded* coil so produced will be shown to generate a field space-harmonic spectrum which includes even harmonics, these being absent for the full-pitch or *unchorded* coil.

Consider Fig. 5.5 in which θ defines position in the airgap with respect to a reference point corresponding to the location of an infinitely thin conductor, at which H_n crosses zero, and is positive-going as θ increases from zero. For simplicity a 2-pole field is shown. We also make the usual assumptions of infinitely permeable iron and a short, uniform airgap of radial length g.

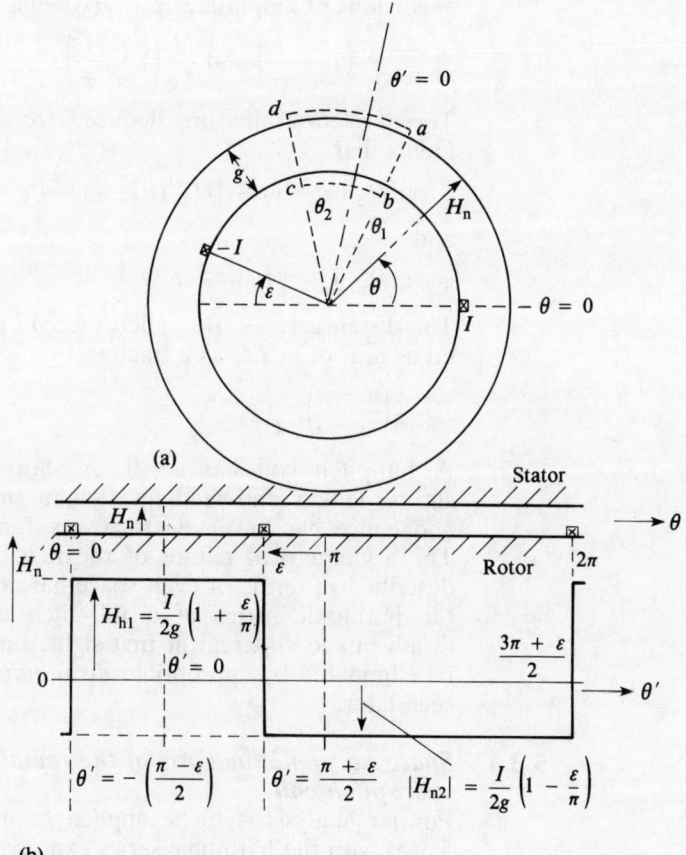

Figure 5.5 Radial two-pole field distribution in uniform airgap due to a single-turn short-pitch coil.

The radial field distribution is again determined using Ampère's law. The line integral

$$\oint \mathbf{H} \cdot \mathrm{d}l$$

is zero for any closed path which crosses the airgap twice and which does not embrace a conductor. The line integral equals $\pm I$ for such closed paths which enclose a conductor carrying current I. By considering one gap-crossing of an elemental closed path to be fixed [$a \to b$ at $\theta = \theta_1$] in Fig. 5.5(a) and allowing the return gap traverse [$c \to d$ at $\theta = \theta_2$] to vary its position, the field at $\theta = \theta_2$ *relative* to that at $\theta = \theta_1$ is readily established. Absolute determination of $H_n(\theta)$ requires the application of Gauss's law, eqn. [2.1], which states that the *total magnetic flux* entering (or leaving) a closed surface must be zero, i.e.

$$\Phi_{total} = \oint B \cdot da = \mu_0 \oint H \cdot da = 0.$$

This means that the total flux entering or leaving the entire rotor (or stator) surface is zero. Thus the *average* value of $H_n(\theta)$ must be zero. Referring to Fig. 5.5(b), the area under the waveform of $H_n(\theta)$ from $\theta = 0$ to $\theta = 2\pi$ is therefore zero. The rectangular waveform is asymmetrical for $\varepsilon \neq 0$, with positive and negative excursions of amplitude H_{n1}, H_{n2} being respectively equal to

$$\frac{I}{2g}\left(1 + \frac{\varepsilon}{\pi}\right) \text{ and } \frac{I}{2g}\left(1 - \frac{\varepsilon}{\pi}\right).$$

These relationships are deduced from the simultaneous requirements that

$$H_{n1}(\pi - \varepsilon) + |H_{n2}|(\pi + \varepsilon) = 0$$

and

$$H_{n1}g + |H_{n2}|g = I.$$

The distance $(\pi - \varepsilon)$ is called the *coil pitch* and is usually expressed as p.u. of π, i.e. as a fraction

$$\left(\frac{\pi - \varepsilon}{\pi}\right).$$

A full-pitch coil has $\varepsilon = 0$. A short-pitch or *chorded* coil has shorter-length end-windings than a full-pitch or unchorded coil, and hence has lower electrical resistance, which is an advantage. The asymmetrical nature of the field distribution, which may be described in terms of even space-harmonics which are absent from the harmonic series of a full-pitch coil, is not necessarily the disadvantage apparent at first sight, since the even harmonics may be eliminated by appropriate *distribution* of the winding, as will be seen later.

5.3.1 Space harmonic analysis of the radial field developed by a short-pitch coil

Fourier analysis is to be applied to the field distribution of Fig. 5.5(b) with the harmonic series expressed with reference to $\theta' = 0$,

where $\theta' = \theta - \left(\frac{\pi - \varepsilon}{2}\right).$

Since $\theta' = 0$ represents an axis of symmetry for the waveform, i.e. $f(-\theta') = f(\theta')$, only cosine terms will be present in the harmonic series. ($\theta' = 0$ represents the centre-line of the fundamental magnetic field component produced by the coil.)
Hence

$$f(\theta') = \sum_{v=1}^{v=\infty} a_v \cos v\theta'$$

where

$$a_v = \frac{1}{\pi}\int_{-\pi}^{\pi} f(\theta')\cos v\theta'\,d\theta'$$

$$= \frac{1}{\pi}\int_{-(\pi-\varepsilon)/2}^{(3\pi+\varepsilon)/2} f(\theta')\cos v\theta'\,d\theta'$$

$$= \frac{1}{\pi}\frac{I}{2g}\left[\int_{-(\pi-\varepsilon)/2}^{(\pi-\varepsilon)/2}\left(1+\frac{\varepsilon}{\pi}\right)\cos v\theta'\,d\theta' \right.$$

$$\left. +\int_{(\pi-\varepsilon)/2}^{(3\pi+\varepsilon)/2}-\left(1-\frac{\varepsilon}{\pi}\right)\cos v\theta'\,d\theta'\right]$$

giving, for $v = 1$ which pertains to the fundamental component of the radial field,

$$a_1 = \frac{I}{2g}\frac{2}{\pi}\left(\cos\frac{\varepsilon}{2}+\cos\frac{\varepsilon}{2}\right) = \frac{I}{2g}\frac{4}{\pi}\cos\frac{\varepsilon}{2}.$$

Thus the fundamental component of the radial field produced by a single-coil short-pitched by angle ε is

$$H_{n1}(\theta') = \frac{I}{2g}\frac{4}{\pi}\cos\frac{\varepsilon}{2}\cos\theta'. \qquad [5.2]$$

The factor

$$\cos\frac{\varepsilon}{2}$$

in eqn [5.2] is called the *coil-span factor* K_{p1} for the fundamental field and, being of value less than unity, it is the factor by which the fundamental field is reduced from that pertaining to a full-pitch coil carrying the same current.

Investigating now the harmonic components of radial field

$$a_v = \frac{1}{\pi}\frac{I}{2g}\left[\left(1+\frac{\varepsilon}{\pi}\right)\left|\frac{\sin v\theta'}{v}\right|_{-(\pi-\varepsilon)/2}^{(\pi-\varepsilon)/2}\right.$$

$$\left.-\left(1-\frac{\varepsilon}{\pi}\right)\left|\sin\frac{v\theta'}{v}\right|_{(\pi-\varepsilon)/2}^{(3\pi+\varepsilon)/2}\right]$$

$$= \frac{1}{\pi}\frac{I}{2g}\frac{2|\sin v\theta'|_{-(\pi-\varepsilon)/2}^{(\pi-\varepsilon)/2}}{v}$$

$$= \frac{I}{2g}\frac{4}{\pi}\frac{1}{v}\sin v\left(\frac{\pi}{2}-\frac{\varepsilon}{2}\right).$$

Hence, for $v = 2$,

$$a_2 = \frac{I}{2g}\frac{4}{\pi}\frac{1}{2}\sin(\pi-\varepsilon) = \frac{I}{2g}\frac{4}{\pi}\frac{1}{2}\sin\frac{2\varepsilon}{2},$$

for $v = 3$,

$$a_3 = \frac{I}{2g}\frac{4}{\pi}\frac{1}{3}\sin\left(\frac{3\pi}{2}-\frac{3\varepsilon}{2}\right) = \frac{I}{2g}\frac{4}{\pi}\frac{1}{3}\left(-\cos\frac{3\varepsilon}{2}\right),$$

and for $v = 4$,

$$a_4 = \frac{I}{2g}\frac{4}{\pi}\frac{1}{4}\sin(2\pi-2\varepsilon) = \frac{I}{2g}\frac{4}{\pi}\frac{1}{4}\sin\frac{4\varepsilon}{2}.$$

Thus the *coil-span factor* for the vth harmonic, K_{pv} is given generally by

$$K_{pv} = \sin v\left(\frac{\pi}{2} - \frac{\varepsilon}{2}\right). \quad [5.3]$$

The radial field of a single-turn coil, short-pitched by ε, is expressed as a harmonic series thus

$$H_n(\theta') = \frac{I}{2g}\frac{4}{\pi}\left(\cos\frac{\varepsilon}{2}\cos\theta' + \frac{1}{2}\sin\frac{2\varepsilon}{2}\cos 2\theta'\right.$$
$$-\frac{1}{3}\cos\frac{3\varepsilon}{2}\cos 3\theta' + \frac{1}{4}\sin\frac{4\varepsilon}{2}\cos 4\theta'$$
$$\left.+\frac{1}{5}\cos\frac{5\varepsilon}{2}\cos 5\theta' + \frac{1}{6}\sin\frac{6\varepsilon}{2}\cos 6\theta' - \ldots\right). \quad [5.4]$$

If the coil is full-pitch, $\varepsilon = 0$ and eqn [5.4] reduces to

$$H_n(\theta') = \frac{I}{2g}\frac{4}{\pi}\left(\cos\theta' - \frac{1}{3}\cos 3\theta' + \frac{1}{5}\cos 5\theta' - \frac{1}{7}\cos 7\theta'\right.$$
$$\left.+\frac{1}{9}\cos 9\theta' - \ldots\right). \quad [5.5]$$

The absence of even harmonic terms in eqn [5.5] is apparent. The radial field of a full-pitch, single-turn coil is shown in Fig. 5.6, together with the fundamental and the two principal harmonic components. Note that the amplitude of each harmonic term is inversely proportional to its harmonic number.

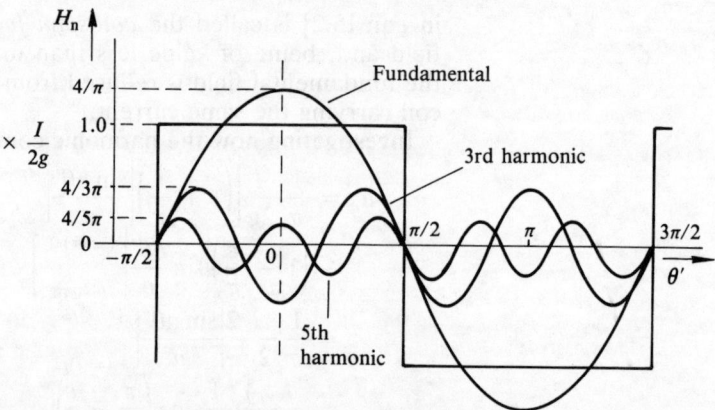

Figure 5.6 Principal harmonic components of the radial field due to a single-turn, full-pitch coil.

Short-pitch or chorded coils employed in polyphase windings do not necessarily produce even harmonics in the resultant field. Armature windings are usually double-layer windings with each slot occupied by the active coil sides of two different coils, not necessarily of the same phase grouping. Consider the two-pole, three-phase winding of Fig. 5.7 having one slot/pole/phase, two coils/phase and $n/2$ turns per coil.

The radial field distribution pertaining to phase a is seen to be symmetrical about the horizontal axis, and hence to contain no even-harmonic components. It could be produced equally well by either (i) locating full-pitch coils in the upper layers of slots 1 and 4 and the lower layers of slots 3 and 6 or (ii) locating coils short-pitched by 60° in slots 1 and 3 for one coil and slots 4 and 6 for the other coil. The former alternative inherently produces two component sets identifiable with the coils, mutually displaced in space by one pole-pitch *of the fundamental* and of opposite polarities. This leads to the cancellation of all even-harmonic components and a doubling of the fundamental and odd-harmonic components – the effect of the latter may be seen by reference to Fig. 5.6, comparing field conditions at $\theta' = 0$ with those at $\theta' = \pi$ radians.

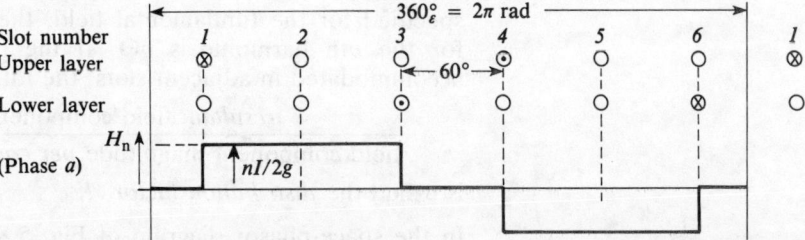

Figure 5.7 Radial field distribution in a uniform airgap due to a double-layer, two-pole, three-phase winding with one slot/pole/phase. Phase *a* only is energised.

5.3.2 Winding distribution, distribution factor evaluation

The winding of Fig. 5.7 may alternatively be represented as a *distributed* winding, since the current/pole of the fundamental field is equally shared between (two) adjacent slots, spanning 60°. If full-pitch coils were used to produce the field of phase *a*, one coil might occupy the *upper* layer of slot 6 and the *lower* layer of slot 3, whilst the other coil would occupy the *upper* layer of slot 1 and the *lower* layer of slot 4. Each excited coil generates a rectangular radial field distribution which may be expressed as a harmonic series, displaced *in space* relative to its neighbour by a distance along the airgap periphery expressed as 60° in terms of the *fundamental* field component. The space phase shift appropriate to each harmonic field component is $v60°$, where v is the harmonic number. The resultant field of each phase is then obtained by adding vectorially the fundamental and harmonic field components of each coil. For the two-coil/phase arrangement of Fig. 5.7 it is easily shown that the resultant field will have fundamental and (odd) harmonic components of magnitude *relative to those of a single coil* given by the factor $2\cos v30°$. Were the two coils concentrated in the same pair of slots the appropriate factor would be 2. The extra multiplier of value $\cos v30°$ is called the *coil-spread factor* or *distribution factor* appropriate to accommodating the two-coil winding in adjacent slots 60° (fundamental) apart.

Since the distribution factor is dependent upon the harmonic number, it follows that the relative significance of particular harmonic fields may be reduced by suitable winding distribution. The same property was evident with chording (short-pitching) the winding. Practical armature windings are distributed in a large number of slots so that the number of coils/phase is large and the slot angle (60° in the above example) low. In polyphase windings it is usual to wind all the available slots, but with single-phase windings the distribution factor for the fundamental field becomes unacceptably low if more than, say, 2/3 of the available slots are wound. This is because the (space) phase-angle between the fundamental fields of the coils at the ends of each phase grouping becomes excessive – approaching 180° in the limit.

5.3.2.1 Distribution factor evaluation

In determining quantitatively the effect of winding distribution, it is convenient to consider a (space-)phasor representation of a sinusoidally distributed harmonic component of the field produced by each similar coil making up the phase winding. As noted above, if each slot is separated from its neighbour by slot angle Θ specified for the fundamental field, the effective phase separation for the vth harmonic is $v\Theta$. If the phase winding has Q coils accommodated in adjacent slots, the ratio

$$\frac{\text{\textit{resultant} field component magnitude}}{\text{field component magnitude \textit{per coil} × number of coils } Q}$$

is called the *distribution factor*, K_{dv} [5.6]

In the space-phasor diagram of Fig. 5.8 the phasors **AB**, **BC**, **CD**, etc. represent harmonic field components of adjacent coils, each of magnitude $|H_{nc}|$. The resultant harmonic field **AF** has magnitude $|H_{nr}|$. Appreciating that a circle may be drawn through the points A, B, C, D, ..., F, the lengths of the phasors and their resultant, as *chords*, may be expressed in terms of the angles subtended at the circle centre.

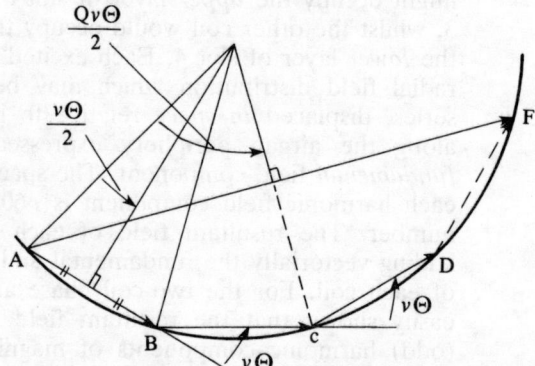

Figure 5.8 Space-phasor diagram for harmonic-field components of a distributed winding with Q coils set in adjacent slots separated by slot angle Θ.

Thus $$K_{dv} = \frac{|H_{nr}|}{Q|H_{nc}|} = \frac{|\mathbf{AF}|}{Q|\mathbf{AB}|} = \frac{\sin\dfrac{Qv\Theta}{2}}{Q\sin\dfrac{v\Theta}{2}}.$$ [5.7]

Considering the winding of Fig. 5.7 as equivalent to a full-pitch, two-pole, distributed winding with two coils/phase and slot angle $\Theta = 60°$, we get for the fundamental component

$$K_{d1} = \frac{\sin 2 \times 30°}{2 \sin 30°} = \sin 60° \, (= \cos 30°).$$

Even harmonics are absent. For the *third* harmonic

$$K_{d3} = \frac{\sin 2 \times 3 \times 30°}{2 \sin 3 \times 30°} = \frac{\sin 180°}{2} = 0.$$

For the fifth harmonic

$$K_{d5} = \frac{\sin 2 \times 5 \times 30°}{2 \sin 5 \times 30°} = \sin 300°(= \cos 150°).$$

for the seventh harmonic

$$K_{d7} = \frac{\sin 2 \times 7 \times 30°}{2 \sin 7 \times 30°} = -\sin 420°(= \cos 210°).$$

for the *ninth* harmonic

$$K_{d9} = \frac{\sin 2 \times 9 \times 30°}{2 \sin 9 \times 30°} = \frac{\sin 540°}{-2} = 0, \text{ etc.}$$

We may note with interest that distribution of the winding between adjacent slot pairs has eliminated the *triplen* harmonics (odd multiples of three) from the resultant field. The magnitude of the harmonic distribution factor has the value 0.866 for all remaining harmonics, and also for the fundamental. Such a winding has little to commend it on account of its relatively low fundamental distribution factor and the lack of any selective reduction of harmonics – except *triplens*.

In three-phase machines triplen field-harmonic effects may be eliminated at source, or by phase interconnection in star or delta, therefore the harmonic reduction capability of winding distribution and/or pitching is usually brought to bear principally on the significant fifth and seventh harmonics.

Distributing the winding in a larger number of slots than the six available for the two-pole (fundamental) winding of Fig. 5.7 is generally advantageous, in the sense that the fundamental fields of adjacent coils carrying the same phase current are more nearly in phase with each other. To keep the distribution factor for the fundamental high, however, it is necessary that the angular *spread* or *span* of a grouping of adjacent coils associated with the same phase winding is kept small. Polyphase windings have an inherent advantage over single-phase windings in this respect.

Worked example 5.1 As an example, let us consider alternative three-phase winding arrangements for accommodation in 12 slots over a pair of (fundamental) poles. The number of slots/pole/phase available is 2. If a full-pitch, single-layer winding is to be used, phase a could comprise two coils located in slots 1 and 7, and slots 2 and 8 as shown in Fig. 5.9(a), with each coil-side group spanning 60° of the fundamental field. The slot angle Θ is 30° and the harmonic distribution factor given by

$$K_{dv} = \frac{\sin 2v\, 15°}{2 \sin v\, 15°} \tag{5.8}$$

giving

$K_{d1} = 0.966, K_{d3} = 0.707, K_{d5} = 0.259,$

$K_{d7} = -0.259, K_{d9} = -0.707,$

$K_{d11} = -0.966, K_{d13} = -0.966.$

(A negative sign indicates that the harmonic field component is in 'antiphase' to the fundamental near $\theta' = 0$.)

Figure 5.9 Three-phase windings accommodated in 12 slots per pole-pair. One double pole-pitch is shown: (a) single-layer winding with 60° groups, (b) double-layer winding with 120° groups, and (c) double-layer winding with 60° groups, short pitched by 30°.

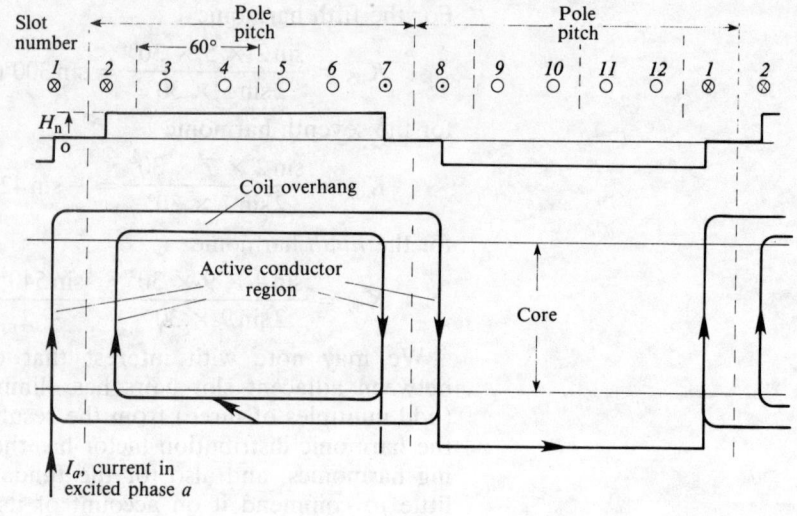

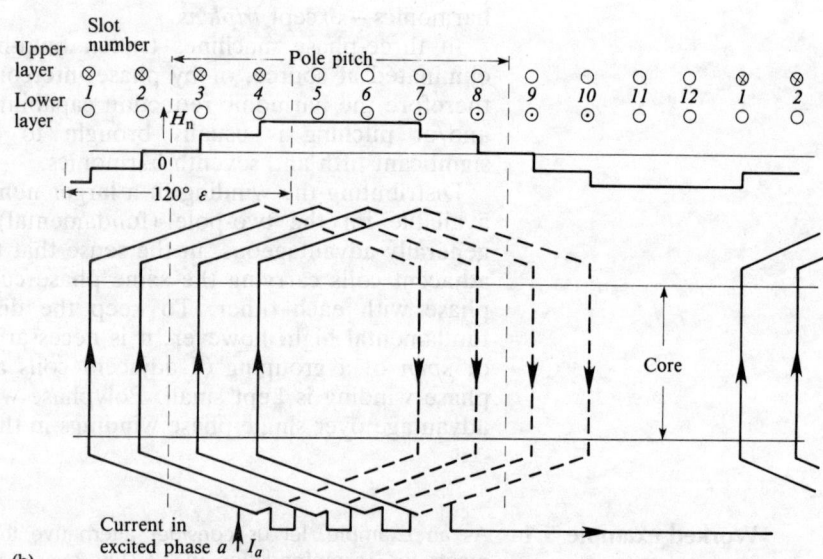

The four slots/phase per (fundamental) pole pair available may alternatively be used to accommodate the upper sides (say) of four coils, series connected to form phase a; this grouping spans 120° of the fundamental field. The lower coil sides would be placed a pole pitch away, if full-pitch coils are again used – alternatively the coil pitch may be reduced to invoke the advantages of short pitch. Thus Fig. 5.9(b) shows the upper coil sides of phase a in slots 1–4. The harmonic distribution factor is now given by

$$K_{dv} = \frac{\sin 4v\,15°}{4\sin v\,15°}$$

which gives

$K_{d1} = 0.837$, $K_{d3} = 0$, $K_{d5} = -0.224$, $K_{d7} = 0.224$, $K_{d9} = 0$,
$K_{d11} = -0.837$, $K_{d13} = -0.837$.

A *third* alternative is to use a two-layer winding with 60° groups. Coils short

Windings for rotating machines 149

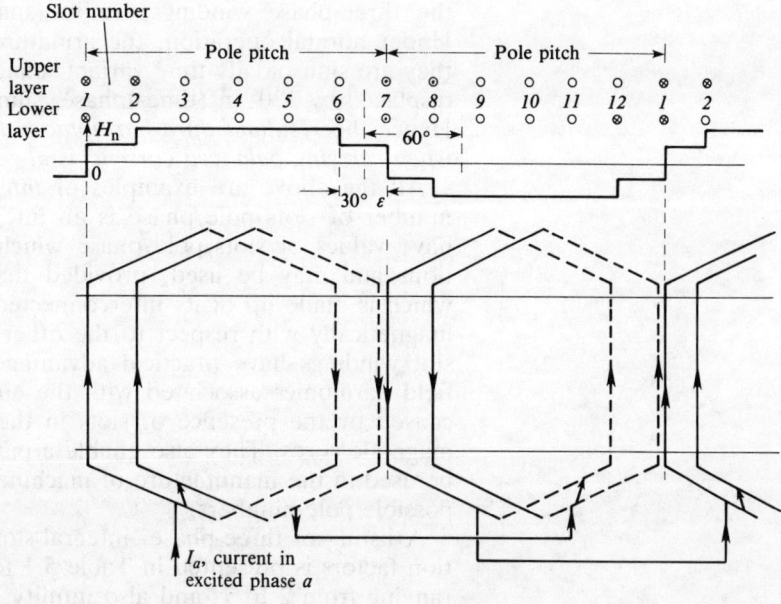

(c)

pitched by $\varepsilon = 30°$ may be incorporated by placing the upper coil sides for the phase a winding in slots, 1, 2, 7 and 8, with the lower coil sides respectively in slots 6, 7, 12 and 1. (See Fig. 5.9(c).)

The overall harmonic winding factor K_{dpv} is then given by the product of K_{dv} as eqn [5.8] and K_{pv} as eqn [5.3], giving

$K_{p1} = 0.966$, $K_{p3} = -0.707$, $K_{p5} = 0.259$,

$K_{p7} = 0.259$, $K_{p9} = -0.707$,

$K_{p11} = 0.966$, $K_{p13} = 0.966$.

Further,

$K_{dp1} = 0.933$, $K_{dp3} = -0.500$, $K_{dp5} = 0.067$, $K_{dp7} = -0.067$,

$K_{dp9} = 0.500$, $K_{dp11} = -0.933$, $K_{dp13} = 0.993$.

Now, incorporating in eqn [5.4] the effect of the winding factors implicit in eqns [5.3] and [5.6], the vth harmonic component of the radial field produced by a short-pitched, distributed winding with nI the total current/pole may generally be stated as follows.

$$H_{nv}(\theta') = n \frac{I}{2g} \frac{4}{\pi} \frac{K_{dpv}}{v} \cos v\theta'. \qquad [5.9]$$

Thus, examination of the values for K_{dpv} derived above for the short-pitched two-layer winding with 60° groups shows that, *providing the effects of the third harmonic can be contained*, the most significant radial-field space harmonics are the 11th and 13th, with amplitude about 9% of that of the fundamental, for which $v = 1$. Effective elimination of the third-harmonic field does indeed occur, due to the fact that the three phase-windings are mutually displaced by 120° in terms of the fundamental field developed. The third-harmonic field has a pole pitch equal to 1/3 that of the fundamental field, hence the third-harmonic field components of the three-phase windings are in (space-)phase with each other. Under normal operation, the armature currents are balanced, i.e. they are sinusoidally time variant, equal in amplitude and mutually displaced by 120° in (time-)phase, summing to zero at all instants. Hence the *resultant third-harmonic field of the three-phase winding when carrying balanced currents is always zero*.

All the above are examples of *integral-slot* windings, since the number of slots/pole/phase is an integer. *Fractional-slot* windings have values of slots/pole/phase which are usually improper fractions and may be used, provided that the entire phase winding which is made up of its interconnected groups of coils is balanced magnetically with respect to the other phase windings. Fractional-slot windings have practical advantages in attenuating the higher field harmonics associated with the airgap irregularities which are caused by the presence of slots in the rotor and stator laminated magnetic cores. They also enable a particular slotted lamination to be used in the manufacture of machines with an extended range of possible pole numbers.

A listing of three-phase, integral-slot winding harmonic distribution factors is presented in Table 5.1 for values of slots/pole/phase ranging from 2 to 7, and also infinity. Windings with the common phase-spread of 60° are assumed. For $Q = \infty$, the phasor summation path equivalent to A B C D E F in Fig. 5.8 is coincident with the arc AF of the circle shown. Since $v\Theta/2$ is then small, eqn [5.7] may be approximated to

$$K_{dv} \simeq \frac{\sin Q \frac{v\Theta}{2}}{Q \frac{v\Theta}{2}}$$

with Θ necessarily in radians in the denominator.

But $Q\Theta = 60°$ for 60° groups (i.e. phase-spread), therefore

$$K_{dv} = \frac{\sin v\, 30°}{Q \sin \frac{v\, 30°}{Q}}, \qquad [5.10]$$

generally, and

$$K_{dv} \simeq \frac{\sin v\, 30°}{v \frac{\pi}{6}} \qquad [5.11]$$

for $Q \to \infty$.

It may be noted from Table 5.1 that the effect of an increase in the number of slots/pole/phase from 2 to 3 on the mid-range 11th and 13th harmonics is quite dramatic. A useful exercise for the reader is to demonstrate that the three-phase winding having 18 slots per pole/pair, short/pitched by one slot ($\equiv 20°$ fundamental) and arranged in 60° groups should theoretically produce a distributed radial field in which no space-harmonic component exceeds 3% of the fundamental. It is evidently the case that a relatively simple phase-winding configuration may generate an effectively sinusoidally distributed radial field in the airgap of the machine,

Table 5.1 Harmonic distribution factors $K_{d\nu}$ for three-phase windings with 60° groups

Harmonic number	Q, slots per pole per phase						
	2	3	4	5	6	7	∞
$\nu = 1$	0.966	0.960	0.958	0.957	0.957	0.957	0.955
3	0.707	0.667	0.654	0.646	0.644	0.642	0.636
5	0.259	0.217	0.205	0.200	0.197	0.195	0.191
7	−0.259	−0.177	−0.158	−0.149	−0.145	−0.143	−0.136
9	−0.707	−0.333	−0.270	−0.247	−0.236	−0.229	−0.212
11	−0.966	−0.177	−0.126	−0.110	−0.102	−0.097	−0.087
13	−0.966	0.217	0.126	0.102	0.092	0.086	0.073
15	−0.707	0.667	0.270	0.200	0.172	0.158	0.127
17	0.259	0.960	0.158	0.102	0.084	0.075	0.056
19	0.259	0.960	−0.205	−0.110	−0.084	−0.072	−0.059
21	0.707	0.667	−0.654	−0.247	−0.172	−0.143	−0.091

the amplitude of which is related to the total phase current/pole nI, thus

$$\hat{H}_{n1} = \frac{nI}{2g}\frac{4}{\pi}K_{dp1}. \qquad [5.12]$$

Equation [5.12] follows from eqn [5.9] with g the airgap length. H_{n1} is sinusoidally distributed about an axis with respect to which the phase groups of active coil sides are symmetrical.

5.4 The Tangential Magnetic Field of Armature Windings

As pointed out in Section 5.2 an armature winding – indeed any machine winding – produces, when energised with current, a *tangential* magnetic field in the airgap in addition to the familiar radial field. This tangential field is most important – without it the machine *would not work*. The radial field alone may be regarded as necessary for the development of armature e.m.f., if relative motion between the armature conductors and the radial flux is arranged but, for the development of mechanical forces on the conductors and the complementary flow of power across the armature conductor/airgap interface, the tangential field H_t is an absolute necessity.

The tangential field H_t of a machine winding differs from the radial field H_n in many important respects. First of all, H_n at any particular airgap location θ, say, is virtually constant over the length of the gap g between stator and rotor, if we assume a uniform gap and $g \ll d$, the rotor diameter. H_t, however, is not constant over the gap but varies approximately linearly from a maximum value at the side of the gap where the current is located to zero at the other. Within the general volume of the airgap the resultant tangential field is established by adding components derived from the current flowing on both sides of the airgap, if the machine is doubly excited. At either airgap/iron boundary, however, only the local source of m.m.f. contributes to the resultant tangential field and this makes machine performance much easier to evaluate because, as will be shown in Section 5.4.1 below, the tangential field of a winding placed coincidentally with the airgap/iron boundary has a magnitude at that boundary which is given simply by the *linear current density* (A/m) of that winding. The essential performance of a rotating electrical machine may be described fully in terms of the tangential field distribution and the radial flux distribution at the armature/airgap surface, together with the velocity of the armature conductors with respect to the radial flux.

Referring back to Fig. 5.2(b) the variation of H_t *over the surface* of the rotor carrying the currents establishing the four-pole field is seen to consist of 'pulses' of alternating polarity separated by a pole pitch. The *area* of each pulse is in fact the current per conductor, and the pulse width is the conductor width. In arriving at this result we assumed conductors of finite width but of zero thickness. In our consideration of the *radial* field we have generally assumed conductors of negligible width (except for the infinitely distributed winding with $Q = \infty$), but we cannot reasonably make the same assumption with regard to H_t. Instead, we shall assume an appropriate, controlled distribution of current even where the number of slots per pole (or slots/pole/phase for an a.c. winding) is much nearer 1 than ∞. For d.c. armature windings the assumed current distribution will be uniform, whilst for a.c. armature windings it will be sinusoidal.

Some justification for these dramatically simplifying assumptions is required, however, perhaps particularly so in the case of a.c. windings where the number of slots/pole/phase is necessarily less than the number of slots/pole. Considering the a.c. winding case, therefore, we have seen (Sect. 5.3) that one phase of a practical three-phase winding can readily produce a near-sinusoidal distribution of *radial* field in the airgap – with no space harmonic exceeding, say 4% of the fundamental amplitude. Net torque is produced by the interaction of radial flux-density and a tangential field of like 'frequency' hence, to the degree that the B_n distribution (to which the 'field' winding located on the other side of the airgap also contributes) is kept nearly sinusoidal, signficant harmonic components in the H_t field will produce only local perturbations in the *stress* distribution over the armature/airgap surfaces – an effect insignificant compared with the effect of slotting on the assumption of uniform airgap permeance. For the a.c. machine, therefore, we are effectively saying that space harmonics in the H_t field can be ignored, as far as machine performance is concerned, as long as the radial flux is maintained nearly sinusoidal. As will be confirmed in Section 5.4.2 below, the sinusoidal current distribution assumed in our model gives rise to a sinusoidal distribution of H_t, naturally precluding any harmonic elements of H_t whose effects we are disposed to discount. A further consequence of an assumed sinusoidal current distribution is the elimination of harmonics in the *radial* field component produced. Superficially, this appears to be an unacceptable simplification, pursued on account of our desire to generate a sinusoidally time-variant armature e.m.f. when the machine operates in the steady state. As will be shown in Section 5.7.2, however, a further effect of distributing

Windings for rotating machines   153

and short-pitching the armature winding is to selectively reduce the influence of harmonic e.m.f.s induced in individual coil-sides or conductors by their relative motion normal to the radial airgap flux, when summed to provide the net e.m.f. of a phase winding.

5.4.1 *General procedures for airgap radial and tangential field evaluation*

The *radial* field distribution is readily calculated by applying Ampère's law to an elementary closed loop which involves two traverses of the uniform airgap of length g, with one gap-crossing conveniently at $\theta_m = \theta_{mo}$ (Fig. 5.10), where the field is zero. This occurs naturally at the mid-points of each band of current of particular polarity. The return gap-crossing occurs alternatively either at $\theta_m = \theta_{m1}$, where the radial field (directed from rotor to stator by convention) has the value $H_{n\theta_{m1}}$, or at $\theta_m = \theta_{m1} + \Delta\theta_m$, where the radial field is $H_{n\theta_{m1}+\Delta\theta_m}$.

Since, $H_{n\theta_{mo}} = 0$, and the stator and rotor iron is assumed to be of infinite permeability, it follows that $gH_{n\theta_{m1}} =$ the total current (into the paper) enclosed within limits $\theta = \theta_{mo}$ and $\theta = \theta_{m1}$ and

$$g(H_{n\theta_{m1} + \Delta\theta_m} - H_{n\theta_{m1}}) = \Delta I,$$

where $\Delta I =$ the total current (into paper) enclosed within limits $\theta = \theta_{m1}$ and $\theta = \theta_{m1} + \Delta\theta_m$.

For θ_{m1} as depicted in Fig. 5.10,

$$\Delta I = (-)A(\theta_m)b\Delta\theta_m$$

because, in the region $\Delta\theta_m$, the current flow is out of the paper.

More generally, $A(\theta_m)$ is the current 'density' in A m^{-1} directed into the paper, b is the rotor radius and $b\Delta\theta_m$ is the element of armature winding on the rotor periphery which subtends angle $\Delta\theta_m$ about the axis of rotation. Thus generally

$$\frac{dH_n}{d\theta_m} = \left[\frac{H_{n\theta_m+\Delta\theta_m} - H_{n\theta_m}}{\Delta\theta_m}\right]_{\Delta\theta_m \to 0} = A(\theta_m)\frac{b}{g} \text{ A m}^{-1}. \quad [5.13]$$

Equation [5.13] shows that the *slope* of $H_n(\theta_m)$ plotted against θ_m is proportional to the local current density. With uniform current density of $(+)A$ A m^{-1} effective between limits $\theta_m = \theta_{mo}$ (at which $H_n = 0$) and $\theta_m = \pi/2$ in Fig. 5.10, it is apparent that in consequence H_n increases linearly from zero at

$$\theta_m = \theta_{mo} = \frac{\pi}{4}$$

to

$$\hat{H}_n = \int_{\pi/4}^{\pi/2} \frac{Ab}{g} d\theta_m$$

at

$$\theta_m = \frac{\pi}{2}.$$

Thus

$$\hat{H}_n = \frac{1}{g}A\left(\frac{b\pi}{4}\right) = \frac{\text{current/pole}}{2g}. \quad [5.14]$$

Figure 5.10 Uniform airgap radial and tangential magnetic field distributions due to a uniform current located on the rotor surface, four-pole case: (a) section at right-angles to the axis of rotation; (b) developed section of airgap (H_n and H_t are not to scale); and (c) developed field distributions. A = linear current density, $A\,m^{-1}$.

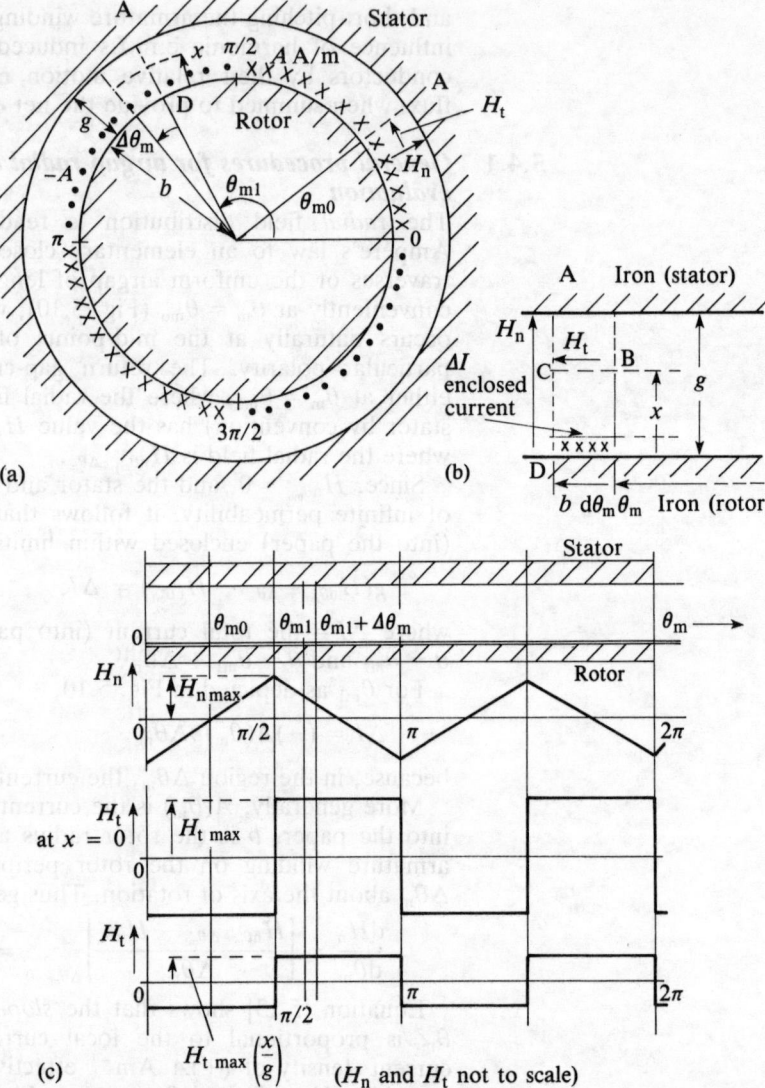

For $\pi/2 < \theta_m < \pi$, A has negative polarity and so the plot of $H_n(\theta_m)$ against θ_m has negative slope. Hence one can justify the triangular variation of $H_n(\theta_m)$ with (θ_m) for the finely distributed winding of Fig. 5.10 represented as a uniform current sheet. This models ideally the current distribution of the armature winding of a d.c. machine. It is noteworthy that the precise location of the current in the airgap is not important, provided that its θ_m co-ordinate is fixed. It is also of interest to evaluate $H_n(\theta_m)$ as a harmonic series.

Fourier analysis of the triangular waveform for $H_n(\theta_m)$ in Fig. 5.10(b) with H_n given by eqn [5.14] yields

$$H_n(\theta_m) = \frac{8}{\pi^2} \frac{\text{current/pole}}{2g} \left(\sin\theta - \frac{1}{3^2}\sin 3\theta + \frac{1}{5^2}\sin 5\theta - \ldots \right). \quad [5.15]$$

In eqn [5.15] $\theta = 2\theta_m$, with θ_m as a *mechanical* measure of angle, whereas θ is an *electrical* or *magnetic* measure such that adjacent like poles of the fundamental field are 2π (elec) radians apart. In general, $\theta = p\theta_m$ where p is the number of (fundamental) pole pairs developed by the winding.

We can compare the amplitude of the *fundamental component* of $H_n(\theta_m)$ with that deduced from eqn [5.12] using the distribution factor for an infinite number of slots/pole Q and a phase-spread of π radians (elec).

Thus $$\hat{H}_{n1} = \frac{\text{(current/pole)}}{2g} \frac{4}{\pi} \frac{\sin \pi/2}{\pi/2} = \frac{8}{\pi^2} \frac{\text{(current/pole)}}{2g}.$$

Agreement of the above expression with the fundamental term of eqn [5.15] is confirmed, but it should also be noted that whereas this equation relates to the particular case of a uniform current sheet, eqns [5.13] and [5.14] are quite general and apply to *any* current distribution. We shall use them in Section 5.4.2 when considering a sinusoidal current density variation with θ – the 'sinusoidal current sheet'.

For the present, however, we will investigate the *tangential* field distribution in the airgap of Fig. 5.10 considering the current to be attached to the rotor/airgap surface. The region of interest around an element of rotor periphery $b\Delta\theta_m$ is shown enlarged in Fig. 5.10(b), which also shows clearly the reference directions for current polarity, field direction and angular position.

Two alternative routes for a closed path linking current element ΔI and additional current complete the path: either (i) from A to D along a line of normal field intensity $H_{n\theta + \Delta\theta}$ or (ii) from A to D via B and C. Since the magnetic potential difference between A and C must be the same whether one moves from C to A directly or via B, magnetic potential differences may be equated as follows

$$H_{n\theta_m + \Delta\theta_m}(g - x) = H_{n\theta_m}(g - x) - H_{t\theta_m} b\Delta\theta_m,$$

assuming that $b\Delta\theta_m$ is small so that H_t may be considered constant at $H_{t\theta_m}$ between θ_m and $\theta_m + \Delta\theta_m$. Hence

$$H_{t\theta_m} b\Delta\theta_m = [H_{n\theta_m} - H_{n\theta_m + \Delta\theta_m}](g - x)$$

$$= -\frac{\Delta I}{g}(g - x)$$

$$= -A(\theta_m)\frac{b\Delta_m}{g}(g - x)$$

$$\therefore H_{t\theta_m} = -A(\theta_m)\frac{(g - x)}{g}. \quad [5.16]$$

From eqn [5.16] it is apparent that H_t is a function of x as well as of θ_m. (x defines position in the airgap space between rotor and stator surfaces.) At the rotor surface where the current is located, $x = 0$

$$[H_t]_{x=0} = -A(\theta_m). \quad [5.17]$$

Equation [5.17] shows that the tangential magnetic field intensity distribution is identical with the rotor current distribution, except for sign, at the current surface. This is depicted in Fig. 5.10(c). We may usefully incorporate eqn [5.17] within eqn [5.13] to show

$$\frac{dH_n}{d\theta_m} = -\left[\frac{b}{g}H_t\right]_{x=0} = A(\theta_m)\frac{b}{g}. \qquad [5.18]$$

Equation [5.18] is a useful general relationship which associates radial and tangential components of *H* *at the rotor surface* with the responsible rotor linear current density *A* and the airgap dimensions. At the *stator* surface $x = g$. Equation [5.16] gives

$$[H_t]_{x=g} = 0. \qquad [5.19]$$

Thus rotor current produces no tangential magnetic field at the stator surface boundary of the airgap. This is a most important result.

Expressions analogous to eqns [5.13] through [5.19] apply also to radial and tangential field components due to current on the *stator*/airgap boundary. Throughout the airgap generally the *radial* field component H_n is independent of *x*, whereas H_t at *x* is critically dependent upon the location of current in the airgap – whether at stator or rotor bounding surfaces or at both. The variation of H_t with θ_m is complementary to that of the current-density distribution $A(\theta_m)$. It is readily shown that the negative sign appropriately included in eqns [5.16] to [5.18] for the tangential field due to *rotor* current is absent when the current is located on the *stator* surface, as the right-hand or 'corkscrew' rule confirms.

This has important consequences for doubly excited machine systems with energised windings on both sides of the airgap. When operated over a range of load or torque values at nominally constant *radial* flux, the rotor and stator *radial* field components must add to provide a constant resultant. This means that the currents flowing in conductors facing each other across the gap will generally be of opposite polarity, although local exceptions will occur. Thus the *tangential* magnetic field in the body of the airgap volume generally has stator and rotor components augmenting each other, the *resultant* tangential field increasing with increasing load current. This significantly influences the flow of power in the airgap, as given by the Poynting vector.

Even under full-load conditions, a practical machine having an airgap length small compared with its rotor radius has H_n very much greater than H_t. Increasing the airgap length has the effect of making H_t relatively less small and more significant, due to the increased cross-section of its flux path, with consequences for the operation of the machine. The *tangential* field of a winding produces flux which links only that winding responsible for its generation, whereas the *radial* field produced by a winding extends its influence over the entire surfaces bounding both sides of the airgap, linking other circuits. By analogy with the transformer, it may be recognised that the tangential field contributes the *leakage* flux within the machine although, as has been noted, its presence is necessary for the machine to function as an energy conversion device.

5.4.2 Radial and tangential magnetic fields of a sinusoidal winding (sinusoidal current sheet)

In the preceding section general expressions for the tangential and radial airgap magnetic field components due to current distributed over rotor (and stator) surfaces have been developed. If we stipulate a sinusoidal surface distribution of rotor current, the expression for current density becomes

$$A(\theta_m) = \hat{A} \sin p\theta_m \qquad [5.20]$$

where p is the number of pairs of magnetic poles which the energised winding will produce over the complete rotor surface and $\hat{A}$ is the maximum linear current density.

From eqn [5.16], therefore, for locations in the airgap at distance x from the rotor surface, with $x < g$

$$H_t(\theta_m) = -\frac{(g-x)}{g}\hat{A} \sin p\theta_m \qquad [5.21]$$

and from eqn [5.18]

$$\frac{dH_n(\theta_m)}{d\theta_m} = \frac{b}{g}\hat{A} \sin p\theta_m. \qquad [5.22]$$

Integrating eqn [5.22] with respect to θ_m and appreciating that the integral of $H_n(\theta_m)$ over 2π radians must be zero in order that the net radial flux crossing the airgap is zero, we get a value of zero for the constant of integration and

$$H_n(\theta_m) = -\frac{b}{pg}\hat{A} \cos p\theta_m. \qquad [5.23]$$

Thus we note from eqns [5.21] and [5.23] that a sinusoidal current distribution at the *rotor*/airgap boundary gives rise to complementary sinusoidal and cosinusoidal distributions of tangential and radial field in the airgap. The tangential field lags the radial field by 90° (elec) in space and, at the *rotor* surface where $x = 0$, has relative magnitude pg/b. Corresponding waveforms are shown in Fig. 5.11.

Recalling that an objective in a.c. machine winding design is to produce a near-sinusoidal radial field distribution, the advantages of adopting a sinusoidal current distribution in modelling the winding are apparent in that harmonic field components, both normal and tangential, are automatically excluded in this simple model. Refinements to incorporate the effects of space harmonics may be made subsequently if required but, for most purposes, the sinusoidal model is adequate.

For a sinusoidal current distribution, the current 'density' A is simply expressed in terms of the current/pole, which is equivalent to the space integral over the current-carrying surface of the current density between adjacent points of zero value. Thus, for the general case with p pole pairs

$$\text{current/pole} = \int_{\theta_m=0}^{\theta_m=\pi/p} \hat{A} \sin p\theta_m b \, d\theta_m$$

$$= \left|\frac{-\hat{A}b}{p}\cos p\theta_m\right|_0^{\pi/p}$$

$$= \frac{2}{p}\hat{A}b.$$

158 Electrical machines and drive systems

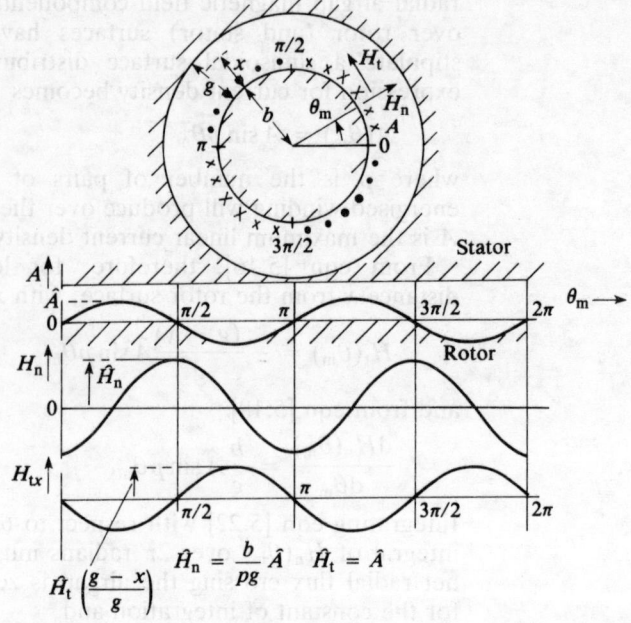

Figure 5.11 As for Fig. 5.10 (a) and (c) with the sinusoidal current sheet located on the rotor surface (four-pole case).

$$\hat{H}_n = \frac{b}{pg}\hat{A} \qquad \hat{H}_t = \hat{A}$$

$$\hat{H}_t\left(\frac{g-x}{g}\right)$$

Substituting for

$$\frac{\hat{A}b}{p}$$

in eqn [5.23]

$$H_n(\theta_m) = -\frac{\text{current/pole}}{2g}\cos p\theta_m. \qquad [5.24]$$

Also in eqn [5.21] with $x = 0$, i.e. at the rotor surface,

$$H_t(\theta_m) = -\frac{\text{current/pole}}{2g}\frac{pg}{b}\sin p\theta_m. \qquad [5.25]$$

The expression

$$\frac{\text{current/pole}}{2g}$$

for the peak value of the radial field produced by a winding has been encountered before – for example, in eqn [5.14] relating to a uniform current sheet. It appears also in expressions for the radial field produced by windings when expressed as a harmonic series – e.g. eqns [5.12] and [5.15], in which the respective additional constants

$$\frac{4}{\pi}$$

and

$$\frac{8}{\pi^2}$$

result from the Fourier analysis.

5.5 Practical Winding Considerations

For the sinusoidal current modelling of a.c. machine windings, therefore, it is apparent that simple magnitude and phase relationships exist between the distributions of radial and tangential fields in the uniform airgap, with harmonic field components completely absent. We have seen (Sect. 5.3) that a practical a.c. armature winding can indeed produce a radial field distribution for which the harmonic components are not significant in comparison with the fundamental, but doubt must remain about the validity of the sinusoidal current model – unless analysis of the *tangential* field distribution of a practical winding shows the space harmonics to be manageable and, particularly, that the *fundamental bears the same magnitude and phase relationship to that of the radial field as does the sinusoidal current model*.

Let us therefore analyse the tangential field produced by one phase of a full-pitch, three-phase winding having 60° groups and an infinite number of slots/pole/phase. From eqns [5.11] and [5.12] the amplitude of the fundamental component of the *radial* field is

$$\hat{H}_{n1} = \frac{\text{current/pole}}{2g} \frac{4}{\pi} \frac{\sin 30°}{\pi/6}$$

$$= \frac{\text{current/pole}}{2g} \frac{12}{\pi^2}. \qquad [5.26]$$

The corresponding *tangential* field distribution is shown in Fig. 5.12 and, expressed with respect to $\theta = 0$ located at the position where the fundamental component of the radial field has its peak positive value, Fourier analysis describes the tangential field at the winding surface on the rotor/airgap boundary in terms of the linear current density A A m^{-1} thus

$$H_t(\theta) = \sum_{v=1}^{v=\infty} b_v \sin v\theta,$$

where

$$b_v = \frac{1}{\pi} \int_{-\pi}^{\pi} H_t(\theta) \sin v\theta \, d\theta.$$

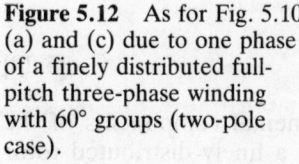
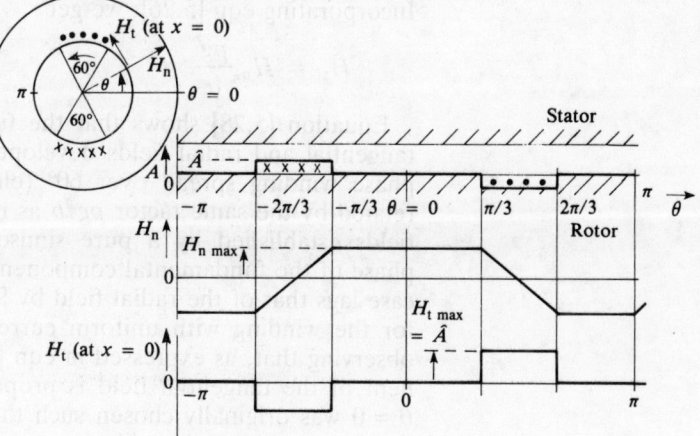

Figure 5.12 As for Fig. 5.10 (a) and (c) due to one phase of a finely distributed full-pitch three-phase winding with 60° groups (two-pole case).

Substituting eqn [5.17] and appreciating that $H_t(\theta)$ is discontinuous,

$$b_v = \frac{1}{\pi}\hat{A}\left[\int_{-2\pi/3}^{-\pi/3} -\sin v\theta\, d\theta + \int_{\pi/3}^{2\pi/3} \sin v\theta\, d\theta\right]$$

$$= \frac{2}{\pi}\frac{\hat{A}}{v}\left[\cos\frac{v\pi}{3} - \cos\frac{v2\pi}{3}\right]$$

$$= \frac{4}{\pi}\frac{\hat{A}}{v}\sin\frac{v\pi}{6}\sin\frac{v\pi}{2}.$$

For $v = 1$,

$$b_1 = \frac{2}{\pi}\hat{A},$$

the amplitude of the fundamental component. The complete harmonic series is

$$H_t(\theta) = \frac{2}{\pi}\hat{A}\left[\sin\theta - \frac{2}{3}\sin 3\theta + \frac{1}{5}\sin 5\theta\right.$$

$$\left. + \frac{1}{7}\sin 7\theta - \frac{2}{9}\sin 9\theta + \frac{1}{11}\sin 11\theta + \ldots\right]. \quad [5.27]$$

We need to express the amplitude of the fundamental component of the tangential field (at the rotor surface) in terms of the current/pole for the general case with p pole pairs and rotor radius b.

$$\text{Current/pole} = \int_{\theta=\pi/3p}^{\theta=2\pi/3p} \hat{A} b\, d\theta$$

$$= \hat{A}\frac{\pi}{3}\frac{b}{p}.$$

Substituting for $\hat{A}$ in eqn [5.27]

$$H_{t1} = \frac{2}{\pi}(\text{current/pole})\frac{3}{\pi}\frac{p}{b}$$

$$= \frac{\text{current/pole}}{2g}\frac{12}{\pi^2}\frac{pg}{b}.$$

Incorporating eqn [5.26], we get

$$\hat{H}_{t1} = \hat{H}_{n1}\frac{pg}{b}. \quad [5.28]$$

Equation [5.28] shows that the fundamental components of the tangential and radial fields developed by a finely distributed rotor phase winding spread over 60° (elec) per pole have magnitudes related by the same factor pg/b as relates the tangential and radial fields established by a pure sinusoidal current distribution. The phase of the fundamental component of the tangential field in each case lags that of the radial field by 90° (elec). This is demonstrated for the winding with uniform current density spread over 60° by observing that, as expressed in eqn [5.27], the fundamental component of the tangential field is proportional to $\sin\theta$. The position $\theta = 0$ was originally chosen such that the radial field fundamental component would have its peak positive value there and hence have its distribution described in terms of cos θ.

We have yet to check the significance of the harmonic content of the tangential field given by eqn [5.27]. The apparently large third harmonic produced by a phase winding is, in practice, no embarrassment in three-phase applications.

The resultant third-harmonic field will be zero always and everywhere, when three-phase windings mutually displaced in space by 120° (elec) for the fundamental (equivalent to a double pole pitch for the third space harmonic) carry balanced polyphase currents which sum instantaneously to zero. All other triplen harmonics (odd multiples of three) similarly cancel out.

The fifth, seventh, eleventh, thirteenth, etc., harmonics remain, however, and are capable of developing harmonic torque components if space harmonics of the same harmonic number exist in the radial flux distribution to which both stator and rotor windings contribute. If, as has been noted in Section 5.4, the radial flux distribution is essentially sinusoidal, such harmonic torques will be negligible.

5.6 Resultant Magnetic Fields of Polyphase Windings (Rotating Field)

Balanced three-phase armature windings have identical phase windings, mutually displaced from each other by 1/3 of a double pole pitch of the fundamental magnetic field they each produce when excited with current (i.e. 120° (elec)). When energised with balanced three-phase currents (sinusoidally time variant, of equal magnitude and of 120° phase difference), the resultant fields, *whether radial or tangential*, may be described in terms of rotating, constant-amplitude components.

It should be apparent from Fig. 5.13 that, as far as the *fundamental* component of radial field is concerned, the resultant due to the three-phase windings, considered at the particular instants in time when the currents in the respective phases have their peak positive values, adopts the distribution of the dominant phase and has an amplitude 1.5 times that of the dominant phase. The analysis in Section 5.6.1 below shows that the resultant fundamental field retains its sinusoidal distribution *at all times*, appearing to *move* with respect to the winding conductors at a *synchronous* speed appropriate to the supply frequency. Also, resultant space-harmonic fields other than triplens, sinusoidally distributed with appropriate pole-pitch, appear to *rotate* at correspondingly reduced speeds in different directions. Triplen harmonic fields sum to zero.

5.6.1 *Resultant radial magnetic field of a balanced, three-phase winding energised with balanced, sinusoidal currents*

The three-phase winding of Fig 5.13 has 120° (elec) separation between the phases, each of which gives rise to an airgap field distribution, tangential or radial, which may be expressed as a harmonic series. From eqn [5.5], with $\theta = 0$ defined appropriately for the expression of the radial field as a cosine series with odd

162 Electrical machines and drive systems

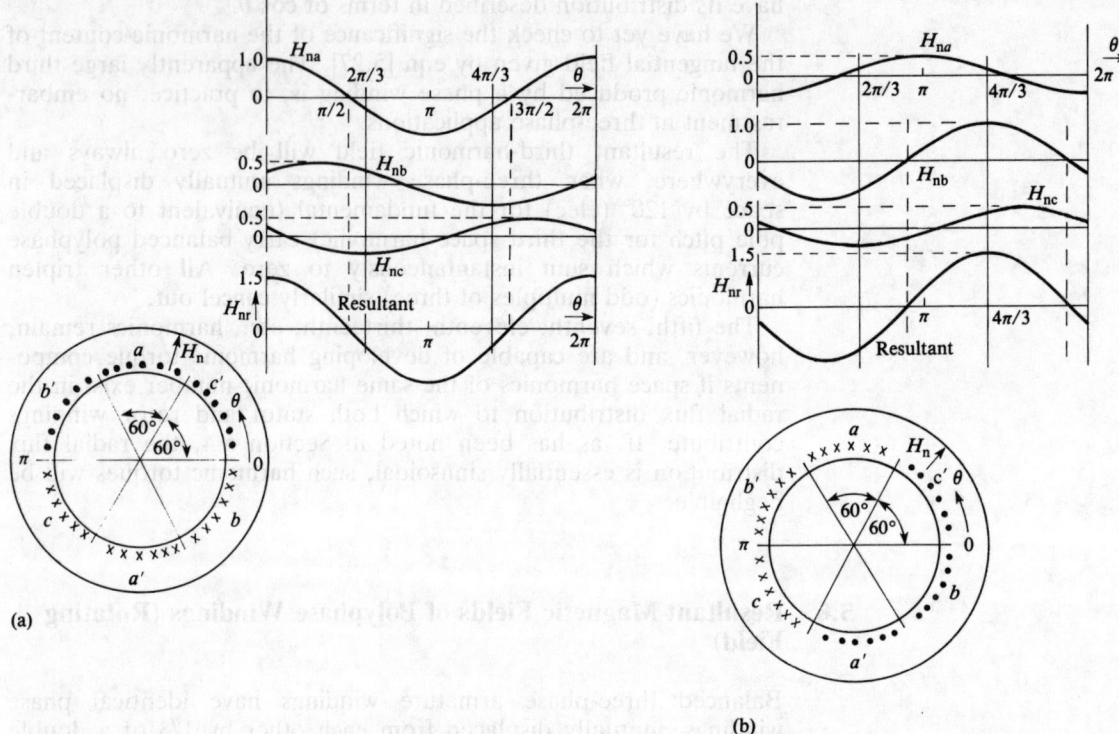

Figure 5.13 Fundamental radial field distribution in a uniform airgap for a three-phase winding with 60° groups. Components due to each phase and resultant are shown: (a) appropriate to instant when $\hat{i}_a = \hat{I}_a$, $i_b = i_c = -\tfrac{1}{2}\hat{I}_a$; and (b) appropriate to instant when $i_b = \hat{I}_a$, $i_c = i_a = -\tfrac{1}{2}\hat{I}_a$ (Two-pole case depicted).

harmonics only, we have for the reference phase a

$$H_{na}(\theta) = i_a[a_1\cos\theta + a_3\cos 3\theta + a_5\cos 5\theta + \ldots + a_\nu\cos\nu\theta + \ldots],$$

where a_ν is a constant and i_a is the instantaneous current in reference phase a.

For the radial fields of phases b and c, similar expressions apply but, since the b phase field leads the a phase in space by 120° and the c phase leads the a phase field by 240°, we write

$$H_{nb}(\theta) = i_b[a_1\cos(\theta + 120°) + a_3\cos 3(\theta + 120°) + a_5\cos 5(\theta + 120°) + a_\nu\cos\nu(\theta + 120°) + \ldots]$$

$$H_{nc}(\theta) = i_c[a_1\cos(\theta + 240°) + a_3\cos 3(\theta + 240°) + a_5\cos 5(\theta + 240°) + a_\nu\cos\nu(\theta + 240°) + \ldots].$$

Substituting for i_a, i_b and i_c as sinusoidal functions of time with i_b lagging i_a by 120° and i_c lagging i_a by 240°, we get for the *fundamental* component of the resultant radial field

$$\begin{aligned}H_{nr1}(\theta, t) &= a_1\,\hat{I}_a[\sin\omega t\cos\theta + \sin(\omega t - 120°)\cos(\theta + 120°) \\ &\quad + \sin(\omega t - 240°)\cos(\theta + 240°)] \\ &= \tfrac{3}{2}a_1\hat{I}_a\sin(\omega t + \theta).\end{aligned} \qquad [5.29]$$

Considering eqn [5.29] at an arbitrary *instant* in time it is apparent that the resultant fundamental field is sinusoidally space-variant with a peak value 3/2 times that of each contributing phase. Considering eqn [5.29] at an arbitrary *position* in the airgap it is apparent that the resultant field is sinusoidally *time*-variant also.

These are the properties of a sinusoidally distributed *travelling wave* of field, with the position of its positive peak value at any instant t located at θ_1, where $\omega t + \theta_1 = 90°$, or $\theta_1 = 90° - \omega t$. Thus θ_1 is a linear function of time and $d\theta/dt = -\omega$, indicating that the resultant radial field in the airgap appears to rotate at constant speed $\omega\,\text{rad}\,\text{s}^{-1}$ in the direction *opposite* to the reference direction of θ. This suggests a clockwise motion for the configuration of Fig. 5.13 employing phase currents of sequence *a b c*.

Considering now the space-harmonic components of the resultant radial field

$$H_{nrv}(\theta, t) = a_v \hat{I}_a[\sin\omega t \cos v\theta + \sin(\omega t - 120°)\cos v(\theta + 120°) + \sin(\omega t - 240°)\cos v(\theta + 240°)]$$

$$= a_v \hat{I}_a \{\sin(\omega t + v\theta) + \sin(\omega t - v\theta) + \sin[\omega t + v\theta + (v-1)120°]$$
$$+ \sin[\omega t - v\theta - (v+1)120°]$$
$$+ \sin[\omega t + v\theta + (v-1)240°] + \sin[\omega t - v\theta - (v+1)240°]\}.$$

If $v = 3, 9, 15, \ldots$ (i.e. v = any odd multiple of 3) then

$$H_{nrv}(\theta, t) = 0.$$

Hence triplen space-harmonics are non-existent in the resultant field.

If $v = 3K + 1$, where K is any positive even integer (or zero) then $(v-1)120° = (v-1)240° = 0°$; $(v+1)120° = 240°$; and $(v+1)240° = 120°$, giving

$$H_{nrv}(\theta, t) = \tfrac{3}{2} a_v \hat{I}_a \sin(\omega t + v\theta). \qquad [5.30]$$

If $v = 3K - 1$, where K is any positive even integer then $(v+1)120° = (v+1)240° = 0°$; $(v-1)120° = 120°$; $(v-1)240° = 240°$, giving

$$H_{nrv}(\theta, t) = \tfrac{3}{2} a_v \hat{I}_a \sin(\omega t - v\theta) \qquad [5.31]$$

Equation [5.30] indicates that the resultant field for the 7th, 13th, 19th, etc., harmonic is of amplitude 3/2 times the harmonic field peak value/phase, apparently rotating in the *same* direction as the fundamental field, but at a reduced speed of $\omega/v\,\text{rad}\,\text{s}^{-1}$.

Equation [5.31] indicates that the resultant field for the 5th, 11th, 17th etc., harmonic is of amplitude 3/2 times the harmonic field peak value/phase, apparently rotating in the *opposite* direction to that of the fundamental field, and at velocity $\omega/v\,\text{rad}\,\text{s}^{-1}$.

5.7 Principal Magnetic Field Properties of Armature Windings (Summary)

Before moving on to consider electromagnetically induced e.m.f in armature windings it may be useful to summarise the main airgap field properties of axially directed current-carrying conductors placed in small, uniform (cylindrical) airgaps bounded by stator and rotor iron surfaces of infinite permeability as follows.

(i) Axially directed current in the airgap gives rise both to radial and tangential components of magnetic field therein.

(ii) The *radial* field is essentially uniform across the gap and is independent of the radial co-ordinate of current in the gap, identifiable in practice with either the stator or the rotor surface. The rate of change with angular position of the radial field is proportional to the local linear current density.

(iii) The *tangential* field is not uniform across the gap but varies essentially linearly from a maximum at the current-carrying surface to zero at the opposite boundary. The maximum value has magnitude equal to the linear current density at the energised surfaces.

(iv) Distributed currents of the same polarity at rotor and stator surfaces develop radial field components which act in the same direction, and tangential field components which mutually oppose.

(v) A sinusoidal current distribution gives rise, respectively, to sinusoidal and cosinusoidal distributions for tangential and radial field components, with amplitudes related by the factor pg/b.

(vi) Practical distributed windings may produce airgap fields which approximate closely to the uniform tangential, triangular radial distributions ideal for d.c. machines, and the cisoidal* tangential and radial distributions ideal for a.c. machines – this is particularly true of polyphase windings.

(vii) Ideal polyphase windings energised with balanced polyphase currents give rise to resultant tangential and radial fields which may be represented as constant-amplitude sinusoidal distributions which move synchronously through one double pole pitch for each complete period of phase current. Non-ideal three-phase windings give rise to space-harmonic resultant field components (triplens excluded) which move in one direction or the other at lower speeds, according to their pole number.

5.8 Electromotive Force Induction in Armature Windings

As noted in Section 5.1 the armature winding of a rotating machine has a dual function to perform. First, the flow of axially directed current in the active coil sides produces radial and tangential components of magnetic field in the airgap which may lead to the development of torque. Second, e.m.f.s induced in the

---

*'Cisoidal' means a general sinusoidal or cosinusoidal variation without specifying the precise functional dependence.

same active coil sides, summated over a path between the winding terminals, may control the flow of electrical power between the winding and the external circuit. It is our purpose now to consider the means by which e.m.f. is induced in these active conductors situated on one or other side of the airgap, and to establish the influence of the winding interconnection in relating such component e.m.f.s to that developed between winding terminals.

The Lorentz equation, eqn [2.39], states the relationship between electromagnetically induced electric field E_e, magnetic vector potential A, conductor velocity v and flux density B appropriate to a conducting material thus

$$E_e = -\frac{\partial A}{\partial t} + v \times B.$$

Knowledge of the magnetic flux distribution in the airgap of a machine allows for the evaluation of magnetic vector potential A via eqn [2.33], i.e.

$$\oint A \cdot dl = \Phi = \oint B \cdot ds,$$

where Φ is the flux enclosed by the line integral of A over a closed circuit.

Thus knowledge of the magnetic flux conditions enables the electric field induced in each conductor to be evaluated, and the winding e.m.f. established by integrating the electric field over the circuit between winding terminals.

Armature-winding active conductors are ideally laid parallel to the axis of rotation such that the apparent motion of each conductor relative to the reference frame adopted is at right-angles to the radial component B_n of the airgap flux density. Any induced electrical field along the length of each conductor due to the $v \times B$ term in eqn [2.39] will correspondingly be maximised at a value equal to the simple product $B_n v$. Under these circumstances the tangential component $B_t = \mu_o H_t$ can make no contribution to the induced electric field except through its contribution to A – which needs time-variance to be effective. Thus the tangential field is conventionally regarded as exhibiting only *leakage* inductance effects in the armature circuit.

As noted earlier, practical armature windings are comprised of interconnected coils (full-pitch or short-pitch) distributed over the armature surface. In consequence, the winding e.m.f. will correspond to the summation of components induced in each coil side. As would be expected, therefore, winding factors attributed to coil pitch and winding distribution apply in relating total winding e.m.f. to that of an individual full-pitch coil, just at they relate the total resultant airgap field to the field of a single full-pitch coil. In both contexts the winding factors have the same essential dependence on slot angle Θ and harmonic number v.

Once again, in this chapter, our primary concern will be with a.c. windings, with particular d.c. armature winding considerations being included in Chapter 6. Alternating-current windings have fixed tapping points connected directly to the external circuit, which feature necessarily identifies the waveform of the winding current with that flowing externally – at least as far as the fundamental component is concerned. Since a.c. supplies are ideally sinusoidal for both voltage and current, one would expect that an a.c. armature winding will carry phase currents which are substantially sinusoidal, and will experience electromagnetically induced e.m.f.s which are also sinusoidal.

5.8.1 Electromotive force induced in an individual rotor coil side moving in a steady radial magnetic field

Consider the elementary full-pitch rotor coil shown in Fig. 5.14 to have coil sides instantaneously located at θ and $(\theta + \pi)$ with respect to a stationary axis about which a two-pole radial magnetic field is co-sinusoidally distributed, i.e.

$$B_n(\theta) = \hat{B}_n \cos \theta. \qquad [5.32]$$

Due to symmetry, the magnetic vector potential A at the two coil side locations due to the radial flux will be axially directed and of equal and opposite magnitudes, such that a loop formed by connecting the coil sides of length l via front and back connectors gives rise to a line integral of A due to the *radial* flux embraced, as follows

$$\oint A \cdot dl = 2A(\theta)l = lb \int_{\theta}^{(\theta+\pi)} \hat{B}_n \cos \rho \, d\rho = -2\hat{B}_n lb \sin \theta$$

giving

$$A(\theta) = -\hat{B}_n b \sin \theta \qquad [5.33]$$

where b is the rotor radius, and the reference directions for A and B_n are as indicated in Fig. 5.14.

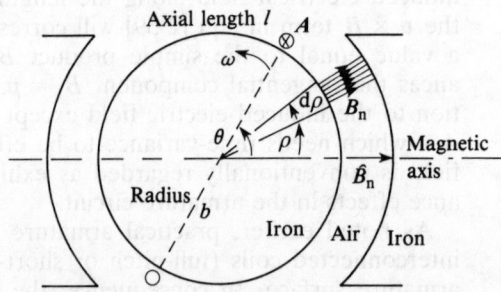

Figure 5.14 Elementary full-pitch rotor coil located in a radial magnetic field.

Although the rotor conductors are in motion, eqn [5.33] does not express A explicitly as a function of time, thus implying that the reference frame for A is stationary. θ in fact defines a particular region in space at which the magnetic field is unchanging and through which the rotor conductors pass. With respect to this stationary reference frame, the velocity v of each rotor conductor is given by

$$v = \omega b.$$

Substituting for A, v and B in eqn [2.39], we get for the electric field along each conductor

$$E_e = \omega b \hat{B}_n \cos \theta. \qquad [5.34]$$

Integrating the electric field over the length of conductor we deduce a familiar expression for the e.m.f. E induced in a conductor of length l moving with velocity v at right-angles to a magnetic field of flux density B_n

$$E = B_n l v. \qquad [5.35]$$

The reference directions for the quantities expressed in eqn [5.35] are given by Fleming's right-hand rule which is illustrated variously in Fig. 5.15.

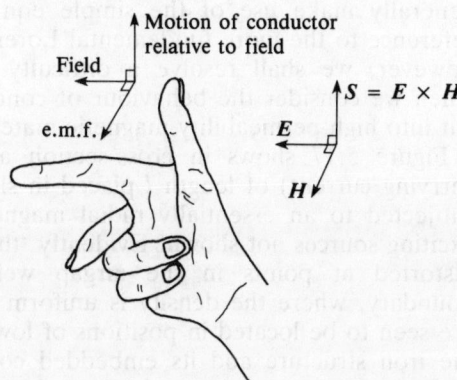

Figure 5.15 Fleming's right-hand rule relating directions of electromagnetic induced electric field (or e.m.f.) to directions of magnetic flux density and motion of a conductor with respect to a magnetic field.

Thu*m*b — *m* for motion,
Fore*f*inger — *F* for field (magnetic),
*C*entre finger — *C* for current flow in response to e.m.f.

The rule also relates vector product terms to their resultant, e.g. $S = E \times H$, $dF = I(dl \times B)$.

Equations [5.34] and [5.35] may also be deduced from eqn [2.39] by restating eqn [5.33] in terms of a space frame rotating with angular velocity ω in synchronism with the rotor conductor motion. Thus $\theta = \omega t + \theta_0$, say, to give

$$A(\theta, t) = -\hat{B}_n b \sin(\omega t + \theta_0).$$

$A(\theta, t)$ now has a non-zero time derivative and, with $v = 0$ since the rotating conductor is stationary with respect to the new space frame, eqn [2.39] yields

$$E_e = -\frac{\partial A}{\partial t} + v \times B$$

$$E_e = \hat{B}_n b \omega \cos(\omega t + \theta_0)$$

which is identical with eqn [5.34] but suggests more explicitly that the conductor e.m.f. will alternate sinusoidally with time as it moves through the field.

Equations [5.34] and [5.35] also apply if the conductor is regarded as stationary and the field system is allowed to rotate as depicted in Fig. 5.16. If we adopt a stationary frame of reference the magnetic field at all points on the stator/airgap boundary will appear to alternate with time, requiring a formulation for A such that, at the reference conductor location,

$$A(\theta, t) = +\hat{B}_n b \sin \theta, \qquad [5.36]$$

where $\theta = \omega t + \theta_0$, say. The positive sign in eqn [5.36] follows from the right-hand rule relating the line integral of A and the enclosed flux. Appreciating that the stator conductors have no velocity relative to a stationary frame of reference, eqn [2.39] gives

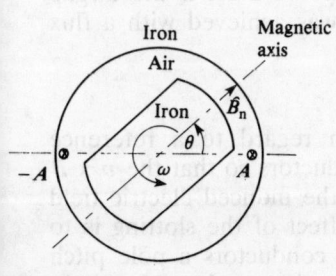

Figure 5.16 Elementary full-pitch stator coil subject to a rotating, radial magnetic field.

$$E_e = -\frac{\partial A}{\partial t} = -\hat{B}_n \omega b \cos(\omega t + \theta_0) \quad [5.37]$$

which is identical with eqn [5.34], except for sign. If, however, v is interpreted as the motion of the conductor *relative to the magnetic field*, eqn [5.35] and its identification with the right-hand rule still apply. In eqn [5.37] ωb represented motion of the field relative to the conductor.

In our subsequent work associated with the e.m.f.s induced in conductors having motion with respect to a magnetic field, we shall generally make use of the simple eqn [5.35] to avoid frequent reference to the more fundamental Lorentz equation. Prior to this, however, we shall resolve a difficulty which becomes apparent when we consider the behaviour of conductors placed within slots cut into high-permeability magnetic materials.

Figure 5.17 shows in cross-section a pair of conductors (not carrying current) of length l placed in slots a pole pitch apart and subjected to an essentially radial magnetic field originating from exciting sources not shown. Evidently, the flux distribution is much distorted at points in the airgap well away from the upper boundary, where the density is uniform at B_n and the conductors are seen to be located in positions of low flux density. Considering the iron structure and its embedded conductors to be in motion with respect to the magnetic flux which is held stationary at the upper surface, the $v \times B$ product in the region of space occupied by the conductors would appear to be small.

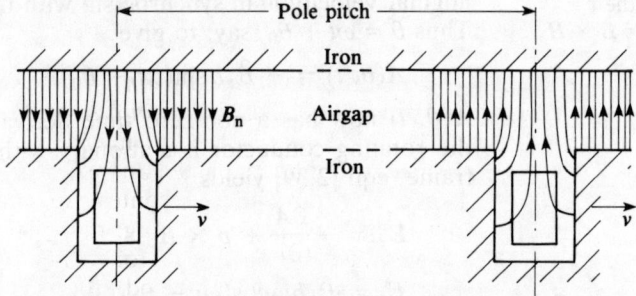

Figure 5.17 Axially directed conductors placed in slots a pole pitch apart, in motion relative to a stationary radial magnetic field.

Experiment would demonstrate, however, that e.m.f. is induced in the conductors of the same value as would exist if the airgap were uniform and the conductor motion was achieved with a flux density held constant at B_n, i.e.

$$E = B_n l v.$$

The phenomenon is best explained with regard to a reference frame moving synchronously with the conductors so that the $v \times B$ contribution to the Lorentz equation for the induced electric field in each conductor is actually zero. The effect of the slotting is to alter the *distribution* of the flux between conductors a pole pitch apart, but not at all to influence its *integral* over the pole pitch. The consequence is that, although the line integral of magnetic vector potential experienced over the full-pitch coil is unchanging as the conductors and the lower iron move, the complementary *local* variations of vector potential with time, through changes in position, at the conductor locations account in total for the induced e.m.f.s. The transfer of the flux lines from one side of each slot to the other as the slotted member incorporating the conductor moves, gives rise to an effect analogous to the 'flux-cutting' concept identified with the $v \times B$ term of the Lorentz equation.

The significant changes in A are brought about by the movement of the lower iron. If both iron surfaces were maintained stationary with a fixed flux distribution, subsequent lateral movement *within the slot* of the conductor would give rise to an e.m.f. identified solely with the low flux density prevailing therein.

The fact that local variations in magnetic field in the vicinity of a conductor *may* have no practical effect on the e.m.f. induced is highly significant. It means that we can ignore slotting effects when developing a simplified model of the machine – as long as we are confident that magnetic field conditions within the central region of the airgap accord reasonably well with our idealised model. This consideration complements the procedure for calculating *torque* through integration of the tangential mechanical stresses over an arbitrary, cylindrical surface placed coaxially in the airgap to surround the rotor (or stator). From our discussions in Section 3.11 we are aware that the effective torque-producing forces are, in practice, established on the sides of the teeth separating the slots which enclose the conductors, but acknowledgement of the airgap to be a mechanically stressed medium in equilibrium enables the resulting boundary forces to be deduced from a knowledge of the stress distribution over appropriate surfaces (Sect. 3.10).

5.8.2 *Coil interconnections within phase windings and their effect on induced e.m.f., distribution and pitch factors*

Phase windings of practical machines are built up from individual, multi-turn coils connected in series or as series/parallel combinations. As noted in the context of evaluating the magnetic field produced by a phase winding, the current flowing in all constituent parts must be identical for series-connected coils, and equal current-sharing for paralleled groups is to be desired. The situation is somewhat different regarding the e.m.f.s induced within individual coils of a group. If the coils occupy different positions in the inducing radial magnetic field, their individual e.m.f.s will be identical in magnitude and waveshape but displaced in time, with a phase shift dependent upon the physical displacement of the coils measured against a pole pitch for each space harmonic. Full-pitch coils have induced e.m.f.s due to the fundamental field oppositely directed along each coil side and acting in the same sense around the coil loop.

Figure 5.18(a) shows the winding layout for two coils comprising a part of one phase of a double-layer, three-phase, four-pole rotor winding having two slots/pole/phase. The vacant locations in the slots shown in full would be used to accommodate the additional six coils of the phase winding considered. Full-pitch coils are shown, but the reader should check that short-pitch coils may be accommodated equally well, by considering the upper coil-sides to remain in the position shown or implied and all lower coil-sides to be displaced by some multiple of a slot pitch in the anticlockwise direction. The effect on the fundamental magnetic field produced by the excited winding is an anticlockwise shift of one-half the angle of short pitch.

Figure 5.18(b) shows a developed diagram over a double pole pitch for the unchorded (full-pitch) winding with the four coils/phase connected in series/parallel. Within each parallel group the currents and summated e.m.f.s should be identical, although the individual conductor e.m.f.s may differ in phase. For the part of the winding shown, each group of coils contains two full-pitch coils, separated by the slot-angle of 15° (mech). This is equivalent to 30° space phase displacement in the fundamental field and $v30°$ for the vth harmonic field. In consequence, the vth harmonic

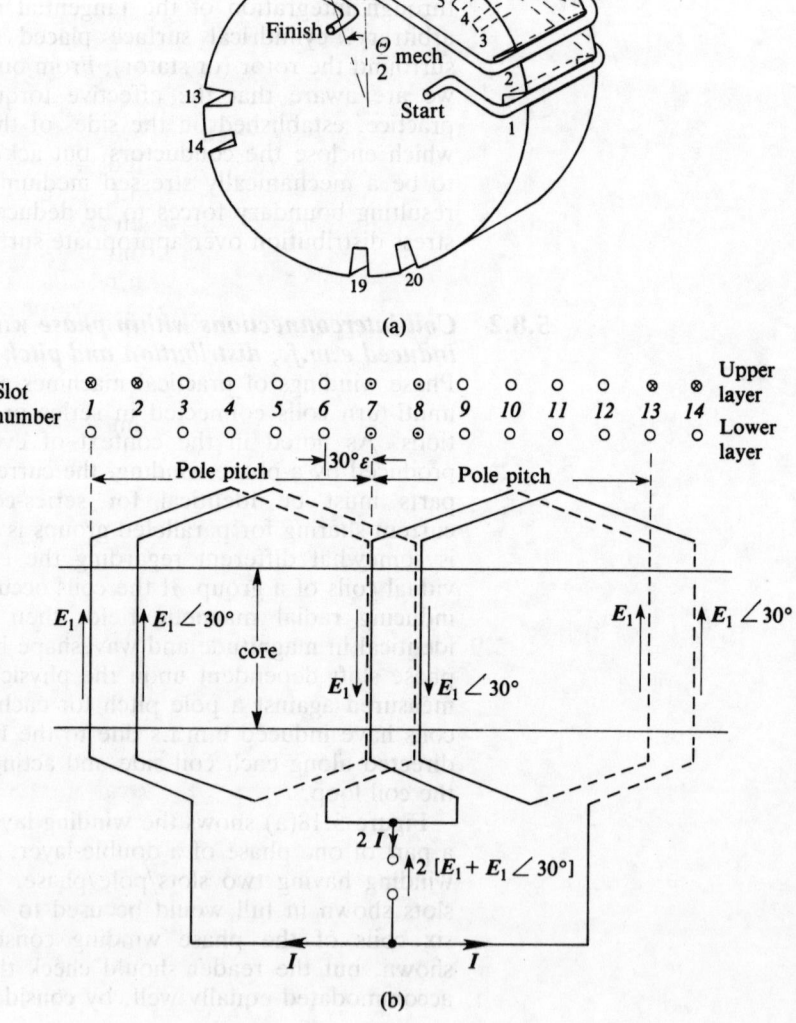

Figure 5.18 Winding layout for one phase of double-layer, three-phase winding with two slots/pole/phase: (a) arrangement appropriate to one-quarter of the four-pole winding with full-pitch coils; and (b) a developed view of a double pole-pitch with parallelin winding connections shown for one phase only.

e.m.f.s per coil within each group have relative time-phase displacements of $v30°$ and the *magnitude* of their resultant is related to the product of the magnitude of the e.m.f. per coil and the number of coils per group by the winding *distribution factor* K_{dv}

$$K_{dv} = \frac{\sin v30°}{2\sin v15°} = \frac{\sin\dfrac{Qv\Theta}{2}}{Q\sin\dfrac{v\Theta}{2}} \qquad [5.38]$$

with $Q = 2$ and $\Theta = 30°$.

The geometrical argument for the validity of eqn [5.38] in which Q is the number of series-connected coils in each group and Θ is the slot angle (elec) is precisely as presented in Fig. 5.8, with the phasors now representing time-phasor e.m.f.s instead of space-phasor magnetic field distributions. Again, for a three-phase winding with all slots wound, a 60° phase spread defining each group of coils and Q alternatively slots/pole/phase, eqn [5.38] reduces to

$$K_{dv} = \frac{\sin v30°}{Q\sin\dfrac{v30°}{Q}} \simeq \frac{\sin v30°}{\dfrac{v\pi}{6}} \quad \text{for } Q \to \infty.$$

The above expressions were presented as eqns [5.10] and [5.11] in the context of distributed magnetic field production. By analogous procedures, the e.m.f. reduction factor due to the use of short-pitch coils is as given by eqn [5.3]. Because even-harmonic magnetic fields do not normally exist in machine airgaps the coil-span or pitch factor in respect of induced e.m.f. may be more simply stated as

$$K_{pv} = \sin v90° \cos\frac{v\varepsilon}{2} \qquad [5.39]$$

in which ε is the (fundamental) angle of short-pitch. If sign is disregarded.

$$|K_{pv}| = \cos\frac{v\varepsilon}{2} \qquad [5.40]$$

5.9 Power Flows Relating to Rotating Machine Windings

The armature winding of an electrical machine is essentially an arrangement of interconnected electrical conductors connected to a source (or sink) of electrical power and interfacing with an air space. In rotating electrical machines this airgap separates the rotor and stator members and has an important bearing on the performance of the machine.

For a rotating machine to achieve electromechanical energy conversion at a steady rate over extended periods of time, it is necessary for the armature winding to continuously radiate or absorb electrical energy at a constant rate. If the machine is a motor, radiated energy is accepted by the airgap, for transmission towards the other airgap bounding surface or for electromechanical conversion within the volume of the airgap. In order that radiation (or absorption) may occur at the armature winding surface, it is necessary that appropriate electromagnetic field conditions exist there.

5.9.1 *Electrical power flow and the Poynting vector*

The flow of electrical energy in a medium such as air is, for most purposes, adequately described by the Poynting vector, defined in Section 2.6.1. This states that the power density S (W m^{-2}) at a point in the medium is given by the vector product of the electric field E and the magnetic field intensity H

$$S = E \times H. \qquad [2.48]$$

The Poynting vector S as defined above is not unique in relating power density to interacting electric and magnetic fields at a particular location in space. This and other formulations have in common the requirement that integration of the normal component of the power density vector over a closed surface equals the net power transfer over that surface.

The general properties of the Poynting vector are dealt with by specialist texts on electromagnetic field theory and need not concern us here, except to note that, since vector analysis is involved, it is convenient to resolve components of S, E and H along axes mutually at right-angles. Appropriate axes for the near-cylindrical structure of a rotating machine airgap are directed axially, radially and tangentially to the rotor or stator surface.

It is apparent, therefore, that if a machine is to maximise the capability of the armature winding to radiate (or absorb) electrical power at its interface with the airgap, it is necessary to facilitate the normal (radial) flow of energy at this surface. Hence we shall be concerned primarily with investigating the interaction at the airgap bounding surfaces of E_a, the axial component of E with H_t, the tangential component of H (Fig. 5.19) or, alternatively, the interaction of the tangential component of E with the axial component of H.

Figure 5.19 Electric and magnetic field components in a machine airgap with axially directed conductors. Radial power density S_n equals the product of the axial component of the electric field and the tangential component of the magnetic field.

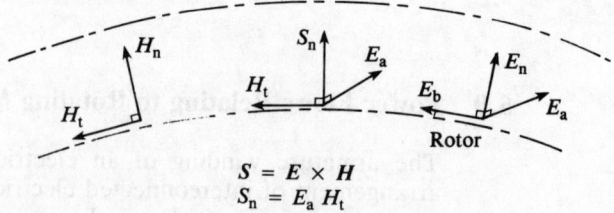

In practical rotating machine structures, the armature conductors are laid out substantially parallel to the rotor axis, which maximises the axial component of E but obliges H_a to be effectively zero. In consequence, the power handling capability of rotating electromechanical machines (the radial power transfer) is fundamentally associated with the *axial* component of the *electric* field and the *tangential* component of the *magnetic* field at the armature/airgap boundary. That is

$$S_n = E_a H_t \qquad [5.41]$$

Within the volume of the airgap H_n and E_n, the radial components of **H** and **E**, respectively, generally exist together with E_t. E_n reacts with H_t and H_n with E_t to give an *axial* power flow, whereas E_a reacts with H_n to give a *tangential* power flow. These internal airgap electrical energy flows may be of some interest in their own right but, as far as machine *performance* in electromechanical terms is concerned, only the electrical energy transfers occurring normally across the airgap boundaries with rotor and stator surfaces are significant. Thus the *net* electrical power transferred to the airgap via interfaces with both stator and rotor windings is accounted for by the mechanical power output and any increase in magnetic field energy stored within the airgap. In many rotating electrical machines which operate in the steady state no significant change occurs in the total magnetic energy stored within the airgap, hence the net electrical power transferred is balanced by the mechanical power developed.

As stated above, it is necessary to establish only the normal (radial) power flows across the stator and rotor bounding surfaces with the airgap to determine machine performance characteristics. This means determining only the distributions of the axial component of **E** and the tangential component of **H** over appropriate surfaces and evaluating the surface integral of the product. It has already been shown in Section 5.4.1 that the tangential component of **H** over either surface is conveniently independent of the current associated with the opposite surface in a model which makes the reasonable assumption of highly permeable iron bounding the airgap. Furthermore, the component of **E** due to motion is axially directed and, for a constant-speed machine, is proportional to the radial component of gap flux density B_n developed by the *resultant* of the normal components of stator and the rotor magnetic fields. It is again convenient that machines in the steady state often operate under conditions of a near-constant resultant radial flux, which property imposes constraints upon the relationship between rotor and stator winding currents.

5.9.2 *Power flow at a smooth cylindrical rotor/stator surface bounding an airgap*

As demonstrated in Section 3.11, torque is readily calculated from a knowledge of the (rotor) surface tangential stress distribution t_t using the tangential magnetic field intensity H_t and the normal (radial) flux density B_n. In fact, from eqn [3.22] $t_t = B_n H_t$. This stress, in association with the rotational speed ω relative to B_n is directly identifiable with the radiated energy flow normal to the (rotor) surface element under consideration.

Proof: Power Transfer Across an Iron/Air Boundary
Consider an elemental area $lb\,d\theta$ of (rotor) surface. From the Poynting vector equation $S = E \times H$ and, specifically,

$$S_n = E_a H_t. \qquad [5.41]$$

The normally directed *electrical* power transfer over the elemental area is

174 Electrical machines and drive systems

given by

$$E_a H_t lb \, d\theta = \omega b \, B_n H_t lb \, d\theta = \omega b^2 l B_n H_t d\theta.$$

From Maxwell's tangential stress equation, the torque contribution due to the same elemental area

$$T = b t_t lb \, d\theta = b^2 B_n H_t l \, d\theta.$$

The corresponding *mechanical* power $= \omega b^2 l B_n H_t \, d\theta$, which equates to the electrical power.

Integration of S_n over the entire (rotor) surface gives the electrical power transferred across the surface available for conversion to or from the mechanical state. This procedure is analogous to the integration of tangential stress to evaluate torque.

With those d.c. and synchronous a.c. machines in which a 'field' winding is d.c. excited, no relative motion occurs between such winding and the radial magnetic field. The axial electric field at the winding/airgap boundary is therefore zero and no electrical power transfer takes place.

5.10 Tutorial Examples

Questions marked with an asterisk * may be deferred until the chapter relating to the particular machine type has been studied.

5.1 A three-phase, eight-pole synchronous machine has a double-layer winding set in 96 slots with a phase spread of 60° (elec) and short-chorded by 1 slot. Draw the winding arrangement over a double pole pitch and sketch the radial field distribution across a uniform airgap as developed by current flowing in one phase only. Calculate

(i) the distribution factors for the winding to the ninth space harmonic,
(ii) the pitch factors to the ninth harmonic,
(iii) the overall winding factors to the ninth harmonic,
(iv) the amplitude of the fundamental component of airgap radial field set up by one phase of the winding, if each slot contains 10 conductors, with all the conductors connected in series and carrying 20 A,
(v) as (iv) but for the fifth space harmonic,
(vi) the amplitude of the resultant fundamental radial field when phases a, b and c are energised with sinusoidal currents mutually displaced in time phase by 120° and of peak value 20 A, and also the apparent distance moved through (° mech) in one period of current supply,
(vii) the amplitude of the resultant third harmonic field under conditions of (vi) and
(viii) the amplitude of the resultant fifth harmonic field under conditions of (vi) and the apparent distance moved through (°mech) in one period of current supply.

Ignore slot effects; assume iron of infinite permeability; and let the airgap length be g.

[(i) 0.958, 0.654, 0.205, −0.158, −0.270; (ii) 0.991, −0.924, 0.793, −0.609, 0.383; (iii) 0.950, −0.604, 0.163, 0.096, −0.103; (iv) 483/g A m^{-1}; (v) 16.6/g A m^{-1}; (vi) 724/g A m^{-1}, (vii) 90°; 0; (viii) 24.9/g A m^{-1}, −18°]

5.2 Taking the amplitude of the fundamental component of the radial airgap field established by the three-phase winding of **5.1** as 724/g A m^{-1}, determine the amplitude of the related tangential component at the energised iron/airgap boundary, and also the space-phase difference between tangential and radial field components. Make the same assumptions and assume a rotor radius $b \gg g$.

[2 896/b A m^{-1}, 22.5°(mech)]

5.3 Show that the r.m.s. value of e.m.f. induced in a full-pitch rotor coil of N turns having relative velocity $v = \omega b$ at right-angles to a sinusoidally distributed radial two-pole flux described by $B_n = \hat{B}_n \sin \theta$, where $\theta = \omega t$ defines the instantaneous position of one coil side, is given by

$$E = \sqrt{2} N \omega b l \hat{B}_n = \sqrt{2} \pi f N \Phi,$$

where the active conductors are of length l, $\omega = 2\pi f$, the rotor radius is b and Φ is the fundamental flux/pole.

Hence show that the fundamental e.m.f./phase with frequency f, induced by the fundamental radial flux in a winding subjected to distribution and pitching, may be expressed by the 'transformer e.m.f. equation' [4.14] i.e.

$$E_{\text{r.m.s.}} = 4.44 f N' \Phi,$$

where N', the effective turns/phase equals the product of actual turns/phase and the winding factor K_{dp1}.

5.4 Draw a diagram of the arrangement of coils in a double pole pitch of a two-layer, three-phase, four-pole, 36-slot stator winding with a phase spread of 60° (elec) and a coil span of seven slots. Calculate the winding factor for the fundamental.

Each coil has 10 turns, and all coils of a phase are connected in series. Estimate the fundamental radial flux per pole when the winding is star-connected to a 380 V 50 Hz supply. How could the winding be reconnected to give approximately the same flux/pole with a 110 V, 50 Hz supply?

E.C. [0.902; 0.00913 Wb; delta with two parallel paths/phase]

5.5 Repeat 5.3 in respect of sinusoidally distributed harmonic radial fluxes expressed as $B_{vn} = \hat{B}_{vn} \sin v\theta$ to show that the r.m.s. value of the vth harmonic e.m.f. is given by

$$E_v = \sqrt{2} N' \omega b l \hat{B}_{nv} = \sqrt{2} \pi f N v \Phi_v, \qquad [5.42]$$

where $N' = K_{\text{dp}v} N$, the fundamental frequency $f = \omega/2\pi$ and Φ_v is the vth harmonic flux/pole.

A 50 Hz synchronous machine has a field system which develops a fundamental radial flux/pole of 0.1 Wb with a third harmonic flux density component of amplitude $\hat{B}_{n3}$ equal to 20% of that of the fundamental, and a fifth harmonic component $\hat{B}_{n5}$ equal to 10% of $\hat{B}_{n1}$. Each armature phase winding has 320 series connected conductors (160 turns) with winding factors $K_{\text{dp1}} = 0.950$, $K_{\text{dp3}} = -0.604$, and $K_{\text{dp5}} = 0.163$. Determine the fundamental r.m.s. open-circuit armature voltage/phase and the harmonic components p.u. of the fundamental. Repeat for line–line open-circuit voltages if the winding is star connected.

What are the consequences of reconnecting the phase windings in delta?
[3374 V, 0.127, 0.017; 5845 V, 0, 0.017; Circulating third harmonic current, developing third harmonic voltage/phase to balance third harmonic e.m.f./phase]

5.6* An idealised polyphase synchronous machine on load has armature conductors on the stator side of a uniform airgap, carrying current such that the tangential magnetic field in the airgap at the stator boundary H_t is represented by a distributed constant-amplitude sinusoid rotating at synchronous speed. The radial magnetic flux density B_n due to the combined influences of the armature current and the (rotor) field winding current is also sinusoidally distributed in the airgap and is synchronously rotating.

Show that the integral of the radial component of electrical power density over a pole pitch of armature peripheral surface (equivalent to the developed [torque/pole] × ω) is proportional to the product of H_t, B_n and the cosine of the space phase angle between them.

5.7* Show on a developed diagram the radial magnetic field intensity distribution produced in a machine airgap by a single full-pitch coil over a double pole pitch, identifying the current locations and polarity.

The radial field distribution produced by a winding consisting of interconnected full-pitch coils is given generally by the following algebraic expression

$$H_n(\theta) = \frac{4}{\pi} \frac{\text{(current/pole)}}{2g} \left[K_{d1} \sin \theta + \frac{K_{d3} \sin 3\theta}{3} + \frac{K_{d5} \sin 5\theta}{5} + \ldots \right].$$

Explain what is meant by the 'distribution factor' of a winding with regard to the magnetic field it develops in the airgap, and show that the distribution factor for a finely distributed d.c. winding is given by

$$K_{dv} = \frac{\sin \frac{v\pi}{2}}{\frac{v\pi}{2}},$$

where v is the (space-) harmonic number. Thus show that the radial field distribution produced by a finely distributed d.c. winding may be given by the expression

$$H_n(\theta) = \frac{8}{\pi^2} \frac{\text{(current/pole)}}{2g} \left[\sin \theta - \frac{1}{3^2} \sin 3\theta + \frac{1}{5^2} \sin 5\theta - \ldots \right]$$

Identify the wave-shape described by the above harmonic series. What type of waveshape will the *tangential* field distribution have at the same winding surface?
U. of B. [Triangular, Rectangular]

5.8* An idealised polyphase induction machine on no-load carries current in the primary (stator) windings only. This establishes a sinusoidally distributed synchronously rotating field in the uniform airgap, with a corresponding circulation of energy therein and across the airgap/primary winding interface. When the machine is on load the constant-amplitude radial flux density wave in the airgap has motion relative to the *secondary* winding conductors at s times the synchronous value, where s is called the p.u. *slip*. Also, the load current density (A/m) induced in the secondary conductors bounding the airgap on the rotor side complements the primary current distribution at the stator/airgap boundary such that the resultant *radial* field component along any radius is independent of load.

Show that the mean value of electrical power absorbed by the airgap for conversion into mechanical power, accounting for power transfers across boundaries with both primary and secondary windings, is proportional to $(1-s) E_1 I_2' \cos \phi$, where ϕ is the phase angle between the e.m.f. E_1 induced in the primary reference phase winding by the radial flux and I_2' the primary reference phase current equivalent of the secondary load current.

5.9* (i) Explain what is meant by the chord (pitch) factor of an a.c. machine winding, illustrating your answer with reference to the e.m.f. induced by the fundamental radial flux wave. Derive an expression for the fundamental chord factor in terms of the coil-span angle (2θ).

(ii) List the orders and directions of the principal harmonics m.m.f.s of a balanced three-phase winding. Explain why some harmonics which are present in the phase m.m.f. do not appear in the resultant three-phase m.m.f.

(iii) What frequencies will be induced in (a) the *stator* and (b) the *rotor* of a 50 Hz synchronous machine by the fifth and seventh stator m.m.f. harmonics? Assume that the rotor is synchronised with the fundamental wave.

E.C.   [(i) $\sin\theta$; (iii) (a) 50 Hz, (b) 300 Hz]

5.10* (i) Explain what is meant by the chord (pitch) factor of a two-layer a.c. winding. What considerations have to be taken into account by the designer in choosing a suitable value for the coil pitch?

(ii) Two three-phase cage induction motors are identical except that the first has a coil pitch of 120° while the second has a coil pitch of 150°. The first motor develops a torque of 50 N m at a slip of 4%. Estimate the torque of the second motor at a slip of 4% when it is operated from the same supply. Make clear any assumptions that you consider necessary.

(*Author's note*: Show first that the effect of harmonic torques is negligible.)

E.C.   [(ii) 40.2 N m]

6 Direct-current machines

6.1 General Arrangement of d.c. Machines

Most d.c. machines are rotating heteropolar machines with the *armature* winding located in slots on the periphery of a cylindrical core assembled from iron laminations, insulated from each other though clamped together, and either keyed directly to the rotor shaft or mounted on a cast-iron 'spider' for large machines. The use of a laminated armature core reduces eddy-current loss in the rotor iron (Sect. 4.7). The main flux carried by the rotor in the region of the teeth pulsates along the length of the teeth as the rotor moves under the influence of the magnetic field created by the concentrated windings surrounding the *salient* main 'field' poles which are bolted to the iron stator *yoke*. The general arrangement of the stator is shown in Fig. 6.1 which corresponds to a four-pole machine. The number of main stator poles is always a multiple of 2, alternating in polarity around the armature surface. The main pole cores are generally unlaminated except possibly in the region of the *pole shoe* adjacent to the airgap where subjected to fluctuations in the flux path as the rotor teeth move past. Frequent flux changes within the main body of the stator pole cores and yoke are not required in most d.c. machine applications.

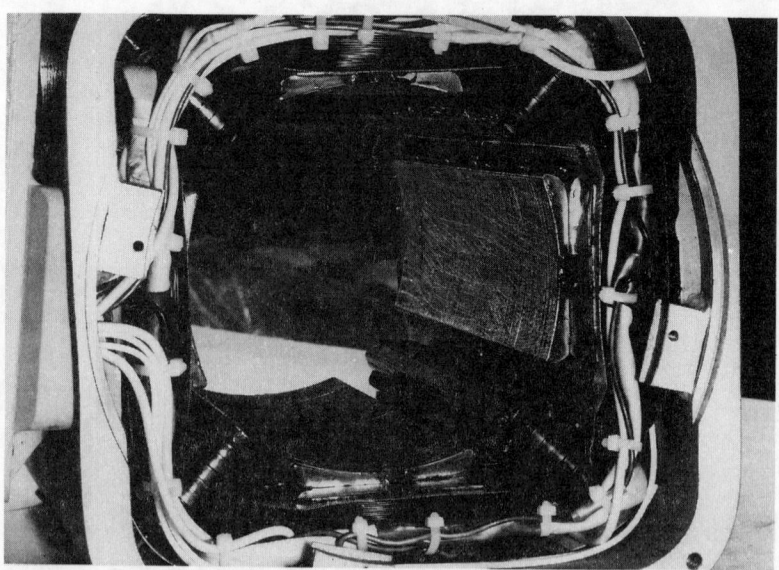

Figure 6.1 Stator-mounted pole detail of 50 kW, 460 V, 1000 rev/min, four-pole d.c. machine with interpoles. (*Photograph by courtesy of Mawdsley's Ltd, Dursley, UK.*)

Figure 5.1 illustrates the distribution of flux produced by the main field poles within the stator and rotor iron and also within the airgap. Ignoring the local effects of the rotor slots, the energised (excited) 'field' windings produce flux which crosses the airgap radially at the rotor/airgap boundary and emerges from (or enters) the stator pole shoes normally.

Much space is wasted on the stator side of the airgap in the region between adjacent poles if the number of poles employed is small. Figure 6.1 shows this interpolar space to be occupied by *interpoles* or *commutating poles*, whose flux crossing the airgap radially at the rotor surface is very much more localised than that of the main stator poles. The function of the latter is to provide a radial flux density distributed over much of the rotor surface, directed at right-angles to the motion of the axially directed armature conductors. Equation [2.39], from which was derived eqn [5.35], indicates that this configuration will maximise the electric field or e.m.f. induced in the armature conductors, whilst eqns [3.3] and [3.2] confirm this in respect of the tangential mechanical stress or torque due to an element of armature winding current flowing axially. Radial flux density is a proportionality factor in both considerations. The function of the interpoles is essentially a secondary one in facilitating *commutation*, the necessary frequent reversal of the current carried by the armature conductors employed in heteropolar d.c. machines.

The magnetic circuits relating to the main-pole field and the interpole field are essentially similar in being an effective series combination of iron and air paths. The objectives are rather different, however. Since the interpole flux should be a linear function of the current carried by the armature conductors, a large airgap is included in order to render the effect of variable iron permeability insignificant. To maximise torque and armature e.m.f. the average value of the main-pole radial flux density should be large, hence the airgap element of the main flux circuit is kept small and those parts of the iron circuit where the flux is concentrated, i.e. the teeth, may well be operated in the region of saturation.

The *commutator* is another characteristic feature of the heteropolar d.c. machine, mounted on the rotor shaft as shown in Fig. 6.2. This assembly of insulated copper bars connected individually to the armature coils, each in cyclical contact with stationary carbon (graphite) brushes as the rotor rotates, is necessary to translate the essentially alternating e.m.f. and current variation with time within the coils into the steady values appropriate to a d.c. supply.

6.2 Generation of e.m.f. between the Armature Terminals of Rotating d.c. Machines

Rotating d.c. machines normally have a heteropolar field system on the stator which is derived electromagnetically via field coils energised with direct current or, alternatively, formed from permanent-magnet material. The rotating axially directed armature conductors experience the radial magnetic field due to the stator poles and have induced therein an electric field which gives rise to an e.m.f. between the ends of each conductor which is given by eqn [5.35]

$$E = B_n l v. \qquad [6.1]$$

Figure 6.2 Direct-current machine shaft-mounted commutator which connects the armature winding tapping points to an external circuit via fixed carbon brushes. (*Photograph by courtesy of Mawdsley's Ltd, Dursley, UK.*)

The above equation has been derived in Section 5.7.1 from the more fundamental Lorentz equation [2.39] for the electric field induced in a conductor moving in a magnetic field. The simple expression [6.1] has conductor length l and v, the velocity of the conductor, at right-angles to each other and also to the radial flux density B_n. The relationship is valid also for stationary conductors if v is interpreted as the motion of the conductor relative to the flux, and holds when, as is usual, the armature conductors are placed in a slotted magnetic core. Relative directions of e.m.f., flux density and motion of conductor with respect to the flux are given by Fleming's right-hand rule illustrated in Fig. 5.15. For armature conductors of constant length, rotating at constant speed under the influence of a steady radial flux derived from the stator pole system, the e.m.f. E will vary with time in accordance with the instantaneous value of radial flux density at the conductor location. Since the field poles alternate in polarity, the induced e.m.f. will correspondingly alternate in time, with a *waveform* precisely matching the angular distribution of radial flux in the airgap at the rotor surface. Such a variation will generally not be sinusoidal – although the assumption of a sinusoidal distribution may prove convenient in certain analyses. Ideally, the distribution of flux should be uniform over the pole face, giving rise to a square waveform for the induced e.m.f.

In order to convert the alternating e.m.f. induced in each armature conductor into unidirectional d.c. voltage, *rectification* is required. Conventional d.c. machines employ a mechanical commutator to achieve this function – the external circuit is connected to the armature conductor via a carbon brush which bears on a wedge-shaped copper segment connected permanently to one side of the armature conductor and insulated from adjacent segments. The complete commutator assembly forms a cylinder concentric with the rotor shaft, as illustrated in Fig. 6.2.

In its most elementary form involving a single-turn armature coil the d.c. machine is as portrayed in Fig. 6.3(a). Inspection of this figure shows that the fixed brushes overlap the two commutator segments at the instant when the two coil-sides are located in the interpolar spaces and hence have values of induced e.m.f. which are instantaneously zero. Figure 6.3(b) shows that the e.m.f. per turn is always twice that of each conductor because each conductor occupies identical positions under adjacent poles – the single-turn coil is said to be 'full-pitch'. Figure 6.3(c) shows that the e.m.f. appearing between the brushes is a rectified version of the coil e.m.f.

The position of the fixed brushes with respect to the stator poles is critical. For that illustrated in Fig. 6.3(a), the brushes are said to commutate the coil 'in the quadrature axis' as the coil sides are located in the centre of the interpolar space such that $\theta = \pm 90°$ when short-circuited by the brushes. This position is displaced by 90° from the magnetic field axis due to the stator poles, known as the 'direct axis', on the basis of a cosinusoidal distribution of flux about this axis. Such a flux distribution has, for simplicity, been assumed in deriving the waveforms of Fig. 6.3.

In the idealised waveforms of Fig. 6.3(b) and (c), zero time has been allowed for the commutation process. In fact, the insulation between commutator segments and also the brushes will both be of finite width, with the latter exceeding the former. During the *commutation time* t_c, when the single armature coil is short circuited by the brushes, the e.m.f. between armature terminals of this simple d.c. machine is zero.

If, however, the brush axis is shifted in the anticlockwise direction to make an angle ψ with the stator field axis (the 'direct' axis), the commutation of the coil will be delayed by a period of time equivalent to the phase angle ψ, with consequences for the 'rectified' terminal e.m.f. waveform illustrated in Fig. 6.3(d).

The instantaneous e.m.f. between terminals now includes negative excursions and has a reduced mean d.c. value. On the

Figure 6.3 Elementary two-pole d.c. machine with single-turn armature winding: (a) normal configuration with fixed brushes commutating the armature coil in quadrature axis; (b) e.m.f.s per conductor and per turn; (c) e.m.f. between brushes commutating the armature coil in quadrature axis; and (d) e.m.f. between the brushes, with brush axis shifted by ψ.

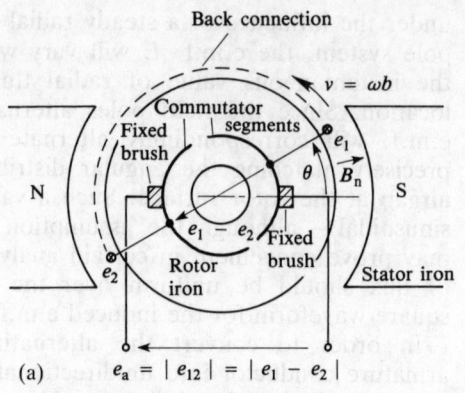
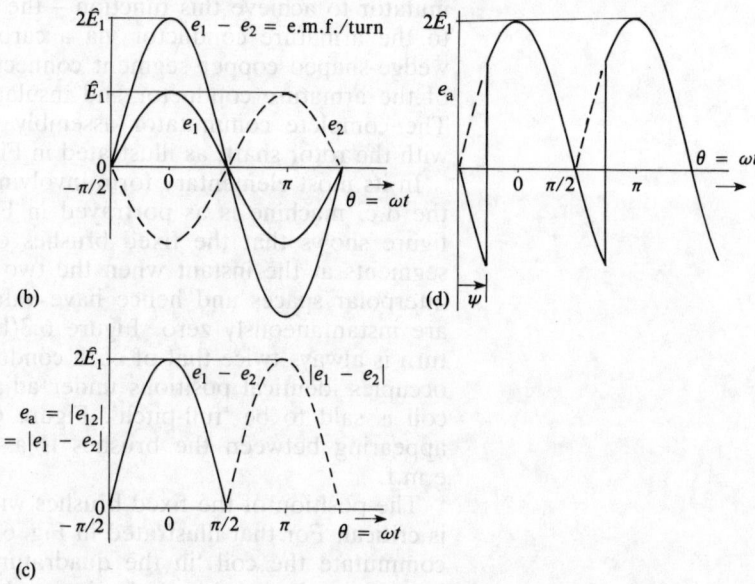

assumption of a sinusoidally distributed radial flux and a correspondingly time-dependent coil e.m.f., a simple calculation shows that the *mean* d.c. value of the terminal e.m.f. requires a multiplying factor $\cos\psi$, with polarity reversed for $90° < \psi < 270°$. Thus, appreciating that the factor $2/\pi$ relates average value to peak value for a sinusoidal waveform, and with $\hat{E}_1$ equal to the maximum value of coil-side e.m.f.

$$E_a = e_{a_{av}} = \frac{4}{\pi}\hat{E}_1 \cos\psi. \qquad [6.2]$$

Before progressing to a consideration of more practical arrangements let us consider the elementary d.c. machine with a single armature coil to be connected to a load resistance of value R_L. Current will flow in the armature coil in the same direction as, and proportional to, the induced e.m.f. alternating *within the coil* but, like the e.m.f., maintaining a constant *direction* as each coil side sweeps past each stator pole in turn. Equation [3.3] may be used to evaluate the direction of the force on each conductor. Thus, referring to Fig. 6.3(a), conductor 1 at the instant depicted has e_1 of positive polarity. With i_1 having the same direction as e_1 its flow will be out of the paper. Use of eqn [3.3]

$$\mathrm{d}\mathbf{F} = I(\mathrm{d}\mathbf{l} \times \mathbf{B})$$

indicates that the force on the conductor at right-angles to both the current flow and the radially directed flux will be tangentially directed and such as to oppose the anticlockwise motion of the coil-carrying rotor. An identical situation holds for conductor 2, hence the coil-side currents are responsible, together with the radial flux, for the rotor torque opposing the motion of the coil. This is to be expected as the machine is an elementary *generator* of electrical power. The torque development is unidirectional, but not constant. With assumed constant speed and the radial flux sinusoidally distributed in the airgap the torque will vary sinusoidally with time about a mean value equidistant between zero and twice the mean value, completing one cycle during the time taken for one coil side to traverse one pole. Shifting the brush axis has no effect on the electrical *power* output of this generator, surprisingly at first sight perhaps, as the *mean* values of terminal e.m.f. and current are influenced by brush shift. The machine is essentially an elementary a.c. generator supplying a resistive load via a full-wave rectifier.

Clearly, the arrangement of Fig. 6.3 is unsatisfactory. As a d.c. generator the armature terminal e.m.f., though invariably unidirectional for brushes commutating the rotor conductors in the quadrature axis, has excessive *ripple* (variation about the *mean* (d.c.) value) whilst the torque is even more variable with time. Matters may be significantly improved by adding a further armature coil, displaced by 90° from the first and connected in like fashion to a similar, two-segment commutator mounted on the rotor shaft and having sliding contact with a further pair of brushes. The brush position with respect to the stator pole axis is identical with that of the original coil. Such an enhanced arrangement is shown in Fig. 6.4(a), with the corresponding e.m.f. waveforms appearing between each brush pair (illustrated in Fig. 6.4(b)) being seen to correspond to a pair of full-wave rectified sinusoids with a relative phase displacement of 90°. If the two sets of brushes are then connected *in series*, the total armature e.m.f. is shown in Fig. 6.4(c) to have an improved ratio of mean value to peak value.

Over the first 1/8 period of one revolution of the rotor from the position $\theta = 0$, the total armature e.m.f. given by the sum of the component e.m.f.s has the instantaneous value

$$e_\mathrm{a} = e_{a_{12}} + e_{a_{34}} = 2\hat{E}_1 \cos \omega t + 2\hat{E}_1 \sin \omega t$$

with average value

$$E_\mathrm{a} = \frac{2\hat{E}_1}{\pi/4} \int_0^{\pi/4} (\cos \omega t + \sin \omega t) \, \mathrm{d}\omega t = \frac{8}{\pi} \hat{E}_1. \qquad [6.3]$$

Now the average value of e.m.f. induced in any one coil side as it sweeps across one pole face

$$\frac{\hat{E}_1}{\pi} \int_{-\pi/2}^{\pi/2} \cos \omega t \, \mathrm{d}\omega t = \frac{2}{\pi} \hat{E}_1.$$

Figure 6.4 Elementary two-pole d.c. machine with two-turn armature winding: (a) normal configuration with brushes commutating the coils in the q-axis; (b) e.m.f.s per conductor and between brushes; and (c) total armature e.m.f. e_a between brush sets in series.

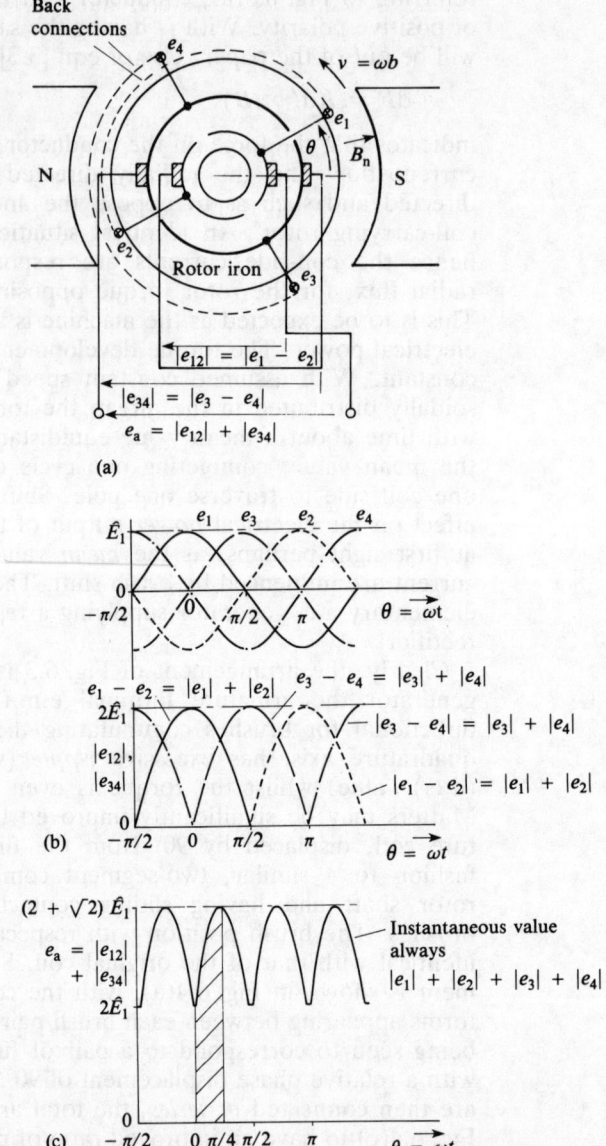

Thus the *total* e.m.f. across both pairs of brushes in series (eqn [6.3]) has an average value equal to 4 times that of an individual coil side. Further, examination of the waveforms of Fig. 6.4 demonstrates that, at all instants in time, the total e.m.f. developed across both pairs of brushes in series is equal to the sum of the rectified instantaneous e.m.f.s of each individual coil side. This result is most significant.

On proceeding to evaluate the contribution of each current-carrying coil to the development of rotor torque when the two-coil generator is loaded on resistance, use is made of the fact that the series-connected coils carry the same current, having waveshape identical with that of the instantaneous armature e.m.f. e_a.

Clearly, this two-coil machine is a more satisfactory energy converter than the single-coil device. If shifting of the common brush axis relative to the stator pole axis is applied, it is readily shown that the mean d.c. e.m.f. E_ψ, expressed instantaneously as $e_{a_{12}} + e_{a_{34}}$ varies as the cosine of the brush-shift angle ψ – assuming a sinusoidal distribution of radial flux in the airgap.

Since $\hat{E}_1$ is proportional to $\hat{B}_n$ and ω, we can make the following general statement about the e.m.f. appearing at the armature terminals of a rotating d.c. machine. The mean d.c. value is proportional to rotor speed and radial flux, and varies sinusoidally with brush position if the flux is sinusoidally distributed, with maximum value occurring when the brushes are set to commutate coils located with their sides along the quadrature axis and zero when the brush positions are such as to switch coil sides along the direct (or polar) axis. Evidently, a sinusoidal alternating 'mean d.c. e.m.f.' is developed if the brush sets are in constant motion relative to the fixed-pole system – ideally at a low rate compared with the motion of the coil sides with respect to the stationary field. A standard d.c. machine has brushes fixed 'in the quadrature axis' to maximise e.m.f. and torque for given values of pole flux, speed and armature current.

It is not, in fact, desirable to arrange for a sinusoidal radial flux distribution in practical d.c. machines. A higher mean d.c. e.m.f. with lower ripple and torque variation is obtained if the radial field approaches a uniform, square-wave distribution. As will be proved formally in Section 6.3 the mean d.c. e.m.f. is proportional to the total radial flux/pole, *regardless of its distribution*.

The improved performance of the two-coil machine featured in Fig. 6.4 over the single-coil arrangement of Fig. 6.3 is due to the *distribution of the armature winding* between the two coils separated by half a pole pitch (90°). Presumably, further improvement regarding e.m.f. waveform and torque ripple is likely if the number of armature coils is further increased by *spreading* the winding over the rotor surface. In practice, the winding will be accommodated in slots around the rotor periphery. Commutation would be difficult to arrange, however, if it proved necessary to have a two-segment commutator for each coil, with each brush pair connected in series with the next!

Fortunately, an alternative *double-layer* winding configuration enables the number of brushes to be reduced to the minimum of two, such that the brushes are required to be connected only to those coils undergoing commutation in sequence as the coil system rotates. All the other armature coils not involved in commutation at any instant are connected in series/parallel groupings between brush pairs, carrying identical currents and contributing to the total armature e.m.f. Thus each coil is shown in Fig. 6.5 to be connected in series with adjacent coils and also to one commutator segment. The complete winding is continuous with each coil usually spanning a full pole pitch. The brushes are also separated by a pole pitch and are fixed relative to the stator pole system, such that any coil short circuited by a brush bridging adjacent commutator segments has its sides instantaneously in the centre of the space where the radial flux density due to the stator poles is ideally zero – the quadrature axis.

Figure 6.5 shows a view of a d.c. machine armature winding with the individual coils connected to insulated commutator segments. Each rotor slot may accommodate more than one pair of coil sides. Each armature coil has the characteristic diamond shape illustrated in Fig. 5.4, formed with a kink in the inactive regions to permit overlapping of adjacent coils. Each coil has one side in the upper slot position and the other side located in the bottom of a slot about a pole pitch away.

Figure 6.5 Elementary two-pole double-layer winding with four coils set in four slots: (a) general configuration with four-section commutator and one brush pair; (b) corresponding developed diagram with reference directions defined for conductor e.m.f.s; and (c) e.m.f.s per conductor, per coil and between brushes.

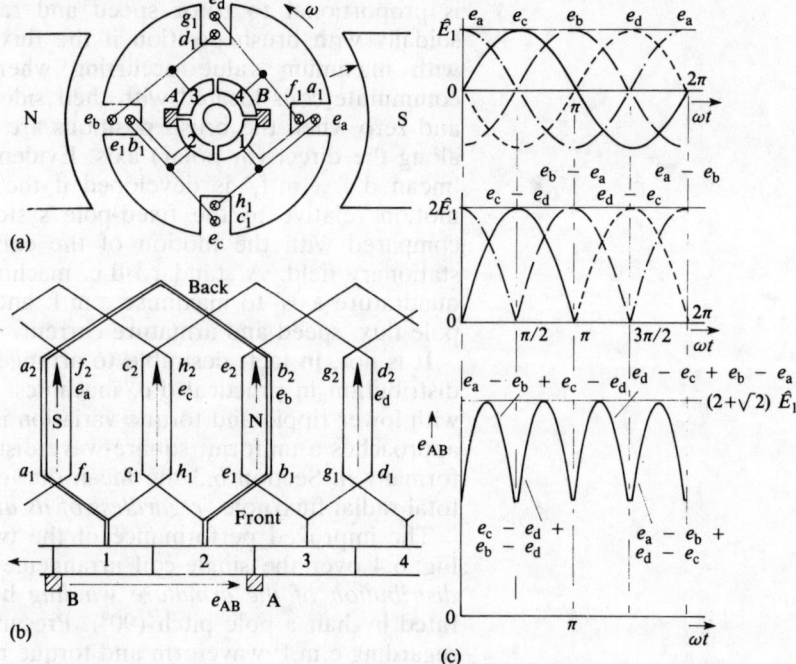

Figure 6.5 shows a four-coil, two-pole, double-layer armature winding set in four slots. Figure 6.5(a) has the back connections interconnecting upper and lower coil sides separated by a pole pitch. At the front, the same upper-coil-side to lower-coil-side sequence is maintained, but the connection spans less than a pole pitch to enable the active winding to be interconnected. Connections are made from each commutator segment to one front connection so that the number of commutator segments equals the number of coils, with each identifiable coil having active sides connected at the front to adjacent commutator segments. When the segments are bridged by a brush during commutation, an individual coil is itself short circuited, but retains its series connection to adjacent coils at either side. These coils provide the terminal elements of a pair of parallel paths through the armature winding for the armature current as it crosses the brush/commutator contact.

Figure 6.5(a) shows also the brush locations to be in line with the radial field axis due to the stator poles. This ensures that each coil when short circuited by a brush has its sides located in the interpolar space and consequently is not subject to an e.m.f. resulting from its motion in the airgap field due to the stator poles.

It also ensures that the maximum possible e.m.f. appears between brushes for a given rotational speed and radial flux/pole.

The waveform of the e.m.f. appearing between brushes is explained with reference to the developed diagram of Fig. 6.5(b) which shows the position of the fixed stator poles and brushes relative to the moving armature coil sides and commutator segments. Electromotive forces induced in each coil side and directed from front to back are defined, respectively, as e_a, e_b, e_c and e_d. Assuming constant speed and sinusoidal radial flux distribution, the e.m.f.s e_a and e_b will be sinusoidal and in antiphase. Electromotive force e_a will apply to both coil sides a and f located in the same slots. Waveforms appropriate to e_a, e_b, e_c and e_d are shown in Fig. 6.5(c), for which the instant $t = 0$ corresponds to the configuration of armature conductors and field poles shown in Fig. 6.5(a) and (b). At $t = 0$. e_a and e_b have peak positive and negative values whilst e_c and e_d are both zero, respectively positive- and negative-going. The pairs of coil sides short-circuited by the brushes, g and h, and c and d, have at that instant zero e.m.f.

As ωt increases from $0°$ to $90°$ the e.m.f. e_{AB} between the brushes is instantaneously given by $e_a - e_b + e_c - e_d$, with each component making a positive contribution to the total. During this interval e_{AB} follows a sinusoidal waveform, rising and falling symmetrically about a peak value equal to $\sqrt{2}$ times the maximum e.m.f. *per coil*, $2\hat{E}_1$. The pattern is repeated for each successive $90°$ displacement and it may be recognised from Fig. 6.5(c) that, at all instants in time, the instantaneous e.m.f. appearing between the brushes is equal to the sum of the instantaneous recitified e.m.f.s induced in each of the four coil sides. The waveform of the brush e.m.f. is completely described between any two successive instants of commutation. Increasing the number of coils and commutator segments, whilst maintaining symmetry regarding their distribution over the rotor and commutator surfaces, will increase the average value of the e.m.f. between brushes and will reduce the ripple content.

6.3 Calculation of Armature e.m.f., the e.m.f. Equation of a Two-pole d.c. Machine

In order to calculate the mean or average value of the e.m.f. developed between a brush pair it is necessary simply to evaluate the sum of the mean values of the e.m.f.s in each of the conductors connected in series between the brushes during the interval separating one instant of commutation from the next. During this period each conductor or coil side sweeps past a part of one pole face such that the sum of all such areas identified with the series-connected conductors is the total area associated with the radial flux of the two poles. After the next commutation instant the procedure is repeated, the actual configuration of the active conductors making up the complement between brushes being different, but summing identical instantaneous e.m.f. components.

Thus, during the interval between any successive pair of commutation instants, the average e.m.f. between brushes due to z series-connected conductors labelled 1 to z is

$$E_a = E_{av} = \sum (e_{1_{av}} + e_{2_{av}} + e_{3_{av}} + \ldots + e_{z_{av}}).$$

Using eqn [6.1] with l being the axial length of each armature conductor having peripheral velocity ωb and $B_{1_{av}}$ being the average value of radial flux density traversed by conductor 1 we get

$$E_a = l\omega b \sum (B_{1_{av}} + B_{2_{av}} + B_{3_{av}} + \ldots + B_{z_{av}})$$

$$= l\omega b \sum \left(\frac{z\Phi_{12}}{2\pi lb} + \frac{z\Phi_{23}}{2\pi lb} + \frac{z\Phi_{34}}{2\pi lb} + \ldots \frac{z\Phi_{z1}}{2\pi lb} \right)$$

where Φ_{12} is that portion of radial flux swept by conductor 1 as it moves between the arbitrary pair of adjacent commutation instants and Φ_{z1} is the corresponding flux swept by conductor z.

Now, $\Sigma(\Phi_{12} + \Phi_{23} + \Phi_{34} + \ldots + \Phi_{z1})$ represents the *algebraic* sum of the radial fluxes encountered by the z conductors in series between the brushes which, for brushes set to commutate coil sides located on the quadrature axis, is equal to twice the total radial flux per pole Φ. Hence

$$E_a = \frac{l\omega bz}{2\pi lb} 2\Phi.$$

For a rotational speed of n rev/s, $\omega = 2\pi n$ giving

$$E_a = 2nz\Phi.$$

Since there are two parallel paths through the armature between brushes

$$z = \frac{Z_a}{2},$$

where Z_a is the total number of armature conductors, hence

$$E_a = Z_a n \Phi. \qquad [6.4]$$

The above expression is valid for an armature winding having two parallel circuits. It has been derived with no assumptions being made regarding the distribution of the radial flux/pole Φ. Φ is strictly the *net* radial flux crossing the armature surface in an arbitrary direction between limits defined by the brush positions. Movement of the brushes from the position at which the *quadrature axis* coils are commutated will reduce Φ and consequently the mean armature e.m.f. E_a. For a symmetrical distribution of radial flux about the direct axis, brush shift in excess of 90° will reverse the polarity of E_a.

Equation [6.4] may alternatively be derived by recognising that the average armature e.m.f. between brushes equates to that developed in any single conductor as it moves over a pole pitch between the brushes, multiplied by the number of series-connected conductors in the path. Thus

$$E_a = l\omega bz B_{av}$$

where

$$B_{av} = \frac{\text{radial flux/pole}}{\text{pole area}} = \frac{\Phi}{\pi bl} = \frac{1}{\pi} \int_{-\pi/2+\psi}^{\pi/2+\psi} B_n(\theta) \, d\theta,$$

giving

$$E_a = 2nz\Phi = Z_a n\Phi,$$

as required

For a cosinusoidal distribution of radial flux about $\theta = 0$ and a brush shift of ψ in the reference direction of θ measurement (Fig. 6.4)

$$\Phi = bl \int_{-\pi/2+\psi}^{\pi/2+\psi} \hat{B}_n \cos\theta \, d\theta = 2bl\hat{B}_n \cos\psi$$

which, for the four-coil, eight-conductor arrangement of Fig. 6.5 with two parallel paths gives

$$E_a = \frac{8}{\pi} \hat{E}_1 \cos\psi, \qquad [6.5]$$

where $\hat{E}_1 = 2\pi nbl\hat{B}_n$ equals the maximum value of e.m.f. per conductor. Equation [6.5] compares with eqn [6.3] and quantifies the effect of brush shift.

6.3.1 *Multipole machine windings*

There are advantages to be gained in designing rotating electrical machines with the number of magnetic poles greater than two. Less space is wasted on the stator between the projecting or *salient* field poles, and the end connections of the armature coils are shorter. The number of poles must be a multiple of two with N–S alternation on the stator. If the procedure for winding the four-coil, two-pole arrangement of Fig. 6.5 is continued over the rotor surface of a multipole machine, maintaining a pole-pitch separation between armature coil sides at the non-commutator end, and with a similar configuration of poles, brushes and conductors for each pole pair, an armature winding will result which has an identical number of conductors connected in series in any path between any pair of adjacent brushes. Such a winding is called a *lap* winding and has, ideally, identical e.m.f.s developed between adjacent brush pairs – assuming identical distributions of flux/pole. Alternate brushes may, therefore, be connected together so that the complete armature wound for p pole pairs offers p times the number (2) of parallel circuits inherently provided by the two-pole arrangement. Figure 6.6 illustrates a lap winding for a four-pole machine having 12 rotor slots to accommodate the armature winding. The armature conductors and commutator segments are normally in motion with respect to the fixed stator poles and brushes. At the instant depicted in Fig. 6.6, the armature coil connected between commutator segments 1 and 2 embraces one of the four stator poles and is short circuited by brush A. The brushes are set, as normal, to commutate coils 'in the quadrature axis'. A short time prior to the instant shown, brush A made contact only with segment 1, a short time later the same brush will contact only segment 2. If the speed of rotation, brush width and commutator dimensions (including insulating spacer thickness) are known, the *time* during which each coil in turn is short circuited by a brush may be calculated. Inspection of the sides of the coil undergoing commutation confirms these to be instantaneously located in the interpolar space where the radial flux is ideally non-existent and so cannot contribute to the e.m.f. developed in the short-circuited coil.

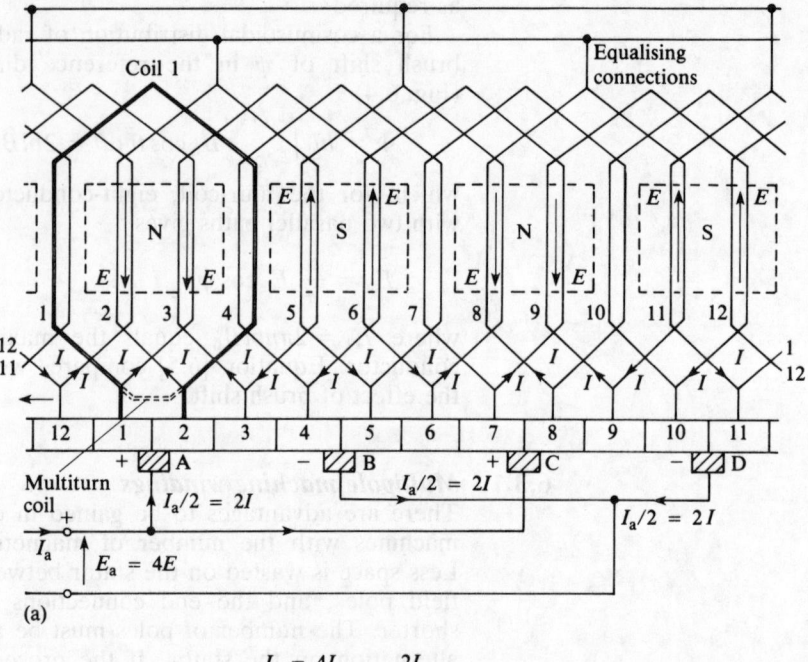

Figure 6.6 Armature lap winding for four-pole machine with 12 slots in total: (a) developed diagram; and (b) an equivalent source representation illustrating four parallel circuits.

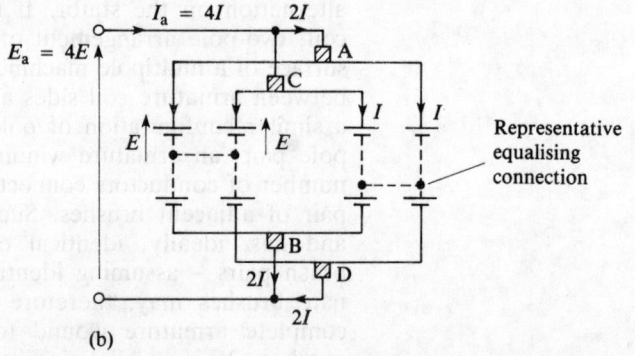

The e.m.f.s due to the radial flux experienced by the effective conductors account for the total e.m.f. developed between a pair of adjacent brushes. If one traces a route through the winding via the particular coil side connected to commutator segment 2 and not associated with the short-circuited coil, e.m.f. is encountered via the upper conductor in slot 2, the lower conductor in slot 5, the upper conductor in slot 3 and the lower conductor in slot 6. The latter conductor is in contact with brush B via commutator segment 4. This brush short circuits the coil which embraces the stator pole adjacent to the first one considered. The components of e.m.f. induced in the four effective conductors or coil sides are seen to act in the same direction, such as to make brush A positive with respect to brush B. On proceeding further through the winding in the same direction as before, prior to meeting a third coil short circuited by brush C and embracing the third stator pole, e.m.f.s induced by radial flux in four conductors are again encountered, of polarity such as to make brush C positive with respect to brush B. The pattern identified, of *passive* coils short-circuited by a brush bridging adjacent commutator segments which are interconnected by four *active* coil sides with e.m.f.s co-directed within each group (but of opposite polarity between adjacent groups) is repeated four times as each polar region is traversed, until the starting point is regained.

If, now, *alternate* brushes are connected together, four parallel circuits within the armature winding may be distinguished, analogous to the parallel connection of four banks of cells to constitute a battery, as shown in Fig. 6.6(b). The armature current I_a is then shared, ideally equally, by the four parallel paths within the armature winding. All conductors, *except those short circuited by the brushes*, carry the same fraction (1/4) of the total armature current and, significantly, all conductors situated in the radial magnetic flux due to the stator poles carry current in a direction consistent with that of the flux. The relative directions of conductor e.m.f. and current depend upon whether the armature winding is absorbing or supplying electrical power, i.e. whether the machine is motoring or generating. Current flow in these active conductors pertaining to motor operation is depicted in Fig. 6.6(a), with the current in the four coils short circuited by the brushes being instantaneously zero.

As the winding moves with respect to the fixed pole and brush systems, the armature e.m.f. and current pattern remains essentially unchanged, although the disposition of actual conductors changes. If the situation of coil 1 is considered just before and just after its short circuit by brush A, the current it carries will be found to change from that of the circuit between brushes A and D to that of the parallel circuit between brushes A and C. In so doing, the *direction* of current flow in the short-circuited coil has to reverse. Since the coil possesses self inductance due to the magnetic fields developed by its active portion within the airgap and the inactive end windings, this reversal of current is opposed by an e.m.f. proportional to the rate of change of current. A finite time is therefore necessary for this reversal or *commutation* of the armature coil current, which is related to the magnitude of the current in each parallel path through the armature winding. The time available is the period of coil short circuit by the brush. The process of commutation is further discussed in Section 6.8 as it constitutes an important feature regarding the design and operation of a d.c. machine. Another aspect for later consideration relates to the airgap magnetic field produced by armature current. At this juncture we will simply note that the *direction* of current flow along each coil side reverses at each location of a short-circuited coil as one progresses along the winding. If the armature conductors are many, and are thus finely distributed over the rotor surface, the armature current distribution may be approximated to a *current sheet* of uniform 'linear density' measured in amperes per metre of rotor periphery, with polarity changes where connection to a brush occurs at the centre of the space between the stator field poles. If the brush positions are shifted with respect to the stator poles the armature current distribution will effect a similar displacement, with consequences for the *torque* developed on the stator and the rotor which complement the influence of brush shift on armature e.m.f.

6.3.2 Wave windings

A *lap* configuration is not the only method available for interconnecting the armature conductors of multipole d.c. machines. The *wave* winding interconnects coil sides in such a manner that a pair of series-connected coil sides associated with adjacent stator poles is connected directly in series with coil sides lying under the remaining stator poles in sequence, rather than to adjacent coil sides under the original pair of poles. To ensure the complete incorporation of all armature conductors, it is necessary for the number of slots (and commutator segments) to be such that, having traversed the armature periphery via the series connection of coil sides from any arbitrary position to the immediate vicinity of the original starting point, the second transit would naturally connect in sequence a coil side adjacent to the initial conductor and lying in a similar location within its slot – whether top or bottom. On continuing the sequence, several passes over the armature surface will be completed in a return to the initial starting point with all conductors interconnected.

With such an arrangement, it is necessary to provide only one pair of brushes, separated by a pole pitch. Whatever the number of poles, only two parallel circuits exist, between which the total armature current I_a should divide equally. In practice, additional brush pairs may be provided to reduce the current flowing across each commutator segment/brush contact surface and generally ease problems of commutation. For given speed, power rating, pole flux, airgap dimensions and number of armature conductors, however, the e.m.f. developed between brushes for a wave winding exceeds that for the lap winding by a factor equal to the number of pole pairs. This is the factor by which the total armature current I_a is reduced. The current per parallel path, and hence per conductor, is basically unchanged. Wave windings are more suitable for use on high-armature-voltage, low-armature-current applications; lap windings for high-current, low-voltage machines.

A wave winding set in 13 slots for a four-pole machine is shown in Fig. 6.7(a). Each slot accommodates conductors associated with the top and bottom sides of different coils. Each coil spans approximately one pole with the *back pitch*, y_b (the separation of the coil sides at the non-commutator end) and the *front pitch*, y_f, such that the product $p(y_b + y_f)$ is equal to the total number of coil sides ±2. p is the number of pole pairs. The winding shown is described as *retrogressive*, as each transit over the armature core periphery via the series-connected coil sides falls short of the starting point. A *progressive* wave winding overlaps the start position, usually by one slot. Such a winding results from a choice of $y_b = y_f = 5$ for the 13 slot, four-pole combination, as the reader may usefully confirm.

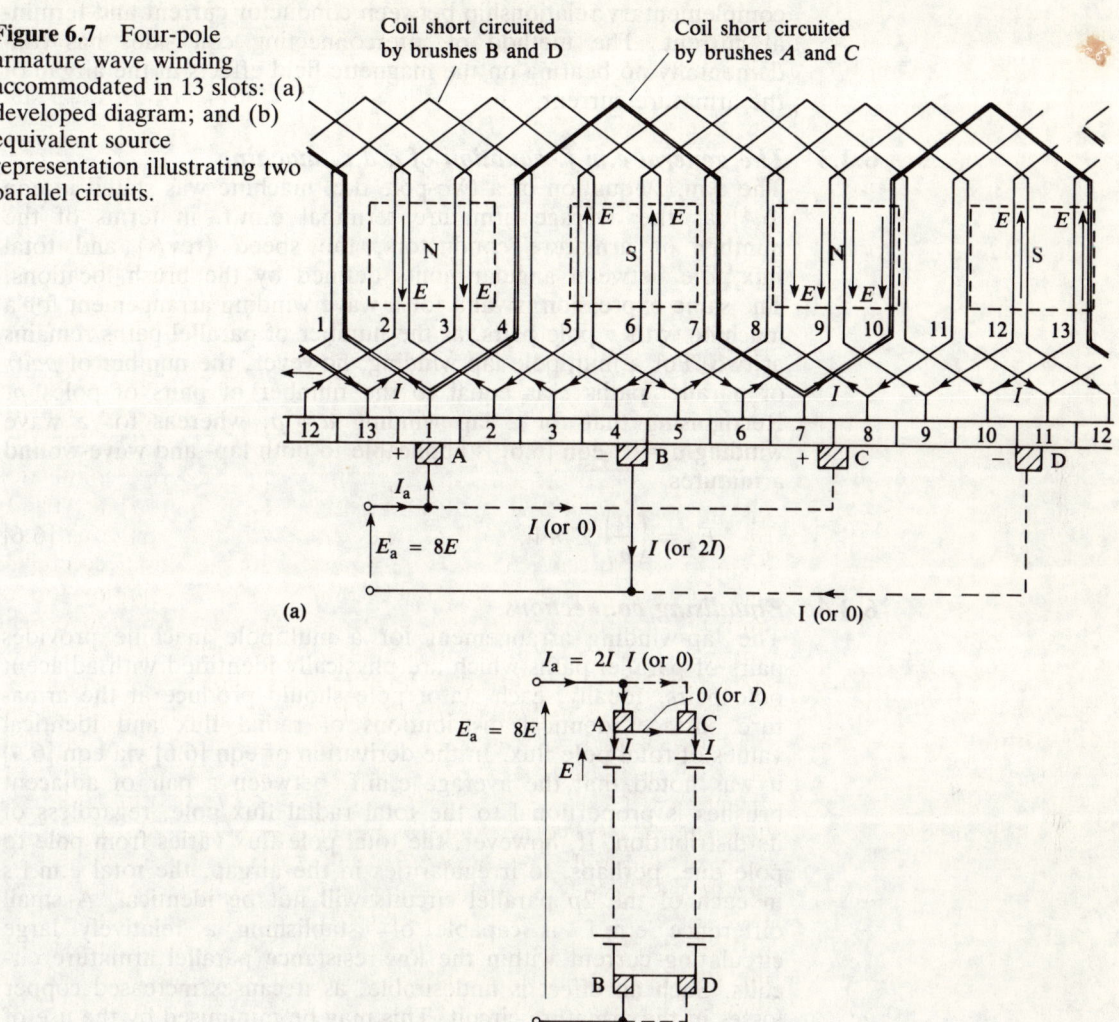

Figure 6.7 Four-pole armature wave winding accommodated in 13 slots: (a) developed diagram; and (b) equivalent source representation illustrating two parallel circuits.

As with a lap winding the essential brush-pair A and B, say, in Fig. 6.7(a), is normally aligned with the centre line of adjacent stator poles, so that the coils commutated have sides in the region of low radial flux density between the poles and the e.m.f. between brushes is maximised. With no additional brush pairs provided, the p coils connected in series between adjacent commutator segments require commutation *as a group* during the time the single brush overlaps the segments. The installation of additional brush pairs, e.g. brushes C and D in Fig. 6.7, eases the transfer of total armature current between the external circuit and the winding conductors, the number of parallel paths through the winding remaining at two.

Evidently, the relationship between conductor e.m.f. and terminal e.m.f. is dependent upon the armature winding configuration employed – whether lap or wave – via the number of parallel paths each alternative provides. The effect is counterbalanced by a complementary relationship between conductor current and terminal current. The method of interconnecting coil sides has fundamentally no bearing on the magnetic field effects in the airgap of the armature current.

6.3.3 The general e.m.f. equation of a d.c. machine

The e.m.f. equation of a two-pole d.c. machine was stated in eqn [6.4] as the average armature terminal e.m.f. in terms of the number of armature conductors, the speed (rev/s) and total flux/pole between angular limits defined by the brush locations. The same expression is valid for a wave winding arrangement for a machine with p pole pairs, as the number of parallel paths remains at two. For a multipole lap winding, however, the number of *pairs* of parallel paths a is equal to the number of pairs of poles p. Recognising that for a lap winding $a = p$, whereas for a wave winding $a = 1$, eqn [6.6] is applicable to both lap- and wave-wound armatures

$$E_a = \left(\frac{p}{a}\right) Z_a n \Phi. \qquad [6.6]$$

6.3.4 Equalising connections

The lap-winding arrangement for a multipole machine provides pairs of parallel paths which are physically identified with adjacent pole pairs. Ideally, each stator pole should produce at the armature surface identical distributions of radial flux and identical values of total pole flux. In the derivation of eqn [6.6] via eqn [6.4] it was noted that the average e.m.f. between a pair of adjacent brushes is proportional to the total radial flux/pole, regardless of its distribution. If, however, the total pole flux varies from pole to pole due, perhaps, to irregularities in the airgap, the total e.m.f.s in each of the $2p$ parallel circuits will not be identical. A small difference e.m.f. is capable of establishing a relatively large circulating current within the low-resistance parallel armature circuits. Such an effect is undesirable, as it causes increased copper losses in the armature circuit. This may be minimised by the use of low-resistance *equalising rings*, fitted concentrically with the rotor shaft at the non-commutator end, and each connected to appropriate points on each parallel path which *should* have equal potentials provided each pole has identical flux distribution and value. If magnetic unbalance is present, out-of-balance currents circulate in shorter paths involving the equalising connections, thus exerting a less disturbing influence than without equalisers. Two such equalising connections are shown in Fig. 6.6

6.4 Torque Development within a d.c. Machine

The torque manifested at the shaft of a rotating electrical machine may be evaluated in several different ways. Electromagnetic torque is developed to balance prime-mover or external load torques, bearing friction and windage effects, and that due to the inertia of the rotating mass when speed changes occur. Practical machine designers need to know the distribution of force on the stator and rotor core teeth, the *tangential* component of which contributes a component of the net torque. The distribution of *radial* force, or stress, is also significant in that out-of-balance radial forces may tend to distort the rotor shaft. The use of Maxwell's stress eqns [3.21] and [3.22] enables the distribution of force to be calculated if the magnetic-field distribution over the iron surfaces bounding the airgap is known. From a knowledge of the tangential stress distribution the corresponding torque may be deduced, in a particularly convenient way if the tangential forces may be considered to act at a constant radius. In order to assess the magnetic field in the airgap of the d.c. machine on load, the modifying influence of the armature current on the field due to the *stator* poles must first be evaluated.

6.5 The Airgap Field of a d.c. Armature Winding – Armature Reaction

Current is caused to flow in the armature conductors of value determined (in the steady state) by the ratio of the net circuit e.m.f. and the circuit resistance. Armature conductors are located in open slots on one side of the airgap and, when carrying current, they produce radial and tangential components of magnetic field which, in conjunction with similar field components pre-existing within the airgap due to the so-called 'field' pole system located on the other side of the airgap, subject the magnetic medium in the (air) gap to mechanical stresses.

Note must be taken of the fact that the armature current is contained within slots, and also that magnetic flux must traverse an iron/airgap boundary virtually at right-angles to the surface, due to the high relative permeability of the iron laminations employed. The flux distribution over a slot as set up by *external* field sources is disturbed by the field effects of current within the slot to create a differential mechanical stress on the sides of the slot. This stress, on integration over its effective area, gives rise to an element of tangential force.

In this section we shall seek to evaluate the effective airgap magnetic field components due to a current-carrying armature winding, with a view to establishing how such current affects the e.m.f. and torque developed by the machine. Any effect of the armature current on the *radial* flux density at the armature surface will modify the e.m.f. induced in individual conductors, as given by eqn [6.1], and will modify the mean e.m.f. developed between brushes in accordance with eqn [6.6].

It has been shown in Section 5.8.1 that, with regard to the evaluation of armature e.m.f., it matters not whether the conductors are regarded as contained within slots or attached to a smooth surface concentric with the axis of rotation. The same feature proves to be the case with torque evaluation. The alternative scenarios shown in Fig. 6.8 relate to configurations of the iron/airgap surface bounding the current-carrying armature conductors as follows:

(a) a round conductor of finite diameter placed within the airgap and secured through adhesive bonding to a parallel, smooth iron surface concentric with the axis of rotation;
(b) an axially directed current finely distributed over the same smooth iron surface; and
(c) a current flowing in conductors laid within open slots separated by iron teeth projecting radially from same smooth iron surface.

In each case, general symmetry about the axis of rotation is assumed and identical expressions deduced for the force per unit axial length of conductor, acting at right-angles to both radial flux density and current flow

$$\frac{F}{l} = B_n I \qquad [6.7]$$

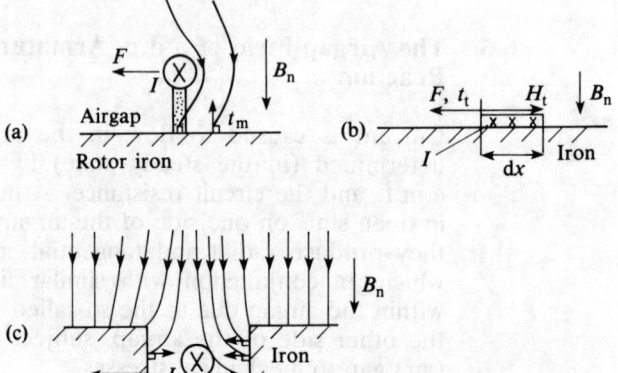

Figure 6.8 Current-carrying conductor adjacent to an iron/airgap bounding: (a) conductor within airgap; (b) conductor current distributed over the surface; and (c) current within slot.

For configuration (a) eqn [6.7] follows immediately from eqn [3.3], whilst for configuration (b) Maxwell's stress equation for *tangential* stress, eqn [3.22], may be applied at the surface of an iron/air boundary with surface current, thus

$$t_t = B_n H_t = B_n \frac{I}{dx}$$

hence

$$\frac{F}{l} = t_t dx = B_n I.$$

The differential force on the sides of the slot in configuration (c) is derived fundamentally from Maxwell's stress equation for *normal* stress, eqn [3.21]. The force may be equated to that derived for (b), however, on recognition that the integrated stress components which contribute to torque over *any* closed surface surrounding the rotor iron give rise to an identical value of resultant torque (Sect. 3.10).

For ease in calculating resultant torque we shall adopt the armature current configuration of Fig. 6.8(b) as it avoids the complicating effect on the magnetic circuit of local variations in airgap length due to slotting, whilst giving a good *general* indication of tangential mechanical stress distribution over the armature surface. The adoption of a *current sheet* at the smooth cylindrical surface bounding the airgap and concentric with the axis of rotation enables both radial and tangential field components due to simple distributions of current to be calculated very easily through direct application of Ampère's law, eqn [2.27].

6.5.1 *Radial and tangential airgap field components of a d.c. armature current sheet*

Consider a two-pole machine with a double-layer armature winding having a uniform current distribution as shown in Fig. 6.9. The brushes are depicted as making direct contact with conductors on the quadrature axis. The key features are retained in the current sheet representation of Fig. 6.10(a) for which the linear current density $(A\,m^{-1})$ over the periphery of the rotor iron is constant, reversing the direction of axial flow at each brush location. The uniform airgap implied in Fig. 6.10(a) is not strictly applicable, however, and will require that predictions of field values in the region of the quadrature axis are subjected to scrutiny.

Figure 6.9 Current distribution due to a double-layer armature winding.

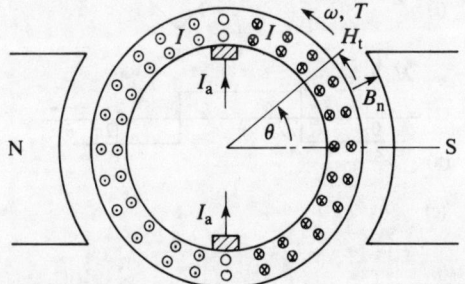

For the uniform armature current distribution and airgap shown in Fig. 6.10(a), the radial magnetic field intensity H_n across the airgap has the triangular distribution shown in Fig. 6.10(c). The radial field distribution is calculated by applying Ampère's law $\oint H \cdot dl = I$ to a closed path involving two crossings of the airgap – one at a fixed location $\theta = 0$, say, and the other at an adjustable location $\theta = \theta_1$. For $-\pi/2 < \theta_1 < \pi/2$, a small increase in the upper limit θ_1 by $\Delta\theta_1$, say, gives rise to an *increase* in the current enclosed and hence to an increase in the value of H_n (with a reference direction from rotor to stator) between $\theta = \theta_1$ and $\theta = \theta_1 + \Delta\theta$. For $\pi/2 < \theta_1 < 3\pi/2$, an increase in θ_1 by $\Delta\theta$ gives rise to a *reduction* in the enclosed current and hence in the value of H_n. Due to symmetry and the requirement that the total flux entering or leaving stator and rotor surfaces must be zero, it follows that the positive and negative peak values for H_n occur respectively at $\theta = \pi/2$ and $\theta = -\pi/2$.

For an elementary flux path crossing the airgap from stator to rotor at $\pi/2$ and from rotor to stator at $\theta = -\pi/2$ the total current enclosed $= \pi b A$, where b is the rotor radius and A is the linear current density $(A\,m^{-1})$. Ignoring the reluctance of the high-permeability iron, the application of Ampère's law gives

Figure 6.10 Current-sheet representation of a d.c. armature winding mounted on a rotor: (a) cross-sectional view; (b) detail at airgap location θ; and (c) developed views of (i) radial field distribution H_n, uniform across the airgap, (ii) tangential field distribution H_{tr} at the energised rotor surface bounding the airgap. Linear current density is A A m^{-1}, with reference direction into the paper.

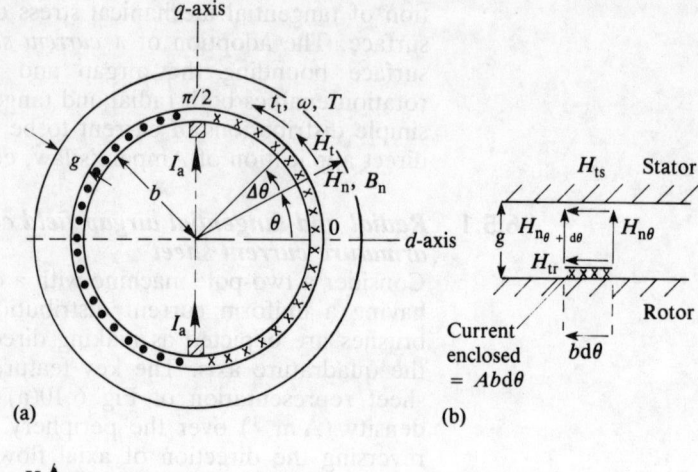

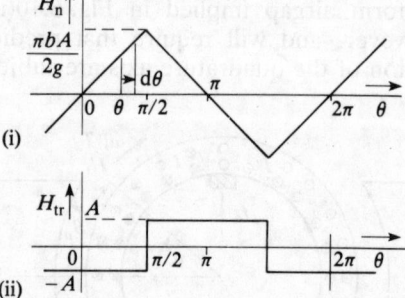

$$\oint \mathbf{H} \cdot d\mathbf{l} = \hat{H}_n g - (-\hat{H}_n)g = 2\hat{H}_n g = \pi b A$$

$$\hat{H}_n = \frac{\pi b A}{2g}. \qquad [6.8]$$

Thus the distribution of the radial field in the airgap, due to the armature current, is uniform across the gap and of *triangular* shape, with peak values located along the quadrature axis where the armature current is commutated. This contrasts significantly with the radial airgap field due to the *stator* poles which tends to a *rectangular* form, centred about the direct (or polar) axis at $\theta = 0$.

By simple application of Ampère's law, the tangential field distribution at the current-carrying boundary between the armature iron and the airgap is readily evaluated. Referring to the element shown in Fig. 6.8(b), the line integral of $\mathbf{H}$ over the surface of the infinitely thin conductor bounded on one side by the infinitely permeable rotor iron yields

$$\oint \mathbf{H} \cdot d\mathbf{l} = -H_{tr} b \, d\theta = A b \, d\theta$$

hence

$$H_{tr} = -A. \qquad [6.9]$$

Since the airgap is bounded on the *stator* side by infinitely permeable iron, $H_{ts} = 0$ due to current on the *rotor*/airgap boundary. The general relationship between tangential field intensity and radial position in the airgap due to current located on one boundary is shown in Section 5.4.1 to be a linear one, but this information is irrelevant to stress evaluation over the airgap bounding surfaces.

6.5.2 *Effect of a non-uniform airgap*

In practice, d.c. machines whose slotting effects are ignored do not have an airgap of uniform length. The rotor armature is indeed cylindrical but the stator poles are *salient*. Excited by embracing, concentrated, rectangular coils supplied with d.c. current (unless employing permanent-magnet material), they project from the iron yoke which completes the magnetic circuit on the stator side of the airgap. An objective in d.c. machine design is the provision of a high, uniform radial field in the airgap over as large a pole area as possible, keeping small the interpolar region in the vicinity of the field-coil sides parallel to the rotational axis. Such an arrangement maximises the torque for given values of armature current and maximum allowable radial gap flux density. Because of the need to commutate armature coils whose sides lie in the interpolar region, the radial flux density should reduce rapidly as the quadrature axis is approached. Control is achieved by shaping the stator *pole shoes*, thus increasing the airgap length in the region of the pole tips.

The high reluctance presented to radial flux in the interpolar region has a marked effect also on the radial field component in the airgap due to *armature* current on the rotor. Thus the distribution of H_n, allowing for increased reluctance near the quadrature axis, is modified as shown in Fig. 6.11(a), with the prospective triangular peaks appropriate to a uniform gap length never materialising.

Figure 6.11(a) corresponds to the case where the brushes are set to commutate armature coils on the quadrature axis with the armature winding alone carrying current. If the stator poles are also energised they contribute significantly to the *radial* field across the gap whilst not affecting the *tangential* airgap field at the rotor surface. As the radial field components due to the stator and rotor are distributed about axes mutally displaced by 90°, the *resultant* radial flux density across the gap adopts the form shown in Fig. 6.11(b).

In this figure B_{ns} and B_{nr} represent radial flux densities produced in the airgap when the stator and rotor windings are energised separately. The *resultant* flux density B_n developed in a practical machine when windings on both sides of the airgap are energised together is not given by direct addition of stator and rotor components of flux density, except in the interpolar region where the airgap is large. Elsewhere, the magnetisation characteristics of the iron dominate to enforce a nonlinear B/H relationship. The magnetic circuit is normally operated such that the iron approaches saturation when the stator field current approaches the operational limit. Thus, in iron regions where incremental increases in H_n occur due to superimposed armature current, the increases in B_n are less than pro-rata, whilst prospective reductions in B_n due to reductions in H_n are relatively enhanced. The effect of armature current is significant in reducing the total radial flux/pole below the no-load value, with important consequences for machine operation.

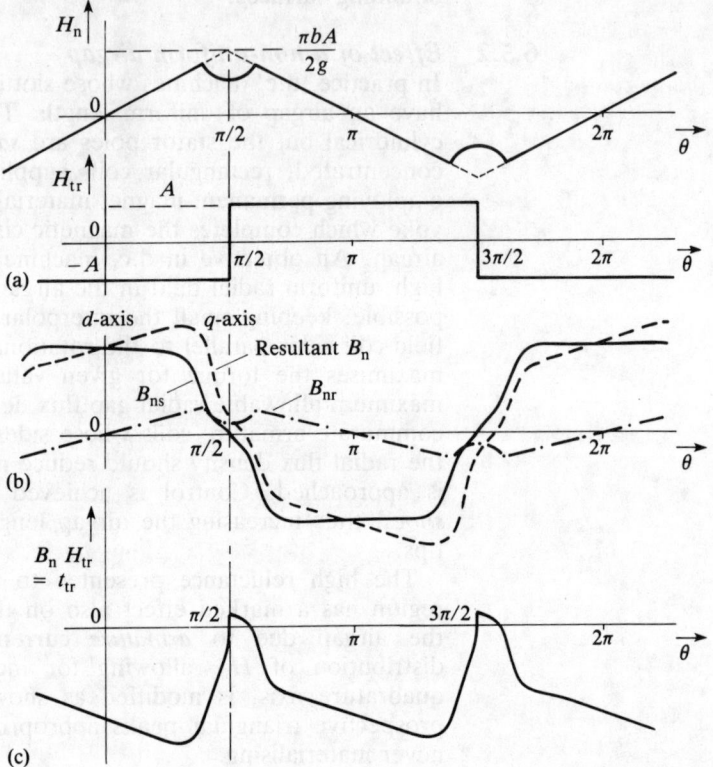

Figure 6.11 Airgap field, flux density and armature-surface tangential stress distribution over a double pole pitch of a d.c. machine with uniform armature current sheet and brushes set to commutator coils in the q-axis: (a) the radial field and the tangential field at the armature surface due to the armature current alone; (b) the resultant radial flux density with the stator field poles energised also; and (c) the tangential stress over the armature surface.

A further influence of the radial field due to armature current is on commutation. The commutated coil sides move relative to the field, and in consequence experience an e.m.f. proportional to the armature current value, invariably directed in such a sense as to oppose the required reversal of current in the coil when short circuited by the brushes.

6.5.3 Provision of compensating winding

The magnetic effect of the armature winding on load is generally called *armature reaction*, the undesirable influences of which may be ameliorated by the simple, though expensive, procedure of providing a distribution of current on the *stator* side of the airgap to counter that of the armature conductors. Such a *compensating winding* consists of interconnected conductors placed in open slots in the stator pole shoes, as shown in Fig. 6.12. The linear current density of the compensating winding should match that of the armature winding and be of opposite polarity at a particular airgap location, in order that the total current, due to both windings and enclosed by any path which involves two radial crossings of the airgap, is zero. Such a facility eliminates the effect of the armature current on the *radial* flux density B_n across the gap, whilst leaving the *tangential* field intensity H_t at the armature surface bounding the airgap unchanged.

Figure 6.12 Close-up view of a compensating winding located in slots in pole shoes of a d.c. machine. (*Photograph by courtesy of Mawdsley's Ltd, Dursley, UK.*)

Thus the rotor-surface tangential field intensity remains at the value $-A$ A m^{-1}, with A being the linear current density due to the armature winding and the reference directions for field and current being defined in Fig. 6.10(a) and (b). The presence of the compensating winding modifies the tangential field conditions at the stator side of the airgap and consequently the distribution of mechanical stress over that surface.

6.6 Tangential Mechanical Stress and Torque Development at the Armature Surface of a d.c. Machine

The tangential mechanical stress at an iron/air boundary is simply given by eqn [3.22] which may be specified in terms of radial flux density and tangential field intensity at the rotor surface as

$$t_{tr} = B_n H_{tr}.$$

For armature current represented as a uniform current sheet flowing axially on the rotor surface with the notation of Fig. 6.10, from eqn [6.9]

$$H_{tr} = -A$$

hence

$$t_{tr} = -AB_n \qquad [6.10]$$

Tangential mechanical stress is evaluated in Fig. 6.11(c) as a function of position θ on the rotor periphery. Since the distribution of A is uniform, with polarity changes at brush positions, it is apparent that t_{tr} follows the distribution of B_n with appropriate reversals in polarity. Thus the tangential mechanical stress on the rotor is almost everywhere in the sense opposing rotor rotation, for the condition depicted with B_n and ω being in the reference directions of Fig. 6.10(a), corresponding to machine functioning as a *generator*.

Rotor torque is evaluated via integration of tangential mechanical stress over the surface of the rotor. Thus, for a two-pole machine

$$T = b \int_{-\pi}^{\pi} t_{tr} lb \, d\theta$$

acting in the direction of t_{tr}, with b being the radius and l being the axial length of the cylindrical rotor

$$\therefore T = -b \int_{-\pi}^{\pi} AB_n(\theta) lb \, d\theta$$

Now

$$\int_{-\pi/2}^{\pi/2} B(\theta) lb \, d\theta = \Phi,$$

where Φ is the total radial flux per stator pole between the limits $\pm\pi/2$, regardless of the nature of its distribution, hence

$$T = -2bA\Phi.$$

Further, if the armature winding has Z_a total conductors set in $2a$ parallel paths carrying *total* armature current I_a,

$$A = \frac{I_a}{2a} \frac{Z_a}{2\pi b}$$

$$\therefore T = -\left(\frac{1}{a}\right) \frac{Z_a}{2\pi} I_a \Phi.$$

For a machine with p pole pairs and with Z_a and Φ again defining total armature conductors and (radial) flux per stator pole

$$T = -\left(\frac{p}{a}\right) \frac{Z_a}{2\pi} I_a \Phi. \qquad [6.11]$$

The minus sign in eqn [6.11] indicates that with stator field polarity such that flux crosses from rotor to stator at $\theta = 0$ and armature *conductor* current directed as indicated in Fig. 6.10(a), the torque acts in the direction opposite to the anticlockwise reference. With such a configuration of armature current and radial flux, rotation of the armature in the anticlockwise reference direction should correspond to generator action. Fleming's right-hand rule (Sect. 5.8.1) confirms that such motion of the armature conductors in the radial flux as described accounts for armature conductor e.m.f. in the direction of assumed armature current.

Alternatively, torque may be defined in terms of the linear current density in the armature winding as

$$T = (-)(2p\Phi)Ab \qquad [6.12]$$

= total flux crossing airgap × linear current density × radius.

Another useful formulation of the torque equation of a d.c.

Direct-current machines 203

machine is simply

$$T = K\Phi I_a, \qquad [6.13]$$

where

$$K = \left(\frac{p}{a}\right)\frac{Z_a}{2\pi}$$

is a machine constant. The factor $K\Phi$ which relates torque to armature current may be identified as that relating armature e.m.f. E_a to speed in rad s^{-1}. This relationship, which follows directly from eqn [6.6], may be confirmed by considering electrical and mechanical power balance with 100% conversion efficiency.

6.6.1 Effect on torque of brush shift away from quadrature axis

In developing expressions for the e.m.f. developed between brushes in Section 6.3 it was recognised that, if the brushes separated by a pole pitch were moved away from the position at which the commutated coils were located on the quadrature axis, the *distribution* of e.m.f. within the conductors remained unchanged but the resultant appearing between brushes would reduce. For the particular case of a sinusoidally distributed flux per pole Φ, the multiplying factor is $\cos\psi$, with ψ being the angle of brush shift.

A similar feature applies to the torque developed by a d.c. machine. For a given armature current and pole flux, maximum torque is developed when the brushes are set to commutate coils with sides on the quadrature axis, embracing maximum flux. If the brushes are moved away from this position, Fig. 6.11(b) is modified such that H_{tr} reverses polarity at locations displaced by ψ from the q-axis positions at which the radial flux density changes polarity. The effect of this is to introduce areas of rotor periphery

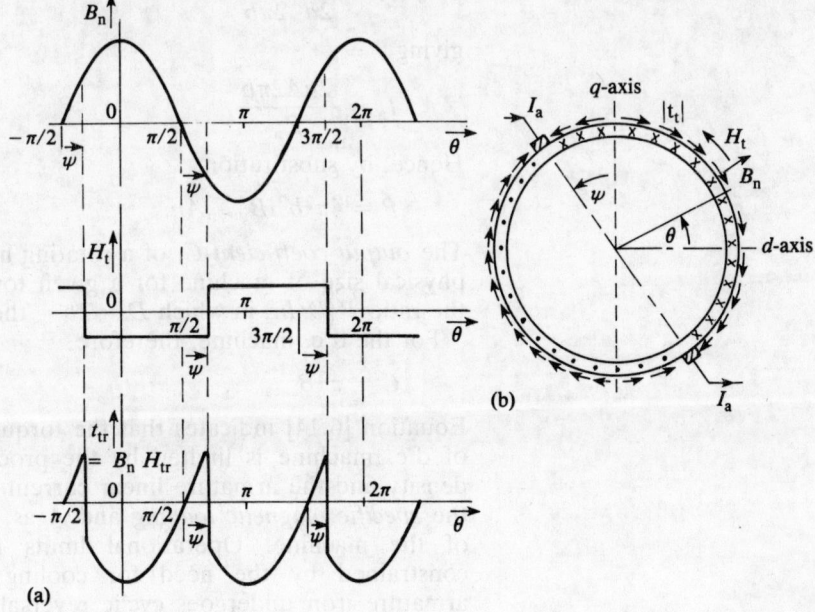

Figure 6.13 Effect of brush shift on tangential stress distribution over (rotor) armature surface: (a) assumed sine-distributed radial flux density but uniform armature current sheet (developed views); and (b) brush position displacement by ψ from the normal q-axis (sectional view).

at which the direction of tangential mechanical stress is reversed but with magnitude unchanged. It is apparent from Fig. 6.13 that integration of the tangential stresses over the rotor periphery gives rise to a net torque which is proportional to the net radial flux between brush positions.

6.7 Output Coefficient and Rating of a d.c. Machine

Direct-current machines are normally operated with brushes set to commutate coil sides along the q-axis in order that both torque and e.m.f. for given values of armature current and speed are maximal. This maximises the power rating for a given pole flux and speed. Further, since the *distribution* of the pole flux is arbitrary as regards output, optimal power for a given frame size will result if the flux distribution is made uniform at a value approaching saturation of the iron over as large a pole area as possible, bearing in mind the need for satisfactory commutation.

The principal dimensions of a d.c. machine armature with q-axis brushes and converting a maximum rated power P may be related to the mean radial gap flux density B_{mean}, and armature current linear density A, as follows.

$$P = E_a I_a = \left(\frac{p}{a}\right) Z_a n \Phi I_a$$

from eqn [6.6]. Now

$$\Phi = B_{mean} \frac{2\pi bl}{2p}$$

and

$$A = \frac{I_a}{2a} \frac{Z_a}{2\pi b}$$

giving

$$I_a = \frac{2aA2\pi b}{Z_a}.$$

Hence, by substitution

$$P = 4\pi^2 b^2 l B_{mean} A n.$$

The *output coefficient* C, of a rotating machine is a measure of the physical size of machine for a given torque capability, defined by the ratio $P/D^2 ln$, in which $D = 2b$ = the rotor diameter.

For the d.c. machine, therefore

$$C = \pi^2 B_{mean} A. \qquad [6.14]$$

Equation [6.14] indicates that the torque capability of a given size of d.c. machine is limited by the product of the mean gap flux density and the armature linear current density. B_{mean} is known as the *specific magnetic loading* and A as the *specific electric loading* of the machine. Operational limits for these parameters are constrained by the need for cooling. The radial flux in the armature iron undergoes cyclic reversals in direction as the rotor moves through each pole pitch, causing hysteresis and eddy current loss in the laminated rotor iron. No such flux reversals are normally required within the stator poles, which may be made of solid steel if the operating conditions do not call for sudden changes of the stator field flux. The *pole shoes* adjacent to the airgap on the stator side may, with advantage, be laminated because of the pulsating nature of flux in this region resulting from the use of slotted armature cores.

6.7.1 *Stator torque*

Calculation of the mechanical stresses which give rise to the reaction torque on the stator is fundamentally difficult on account of the complicated stator pole profile. The presence of compensating windings carrying armature circuit current introduces a tangential component of magnetic field at the stator pole surfaces, which is absent if the sole stator windings are concentrated coils surrounding the poles. Under these latter circumstances, stator torque results from asymmetry in the mechanical stresses developed over the pole faces and sides created by the effect of rotor armature current flow on the complex field pattern existing at the stator pole surfaces.

6.8 Commutation of the d.c. Machine

Commutation is the process by which the external circuit connection to the armature circuit of a heteropolar machine is transferred from one armature winding tapping point to the next, during which period one (or more) armature coil(s) is/are short circuited. Whilst commutation is taking place, the current in a short-circuited coil – initially equal in magnitude to the current flowing in one of the parallel paths through the armature – has to reverse its direction of flow. For good commutation, with electrical discharge unlikely at the trailing edge of the brush, such reversal is required to take place at a substantially constant rate within the commutation time allowed. This period is dependent upon the (equal) number of commutator segments and armature coils, the width of each segment and its insulating spacer, the width of the brushes and the speed of rotation of the rotor shaft upon which the commutator is mounted. Reversal of this current is opposed by the e.m.f. of self induction e_l of the armature coil, which is equal to the product of the rate of current change and the effective self inductance of the coil.

Commutation is a complex procedure. During the time a coil is short circuited by a brush, the coil current i_c flows in a closed circuit which includes the resistance of the coil. The brush itself conveys the circulating current i_c across the width of the carbon brush via the face in contact with adjacent copper commutator segments. The contact area of the *lagging* segment of a pair increases linearly with time from zero at the onset of commutation, whilst that of the *leading* segment decreases linearly at the same rate.

Figure 6.14(b) illustrates the variation with time of (i) the brush contact area for leading and lagging segments and (ii) the current circulating in the commutated coil for ideal or 'straight-line', commutation, which is characterised by a uniform current density ($A m^{-2}$) across the brush/commutator segment contact and a constant e.m.f. of self induction within the commutated coil. Figure 6.14(b)(iii) shows the steady transfer of the constant current carried by a pair of parallel paths within the armature from the leading segment to the lagging segment. Figure 6.14(a) shows the current configuration in the commutated coil and those adjacent coils to which it is connected in a *lap* arrangement. As the commutated coil moves from position AA' through BB' to CC' it is clear that the current therein must reverse – as illustrated by the initial and final values of i_c in Fig. 6.14(b)(ii).

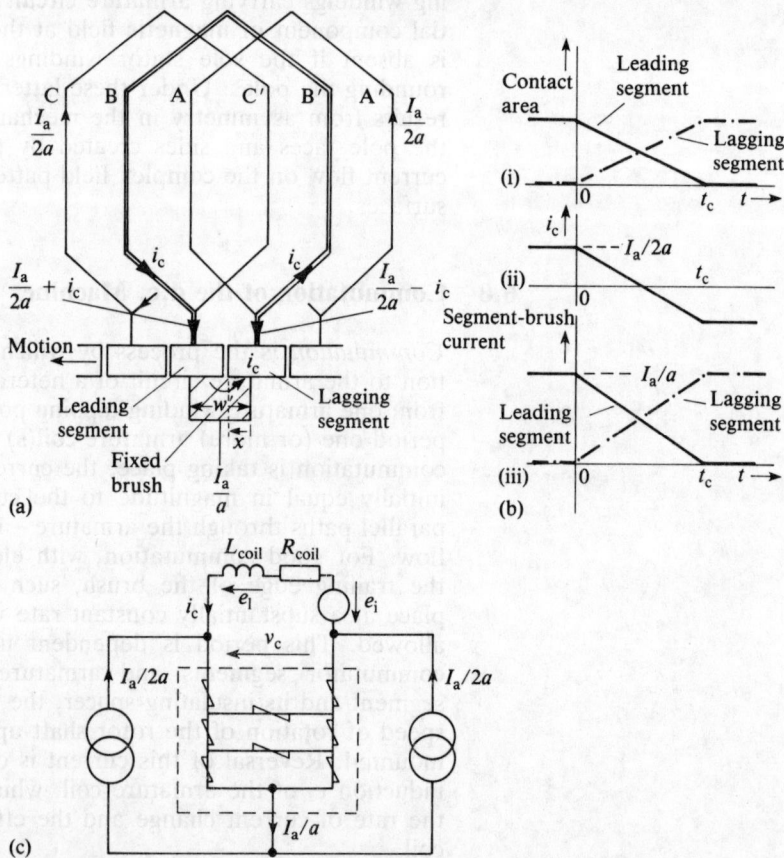

Figure 6.14 Short circuit of an armature coil during commutation: (a) current distribution in commutated and adjacent coils; (b) variation with time during commutation of (i) the brush contact area, (ii) the commutated coil current, and (iii) the leading and lagging segment-brush current; and (c) equivalent circuit model.

In reality, such ideal commutation is difficult to justify by simple theory. It reflects well to the credit of commutator brushgear manufacturers that mechanical commutation has proved to be highly effective and reliable over the years. Straight-line commutation implies a constant e.m.f. of self induction directed to oppose the change in current, whereas the voltage drop across the resistance of the commutated coil is roughly proportional to the instantaneous value of i_c reversing polarity in sympathy at the half-way stage. An implied uniform current density across each commutator segment/brush contact suggests a constant voltage drop in the brush region near the commutator contact – if the brush resistivity is uniform and constant. Around the closed loop formed by the coil and the brush the voltage drops due to flow of current normal to the contact surface should then cancel out, apart from the complicating influences corresponding to the transverse flow of i_c across the width of the brush via a complex route whose effective length and cross-sectional area may change with the value of i_c. This effect, together with the nonlinear nature of brush resistivity and the contact potential differences present at the copper segment/carbon brush boundary combine to enhance the prospects for good commutation, provided that the e.m.f. of self induction is not excessive. The latter increases with the number of winding turns incorporated in each armature coil and the value of the armature current.

An equivalent circuit model appropriate to an armature coil undergoing commutation is shown in Fig. 6.14(c). The carbon brush and its contacts with the commutator segments are represented as a network of nonlinear resistances.

In small d.c. machines the effective inductance/resistance ratio relationship in the commutation circuit is such that satisfactory commutation is achieved up to full load current without the need to employ additional techniques. Poor commutation is indicated by arcing at the trailing edge of the brush. The arc is drawn between the brush and the receding commutator segment whose current transfer has not been completed within the time allowed.

6.8.1 *Methods of improved commutation*

Commutation problems may be eased through the use of several alternative procedures. Commutation time may be increased by extending the brush width such that more coils are short circuited simultaneously than the minimum number. This is done with large machines having a considerable number of coils per pole pair, at the expense of increased loss in the commutation circuits and a possible slight reduction in the armature e.m.f. Rotational speed may be reduced – having its effect on machine rating. High-resistance brushes may be installed, again increasing losses.

More satisfactory methods involve accelerating and controlling current reversal in the short-circuited coil through the induction of e.m.f. as the coil sides move through space. Normally, these pass through the quadrature axis where the *radial* flux density due to the stator field poles is ideally zero, although that due to the armature winding is high, despite increased magnetic circuit reluctance along this axis – unless counteracted by the provision of a compensating winding located on the stator side of the airgap. What appears necessary to assist the reversal of current in the short-circuited coil is the induction of an e.m.f. with polarity similar to that about to be experienced by the coil as it emerges after commutation. This statement assumes that the armature conductor current flow is normally in the direction of e.m.f. induced by motion in the stator pole flux, i.e. the machine is a *generator*. This *commutating e.m.f.* is best achieved through the provision of additional electromagnets, secured to the stator core between the main field poles and having a magnetic circuit designed so that the commutating radial flux is confined to the quadrature axis where commutation is taking place, and provided with a concentrated winding energised with a definite fraction of the armature current. Such *commutating poles*, called *compoles* or *interpoles*, have their magnetic circuit completed by the rotor iron and an airgap, made large to ensure a linear relationship between pole flux density and interpole winding current. Thus the strength and polarity of the commutating poles will vary automatically as required with changes in load or function (whether motor or generator) – provided the direction of rotation and main field polarity remain unchanged.

The roles of compensating winding and interpole windings should not be confused. The *compensating* winding cancels the armature *radial* field along the quadrature axis and elsewhere, enabling the radial flux per main pole and its distribution along the direct axis to be unaffected by armature current changes. Otherwise, with severe duty cycles in which the load current may change radically and quickly, the consequent change in voltage distribution around the commutator periphery may cause flashover between segments. Interpoles assist commutation and have a local effect producing radial flux proportional to armature current confined to the quadrature axis. The radial field of the compensating winding is also along the quadrature axis, with distribution equal and opposite to that of the armature winding.

Large machines required for arduous duties including rapid changes in torque and speed demands, such as steel rolling-mill drives, are invariably provided with both compensating windings and interpoles. Such a machine is represented in Fig. 6.15 which clearly shows the magnetic axes about which the *radial* field components in the airgap due to armature, main field, interpole and compensating windings are directed. In small machines, commutation at higher levels of load current may be improved by moving the brushes away from the position at which they contact coil sides located in the q-axis towards a position at which they make contact with conductors on the fringe of the main pole field. Brush displacement for a generator is in the direction of rotor rotation whereas, for a motor, the shift from the quadrature axis should be in the direction *opposite* to that of rotor rotation.

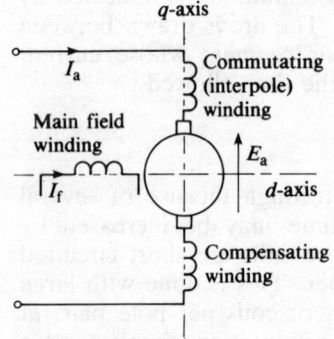

Figure 6.15 Schematic representation of a d.c. machine armature with main field, compensating and commutating (interpole) windings.

6.8.2 *Electronic commutation*

The mechanical commutator limits the electromechanical power conversion potential of the d.c. machine. Centrifugal-force considerations and the need to clamp securely limit the diameter, rotational speed and length of the commutator. The desire to limit the thickness of insulation between adjacent commutator segments restricts the allowable e.m.f. per armature coil. This limits the number of turns per coil to one in large machines and, ultimately, constrains also the armature voltage rating. Unlike polyphase a.c. machines which may operate with line voltages commonly 11 000 V and more, larger d.c. machines are typically 2000 V maximum.

Many of the limitations are mitigated if electronic commutation of the armature coils is applied. One immediate advantage is that the armature winding may then be accommodated on the *stator*, with consequent easing of problems relating to insulating and securing the conductors. The apparent need for slip-rings and brushgear associated with supplying d.c. exciting current to rotating main-field coils may be avoided by mounting the armature of an exciting generator or *exciter* on the common shaft. On small machines permanent-magnet field poles may be employed. If commutation is achievable without the need for interpoles, a very simple rotor assembly is possible.

Electronic commutation eliminates the chance of an arcing brush–commutator contact which may proscribe the use of d.c. machines in hazardous environments. Assuming that semiconductor devices are able to meet the switching duty requirements, constraints on rotor speed and overall dimensions are eased.

The principal objections to electronic commutation relate to the cost and complexity of the control equipment. Semiconductor devices, with the exception of the simple diode, require control circuitry which increases in extent with the number of switching devices required. A d.c. machine should ideally employ a large number of armature coils in order that the effects of coil short circuit during commutation are relatively insignificant. Each commutator segment–brush contact is then required to carry current for only a small proportion of total time – the ratio of peak current to average current is correspondingly very high, implying poor utilisation of a semiconductor switch. Further, when not conducting, such a switch would be required to withstand a variable reverse voltage rising to a peak value equal to the total armature voltage. This contrasts with the mechanical commutator in which the maximum voltage between adjacent segments not involved in commutation is limited to the maximum voltage *per coil*.

Brushless commutation using thyristors is illustrated in Fig. 6.16. Each stator armature coil is connected directly to adjacent coils and also to d.c. positive and negative supply rails. The thyristor permits unidirectional current flow from anode to cathode, providing the anode potential is positive with respect to the cathode, from an instant when the injection of charge carriers into the gate–cathode region enables the device to change state. Having initiated conduction, the gate has no further control over the anode–cathode current which must be brought to zero by a change in circumstances external to the anode and cathode terminals. For the thyristor to recover its current blocking faculty in the forward (anode-to-cathode) direction a small reverse voltage must be maintained between anode and cathode for a brief period ($\sim \mu$s) immediately after switch-off.

In thyristor commutation the devices are turned on in pairs according to the relative position of the appropriate armature conductors and the main field system. A rotor-position sensing device is therefore needed to ensure that the gate-firing circuits of the thyristors are activated at the correct instant in each sequence. Assuming that a previously conducting thyristor is successfully turned off by each newly turned on device the field components in the airgap due to the armature current will move in a series of

Figure 6.16 Brushless commutation of conventional d.c. armature winding. Thyristor turn-off facilitated by interpoles.

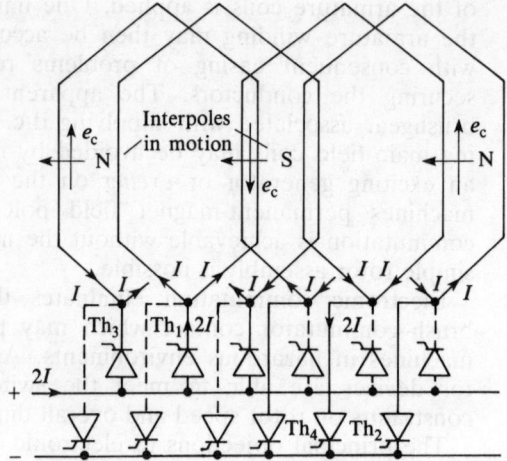

steps, appearing to rotate in synchronism with the field-pole system mounted on the rotor. The firing pulses are normally phased such that the radial field of the armature winding is in space quadrature with that of the rotor field winding – this is the maximum-torque condition for given armature and field currents.

Interpoles are usually necessary to effect the turn-off of the thyristors. Thus, in Fig. 6.16 the interpole positions are shown for Th_1 and Th_2 conducting and Th_3 and Th_4 about to be fired. When this occurs the commutating e.m.f.s e_c induce a circulating current component in each of the two closed loops, forcing the anode–cathode current of Th_1 and Th_2 towards zero and then maintaining a reverse voltage between anode and cathode. Armature current $2I$ is thus transferred from Th_1 to Th_3 and from Th_2 to Th_4. The process is repeated in sequence.

Figure 6.16 depicts an application in which the armature current flow is directed away from the positive terminal of the supply, corresponding to *motoring* action. If the polarity of the supply voltage is reversible, regenerative braking of a motor load is possible without the need to duplicate the thyristors using reversed anode–cathode connections – provided that a relatively slow response time is acceptable.

Commutation without the need for interpoles using the fringing flux of the main field poles is possible if the phasing of the gate-firing pulses is advanced (for a generator) or retarded (for a motor), this effect is analogous to brush shift in the mechanically commutated machine. Commutation problems may be further eased by the use of *gate turn-off thyristors* (GTOs) which have the capability of interrupting the flow of anode–cathode current through the application of a gate–cathode current pulse of polarity opposite to that required to turn the device on. The stored magnetic field energy due to the self inductance of the commutated coil at switch-off is transferred to the capacitive element of a *snubber* circuit connected between the anode and the cathode of the GTO thyristor for the purpose of controlling the rate of rise of the anode–cathode voltage during turn-off. If the latter is allowed to increase too quickly the device reverts to the low-resistance state and effective turn-off is not achieved.

Some small machine requirements are met using very few stator coils and *field-effect transistor* (FET) switches. Rotor position may be sensed by stationary photo-transistors or Hall-effect devices responding to rotor pole fluxes. A single pair of stator coils identified with each of two axes mutually at right-angles in conjunction with a permanent magnet rotor gives rise to a very compact machine. Brushless permanent-magnet machines are discussed in Section 6.21.

6.9 Methods of Stator Pole Energisation

Our main concern to date has been with the armature circuit of the heteropolar d.c. machine, without regard as to the means by which the radial stator flux/pole Φ is provided. Unless permanent magnets are to be installed the electromagnetic alternatives require sources of m.m.f. With magnetic saturation ignored the relationship between pole flux and winding ampere-turns should be a linear one, involving the dimensions and permeability of the airgap and iron which constitute the effective magnetic circuit. As noted in Section 6.5.2, an effect of armature reaction when the armature carries current is to alter the distribution of the flux/pole, and also its value if saturation effects are incorporated.

The induction of steady-state e.m.f. within the rotating armature has no counterpart in the stator field winding because no relative motion exists between the field-winding conductors and the radial airgap field – assuming that the armature brushes are stationary. Since no electric field is electromagnetically induced in the stator field winding in the steady state, it follows from consideration of the Poynting vector (Sect. 2.6.1) that the field winding cannot contribute to the fundamental electromechanical energy-conversion process of a d.c. machine.

Thus the voltage developed across a stator field winding in the steady state is simply the product of current and winding resistance. If, however, the field current is changing, an e.m.f. of self induction will appear across the field coils and an e.m.f. proportional to field current will be developed between fixed brushes set to commutate coil sides located on the quadrature axis as the armature rotates at constant speed. If the armature current is changing, e.m.f.s of self and mutual induction appear across those machine windings which develop their radial fields along the q-axis, i.e. the armature, compensating and interpole windings. The relative winding senses of these three series-connected windings are, in practice, such that the effective self inductance of the composite armature circuit is reduced. Changes in armature current induce no e.m.f. in any field winding, however (and *vice versa*) since the magnetic axis of armature and field windings normally lie at right-angles to each other and consequently experience no magnetic coupling.

The armature circuit (incorporating compensating and interpole windings) and the field circuit are therefore inherently conductively and magnetically uncoupled – for a machine with q-axis brushes. The armature e.m.f. is related through speed to the field flux, but the effect is non-reciprocal and, ideally, the field flux is unaffected by armature current or voltage – unless constraints are placed on the field circuit by connecting it in some way with the armature circuit. Obvious possibilities are *shunt* and *series* interconnections, with *compounding* offering a combination of the two. These, together with independent, *separate* excitation provide the alternative methods of field-winding energisation.

Thus Fig. 6.17(a) shows the circuit arrangement of a *shunt-wound* d.c. machine. The figure also provides an equivalent circuit model which is adequate for many applications. The parallel connection requires that the voltages across the field and armature circuits must be equal to each other and the d.c. supply voltage. Field current and, consequently, radial flux/pole are adjustable by means of the field-circuit rheostat. Voltages and armature e.m.f. throughout the circuit are instantaneously related as follows:

$$R_f i_f + L_f \frac{di_f}{dt} = E_a + i_a r_a + L_a \frac{di_a}{dt} = V_a$$

and in the steady state

$$R_f I_f = E_a + I_a r_a = V_a.$$

Figure 6.17(b) has the field winding connected in *series* with the armature, their currents in consequence necessarily being equal. The field current may be made an adjustable proportion of the armature current by the inclusion of a *diverter resistance* (not shown) in parallel with the field coils.

Figure 6.17 Alternative field/armature circuit interconnections: (a) shunt, (b) series, and (c) compound (long shunt).

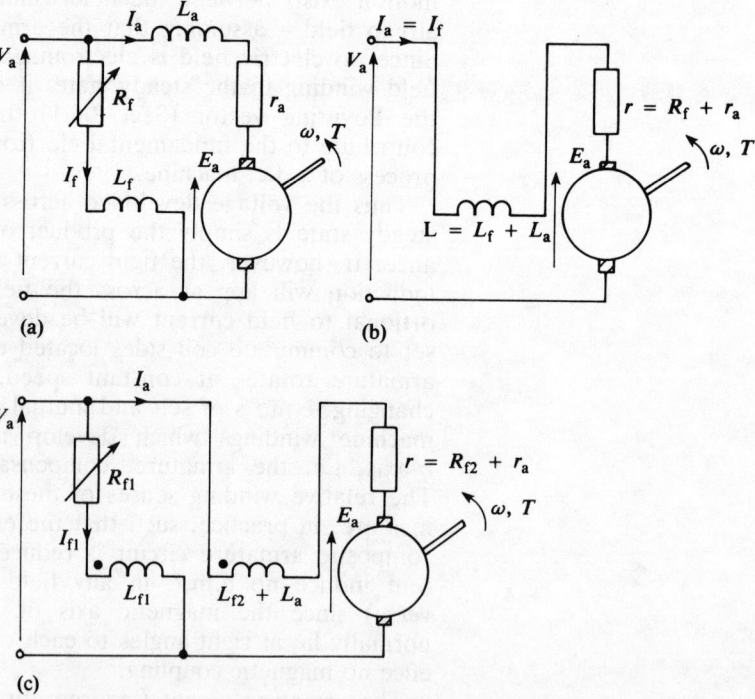

Figure 6.17(c) has one field winding in series with the armature and another in parallel with the series field/armature combination. The alternative 'short shunt' *compound* connection has the shunt field winding connected directly across the armature circuit alone,

excluding the series field winding. Whether the m.m.f.s of the two sets of field coils effectively add or mutually oppose depends on the relative polarities of the windings and the directions of the exciting currents. With I_a and I_f both positive in Fig. 6.17(c), both fields act *cumulatively*. If the machine changes function from motoring to generating with the terminal voltage held constant at V_a, an increase in armature e.m.f. E_a enabling I_a to become negative, the net field will weaken as the armature current increases in magnitude. The excitation condition of the resulting generator is said to be *differentially* compounded.

Shunt-field coils are generally designed to accommodate the rated armature voltage of the machine, developing the required magnetising ampere-turns via a large number of turns of copper wire having small cross-section and correspondingly high resistance, which limits the winding current to typically 3% of the rated armature current. Control over the shunt-field current for a constant terminal voltage may be achieved by means of a rheostat or *field regulator* in series with the field winding. The power dissipated in the field-circuit resistance contributes to the total losses of the energy conversion process.

Series field coils require, conversely, only a few turns of thick wire in order to develop the required m.m.f. A large cross-section is used to provide a low value of coil resistance which, otherwise, would be responsible for an unacceptable copper loss when carrying armature current.

6.9.1 *Magnetisation characteristics; the open-circuit curve (o.c.c)*

An important characteristic of any d.c. machine is the relationship between armature e.m.f. and field current. This includes the essentially nonlinear radial flux/m.m.f. relationship of the magnetic circuit. It is normally expressed as the *open-circuit characteristic* (o.c.c.) which presents graphically the armature terminal voltage on open-circuit – the armature e.m.f. – as a function of field-winding current at a particular rotor speed. If the test is carried out for a particular machine with the field current increasing from zero into the region of saturation and then is reduced to zero, the effects of magnetic hysteresis are demonstrated by the non-coincidence of the rising and falling portions of the curve, with residual magnetism remaining. If the test is repeated at a different speed, the basic shape of the o.c.c. is retained, the scaling of the armature e.m.f. being proportional to speed. Figure 6.18 shows typical open-circuit characteristics for a d.c. machine at two values of speed. The radial flux/pole available under load conditions is reduced below the open-circuit value due to the armature reaction m.m.f. This fact should be borne in mind when predicting machine behaviour.

6.10 Performance of d.c. Machines

Rotating d.c. machines are inherently reversible energy converters. A d.c. generator absorbs mechanical power at its shaft by developing electromagnetic torque opposing rotation, whilst current flows from a positive armature terminal to a negative terminal via the

Figure 6.18 Open-circuit characteristic of a d.c. machine at (a) 1000 rev/min, (b) 750 rev/min.

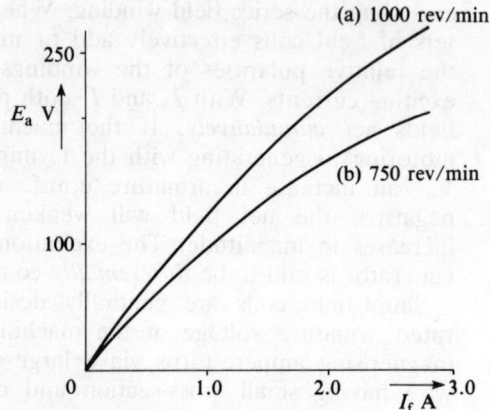

external circuit. The terminal voltage differs from the armature e.m.f. by the resistive and inductive voltage drops developed across the armature circuit which includes the interconnected armature winding, the commutator and brushgear, compensating and interpole windings, if fitted, and any series field winding which contributes to main pole flux.

The function of a d.c. generator is to provide a source of d.c. electrical power for absorption within a load circuit. The nature of the load circuit will influence the way in which the generator responds to changes in the load configuration, as also will changes in the applied torque and speed of the prime mover. A d.c. generator, when supplying a resistive load, will respond differently to a change in its field current, for example, than would be the case if its armature circuit were connected to a constant-voltage busbar system. The characteristics of the prime-mover source are also relevant. In predicting or declaring the load characteristics of a generator, for example, the mechanical power input source is often regarded, for simplicity, as having one of its torque and speed parameters held constant.

Similar simplifying assumptions are often made with regard to motors. The functional requirement of a motor in many common applications is to overcome a steady load torque at a constant speed. It may be a further requirement that the speed is adjustable – or that the speed remains constant over a range of load torques. Dynamic response will also be important. If rapid acceleration of a high-inertia rotating system is required, the motor may be required to develop a torque at low speeds well in excess of the eventual steady load torque. Braking of the connected load may be a required feature of the duty cycle imposed upon the machine. Again, the characteristics of the source require consideration – whether the supply voltage falls as the load current increases due to the regulation of a nominally constant voltage source or, indeed, whether a nominally constant current source might be more appropriate.

6.10.1 *Direct-current generator characteristics*

Although the d.c. machine is essentially a reversible energy converter, and general equations may be set up to relate the electrical and mechanical variables at armature terminals and shaft, via its internal parameters, the nature of the problems requiring solution is such that a generalised analysis is seldom the most appropriate. Most d.c. machines are built for a basic *motoring* duty with the prospect of regeneration into a d.c. mains supply, or dissipating energy within resistance during braking, a useful feature of the drive system. Purpose-built d.c. generating equipment is a relatively uncommon requirement, due to the ready availability of an alternative polyphase supply with an a.c./d.c. static converter.

The simplest d.c. generator is self excited, having an independent field circuit as shown in Fig. 6.19, without the dotted connection. When driven at constant speed the steady-state output terminal voltage equals the armature e.m.f. minus the armature resistance voltage drop. Armature reaction effects and prime-mover speed regulation on load may be compensated for by an increase in field current.

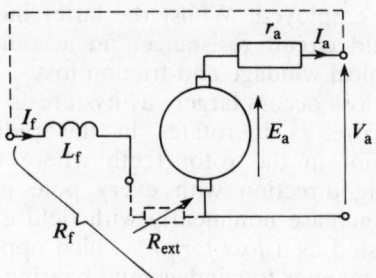

Figure 6.19 Equivalent circuit of a d.c. machine with current and voltage polarities appropriate to generator operation; dotted connections refer to shunt self excitation.

The need for a separate d.c. supply to provide the field current may be avoided by connecting the field winding (having appropriately high values of resistance and number of turns) across the armature terminals. Provided that the field poles retain a modicum of residual flux which has polarity appropriate to the direction of rotation, the machine will build up its magnetisation on no-load to a value of armature e.m.f. defined by the intersection of the open-circuit characteristic E_a/I_f, with the straight-line locus V_f/I_f defined by the field circuit resistance R_f. The open-circuit voltage to which the machine will self excite is critically dependent upon speed, field circuit resistance and the shape of the o.c.c. – which must incorporate a degree of saturation to limit the buildup of stator pole flux. The field resistance (or speed) is said to be 'critical' when a small change in value results in a large change in open-circuit voltage. On load, the inherent reduction in terminal voltage due to the armature resistance and armature reaction is enhanced by a consequential reduction in field excitation. The machine offers a poorly regulated d.c. voltage source unless load compensation is applied by providing a series-connected field winding to supplement the main shunt field. The machine is then described as being cumulatively compound-wound.

6.10.2 *Direct-current motor characteristics*

The considerations noted above (Sect. 6.10) relating source and load characteristics to the performance of a d.c. motor are strictly more relevant to a wider view of the motor as a component within a *drive system* incorporating control facilities. Such will be the subject of further consideration in Section 6.12. For the present we shall be concerned with the basic performance equations of a d.c. machine in a motoring context, relating the external electrical and mechanical parameters of power, i.e. terminal voltage and current, rotor angular velocity and load torque, via such internal parameters as radial flux/pole, armature conductor configuration, etc., by means of the electromagnetic relationships for torque and armature e.m.f. Thus

$$T_{e-m} = -\left(\frac{p}{a}\right)\frac{Z_a}{2\pi} I_a \Phi \qquad [6.11]$$

$$E_a = \left(\frac{p}{a}\right) Z_a n \Phi. \qquad [6.6]$$

It is appropriate to note that a practical machine has components of power loss which are not always explicit in the equivalent circuit usually employed. Whilst the latter incorporates directly armature and field circuit resistance, no account is taken of iron loss, or mechanical windage and friction loss.

Iron loss occurs largely as hysteresis and eddy-current loss in the rotor core as it rotates in the radial flux, being particularly significant in the rotor teeth where the flux density is highest, changing direction with every pole pitch of displacement. Such losses increase nonlinearly with field excitation and speed and are manifested as a *loss torque*, which opposes the rotor rotation, in a similar manner to windage and bearing friction torques.

6.10.2.1 *Torque balance*

The reference direction for rotor torque $T_r = T$ in Fig. 6.17(a) corresponds with that of rotational speed ω. In the figure, T defines a reference direction for rotor torque such that, if T_{e-m} is positive, with ω also positive, the machine might be expected to supply mechanical power via its shaft. T_{e-m} is only one component of rotor torque, however. The total net torque acting on a shaft is always zero, with all components of the resultant torque interpreted in accordance with a common reference direction.

Thus an armature accelerating a load in the direction of rotor rotation, whilst subjected to loss torque due to friction, windage and iron loss, has a torque balance as follows

$$\sum T = T_{e-m} - |T_L| - |T_{loss}| - |T_{accel}| = 0, \qquad [6.15]$$

where

$$T_{accel} = \frac{J d\omega}{dt},$$

with J being the moment of inertia of the rotating system. The use of the modulus sign with a negative term emphasises that the action is in the sense opposing the torque reference.

The same machine operating as an electromagnetic brake to slow the coupled mechanical load will have speed ω positive, as before, and rotor torque components balancing thus

$$\sum T = -|T_{e-m}| - |T_L| - |T_{loss}| + T_{accel} = 0.$$

6.10.2.2 Power balance

In seeking to apply eqns [6.6] and [6.11] for armature e.m.f. and torque it is appropriate to query the relevance of the minus sign appearing in eqn [6.11]. This arose on account of the particular reference directions adopted for radial flux and armature current within the airgap. (See Fig. 6.10 and Sect. 6.5.1.) The current flow within each armature conductor was taken to be in the same direction as the rotationally induced e.m.f., predisposing the function of the machine as a generator. The machine may alternatively be observed from a point of view which excludes detail concerning the airgap, recognising that energy conversion takes place therein at 100% efficiency and, in the steady state, with no change in stored energy. It is then in order to specify only magnitudes for the e.m.f. and torque expressions, interpreting their necessary polarities in the light of the above property.

Thus the reference directions for armature e.m.f. E_a and armature current I_a in Fig. 6.17 correspond to the machine armature absorbing electrical power if both quantities are positive, i.e. directed as shown. The armature is appropriately represented as an ideal two-port device with electrical input power $E_a I_a$ and mechanical output power ωT_{e-m} always in balance, such that $E_a I_a = \omega T_{e-m}$. From eqns [6.11] and [6.6] we may write

$$T_{e-m} = \left(\frac{p}{a}\right) \frac{Z_a}{2\pi} I_a \Phi = K\Phi I_a \qquad [6.13]$$

$$E_a = \left(\frac{p}{a}\right) Z_a n \Phi = K\Phi \omega, \qquad [6.16]$$

with K, a machine constant

$$= \left(\frac{p}{a}\right) \frac{Z_a}{2\pi}.$$

In determining the general performance characteristics of a range of d.c. machine configurations the simplified eqns [6.13] and [6.16] are convenient to use. In many derived expressions the radial flux/pole Φ may not appear explicitly, but caution should be exercised in relating the results to practical machines on account of magnetic saturation and the armature reaction effect of an uncompensated load current which further reduces the available flux/pole.

Figure 6.20 accounts for the flow of power within a d.c. motor; the distribution of the losses is typical for a small industrial machine.

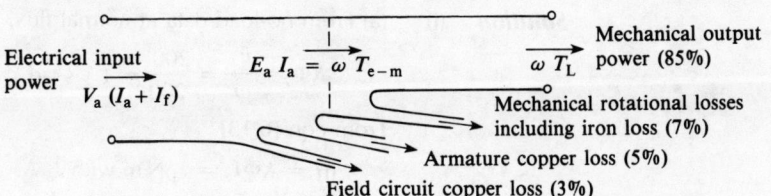

Figure 6.20 Power balance in a small d.c. machine when motoring.

6.10.3 External characteristics of a d.c. motor with separate excitation

The external characteristics of a motor relate the rotation speed and the torque available at the shaft to drive a mechanically

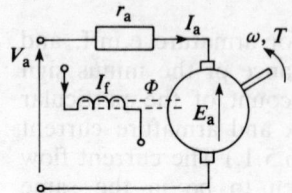

Figure 6.21 Equivalent circuit of a d.c. machine. Reference directions correspond with all variables positive for motor operation.

coupled load. They are usually predicted or determined by test under steady-state conditions for the mechanical and electrical parameters involved. Equations [6.13] and [6.16] are applicable, together with electrical circuit constraints which may define the armature e.m.f. in terms of the armature supply voltage or may define the field flux. For the separately excited machine illustrated in Fig. 6.21 having armature resistance r_a the defining equations are as follows

$$T_{e-m} = K\Phi I_a \qquad [6.13]$$

$$E_a = K\Phi\omega \qquad [6.16]$$

$$V_a - E_a = I_a r_a. \qquad [6.17]$$

The intermediate relationships between (i) torque and armature current and (ii) speed and armature current are of interest; the latter follows from eqns [6.16] and [6.17] to give

$$\omega = \frac{1}{K\Phi}(V_a - I_a r_a). \qquad [6.18]$$

The graph of speed against torque may be deduced from the torque/I_a, speed/I_a relationships by identifying torque and speed for identical values of I_a. Alternatively, the theoretical torque–speed relationship may be deduced by eliminating I_a between eqns [6.13] and [6.18], giving

$$T_{e-m} = \frac{K\Phi}{r_a}(V_a - K\Phi\omega). \qquad [6.19]$$

A pair of examples serves to illustrate the use of the performance equations of a d.c. machine to predict load characteristics.

Worked example 6.1 (i) A 20 kW, separately excited d.c. motor is normally supplied with an armature voltage of 200 V and excitation such that its no-load speed is 20 rad s^{-1}. Its armature resistance is 0.1 Ω. Neglecting losses (other than armature copper loss) and armature reaction effects, plot the steady-state characteristics of torque/armature current, speed/armature current and torque/speed for a constant terminal voltage of 200 V and armature current ranging from zero to 200 A. Values of excitation correspond to (a) normal flux, (b) twice normal flux, (c) one-half normal flux.

(ii) Repeat for flux held constant at its normal value and alternative armature voltages of (a) 300 V, (b) 200 V, (c) 100 V.

Solution (i) (a) From no-load data at normal flux, eqn [6.16] gives

$$K\Phi = \frac{E_a}{\omega} = \frac{200}{200} = 1 \text{ V s rad}^{-1}.$$

From eqn [6.13]

$$T = K\Phi I_a = I_a \text{ N m with } I_a \text{ A}.$$

From eqn [6.18]

$$\omega = \frac{V_a - I_a r_a}{K\Phi} = 200 - 0.1 \, I_a \text{ rad s}^{-1} \text{ with } I_a \text{ A}.$$

(b) At twice normal flux, $K\Phi = 2$, $T = 2I_a$, $\omega = \dfrac{200 - 0.1I_a}{2}$

(c) At half normal flux, $K\Phi = 0.5$, $T = 0.5I_a$, $\omega = \dfrac{200 - 0.1I_a}{0.5}$.

Values for T and ω are listed in Table 6.1 for I_a ranging from 0 to 200 A in increments of 50 A. The results are presented graphically in Fig. 6.22(i), extrapolated into the negative armature current region of generator operation. With terminal voltage and field flux constant, an increase in load torque occasions a small reduction in speed. For a given torque a reduction in flux increases the speed.

Table 6.1 Table for Worked Example 6.1

$V_a = 200$ V	Normal flux $K\Phi = 1$		Twice normal $K\Phi = 2$		Half normal $K\Phi = 0.5$	
I_a A	T Nm	ω rad s^{-1}	T Nm	ω rad s^{-1}	T Nm	ω rad s^{-1}
0	0	200	0	100	0	400
50	50	195	100	97.5	25	390
100	100	190	200	95	50	380
150	150	185	300	92.5	75	370
200	200	180	400	90	100	360

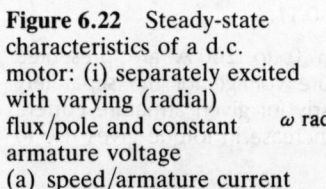
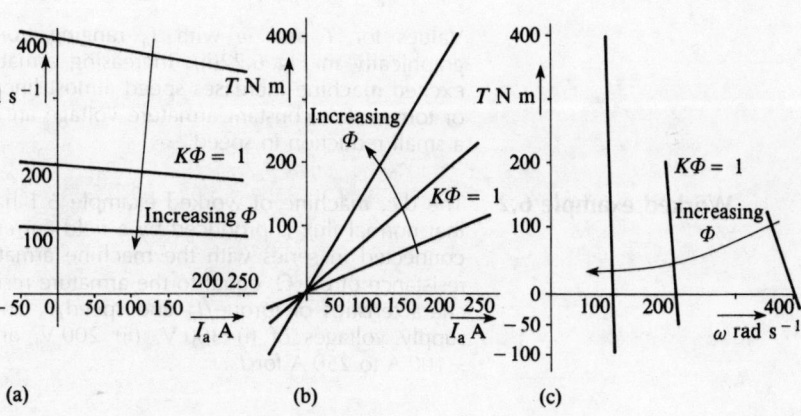
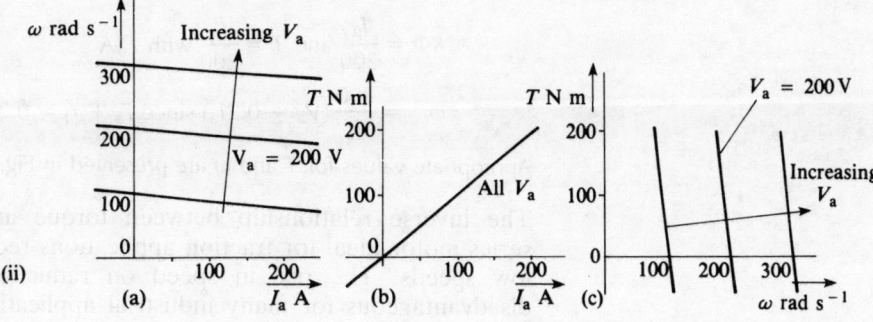

Figure 6.22 Steady-state characteristics of a d.c. motor: (i) separately excited with varying (radial) flux/pole and constant armature voltage (a) speed/armature current (b) torque/armature current (c) torque/speed current (ii) separately excited with varying armature voltage and constant flux;

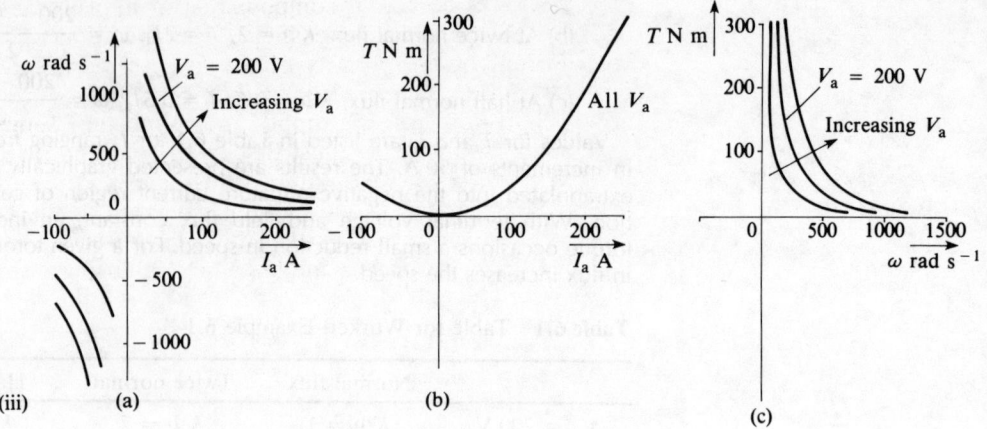

Figure 6.22 (*cont.*)
(iii) series-connected with varying armature voltage.

(ii) At normal flux with variable armature voltage

$K\Phi = 1$, $T = I_a$, $\omega = V_a - 0.1 I_a$

(a) $V_a = 300$ V, $\omega = 300 - 0.1 I_a$ rad s^{-1} with I_a A
(b) $V_a = 200$ V, $\omega = 200 - 0.1 I_a$
(c) $V_a = 100$ V, $\omega = 100 - 0.1 I_a$.

Values for T and ω with I_a ranging from 0 to 200 A are presented graphically in Fig. 6.22(ii). Increasing armature voltage for the separately excited machine increases speed almost linearly for given armature current or torque. For constant armature voltage an increase in torque gives rise to a small reduction in speed.

Worked example 6.2 The d.c. machine of worked example 6.1 has its field coils rewound such that normal flux is produced by a field current of 200 A. The field coils are connected in series with the machine armature winding and have a total resistance of 0.1 Ω, equal to the armature resistance. Determine steady-state characteristics of torque/I_a and speed/I_a and torque/speed for alternative supply voltages of (i) 300 V, (ii) 200 V, and (iii) 100 V over the range -100 A to 250 A for I_a.

Solution $K\Phi = 1$ at $I_a = 200$ A and $\Phi \propto I_a$

$$\therefore K\Phi = \frac{I_a}{200} \text{ and } T = \frac{I_a^2}{200} \text{ with } I_a \text{ A}$$

$$\omega = \frac{200}{I_a}(V_a - 0.2 I_a) \text{ since } r_a + r_f = 0.2 \text{ Ω}.$$

Appropriate values for T and ω are presented in Fig. 6.22(iii).

The inverse relationship between torque and speed makes the series motor ideal for traction applications requiring high torque at low speeds. The rise in speed on reduction in load torque is disadvantageous for many industrial applications, however, unless modified by a degree of compounding using a weak shunt field in addition to the main series field. The direction of the torque is independent of the polarity of the armature current. In order to *brake* a rotating series motor load by permitting generated electrical power to flow into the supply, the connection between armature and series field winding must first be broken and then reconnected in the opposite sense, a feature which is unnecessary with the d.c. shunt or separately excited motor. Armature voltage control is less popular with series motors than field circuit series/parallel switching or diverter resistance methods of influencing the speed/torque characteristic.

6.11 Direct-current Motor Dynamics – Speed Control, Starting and Braking

The load characteristics for separately excited and series d.c. machines described in the previous section indicate how the steady-state speed responds to changes in load torque, armature voltage and field flux. A separately excited motor – or a shunt motor – supplied with a constant armature voltage develops a speed which decreases slightly at a fairly uniform rate as the load torque is increased. Reversal of the load torque whilst maintaining the same direction of rotation but at a speed higher than that corresponding to zero torque (the no-load speed), corresponds to a change of state for the machine from motoring to generating. Such a condition may alternatively be brought about by an increase in the field excitation which raises the e.m.f. of the machine above the supply voltage value, reversing the armature current, torque and electrical power flow in the armature circuit. A drive system with this facility is described as a *two-quadrant* drive, since the torque/speed characteristic is continuous between the first and second quadrants of the speed/torque plane. As illustrated in Fig. 6.23, the development of electromagnetic torque on the rotor in the same direction as that of rotation invariably corresponds to a *motoring* function for the machine.

Operation in the third and fourth quadrants of the speed/torque plane is achievable in a separately excited machine by reversing the base direction of rotation. Two methods are apparently available: (i) changing the polarity of the field flux by reversing field winding connections, and (ii) reducing to zero and then reversing the polarity of the armature voltage supply.

Alternative (i) is often not practical since the field would be reduced to zero during the polarity change, so that the transition from first and second to fourth and third quadrants is made via infinite speed rather than zero speed, as appropriate to armature voltage control. A full four-quadrant drive has the capability of changing state from motoring to generating, or *vice versa*, via torque reversals through zero torque, or speed reversals through zero speed. Such a facility is demonstrated by the traditional *Ward–Leonard* drive system shown in Fig. 6.24(a) to employ three

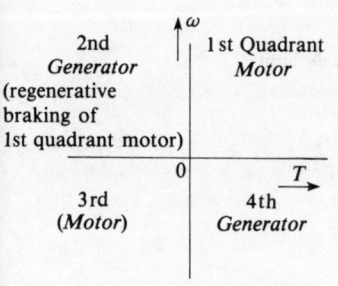

Figure 6.23 The four quadrants of speed/torque drive characteristics.

rotating machines, including two complementary, separately excited d.c. machines with electrically coupled armature circuits. Direct-current machine 1 is mechanically coupled to a nominally constant-speed primary drive and has its e.m.f. controlled by the separately excited field current. Direct-current machine 2 is mechanically coupled to the mechanical load. Absolute and relative adjustments of the field currents of both machines control the speed/torque characteristics of machine 2 presented in Fig. 6.24(c). Whilst offering fine control facilities the Ward–Leonard drive has the disadvantages of requiring three bulky, fully rated machines with low overall efficiency, and response time limited by the time-constants of the field circuits.

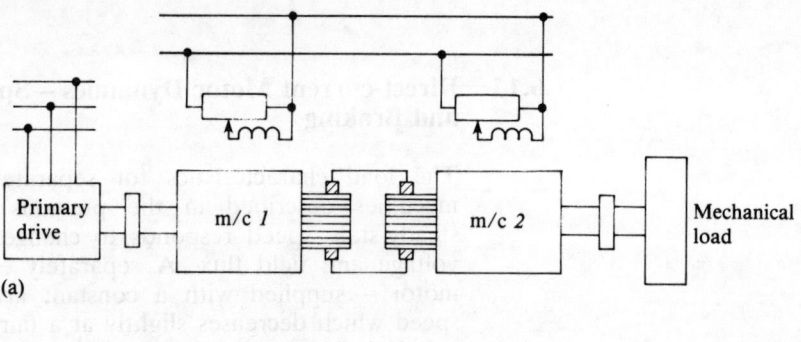

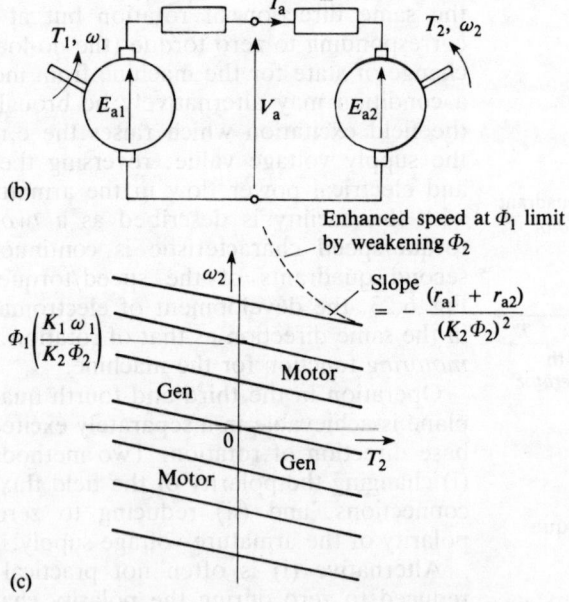

Figure 6.24 Ward–Leonard drive system: (a) general layout, (b) equivalent circuit, and (c) speed/torque characteristics with adjustable field current.

6.11.1 *Direct-current machine dynamic behaviour*

To date we have considered only the steady-state characteristics of d.c. machines, effectively allowing infinite time over which changes in the operating conditions occur. In practice, changes are initiated by some alteration to one or more operating parameters and the machine subsequently undergoes a transient period of adjustment, during which other unconstrained variables alter prior to settling down to a new, hopefully stable, condition of operation.

One important operational change for a motor is the application of a sudden *change* in load torque. If the machine is shunt connected or separately excited at constant armature voltage and field flux, the steady-state characteristics imply that an increase in load torque gives rise to a reduction in speed. This is ultimately the case but, during the transient period, the additional load torque demand is met initially by a sudden retardation of the rotating masses which contribute to the *moment of inertia J* of the shaft. The product of the shaft retardation $(-)d\omega/dt$ and the rotor moment of inertia equals the step increment in torque demand. As the speed falls so does the armature e.m.f., thus increasing the armature current and consequently the electromagnetic torque. If inductance in the armature circuit is neglected the speed change is readily shown to occur exponentially with time-constant $Jr_a(K\Phi)^{-2}$. If inductance in the armature circuit is significant the buildup of armature current lags behind the reduction in speed, delaying the development of balancing electromagnetic torque and causing the speed to fall below the eventual equilibrium value. As the armature current increases further the electromagnetic torque exceeds the new value of load torque and the rotor accelerates once more. Subsequently, the speed exhibits damped oscillations about the steady-state value appropriate to the increased load, the *damping factor* relating to the armature circuit time-constant.

In dealing with d.c. machine dynamics we shall, for simplicity, restrict our considerations to single time-constant electromechanical systems which give rise to first-order differential equations. Many practical drive installations would require more precise modelling and, consequently, more complex mathematical analysis. The general responses of basic systems may be deduced quite simply by ignoring second-order effects. As a general rule, therefore, when considering *mechanical* system time-constants, electrical transients will be disregarded by ignoring machine-field and armature-winding inductance. When investigating *electrical* circuit responses, mechanical system transients will be ignored by considering rotor speeds to be constant over the short term.

6.11.2 *Starting of d.c. motors*

At standstill the d.c. machine develops no armature e.m.f. The armature current and its rate of rise on connection to a constant-voltage source is therefore limited only by the armature circuit resistance and inductance. Excessive armature current during run-up of a d.c. motor is avoided by reducing the voltage actually applied between the armature brushes. The simplest way of achieving this is to incorporate an additional 'starting' resistance in series with the armature circuit, gradually reducing its ohmic value as the motor accelerates. Suitable grading and switching of the external resistance enables armature current, and hence electromagnetic torque, to be held substantially constant at a maximum safe value during the acceleration period. The method is clearly inefficient but is justifiable where frequent starts are not required. If the starting resistors are appropriately rated they may be used subsequently to provide a measure of inefficient and poorly regulated speed control when the machine is on load.

6.11.3 Acceleration and braking of a d.c. motor from standstill with inertia load

It is instructive to consider the particular case of a separately excited or shunt-connected d.c. motor accelerating, from rest, a pure inertia load when energised from a constant-voltage d.c. supply. Armature inductance, frictional and load torques are ignored and the armature circuit resistance is assumed constant throughout. The circuit arrangement is shown in Fig. 6.25(a), the object of the analysis being to establish the rotor speed variation with time and the total energy dissipated in the armature circuit resistance during run-up to steady-state speed ω_∞.

From eqns [6.16], [6.13] and [6.15]

$$E_a = K\Phi\omega, \quad T_{e-m} = K\Phi i_a, \quad T_{e-m} - \frac{J d\omega}{dt} = 0.$$

Also,

$$i_a = \frac{V_a - E_a}{r_a}.$$

Hence

$$\frac{d\omega}{dt} = \frac{K\Phi}{J} \left(\frac{V_a - K\Phi\omega}{r_a} \right)$$

$$\frac{d\omega}{(V_a - K\Phi\omega)} = \frac{K\Phi}{Jr_a} dt.$$

Integrating both sides and evaluating the constant of integration such that $\omega = 0$ at $t = 0$

$$-\frac{(K\Phi)^2 t}{Jr_a} = \log_e\left(\frac{V_a - K\Phi\omega}{V_a} \right).$$

Hence

$$\omega = \frac{V_a}{K\Phi} (1 - e^{-(K\Phi)^2 t/Jr_a}). \qquad [6.20]$$

The exponential variation of speed ω with time is shown in Fig. 6.25(b). The energy W_r dissipated in the armature circuit resistance during run-up is evaluated as follows. At any speed ω

$$E_a = K\Phi\omega = \frac{V_a}{\omega_\infty}\omega,$$

where ω_∞ is the eventual, steady-state speed. Also

$$i_a^2 r_a = i_a(V_a - E_a) = i_a \frac{E_a}{\omega}[\omega_\infty - \omega].$$

Now

$$T_{e-m} = \frac{E_a i_a}{\omega} = J\frac{d\omega}{dt}.$$

Hence

$$W_r = \int_0^\infty i_a^2 r_a \, dt = \int_0^{\omega_\infty} J(\omega_\infty - \omega) d\omega$$

$$= \tfrac{1}{2} J\omega_\infty^2. \qquad [6.21]$$

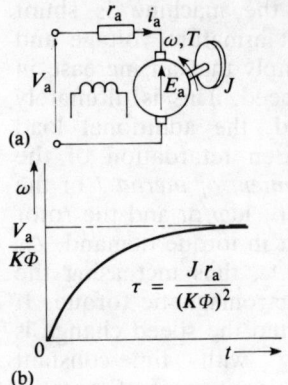

Figure 6.25 Acceleration of a d.c. motor with inertia load: (a) equivalent circuit; and (b) speed/time characteristic.

$\tau = \dfrac{J r_a}{(K\Phi)^2}$

Thus the energy dissipated in rotor resistance during run-up to steady-state speed equals the eventual stored kinetic energy of the rotating mass. The variation of speed with time during run-up and the eventual destinations of energy supplied by the source are analogous to those applicable to the rise of current and energy transfer in an R–L circuit subjected to the sudden application of a constant voltage. By comparison, we should expect that the speed of a d.c. machine, coupled to a load of high inertia, will decay exponentially to zero with time-constant $Jr_a(K\Phi)^{-2}$ if the armature of the d.c. machine is disconnected from the supply and is short circuited. If the armature circuit has incorporated an additional resistance, the armature current will be reduced and the rate of deceleration will be slowed. The process of bringing an excited d.c. machine to rest by dissipating the kinetic energy of rotation within a resistance connected to the armature circuit is called *dynamic or rheostatic braking*. The method is also used to control the rate of descent of hoists and lifts, etc., driven backwards by their connected load.

6.11.3.1 *Plug braking (plugging)*

Yet more rapid braking may be achieved by reversing the connections of the armature circuit to the d.c. supply. Armature e.m.f. and supply voltage then act in concert, reversing the armature current and torque to bring the rotor rapidly to rest. At the appropriate instant the armature must be disconnected from the supply – otherwise the rotor will accelerate in the reverse direction. The method is wasteful of energy, this being dissipated within the armature circuit resistance and supplied in the ratio 1 : 2 from the initial load kinetic energy and the d.c. source.

6.12 Electronically Controlled d.c. Machine Drive Systems

Modern versatile d.c. machine drives invariably incorporate electronic control facilities which are able to exploit efficiently the eminent controllability feature of the d.c. machine. Such methods are brought to bear almost exclusively on the control of the armature voltage supply and fall into one of two general categories:

(i) d.c./d.c. converters for use on fixed-voltage d.c. supplies, and
(ii) phase-controlled a.c./d.c converters, or rectifiers, for use on fixed-voltage/fixed-frequency a.c. supplies.

Within the first category the *chopper* circuit is most commonly employed, although *switch-mode* power supply units are becoming evident at lower ratings. In comparison with a.c. supply provision, d.c. sources are rare. The direct identification of armature voltage with speed, and armature current with torque for the separately excited machine, and alternatively the high speed with low torque, low speed with high torque characteristics of the series machine, render the d.c. machine a most attractive contender for electromechanical drive systems in which close control of speed and torque is required. The prospects for effective and efficient braking, including regeneration, add to the appeal. Machine drive systems may be classified as a.c. or d.c., in accordance with the nature of the primary electrical power supply. The a.c. group then strictly includes the Ward–Leonard three-machine set with a polyphase induction motor as the primary drive. Such a system offers controlled operation throughout the four quadrants of the speed/torque plane of Fig. 6.23, with a relatively slow response to changes in the operating conditions due to the naturally long time constants of the field circuits of the two d.c. machines, and an overall efficiency of about 80%. The primary drive a.c. machine may have a relatively low power factor on reduced loads but invariably draws a favoured sinusoidal current from the supply system. A phase-controlled a.c./d.c. converter alternative with d.c. motor load promises controllability comparable with the Ward–Leonard arrangement at lower cost and higher efficiency, but with disadvantages in respect of its impact on the a.c. supply through power factor and harmonic distortion of the line current.

Direct-current supply systems are relatively rare – except in traction applications for rail transport and battery-driven electric road vehicles. Traditionally, series motors are employed, with speed control effected by large resistor banks in series with armature and field, the speed range being extended by field weakening via 'diverter' resistances connected in parallel with the field coils. Such a method is clearly inefficient and modern methods invariably involve semiconductor switches which 'chop' the constant supply voltage at a fairly high repetition rate, enabling the *average* value of armature voltage to be varied between zero and the supply value. Efficiency is affected by losses in the switching circuitry and enhanced iron losses within the machine.

The chopping function requires that the machine should present significant inductance to the switching circuitry.

6.13 Direct-current Chopper Drive Systems

The essential circuit elements of a d.c./d.c. chopper circuit supplying the armature circuit of a d.c. motor are shown in Fig. 6.26(a). Figure 6.26(b) illustrates typical waveshapes of voltage and current developed at the armature circuit terminals when on load. The constant voltage source V is applied to the armature circuit via a conducting thyristor switch Th_1 over the period $t_1 \rightarrow t_2$ after which the switch is opened. Current continues to flow in the armature circuit, now closed via the previously reverse-biased diode D_1. The flow of armature current is maintained by the release of stored energy through the collapsing magnetic field of the inductance L, declining exponentially at a rate determined by the effective-time constant of the armature circuit, itself affected by the method of exciting the motor.

The fact that the armature current has been diverted away from Th_1 enables the thyristor to revert to a 'blocking' condition towards the end of an appropriate commutation process, but then

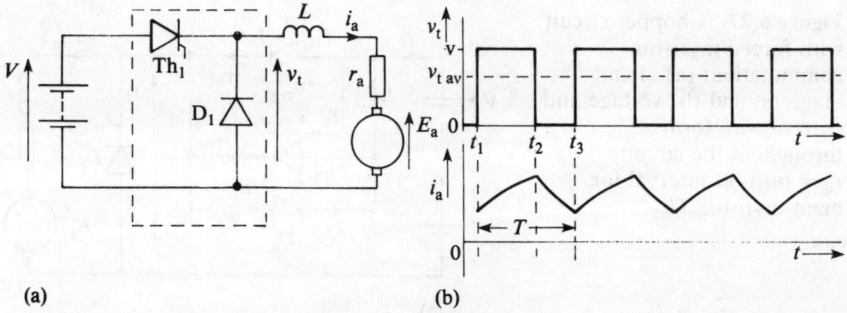

Figure 6.26 Essential d.c./d.c. chopper circuit diagram: (a); with (b) output voltage and current waveforms for d.c. motor load.

to have forward bias restored. With the anode positive with respect to the cathode the thyristor may subsequently respond to a gate trigger pulse applied between the gate and the cathode at $t = t_3$, turning on to permit the re-application of the d.c. supply voltage to the armature circuit. Diode D_1 is immediately reverse-biased, and the armature current proceeds to increase in complementary fashion with the same time-constant, if speed variations about the mean value and saturation effects are ignored. A simple analysis shows that the mean armature current is proportional to the mean armature voltage, both being adjustable by varying the proportion of the repetition period T allocated to thyristor conduction.

In the above discussion no mention has been made as to how thyristor turn-off at instant t_2 might be achieved. This is accomplished by means of *forced commutation* in which the thyristor current flow is halted by the connection of a charged capacitor across the anode–cathode terminals of the thyristor, which has polarity such that the discharging current impulse opposes the pre-existent anode–cathode current, forcing the total current to zero. The partially discharged capacitor then maintains a negative potential difference between anode and cathode for sufficient time to ensure that the thyristor is able to retain its blocking condition before external circumstances make the anode positive with respect to the cathode, prior to the next turn-on in response to gate current applied between gate and cathode terminals.

One of several alternative chopper circuits is shown in Fig. 6.27(a) with the waveforms of Fig. 6.27(b) corresponding to steady-state conditions in which the d.c. machine supplies a constant load torque. The variable nature of the armature current i_a gives rise to torque pulsations and, consequently, speed perturbations about a mean value. The T_{e-m}/i_a relationship depends on the nature of the machine excitation – whether the field winding is separately excited or is series-connected. The latter arrangement is more responsive to armature current fluctuations but inherently offers good smoothing of the armature current through a higher effective time-constant via the influence of a common armature and field current on induced e.m.f.

Electromagnetic torque fluctuations are counterbalanced by rotor acceleration and deceleration which result in speed variation. The period of torque fluctuation is short, however. A typical chopper repetition rate is 500 Hz and, with high-inertia rotors, speed fluctuations are likely to be negligible.

Figure 6.27 Chopper circuit with forced thyristor commutation: (a) circuit diagram, and (b) voltage and current waveforms throughout the circuit. t_q = turn-off interval for main thyristor Th_1.

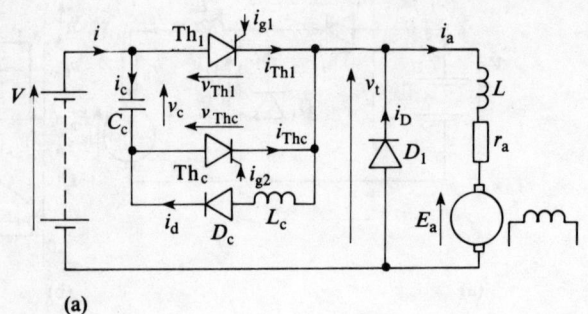

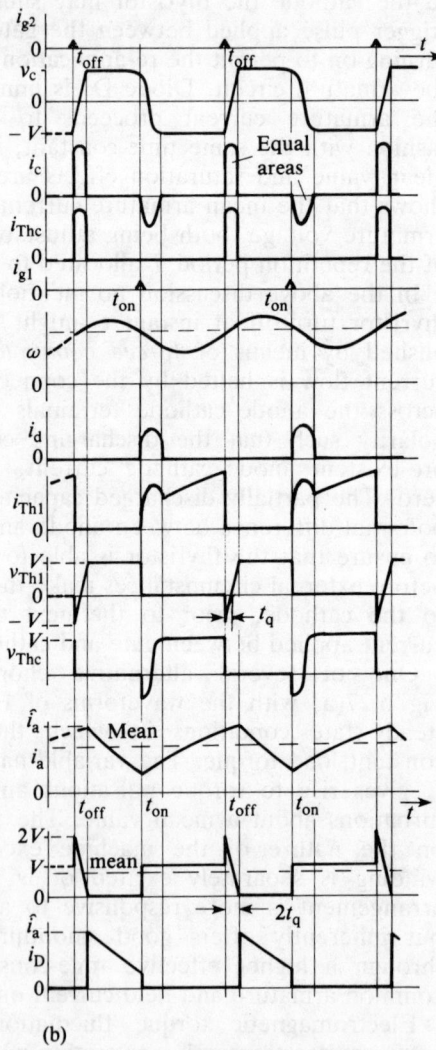

Within the circuit diagram of Fig. 6.27(a) armature inductance L is crucial to effective operation as a chopper. It provides a reservoir of energy, enabling the armature current to be continuous, whereas the d.c. supply current $i_{Th_1} + i_c$ is pulsed. Th_1 is the main series switching thyristor and D_1 is the 'free-wheel' diode. The role of the latter is two-fold – permitting the armature current sustained by the stored energy of series inductance L to circulate and continue the development of torque whilst diverting this current away from Th_1 after its forced commutation has taken place. Th_c is the *commutation thyristor* whose function is to connect the appropriately charged capacitor C_c across the anode and cathode of Th_1, so turning the latter off and maintaining negative anode–cathode voltage for sufficient time to ensure recovery to the blocking condition. The role of D_c and L_c is to form a series-resonant circuit with C_c, oscillatory for half a period only after Th_1 has been fired, thus correcting the polarity of the charged capacitor C_c.

The sequence of events occurring within the circuit corresponding to one complete cycle of operation will now be described, commencing with conditions just before Th_1 is to be turned off. Armature current i_a is approaching its maximum value whilst capacitor C_c has initial charge such that $v_c = -V$. Since Th_1 is conducting, $v_{Th_1} = 0$ and therefore $v_{Th_c} = V$. The commutating thyristor Th_c is thus forward biased and will respond at instant t_{off} to a pulse of gate current i_{g2}, turning on, and permitting C_c to discharge through Th_1, the current pulse being large enough to extinguish anode–cathode current flow through Th_1. At this instant v_t is still positive, maintaining reverse bias for D_1. The armature current flows via Th_c and defines the rate of discharge of C_c, such that $i_c = i_{Thc} = i_a$. The effect of the initial charge on C_c is to increase the common current only slightly as the initial stored energy is located predominantly in armature inductance L rather than commutation capacitor C_c. Hence the capacitor voltage v_c changes subsequently at a near-constant rate as defined by the armature current: positive-going to reverse its polarity.

The flow of i_a through C_c ceases when $v_t = (V - v_c)$ goes negative, having fallen from a transitory peak value of $2V$ at the instant t_{off}. Subsequently, i_a flows via the free-wheel diode D_1 with C_c retaining charge such that $v_c = V$.

The peak value of armature current thus effectively defines the rate of decline of v_t, which complements the rate of rise of v_c. v_{Th1}, which is equal to v_c during the period when v_c is increasing, becomes positive at an instant midway through the positive current pulse i_c. This instant defines the end of the *turn-off interval* for Th_1 as approximately $C_c V / \hat{i}_a$. The effect of C_c charging via the armature current is to raise the *average* value of output voltage by an amount $2V$ (turn-off interval/chopper period).

These conditions with v_c constant at V and $i_D = i_a$ declining exponentially towards zero, are maintained until the instant t_{on} at which the main thyristor Th_1 is subjected to gate–cathode current injection. The forward-biased thyristor turns on, placing the supply voltage across the armature circuit, with i_a increasing exponentially and i_D becoming zero. Capacitor C_c attempts to discharge via Th_1 and Th_c, but the latter is now reverse biased and blocks. C_c does, however, form a series-resonant circuit with L_c via Th_1 and D_c, and a transfer of energy between C_c and L_c takes place as v_c falls to zero and $i_d = -i_c$ builds up sinusoidally to a peak value over

1/4 cycle of resonant frequency $f_0 = (2\pi \sqrt{L_c C_c})^{-1}$. The oscillation continues for the second quarter-cycle with energy transfer back to C_c from L_c, charging C_c with opposite polarity, such that v_c falls to $-V$. Diode D_c prevents further oscillatory interchange, leaving C_c charged appropriately for subsequent turn-off Th_1 when Th_2 is fired to repeat the chopper cycle.

6.13.1 *Operation under a light motoring load*

The waveforms of Fig. 6.27(b) correspond to the machine being fairly heavily loaded as a motor and the mark/space ratio of the square-wave voltage v_t applied to the armature exceeding 1:1. Control of the motor speed is achieved by adjusting the timing of gate-firing pulses i_{g1} and i_{g2} to increase or reduce the average value of voltage applied to the armature circuit. If the mark/space ratio is set to correspond to a low speed and the armature circuit inductance or chopper frequency is relatively low the armature current i_a may become discontinuous. Due to the armature e.m.f., the 'target' value for i_a when free wheeling via inductance L and diode D_1 is negative and the free-wheel diode will not permit i_a to reverse. The armature current having fallen to zero, the stored field energy in the circuit inductance is also zero and the load torque demand is met by deceleration of the rotor.

6.13.2 *Regeneration*

Modification of the circuit enables i_a to reverse polarity and to permit regenerative braking of inertial loads, provided that the armature circuit inductance is adequately large. The basic arrangement excluding the thyristor commutation circuitry is shown in Fig. 6.28(a). Thyristor Th_1 and diode D_1 carried over from Fig. 6.27(a) are complemented by the *anti-parallel* connected diode D_2 and thyristor Th_2, respectively. The key feature to note is that any energy returned to the d.c. source from the braked mechanical load must be accompanied by the simultaneous transfer of energy from the series inductance L. The energy abstracted from the field of L is first built up during an immediately preceding period when the armature is short circuited via Th_2 and the circuit inductance.

For the familiar case of a d.c. machine operating in the steady state as a generator with constant armature voltage, the magnitude of the armature e.m.f. must exceed the terminal voltage V by the voltage drop across the armature circuit resistance.

A chopper circuit, however, is always operating in one or other *transient* mode and, for generator action, the above observation is inappropriate. For both motoring and generating states the armature e.m.f. is normally always *less than* the supply voltage V; the essential difference $V - E_a$ (always positive) being the voltage developed across the armature inductance when the chopper switch is closed.

In the two-quadrant drive chopper circuit four alternative paths exist for the armature current. In Fig. 6.28(b) the chopper switch is open and i_a flows through the inductance L via the external diode D_1 whenever i_a is positive and decreasing with time, so that $v_L = L\, di/dt$ is negative. The inductance is then giving up energy for conversion into mechanical power since E_a and i_a are both positive. Whenever i_a is negative and going more negative with increasing time, v_L is again negative as required by positive E_a and the inductor absorbs energy provided by the braked mechanical load. The negative current i_a is then carried by thyristor Th_2.

In Fig. 6.28(c), the chopper switch is closed and armature current flows in the source, either via diode D_2 when i_a is negative or via thyristor Th_1 when i_a is positive. In the former case i_a is negative but is required to go less negative with increasing time so that $v_L = L\, di_a/dt = V - E_a$, which is a positive quantity. The braked mechanical load is then returning energy to the source, supplemented by energy derived from the collapsing field of inductance L. In the latter case with i_a positive, carried by Th_1 and increasing with time, energy is being supplied by the d.c. source for simultaneous increase in the stored energy of the series inductance and conversion into mechanical power.

Figure 6.28(d) shows waveforms appropriate to a two-quadrant chopper drive providing an intermediate value of average voltage to the d.c. machine armature circuit. The thyristor gate-firing pulses are now necessarily sustained over the entire period during which thyristor conduction might be required. During the first part of each such interval the associated anti-parallel diode conducts and holds a reverse-bias voltage across the thyristor. This will

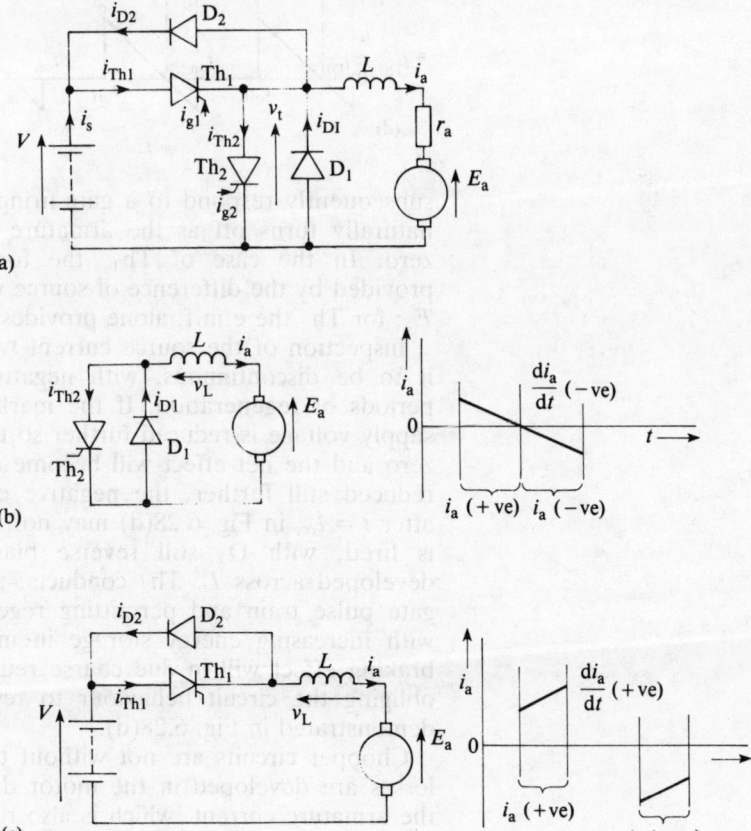

Figure 6.28 Essential two-quadrant chopper circuit: (a) circuit diagram; (b) effective circuitry when armature current decreases; (c) effective circuitry when armature current increases; and (d) voltage and current waveforms throughout the circuit.

Figure 6.28 (*cont.*)

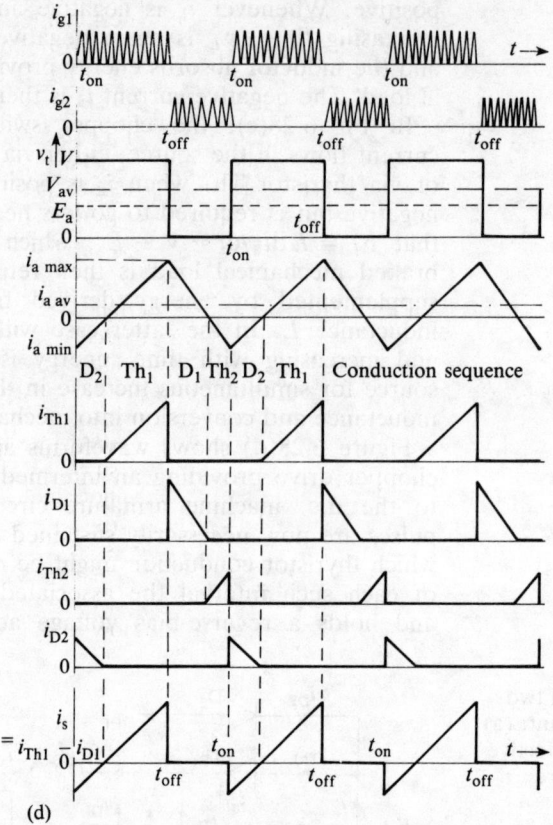

(d)

subsequently respond to a gate-firing pulse immediately the diode naturally turns off as the armature current carried goes through zero. In the case of Th$_1$, the forward anode–cathode bias is provided by the difference of source voltage V and armature e.m.f. E_a; for Th$_2$ the e.m.f. alone provides.

Inspection of the source current waveform $i_s = i_{Th_1} - i_{D2}$ shows it to be discontinuous, with negative portions corresponding to periods of regeneration. If the mark/space ratio of the armature supply voltage is reduced further so that $V_{av} < E_a$, i_{av} will become zero and the net effect will become one of regeneration. If V_{av} is reduced still further, the negative current i_a carried by D$_2$ just after $t = t_{on}$ in Fig. 6.28(d) may not have risen to zero before Th$_2$ is fired, with D$_1$ still reverse biased by E_a and the voltage developed across L. Th$_2$ conducts, responding immediately to its gate pulse train and permitting regenerative braking to continue with increasing energy storage in inductance L. The continuous braking effect will in due course reduce the speed and hence E_a, obliging the circuit behaviour to revert to that characteristically demonstrated in Fig. 6.28(d).

Chopper circuits are not without their disadvantages. Increased losses are developed in the motor due to the pulsating nature of the armature current, which is also reflected in the supply current, giving rise to electromagnetic interference – especially if carried by an overhead line or a railway track. The commutation circuits need

Direct-current machines

to function reliably and efficiently over a wide range of load current conditions, including overload.

Worked example 6.3 A d.c. series motor rotating at a constant speed of 1500 rev/min is supplied via an ideal chopper from a 500 V d.c. source, presenting total resistance and inductance of 0.05 Ω and 0.025 H, respectively. The constant relating armature e.m.f. and field current has the value 12.50 V A^{-1} at 1500 rev/min. If the chopper is operated at a frequency of 500 Hz with a 1:1 mark/space ratio calculate:

   (i)   the mean armature current,
   (ii) the mean torque developed,
   (iii) the maximum and minimum values of armature current during the chopper cycle,
   (iv) the difference between maximum and minimum values of the energy stored in the machine inductance, and
   (v)   the mechanical energy supplied during each 'chopper-off' period.
   (vi) Calculate the commutation circuit capacitor and inductor values on the basis of a main thyristor turn-off time of 5 μs and maximum and minimum load currents as given in (iii) above, i.e. 24.73 A and 14.96 A, restricting the main thyristor peak current to 30 A. Hence determine the minimum value of mean output voltage available from the chopper circuit below which commutation will fail.

Solution (i) Referring to the basic chopper circuit of Fig. 6.26(a) and the waveforms of Fig. 6.26(b), and recognising that the *mean* voltage across the inductance L is zero, for a 1:1 mark-space ratio of chopper output voltage the *mean* armature supply voltage = $500 \times 0.5 = 250$ V. The mean machine e.m.f. at 1500 rev/min = $12.5\ i_{a\,mean}$. Hence $i_{a\,mean}\ (r_a + 12.50) = 250$ V, giving $i_{a\,mean} = 250/12.55 = 19.92$ A.

(ii) From eqn [6.13], (mean) torque = armature constant $K\Phi \times$ (mean) armature current, where $K\Phi$ relates e.m.f. to speed (rad s^{-1}) at the same field current. Hence $K\Phi = 12.50\ i_{a\,mean}/2\pi \times 25$ and mean torque = $12.50 \times 19.92^2/2\pi \times 25 = 31.58$ N m.

(iii) Chopper period = 0.002 s. Chopper 'on' time = chopper 'off' time = 0.001 s. During acceleration

$$V_a = E_a + i_a r_a + L\ di_a/dt$$

$$500 = (12.50 + 0.05)i_a + 0.025\ di_a/dt, \qquad [A]$$

with $i_a = i_{a\,min}$ at $t = 0$ and $i_a = i_{a\,max}$ at $t = 10^{-3}$ s.
During deceleration

$$0 = E_a + i_a r_a + L\ di_a/dt$$

$$0 = (12.50 + 0.05)i_a + 0.025\ di_a/dt, \qquad [B]$$

with $i_a = i_{a\,max}$ at $t = 0$ and $i_a = i_{a\,min}$ at $t = 10^{-3}$ s. By Laplace, eqn [A] becomes $500/s = 12.55\ \bar{i}_a + 0.025\ (s\ \bar{i}_a - i_{a\,min})$, giving

$$\bar{i}_a = \frac{500/0.025}{s(s + 12.55/0.025)} + \frac{i_{a\,min}}{s + 12.55/0.025}$$

$$= \frac{500}{12.55}\left[\frac{1}{s} - \frac{1}{s + 10^3/1.99}\right] + \frac{i_{a\,min}}{s + 10^3/1.99}.$$

Hence

$$i_a(t) = 39.84[1 - \exp(-10^3 t/1.99)] + i_{a\,min} \exp(-10^3 t/1.99).$$

At $t = 0.001$

$$i_{a\,max} = 39.84(1 - 0.605) + i_{a\,min}(0.605)$$

$$i_{a\,max} = 15.74 + 0.605\, i_{a\,min}. \qquad [C]$$

By Laplace, eqn [B] becomes

$$0 = 12.55\, \bar{i}_a + 0.025\,(s\bar{i}_a - i_{a\,max})$$

giving

$$\bar{i}_a = i_{a\,max}/(12.55/0.025 + s)$$

and

$$i_a(t) = i_{a\,max} \exp(-10^3 t/1.99)$$

At $t = 0.001$

$$i_{a\,min} = 0.605\, i_{a\,max} \qquad [D]$$

Substituting eqn [D] in eqn [C] yields

$$i_{a\,max} = 15.74 + (0.605)^2\, i_{a\,max},$$

giving

$$i_{a\,max} = 15.74/(1 - 0.605^2) = 24.73\ \text{A}.$$

From eqn [D],

$$i_{a\,min} = 0.605 \times 24.73 = 14.96\ \text{A}.$$

(iv) The difference between the maximum and minimum values of energy stored in inductance $L = \tfrac{1}{2}L(i_{a\,max}^2 - i_{a\,min}^2) = 0.0125(24.73^2 - 14.96^2) = 4.847$ J.

(v) The mechanical energy supplied during the 'off' period = $T_{mean}\omega \times t_{off} \simeq 31.58 \times 2\pi \times 25 \times 0.001 = 4.961$ J.

(vi) Assuming the load current to be constant at $i_{a\,max}$ when carried by commutation capacitor C_c (Fig. 6.27) after turn-off of main thyristor Th_1, $v_{Th1}(= -v_c)$ rises linearly from $-V$ to zero over a period of time Δt not less than the thyristor turn-off time t_q. Thus $\Delta t = VC_c/i_{a\,max} \not< t_q$ giving $C_c \not< t_q i_{a\,max}/V$. Hence $C_c \not< 5 \times 10^{-6} \times 24.73/500 = 0.25\ \mu\text{F}$, minimum.

During the resetting of initial charge on the commutation capacitor prior to each successive commutation, the main thyristor instantaneously carries peak current in the oscillatory $L_c C_c$ circuit in addition to the load current $i_{a\,min}$. If the total thyristor current is limited to 30 A, the peak current $\hat{i}$ in the resonant circuit must not exceed $30 - 14.96 = 15$ A. Peak oscillatory current occurs when instantaneously $v_c = 0$, the initial energy stored in the capacitor ($\tfrac{1}{2}C_c V^2$) having been transferred to the inductor and evaluated as $\tfrac{1}{2}L_c \hat{i}^2$. Thus $\hat{i}^2 = \tfrac{1}{2}C_c V^2/\tfrac{1}{2}L_c$ giving $\hat{i} = V(C_c/L_c)^{1/2}$. Hence $L_c = C_c(V/\hat{i})^2 = 0.25 \times 10^{-6} \times 500^2/15^2 = 277\ \mu\text{H}$.

The minimum time required to reset C_c is half a period of $L_c C_c$ resonance = $\pi(L_c C_c)^{1/2} = \pi(277 \times 10^{-6} \times 0.25 \times 10^{-6})^{1/2} = 26\ \mu\text{s}$. Allowing for the transient ramp decline from $2V$ to zero in $2t_q$ on switch-off, the minimum mean output voltage over each chopper period of 2 ms is $500(26 \times 10^{-6} + 2 \times 5 \times 10^{-6})/2 \times 10^3 = 9$ V.

Tutorial Example 6.3(A) If the load torque on the d.c. series motor of Worked Example 6.3 is halved, to what value will the rotational speed increase? Ignore magnetic saturation and second-order effects.

[2123 rev/min]

Tutorial Example 6.3(B)   If the d.c. series motor of Worked Example 6.3 is to be regeneratively braked at the original speed of 1500 rev/min, such that the maximum and minimum values of armature current are as evaluated in (iii) above, determine the necessary chopper 'on' and 'off' times, and also the corresponding chopper frequency and duty cycle (mark/space ratio).

[0.001 01 s, 0.000 99 s; 500 Hz, 101/99]

6.14 Phase-controlled Converter-fed Drive Systems

Apart from electric vehicle applications for which battery packs are a convenient power source, constant voltage, constant frequency sinusoidal a.c. supplies (either single-phase or polyphase) are likely to be most readily available for electric drive systems. Simple rectifier circuits may be employed to convert such sources into nominally constant-voltage d.c. supplies with superimposed ripple, to which most standard d.c. motors will respond, developing operational characteristics not significantly different from those obtained using a pure d.c. source, apart from some increase in iron loss. Most rectifier installations, however, take advantage of the opportunity to control the mean d.c. output voltage of the converter to provide motor speed control via armature voltage. The most common method of control is designated 'phase control', in which the controlling series thyristors are 'gated' or 'fired' repetitively at the same instant in each cycle of the supply frequency, the instant being adjustable with respect to a defined reference, for example voltage zero, positive-going, of supply phase voltage. From the operational point of view, a.c./d.c. conversion systems employing polyphase bridge circuits are best suited for this duty, but alternative, simpler configurations have the advantages of reduced cost and complexity which may outweigh certain disadvantages with regard to flexibility and performance.

The desirable features of an a.c. drive system may be listed as follows:

(i)   no-load speed should be easily and efficiently controllable;
(ii)   speed variation with load torque should match the required duty cycle;
(iii)   acceleration and braking functions should be controllable, the latter ideally permitting regeneration; and
(iv)   the a.c. supply current should be alternating and balanced, with acceptable harmonic content and power factor.

Several alternative configurations of increasing complexity will be examined with a view to establishing how these features might be implemented. First, it is appropriate to consider d.c. machine armature representation in the analysis of a.c.-supplied drive systems.

The effective armature circuit of a rotating d.c. machine is as shown in Fig. 6.29, which incorporates a small change in notation to identify nominally d.c. quantities. Under steady conditions the relationship between terminal voltage and current is given by

$$V_d = I_d r_a + E_a,$$

whereas, instantaneously,

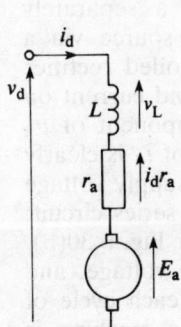

Figure 6.29 Armature circuit representation of a d.c. machine.

$$v_\mathrm{d} = L\frac{\mathrm{d}i_\mathrm{d}}{\mathrm{d}t} + i_\mathrm{d}r_\mathrm{a} + E_\mathrm{a}.$$

Now $E_\mathrm{a} = k\omega I_\mathrm{f}$ and hence will be time invariant only if the field excitation is constant and speed fluctuations are negligible. In solid-state drive applications i_d may have appreciable ripple, causing electromagnetic torque fluctuations with a constant field current. Rotational speed will therefore tend to vary in sympathy unless the rotating mass has large inertia. For convenience, we shall assume this to be the case.

If the armature circuit inductance L is large enough to effectively eliminate ripple in i_d the voltage drop across r_a will be virtually constant and the alternating component of v_d will be developed across inductance L. Since $v_L = L\,\mathrm{d}i_\mathrm{d}/\mathrm{d}t$, it follows that a well-smoothed armature current requires a correspondingly larger value of inductance if the a.c. component of v_d is appreciable. In practical systems the value of L is not such to provide a reservoir of energy so large that the electromechanical energy conversion may proceed at a constant rate whilst permitting wide fluctuations in the power input to the circuit from the a.c. source. Simplified analyses may, however, with little loss of generality, ignore the small a.c. component of voltage across r_a so that the a.c. component of v_d is considered to be developed across L whilst the mean value or d.c. component of v_d, given by $V_\mathrm{d} = v_{\mathrm{dav}}$, is developed across the series combination of armature resistance and e.m.f.

Further degrees of simplification may ignore armature resistance or assume perfect smoothing of the armature current, so that $V_\mathrm{d} \simeq E_\mathrm{a}$, or $i_\mathrm{d} = I_\mathrm{d}$, with $I_\mathrm{d} = (V_\mathrm{d} - E_\mathrm{a})/r_\mathrm{a}$. It is not feasible to ignore armature resistance with a perfectly smoothed armature current unless the current is defined by a knowledge of the torque. The nature of the applied voltage v_d when derived from an a.c. source is ideally one whose mean value is readily adjustable, is capable of polarity reversal and is accompanied by negligible ripple. Such are the properties of complex, polyphase converter systems as will be shown later in this section, but the principal features of d.c. machine drives powered from a.c. sources will first be demonstrated using simple single-phase circuits.

6.14.1 Single-phase, half-wave converter

Thus Fig 6.30(a) shows the armature winding of a separately excited d.c. machine supplied from a single-phase source via a single diode, which constitutes a half-wave, uncontrolled rectifier circuit. Figure 6.30(b) shows waveforms of voltage and current on the assumptions of constant speed and the a.c. component of v_d being equated to $L\,\mathrm{d}i_\mathrm{d}/\mathrm{d}t$. The discontinuous nature of i_d is clearly seen. The armature voltage v_d is defined by the a.c. supply voltage between instants t_1 and t_4 when current flows in the series circuit; at other times v_d is equal to the armature e.m.f. In Fig. 6.30(b), E_a approaches the peak value $\hat{V}_1$ of the a.c. supply voltage, and consequently current flows for only a small part of each cycle of supply frequency. If the mechanical loading of the machine is increased so that its speed is obliged to fall through an imbalance of electromagnetic and load torques, E_a will fall in sympathy and the instant t_1 at which the supply voltage equates to E_a will occur earlier in each cycle. Thus the circuit inductance will be exposed to $v_L \simeq \hat{V}_1 \sin \omega t - E_a$ for a longer period. The consequent current pulse i_a will contribute an increased and extended pulse of electromagnetic torque.

The buildup of current initiated at $t = t_1$ continues until i_d achieves its peak positive value at $t = t_2$ when $v_1 = E_a$ once more and the energy stored in the inductance peaks at $\frac{1}{2} L \hat{i}_d^2$. Thereafter, i_d decreases, the inductance voltage reversing polarity to maintain the diode D_1 in conduction. The inductance maintains the flow of current until its stored energy falls to zero, thus providing energy for electromechanical conversion and permitting power flow from the a.c. supply whilst v_d is positive $(t_2 < t < t_3)$, and returning energy to the supply when v_d is negative $(t_3 < t < t_4)$.

The instant t_4 at which the current falls to zero may be deduced from the voltage/time waveform of Fig. 6.30(b). The area under the curve $v_1 - E_a$, which is equivalent to $L di_d / dt$, plotted to a base of time between limits t_1 and t_2 corresponding to zero and

Figure 6.30 Uncontrolled single-phase, half-wave converter with a d.c. motor load: (a) essential circuit with the free-wheel diode connection dotted; (b) machine terminal voltage and current waveforms with discontinuous current and the freewheel diode disconnected; and (c) waveforms of terminal voltage and distributed current with free-wheel diode, and with a continuous current in the load.

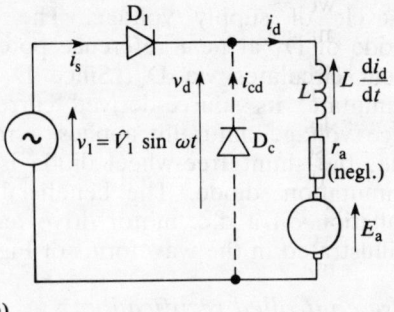
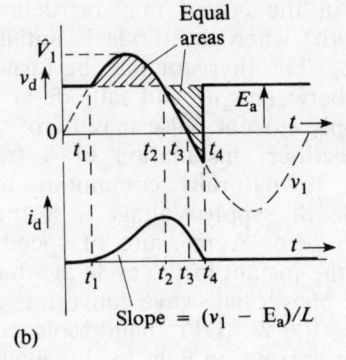
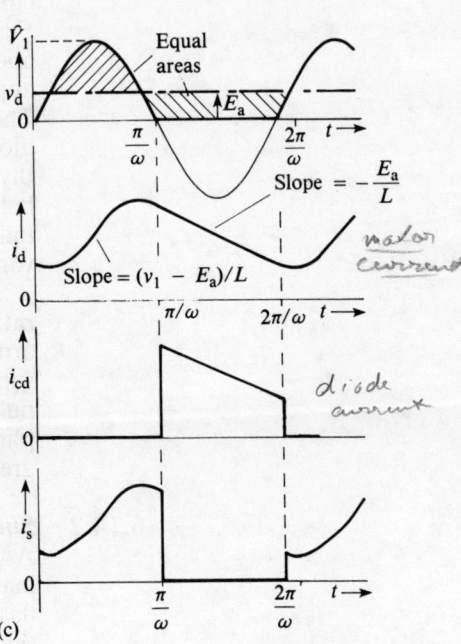

maximum current values, equates to the peak flux linkage developed within the inductance. As the stored energy falls subsequently to zero, the corresponding reduction in flux linkage to zero is given by the area under the curve $E_a - v_1$ against time between $t = t_2$ and $t = t_4$.

The armature current pulse may be encouraged to persist for a longer time, thus increasing the average value of armature current and consequently the torque, by installing a 'free-wheel' diode D_c in parallel with the armature circuit. The diode is of polarity such as to avoid a direct short circuit of the supply but to permit a short circuit of the armature, if current flow continues into the negative-going half-cycle of supply voltage. The current in the armature circuit will then decline towards zero at a rate given by the equation $L di_d/dt = -E_a - i_d r_a \simeq -E_a$. If the magnitudes of $\hat{V}_1$, E_a and L are correctly proportioned, the flow of armature current will be continuous, thus providing a more uniform torque development.

Incorporation of the free-wheel diode D_c enables the main series diode D_1 to cease conduction at the end of each positive-going half-cycle of supply voltage. The free-wheel diode holds the cathode of D_1 at near reference potential for as long as armature current circulates via D_c. Since D_c enables D_1 to turn off or 'commutate' its source-derived current at the instant when the source voltage naturally applies a reverse bias across the series diode, the shunt free-wheel diode is alternatively described as a 'commutation' diode. The beneficial effects of free-wheel diode installation on a d.c. motor drive fed from a single-phase supply are illustrated in the waveforms of Fig. 6.30(c).

6.14.1.1 *Phase-controlled rectification*

If the series diode D_1 is replaced by a thyristor, the buildup of current in the circuit may be delayed beyond the instant t_1 in Fig. 6.30(b) when the diode is initially forward biased as v_1 rises above E_a. The thyristor may be turned on by a suitable gate pulse applied between gate and cathode at any instant between t_1 and t_2. The supply current pulse may be of shorter duration than with the diode rectifier, installation of a free-wheel diode enabling the thyristor to naturally commutate at the end of each positive half-cycle of sypply voltage – if the current has not previously fallen to zero. A measure of speed control is available through varying the instant in the cycle at which the thyristor is turned on.

Single-phase, half-wave converters are suitable for motors having ratings ≈ 100 W. DIY hand-tools come into this category. The armature current on light load is likely to be discontinuous and the supply current at all times consists of short duration, unidirectional pulses inappropriate for an a.c. supply system. Full-wave, single-phase bridge converters offer better performance at the expense of greater complexity and higher component costs.

6.14.2 *Single-phase, full-wave thyristor bridge converter*

A fully controlled single-phase bridge converter with d.c. motor load is illustrated in Fig. 6.31(a) with the free-wheel diode, shown dashed, not initially in circuit. Here two alternative paths including the armature circuit are provided, either via Th_1 and Th_2 which

Direct-current machines

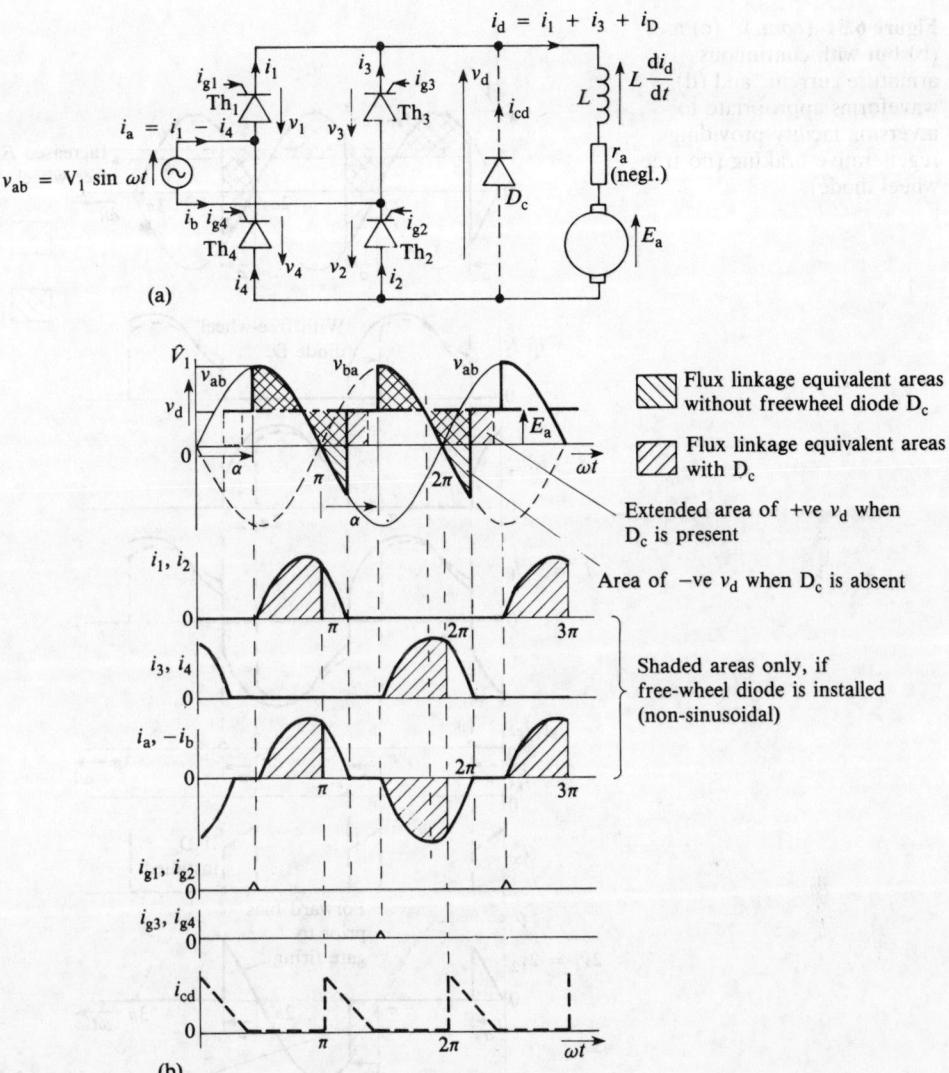

Figure 6.31 Fully controlled single-phase (full-wave) thyristor bridge converter with d.c. motor load: (a) essential circuit with optional free-wheel diode; (b) terminal voltage and distributed current waveforms with and without free-wheel diode in circuit (discontinuous armature current);

are both naturally forward biased during any period when $v_{ab} - E_a$ is positive, or via Th$_3$ and Th$_4$ – which are naturally forward biased when $v_{ba} - E_a$ is positive. Such periods of forward bias extend to a half period of supply frequency if $E_a = 0$. At the end of each period of forward bias defined above the appropriate pair of thyristors, if previously gated to permit current flow, will be held in conduction by the armature current sustained by the armature inductance, until either the current declines to zero or the complementary pair of thyristors is fired.

The first-mentioned condition of discontinuous current is shown for a lightly loaded motor in Fig. 6.31(b), corresponding to a value of E_a which exceeds the mean value of armature voltage v_d, neglecting the armature resistance voltage drop. Evidence of discontinuous armature current (and torque) is provided in the armature voltage waveform by the intervals over which $v_d = E_a$. If the armature current does not fall to zero before the next pair of

Figure 6.31 (*cont.*) (c) as (b) but with continuous armature current; and (d) waveforms appropriate to inversion facility providing regenerative braking (no free-wheel diode).

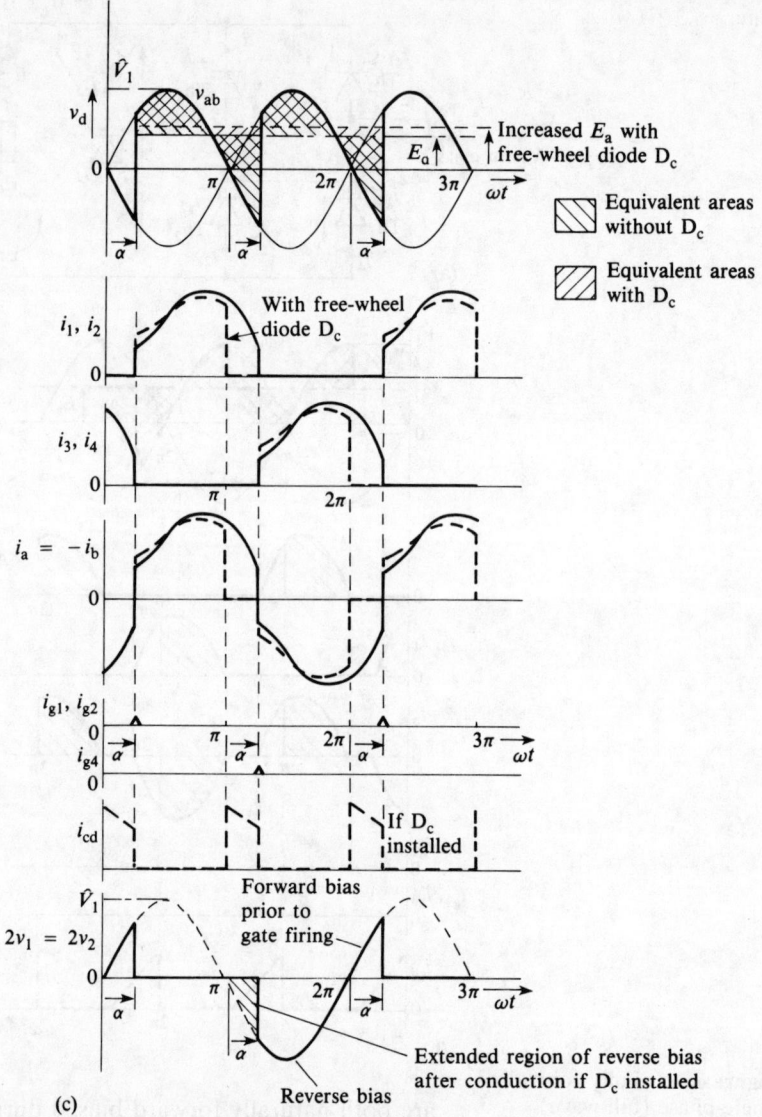

thyristors is gated the armature current is then naturally transferred to the alternative path, Th_3 and Th_4, say. The previously conducting pair Th_1 and Th_2 revert to the blocking state, reverse biased until $v_{ab} - E_a$ is again positive and either the armature current falls to zero or Th_1 and Th_2 are fired. Appropriate waveforms are shown in Fig. 6.31(c).

The continuous armature current condition is most likely with a heavily loaded or accelerating motion for which the delay angle α measured from voltage zero is relatively low. Neglecting resistance, the mean value of v_d is again equal to the armature e.m.f. In terms of the a.c. supply voltage and the delay angle, the mean value v_{dav} without the free-wheel diode is given simply by

Figure 6.31 (*cont.*)

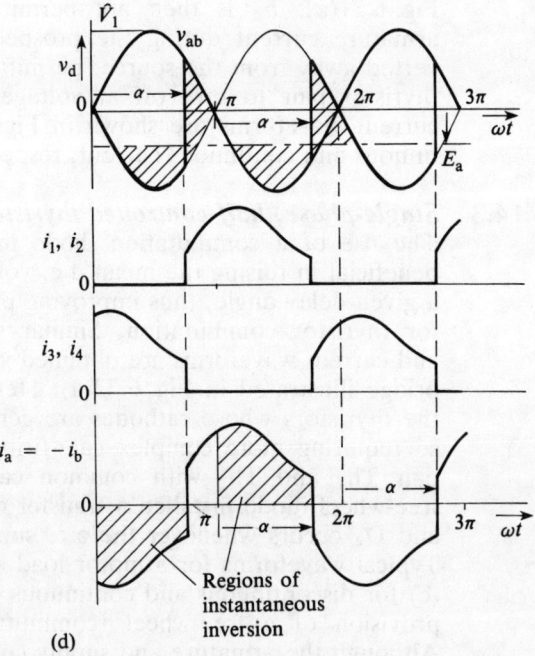

(d)

Regions of instantaneous inversion

$$v_{\text{dav}} = V_{\text{d}} = \frac{1}{\pi} \int_{\alpha}^{\alpha+\pi} \hat{V}_{\text{a}} \sin \omega t \, d\omega t$$

$$= \frac{2}{\pi} \hat{V}_{\text{a}} \cos \alpha. \qquad [6.22]$$

The derivation of an analogous expression for v_{dav} under discontinuous current conditions is difficult, involving rather complicated mathematics. Unlike eqn [6.22] the result is not a simple function of the a.c. voltage amplitude $\hat{V}_{\text{a}}$ and the delay angle α.

Inversion, electrical power flow into the a.c. supply network, is an available feature of the fully controlled bridge converter. Instantaneous inversion occurs at the beginning of each half cycle of supply voltage whilst the armature current maintains its previous unidirectional flow between the source terminals, the source voltage having undergone a reversal of polarity. The energy returned to the supply is derived from the armature inductance. For straight motor drives such periods of instantaneous inversion are an embarrassment, reducing the average power converted by the machine for a given mean value of armature current through a reduction in the mean value of armature voltage v_{dav} for given values of delay angle and supply voltage. The effect is reflected as a reduction in the power factor of the load on the a.c. supply. The inherent inversion capability may be exploited, however, with the regenerative braking conditions illustrated in Fig. 6.31(d) for which $90° < \alpha < 180°$ and E_{a} is negative, being balanced by the negative value of v_{dav} under continuous current conditions.

If the facility for regenerative braking is not required, the inversion intervals may be avoided by incorporating a free-wheel or commutating diode D_{c} across the armature as shown dashed in

Fig. 6.31(a). v_d is then not permitted to go negative and the armature current during the prospective inversion periods is diverted away from the source, permitting the previously conducting thyristor pair to turn off at voltage zero. Modified voltage and current waveforms are shown in Figs. 6.31(b) and (c) for discontinuous and continuous current, respectively.

6.14.3 Single-phase, half-controlled thyristor bridge converter

The use of a commutation diode in a fully controlled bridge is beneficial in raising the mean d.c. voltage of the converter v_{dav} for a given delay angle, thus improving power factor and the prospects for thyristor commutation. Similar source and armature voltage and current waveforms are obtained via the use of a half-controlled bridge illustrated in Fig. 6.32(a). Here diodes D_2 and D_4 replace the thyristors whose cathodes are generally at different potentials, so requiring more complex gate-firing circuits than the remaining pair Th_1 and Th_2 with common cathode connections. With no free-wheel diode installed a transfer of current between diodes D_2 and D_4 occurs whenever the a.c. supply voltage changes polarity. Typical waveforms for a motor load are shown in Fig. 6.32(b) and (c) for discontinuous and continuous current respectively, with the provision of a free-wheel (commutating) diode being optional. Although the armature and supply current waveforms are basically the same as for the fully controlled bridge provided with a commutation diode, the thyristor conduction periods for the basic half-controlled bridge extend over an entire half-cycle of supply frequency, which is a disadvantage from the commutation viewpoint. Installation of a specific commutation diode will, however, divert current away from each thyristor/diode pair at the beginning of each half cycle of source voltage, thus enabling the thyristor to revert to the blocking state earlier than would otherwise be the case and easing practical difficulties with the sequential commutation of continuous armature current from one thyristor to the other. Such a problem is more likely during operation with large delay angles, particularly where significant a.c. source inductance delays the transfer of load current between thyristors. This 'overlap' effect is discusssed in Section 6.16.

Since the armature voltage cannot reverse polarity, whether or not the free-wheel diode is installed, inversion with regenerative braking is not possible.

Referring to Fig. 6.32(c) for the case of continuous current, the mean d.c. voltage derived from the converter is given by

$$v_{dav} = V_d = \frac{1}{\pi} \int_\alpha^\pi \hat{V}_a \sin \omega t \, d\omega t$$

$$= \frac{1}{\pi} \hat{V}_a [1 + \cos \alpha]. \qquad [6.23]$$

The inclusion of a separate commutating diode is unnecessary if the half-controlled bridge is reconfigured as in Fig. 6.32(d). The cathodes of the controlled thyristors are at different potentials, however, creating some additional complexity for the gate-firing circuits.

Direct-current machines 243

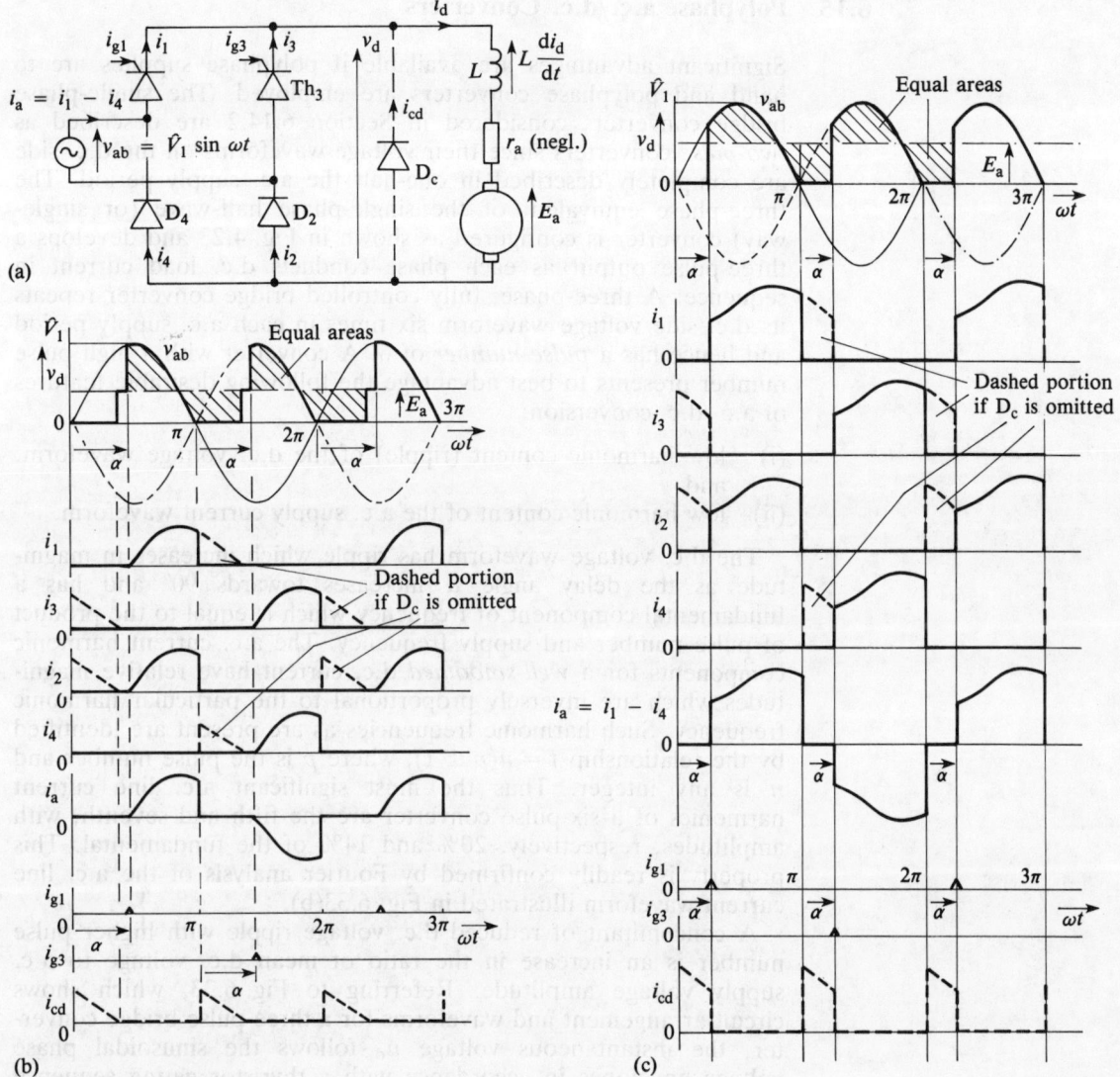

Figure 6.32 Half-controlled single-phase thyristor bridge converter with d.c. motor load: (a) essential circuit with optional free-wheel diode; (b) voltage and current waveforms with discontinuous armature current; (c) as (b) but with continuous armature current; and (d) alternative configuration with inherent free-wheel diode provision. The waveforms are as in (b) and (c) less the dashed portions but incorporating i_{cd} with i_2 and i_4.

6.15 Polyphase a.c./d.c. Converters

Significant advantages are available if polyphase supplies are to hand and polyphase converters are employed. The single-phase bridge converters considered in Section 6.14.2 are described as *two-pulse* converters since their voltage waveforms on the d.c. side are completely described in one-half the a.c. supply period. The three-phase equivalent of the single-phase half-wave (or single-way) converter is configured as shown in Fig. 4.23 and develops a three-pulse output as each phase conducts d.c. load current in sequence. A three-phase, fully controlled bridge converter repeats its d.c. side voltage waveform six times in each a.c. supply period and hence has a *pulse number* of 6. A converter with a high pulse number presents to best advantage the following desirable features of a.c./d.c. conversion:

(i) low harmonic content (ripple) of the d.c. voltage waveform, and
(ii) low harmonic content of the a.c. supply current waveform.

The d.c. voltage waveform has ripple which increases in magnitude as the delay angle α increases towards 90° and has a fundamental component of frequency which is equal to the product of pulse number and supply frequency. The a.c. current harmonic components for a *well-smoothed* d.c. current have relative magnitudes which are inversely proportional to the particular harmonic frequency. Such harmonic frequencies as are present are identified by the relationship $f = n(p \pm 1)$, where p is the pulse number and n is any integer. Thus the most significant a.c. line current harmonics of a six-pulse converter are the fifth and seventh, with amplitudes, respectively, 20% and 14% of the fundamental. This property is readily confirmed by Fourier analysis of the a.c. line current waveform illustrated in Fig. 6.33(b).

A concomitant of reduced d.c. voltage ripple with higher pulse number is an increase in the ratio of mean d.c. voltage to a.c. supply voltage amplitude. Referring to Fig. 6.33, which shows circuit arrangement and waveforms for a three-pulse bridge converter, the instantaneous voltage v_d follows the sinusoidal phase voltage envelopes in accordance with a thyristor gating sequence such that thyristors Th_1, Th_3 and Th_5 with common cathode connections are fired at 120° intervals, in sympathy with the phase sequence, whereas thyristors Th_2, Th_4 and Th_6 with common anodes are fired at the same intervals but such that i_{g2} occurs 60° after i_{g1}. Gate delay angle α is measured from each 'instant of natural commutation', e.g. t_1 in Fig. 6.33(b), which occur at 60° intervals when the pair of source phase voltages involved have identical instantaneous values. The thyristor associated with the lagging voltage of the pair is subsequently forward biased for a duration equivalent to 180°.

With $\alpha = 0°$, the current flow is always from the phase with the most positive voltage to that most negative. Gate delay to $\alpha = 180°$ is theoretically possible, with net inversion for $\alpha > 90°$. In practice, α must be limited to rather less than 180° so that an adequate period of reverse bias will permit a previously conducting

Direct-current machines 245

Figure 6.33 Fully controlled three-phase thyristor bridge converter with d.c. motor load and well-smoothed armature current: (a) essential circuit; (b) voltage and current waveforms for delay angle $\alpha = 30° = \pi/6$ rad; (c) as (b) but for $\alpha = 120° = 2\pi/3$ rad.

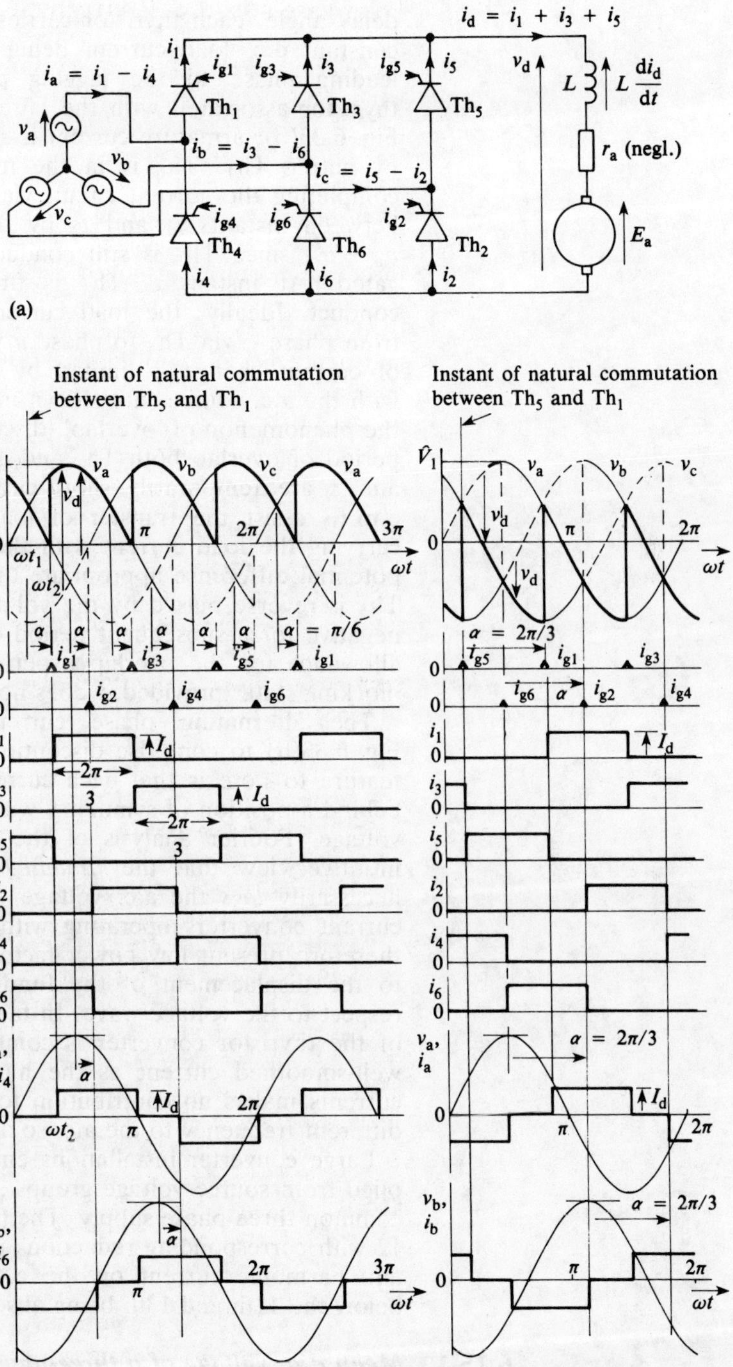

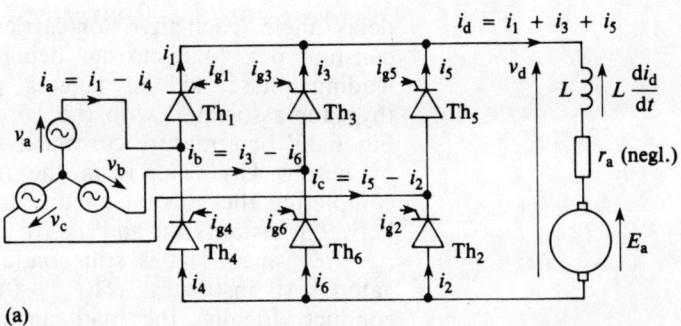

thyristor to recover to the blocking state prior to the re-application of forward bias.

Instantaneous voltage waveforms for v_d with $\alpha = 30°$ and $120°$ are shown in Figs. 6.33(b) and (c), repectively. Regardless of the delay angle, each thyristor carries current over 120° intervals, the constant d.c. load current being naturally commutated from the leading phase to the lagging phase when the forward-biased thyristor associated with the latter is fired. Thus, at instant t_2 in Fig. 6.33(b), armature current is switched from phase c to phase a by gating Th_1. Th_6 is in the middle of its conducting period, completing the current circuit via phase b. Th_1 is forward biased between instants t_1 and t_2 by the increasing voltage difference $v_a - v_c$, since Th_5 is still conducting and Th_1 has not yet been gated. At instant t_2 Th_1 is fired and immediately begins to conduct. Ideally, the load current is instantaneously transferred from phase c via Th_5 to phase a via Th_1 but, in practice, the rate of current change is limited by inductance invariably associated with the a.c. source (not shown in Fig. 6.33(a)), thus giving rise to the phenomenon of 'overlap', discussed in Section 6.16. During the period of overlap both Th_1 and Th_5 are conducting and phases a and c are temporarily short circuited, but the changing source e.m.f.s assist the transfer of load current so that Th_1 emerges carrying the load current with the small, positive, anode–cathode potential difference appropriate to a conducting thyristor, whereas Th_5 is reverse biased by the voltage difference v_{ca} which remains negative for a subsequent period equivalent to $(180° - \alpha)$, less an allowance for the overlap experience. Thus Th_5 recovers to the blocking state, provided α does not approach 180° too closely.

The alternating phase current waveshapes are shown in Fig. 6.33(b) to comprise discontinuous, rectangular blocks. A key feature to note is that each current block is delayed by angle α behind a position of symmetry with respect to the applicable phase voltage. Fourier analysis of the current waveform confirms the intuitive view that the *fundamental* component of a.c. current necessarily *lags* the a.c. voltage by α. Alternating-current/direct-current converters operating with a delay angle approaching 90° therefore present low power factor loading to the a.c. supply due to the displacement of the fundamental current waveform with respect to the voltage wave. In fact, the low power-factor property of the thyristor converter is compounded when operating with a well-smoothed current as the harmonic content of the a.c. line currents makes no contribution to *active* power transfer, being of different frequency to the a.c. source voltage.

Large converter installations employ double-bridge circuits supplied from source voltage groups 30° out of phase, derived from a common three-phase supply. The pulse number is thereby raised to 12 with corresponding reductions in voltage ripple on the d.c. side and harmonic current on the a.c. side, with harmonic currents below the 11th and 13th being absent.

6.15.1 Mean d.c. voltage of a three-phase bridge converter with continuous current flow

In general, the d.c. side voltage of a p-pulse bridge converter is completely described by the envelope of the a.c. phase voltages involved during the common conduction period of one pair of thyristors. In particular, during the 60° period when Th_1 and Th_2 are both conducting in the three-phase bridge converter of Fig. 6.33, the mean voltage developed across the d.c. load, for any delay angle α, is

$$V_d = v_{dav} = \frac{3}{\pi} \int_{\pi/6+\alpha}^{\pi/6+\alpha+\pi/3} (v_a - v_b)\, d\omega t$$

$$= \frac{3}{\pi} \hat{V}_a \int_{\pi/6+\alpha}^{\pi/2+\alpha} \left[\sin \omega t - \sin\left(\omega t - \frac{2\pi}{3}\right)\right] d\omega t$$

$$= 2\left(\frac{3}{\pi}\right) \sin\left(\frac{\pi}{3}\right) \hat{V}_a \cos \alpha \qquad [6.24]$$

$$= 2.34 V_{a_{rms}} \cos \alpha.$$

As with eqn [6.22] which is applicable to a single-phase bridge converter, the mean d.c. voltage of the polyphase converter is proportional to $\cos \alpha$. Net inversion occurs with $\alpha > 90°$, provided that the polarity of the machine e.m.f. is reversed. Allowing for armature circuit resistance and assuming a smooth armature current, the necessary relationship between armature e.m.f. and converter mean d.c. voltage is then

$$|V_d| = |v_{dav}| = |E_a| - I_d r_a.$$

6.16 Overlap

In the analyses of converter systems carried out in the previous sections it has been assumed, not only that the diodes or thyristors are ideal, but also that the transfer of load current from one phase of the a.c. system to another occurs instantaneously. In practice, inductance associated with the a.c. voltage sources obliges the rate of change of current carried by the phases involved to be finite, such source inductance being concentrated largely in the leakage inductance of any transformer directly connected to the converter. Voltage drop across the resistance of the transformer windings is generally small, and is negligible in comparison with the effect of leakage reactance; the reactance/resistance ratio at supply frequency being of the order of 5 to 10 in large installations (> 100 kVA). In many applications it is in order to neglect the resistive voltage drop across the semiconductor devices when conducting, together with the resistance of converter transformer windings.

The effect of a.c. source inductance is to delay commutation and modify the voltage developed on the d.c. side of the converter. During the period of current transfer from one phase to another the current is shared in varying proportion by the two phases involved. The phenomenon is called *overlap* and has the effect of making more negative the mean d.c. voltage of the converter. The duration of overlap and its depression of d.c. voltage is dependent upon load current value.

Referring to Fig. 6.34(a) and (b), a controlled bridge converter is considered to be operating with a small delay angle α and smooth d.c. current constant at I_d. Prior to the attempted transfer of current from leading phase *a* to lagging phase *b*, $i_a = I_d$ and $i_b = 0$. When at $t = t_1$ commutation to phase *b* should naturally occur, Th$_3$ begins to conduct in response to a gate pulse. The inductance L_s in the leading phase resists an immediate collapse of

i_a to zero, developing an e.m.f. of self induction with value $L_s di_a/dt$ directed so as to sustain i_a. Simultaneously, an e.m.f. $L_s di_b/dt$ tends to oppose the immediate establishment of current in phase b.

During the period of overlap, both thyristors are conducting, maintaining common anode and cathode potentials. The a.c. sources of e.m.f. are momentarily short circuited, with the current limited by the source inductances of the two phases involved and driven by the difference of the source e.m.f.s. Within the short-circuited loop the difference in source e.m.f.s is balanced by the difference of the induced voltages. Hence, from Fig. 6.34(a),

$$e_b - e_a = L_s \frac{di_b}{dt} - L_s \frac{di_a}{dt} = 2L_s \frac{di_b}{dt} \quad [6.25]$$

Figure 6.34 Overlap – transfer of load current from leading phase a to lagging phase b: (a) effective circuit; and (b) voltage and current waveforms.

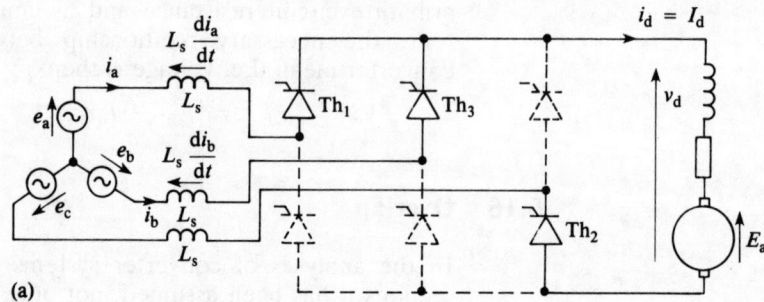

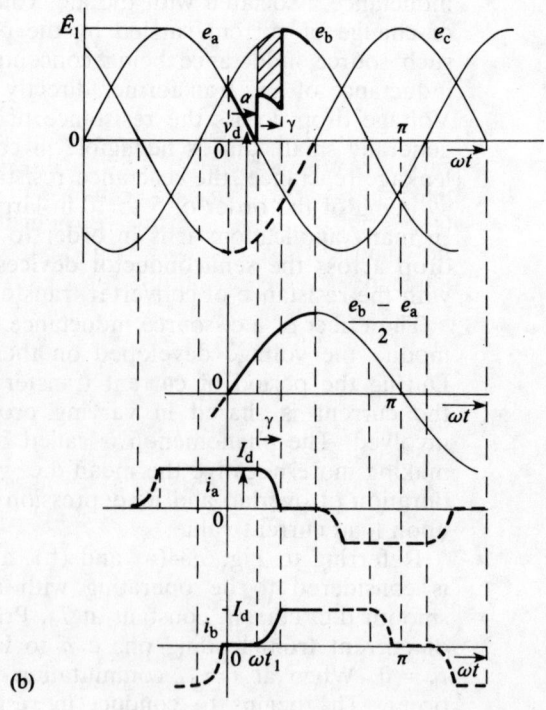

since the rates of change of current in phases a and b are complementary.

Referring to Fig. 6.34(b) the source e.m.f. difference during commutation, expressed with reference to $t = 0$ at the instant of natural commutation between phases a and b via Th_1 and Th_3 (appropriate to $\alpha = 0$), is

$$e_b - e_a = \hat{E}_a \cos\left(\omega t - \frac{\pi}{3}\right) - \hat{E}_a \cos\left(\omega t + \frac{\pi}{3}\right)$$

$$= 2\hat{E}_a \sin\left(\frac{\pi}{3}\right) \sin \omega t. \qquad [6.26]$$

Solution of eqns [6.25] and [6.26] for i_b, and the initial condition that $i_b = 0$ when $\omega t = \omega t_1 = \alpha$ gives

$$i_b = \frac{\hat{E}_a}{\omega L_s} \sin\left(\frac{\pi}{3}\right)[\cos \alpha - \cos \omega t]. \qquad [6.27]$$

Thus i_b takes the form of an offset cosine waveshape, zero-valued at the instant commutation is initiated and attaining the value I_d when $\omega t = \alpha + \gamma$ and i_a has fallen to zero. Appropriate substitution in eqn [6.27] enables the angle of overlap γ to be calculated from

$$\cos(\alpha + \gamma) = \cos \alpha - \frac{I_d \omega L_s}{\hat{E}_a \sin\left(\frac{\pi}{3}\right)}. \qquad [6.28]$$

During overlap v_d is modified, being displaced from the envelope of the lagging phase e.m.f. e_b by the voltage developed across the source inductance of the same phase. Hence, during overlap

$$v_d = e_b - \left(\frac{e_b - e_a}{2}\right) = \frac{e_b + e_a}{2}.$$

The instantaneous d.c. voltage is seen to follow the displaced sinusoidal envelope given by the *instantaneous mean* value of the phase e.m.f.s involved. The change in the waveform of v_d affects its average value. The effective reduction in $v_{dav} = V_d$ is given by the shaded 'lost' area in Fig. 6.34(b), divided by the interval between successive commutations, equivalent to $\pi/3$ radians for the three-phase bridge circuit.

Thus the reduction in mean d.c. voltage is

$$\Delta V_\gamma = \frac{3}{\pi} \int_\alpha^{\alpha+\gamma} \left(\frac{e_b - e_a}{2}\right) d\omega t$$

$$= \frac{3}{\pi} \int_\alpha^{\alpha+\gamma} \hat{E}_a \sin\left(\frac{\pi}{3}\right) \sin \omega t \, d\omega t$$

$$= \left(\frac{3}{\pi}\right) \sin\left(\frac{\pi}{3}\right) \hat{E}_a [\cos \alpha - \cos(\alpha - \gamma)]. \qquad [6.29]$$

Equation [6.24] for the mean value of d.c. voltage developed by a three-phase, fully controlled bridge converter may be modified to take overlap into effect using eqn [6.29]

$$V_{d\gamma} = V_d - \Delta V_\gamma$$

$$= \left(\frac{3}{\pi}\right)\sin\left(\frac{\pi}{3}\right)\hat{E}_a[2\cos\alpha - \cos\alpha + \cos(\alpha + \gamma)]$$

$$= \left(\frac{3}{\pi}\right)\sin\left(\frac{\pi}{3}\right)\hat{E}_a[\cos\alpha + \cos(\alpha + \gamma)]. \qquad [6.30]$$

ΔV_γ, the mean value of converter voltage 'lost' by overlap may alternatively be expressed in terms of the load current I_d.

$$\Delta V_\gamma = \left(\frac{3}{\pi}\right)\omega \int_{\alpha/\omega}^{(\alpha+\gamma)/\omega} \left(\frac{e_b - e_a}{2}\right) dt$$

$$= \left(\frac{3}{\pi}\right)\omega \int_{\alpha/\omega}^{(\alpha+\gamma)/\omega} L_s \frac{di_b}{dt} dt.$$

Changing the variable of integration from time to current i_b, and the limits of integration to correspond

$$\Delta V_\gamma = \left(\frac{3}{\pi}\right)\omega \int_0^{I_d} L_s di_b$$

$$= \left(\frac{3}{\pi}\right)\omega L_s I_d. \qquad [6.31]$$

It is apparent from eqn [6.31] that the reduction in mean d.c. voltage due to overlap is simply proportional to the product of mean d.c. current and the *commutating reactance per phase* ωL_s. It is unaffected by delay angle. A controlled a.c./d.c. converter with overlap may be simply modelled, when viewed from the d.c. side, as a voltage source in series with a source 'resistance'. In such an equivalent circuit shown in Fig. 6.35, R_s accounts for the d.c. voltage difference between $E\cos\alpha$ and $V_{d\gamma}$. Unlike r_a it is not associated with real power loss.

Figure 6.35 Equivalent-circuit model of a controlled converter, with overlap, supplying a d.c. motor load.

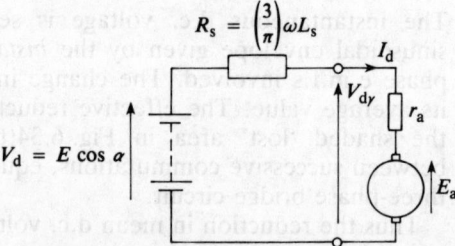

Equations [6.27] to [6.31] are valid for all values of α. With inverter operation the general effect of overlap is to increase the magnitude of $V_{d\gamma}$.

6.17 Anti-parallel Bridge Converters for Four-quadrant, Controlled d.c. Motor Drives

The provision of 'back-to-back' fully controlled converter bridge circuits connected across the armature of a d.c. machine enables operation with the options of rotation in either direction as motor or as generator. Mechanical loads may be braked regeneratively, offering high efficiencies in demanding duty cycles. The essential circuit arrangement is shown in Fig. 6.36 with the armature circuit inductance, crucial to the operation of the system, being prominent. The procedure to be described relates to initial motor operation, with converter 1 providing the supply and having a delay angle α_{10} of around 20°. Converter 2 is initially inactive. The motor load is to be braked, brought to rest and then accelerated against mechanical load torque in the opposite direction. This is achieved without any switching of armature or field circuits, by control of the thyristor gate firing delay angles, whilst enabling the bulk of the kinetic energy released on braking the mechanical load to be returned to the electrical supply over a period of time. Three-phase bridge circuits are shown, for which industrial drive ratings ≈ 1000 kW are commercially available, but the principles of control and operation are equally valid for fully controlled circuits with higher or lower pulse numbers. The a.c. converters of the bridge circuits will normally be paralleled, as shown.

The initial phase of the braking procedure has the object of reducing the armature current to zero. This is achieved by increasing the delay angle of converter 1 to approach 160°, so reversing the polarity of the mean value of v_{d1}, which retains significant ripple. The armature e.m.f. E_a remains unchanged in the short term, maintained by the constant field current and the yet unaffected speed. The armature current i_a, which is well smoothed by the armature circuit inductance L, cannot change immediately but would eventually respond to the polarity reversal of V_{d1}, now co-directed with E_a, by reversing its direction of flow. In the short term, i_a falls towards zero at a rate determined by the circuit parameters. During this period the machine experiences a diminishing electromagnetic torque, still in the direction of rotation but insufficient to support the connected mechanical load torque which thus requires deceleration of the rotating system – hence a fall in speed. The reducing electrical power absorbed by the armature is supplied by the armature inductance which also returns energy to the supply via inverting converter 1.

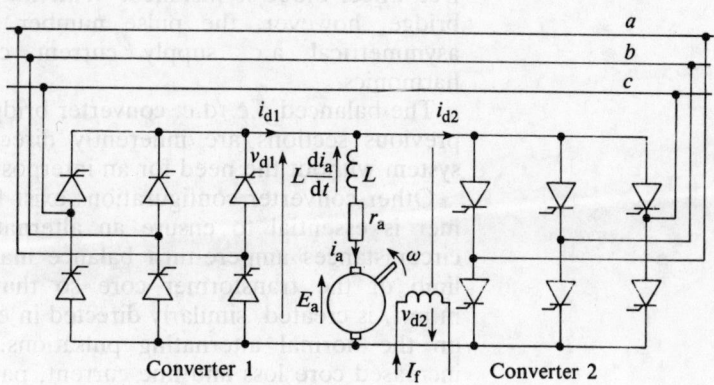

Figure 6.36 Four-quadrant d.c. machine drive employing two anti-parallel bridge converters.

The second phase of the controlled reversal is initiated when the armature current, and electromagnetic torque, have fallen to zero. Converter 1 is turned off and converter 2 is activated with a delay angle in excess of 90°, such that v_{d2} has a negative mean value with modulus less than E_a. $V_{d2} + E_a$ thus has a positive resultant, equal to $Ld(-i_a)/dt$, allowing i_a to build up in the reverse direction, completing its circuit as i_{d2} via converter 2 in Fig. 6.36. Regenerative braking of the connected mechanical load is then taking place, the energy corresponding to an armature current held constant by a progressive reduction of delay angle α_2 being returned to the supply. When the speed has fallen to zero, α_2 is reduced below 90° towards zero, making V_{d2} positive. The reversed direction of i_a is then maintained as the rotor accelerates in the opposite direction to the required final condition – motoring, supplied by converter 2. During these control procedures armature current is continuously monitored, with excessive values being avoided by adjustment of thyristor gate-firing delay angles α_1 and α_2.

6.17.1 Single bridge with switched armature

The scheme described above has the disadvantage of requiring two fully controlled, fully rated bridge circuits. One of these may be dispensed with if, at the instant of current zero the armature connections are reversed. v_{d1}, say, is then reapplied with a negative mean value less than the *magnitude* of E_a. After regenerative braking and speed reversal α_1 is reduced below 90° to achieve motor operation in the opposite direction.

6.18 Half-controlled, Polyphase Bridge Converter

As pointed out in Section 6.14.3 with reference to the single-phase bridge converter there are advantages to be gained by replacing with diodes the set of thyristors having common-anode connections. Provided reverse power flows are not required, this alternative configuration offers simpler control circuits, a higher power factor and a larger mean d.c. voltage with less ripple at a given delay angle, with commutation prospects being improved if a free-wheel diode is installed. With the three-phase half-controlled bridge, however, the pulse number remains at three and the asymmetrical a.c. supply current contains undesirable even harmonics.

The balanced a.c./d.c. converter bridge circuits considered in the previous sections are inherently directly presentable to an a.c. system without the need for an interposing transformer.

Other converter configurations exist for which a supply transformer is essential to ensure an alternating line current. In some circumstances ampere-turn balance may not be achieved in each limb of the transformer core so that a unidirectional residual m.m.f. is created, similarly directed in each limb and superimposed on the normal alternating pulsations. The effect gives rise to increased core loss and line current, particularly if the transformer core saturates at a low value of flux density (see Tutorial Example **4.13.**)

6.19 Closed-loop d.c. Motor Control System

Very tight control of the speed of a d.c. motor drive may be achieved by incorporating the machine within a closed loop, employing feedback of a signal proportional to the motor speed for comparison against a reference quantity. The difference or *error signal* provides appropriate adjustment to the gate-firing delay angles of the converter providing the armature supply. Speed of response is limited by the mechanical time constant and the need to avoid excessive increases in thyristor current levels. Rates of acceleration and deceleration of motor loads are dependent upon the value of armature current – itself a function of armature e.m.f., resistance and inductance, and converter voltage. Under dynamic conditions the current flowing through the armature via sequentially switched thyristors may vary over a wide range and steps must be taken to limit current flow to avoid thyristor damage. Such current-limiting precautions will be incorporated in motor-starting and speed-control procedures.

The current feedback signal is usually a voltage, derived by monitoring the armature current and employed to have the general effect of increasing the gate delay angle of the converter.

Figure 6.37 shows in block diagram form a closed-loop speed-control system for a d.c. machine incorporating current feedback. The need for electrical isolation precludes direct derivation of the current feedback signal from the motor armature circuit. An indirect measure may be derived from conventional current transformers connected in the a.c. supply lines to the converter. Alternatively, a d.c. c.t. may be used with its control winding connected directly in the armature circuit. This device employs two independent saturable magnetic circuits, each provided with two windings – one connected via a low-resistance 'burden' to an a.c. voltage source of constant voltage and frequency, the other *control* winding carrying the d.c. current to be measured. Relative winding senses are such that the a.c. and d.c. m.m.f.s are cumulative on each core in turn during alternate half cycles of a.c. supply current.

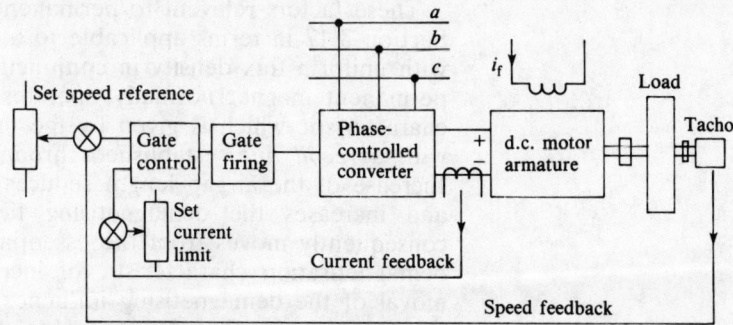

Figure 6.37 Closed-loop speed-control system for a d.c. machine with current feedback.

The d.c. current should be sufficient on its own to cause saturation of each magnetic circuit. As the a.c. supply voltage proceeds to change from a peak value of arbitrary polarity, one particular saturated core develops no opposing e.m.f. in its a.c. winding to limit the a.c. circuit current rise. The a.c. current has the effect of driving this core yet further into saturation but has the opposite effect on the other core, taking it out of saturation such that the rate of change of flux linking its a.c. winding balances the instantaneous a.c. supply voltage (resistance neglected) over the duration from negative peak (say) to positive peak. The current in the a.c. circuit thus rises to the value at which the ampere-turns of a.c. and d.c. windings on the second core are in balance appropriate to a high-permeability magnetic core, and remain thus until the flux in the second core is restored to the saturation value at the end of the half cycle. The a.c. current approximates to a rectangular waveform of amplitude proportional to the d.c. current to be measured.

6.20 Permanent-magnet d.c. Machines

The incorporation of a permanent magnet field system in d.c. machines offers the possibility of economies in volume, weight and manufacturing and running costs. These advantages are at the expense of reduced flexibility in operation as the radial flux density in the airgap is no longer directly controllable by field-winding current. The space-saving benefit associated with the absence of field coils is more pronounced with machines utilising smaller poles. The lack of field-current control may not be important in motor applications with a controlled armature voltage supply, and the absence of field-circuit resistance loss permits increased efficiency. A practical problem is the possibility of loss of permanent magnetisation through misuse or overload in service, or through dismantling during maintenance or repair. Dependent upon the materials employed, there may be a problem in providing the initial magnetisation after construction or reassembly. These considerations assume less importance with modern permanent-magnet materials incorporating rare-earth elements, which possess the twin benefits of high coercive force and a recoil line coincident with the linear demagnetisation characteristics displayed in the second quadrant of the main B/H loop.

These factors relevant to permanent magnetism are discussed in Section 3.17 in terms applicable to a sample of material operated with uniform flux density in conjunction with a series airgap. The permanent magnet normally operates about a point on the B/H characteristic which is given by the intersection of the *airgap line* with a *recoil line*, established through a process of *stabilisation*. Increase of the airgap length reduces the slope of the airgap line and increases the demagnetising field. If the operating point consequently moves from the essentially linear recoil line on to a demagnetisation characteristic of increased slope, subsequent removal of the demagnetising influence will not permit recovery to the original gap flux density value. Section 3.17.4 considers the effect of current linked by the magnetic circuit to move the base position of the airgap line from the origin of the B/H plane – to the left if the linked current is demagnetising. The current carried by the armature conductors of a heteropolar d.c. machine on load can have such an influence under appropriate circumstances.

As noted in Section 6.5.1, the armature current of a heteropolar d.c. machine establishes a triangular radial field distribution over the airgap when the gap is assumed to be bounded by iron of infinite permeability and of uniform length. For brush positions such as to maximise the torque and armature e.m.f. the radial field peaks in the space between the poles – the quadrature axis. At the mid-point of the poles, therefore, the influence of the armature *radial* field is zero. It is greatest at the pole edges acting, at one side, to enhance the radial field due to the 'field' winding – or permanent magnet – and, at the other side, to oppose it. If the armature current is large enough, loss of permanent magnetism may occur in that portion of the permanent-magnet material subjected to reduced flux density. An approximate method of evaluating the demagnetising field due to the armature current equates this to the value developed radially across the airgap on the assumption of infinitely permeable iron boundaries, but accurate assessment requires detailed modelling of the B/H characteristics of the magnetic media.

The likelihood of undesirable permanent-magnet demagnetisation may be reduced by careful design and shaping of the pole profiles, and also by the provision of low-reluctance pole shoes which serve to divert flux established in the pole-face assembly by the armature field along paths directed at right-angles to the permanent-magnet flux.

6.21 Brushless Permanent-magnet d.c. Machines

Direct-current supplied machines with electronically commutated stator armature coils, excited by permanent magnets mounted on the rotor are currently a subject of much interest as a means of meeting the increasing demand for controllable, high-speed, high-efficiency motors in the low-power range extending from tens of watts to several kW.

Although generally classified as d.c. machines since their armature coils are connected via a 'commutator' to a d.c. supply, they may alternatively be viewed as members of the family of synchronous a.c. machines, supplied by the output of an inverter, where the speed of the machine is synchronised to the inverter frequency. With the latter machines, however, which are discussed in Chapter 7, cognisance is not normally required of the instantaneous position of the rotor, whereas, in the d.c. machine, the commutator switching procedures are controlled by sensors which provide this information. The armature coil switching is normally arranged such that the *average* position in space of the tangential airgap field of the armature is in phase with the radial permanent-magnet field.

The usual brushed d.c. machine has a continuous armature winding whose coils carry currents of identical magnitude at all times, except when undergoing commutation, during which time the current reverses, ideally at a constant rate. The number of armature coils is ideally large. The tangential field in the airgap due to the armature is thus of rectangular distribution and the radial field is triangular. The electronic commutation of such a winding is discussed in Section 6.8.2. With an alternative, preferred brushless d.c. machine format, which employs fewer switching devices, the armature current is switched sequentially between groups of coils (or 'phases') and, within a group of coils may be alternating and may be discontinuous. The number of 'phases' may be 2, 4 or commonly 3. The waveform of e.m.f./phase induced by the permanent-magnet field as the rotor rotates is essentially trapezoidal, giving rise to the alternative designation as a *brushless trapezoidal synchronous a.c.* machine.

We shall consider a three-phase(d) *bipolar*-driven motor, so-called because the current in the armature phase windings alternates with time and hence requires a bi-polarity current source to enable continuous torque development as the rotor magnet poles of alternate N and S polarity sweep past. The power circuit arrangement employing a transistor bridge circuit controlled by photo-transistors is shown in Fig. 6.38(a), with, in (b), a cross-section of a six-pole stator having its armature winding accommodated in one slot/pole/phase. A rotating shutter coupled to the shaft ensures that the power transistors are sequentially switched from off (cut-off) to on (saturated). The waveforms presented in Fig. 6.38(c) are appropriate to a switching sequence in which each phase winding carries current for 120° of each half period, and each transistor for 120° of each complete period, before repetition of the switching sequence. Two transistors are 'on' at any time, the load current generally flowing via a pair of phase windings and being *commutated* between transistors of the same group – i.e. common-collector 1, 3, 5 or common-emitter 2, 4, 6 – alternately at 60° intervals. Thus, referring to Fig. 6.38(c), at $\omega t = 60°$, the current flow via T_1 through phase a is sustained whilst being transferred from phase b via T_6 to phase c via T_2. At $\omega t = 120°$, current is transferred from phase a to phase b as T_3 is turned on and T_1 is turned off whilst i_c is maintained. The switching transients incorporated in the phase current waveforms indicate that commutation is not instantaneous but is delayed, due to the effect of e.m.f.s induced within the machine windings. Idealised phase voltage waveforms are also shown – an investigation of the *difference* of terminal voltages appropriate to the current-carrying phases shows this to equate to the source voltage E at all times.

The anti-parallel diodes of Fig. 6.38(a) permit regenerative braking to be carried out automatically if the transistors are turned off and the machine speed has a value at which the mean d.c. voltage equivalent of the phase e.m.f.s exceeds that of the controllable source E. The diodes also have an important role during commutation in permitting the phase current to decay naturally whilst rapid turn-off of the previously conducting transistor is allowed. Thus diode D_4 enables i_a, when positive, to be sustained when T_1 turns of at instant t_1. The pulse of current through D_4 completes its circuit either via saturated transistor T_2 or via source E and diode D_3 with energy recovery by the source.

On the assumption of a uniform radial field (derived principally from the permanent magnet) in the region of airgap over the section of stator surface carrying instantaneous current, the instantaneous torque due to each phase winding approximates to the modulus of the phase current waveform. Ripple in the resultant torque due to all phases is enhanced by the *cogging* effect of the small number of stator teeth per pole and the wide variation locally in airgap length. The influence of cogging may be reduced by skewing the stator slots by up to one slot pitch over the axial length of the lamination pack. The stator conductors are then no longer directed purely axially but have a small tangential component.

It is important to guard against possible demagnetisation of the permanent-magnet field system when the brushless d.c. machine is on load. Because the rotor-mounted magnets maintain an essentially constant motion whereas the armature m.m.f. progresses in discrete steps, an extensive area of the permanent-magnet poles is subject to the demagnetising influence of the armature field. This contrasts with the conventional brushed d.c. motor in which only the trailing pole tips (when motoring) are worst affected.

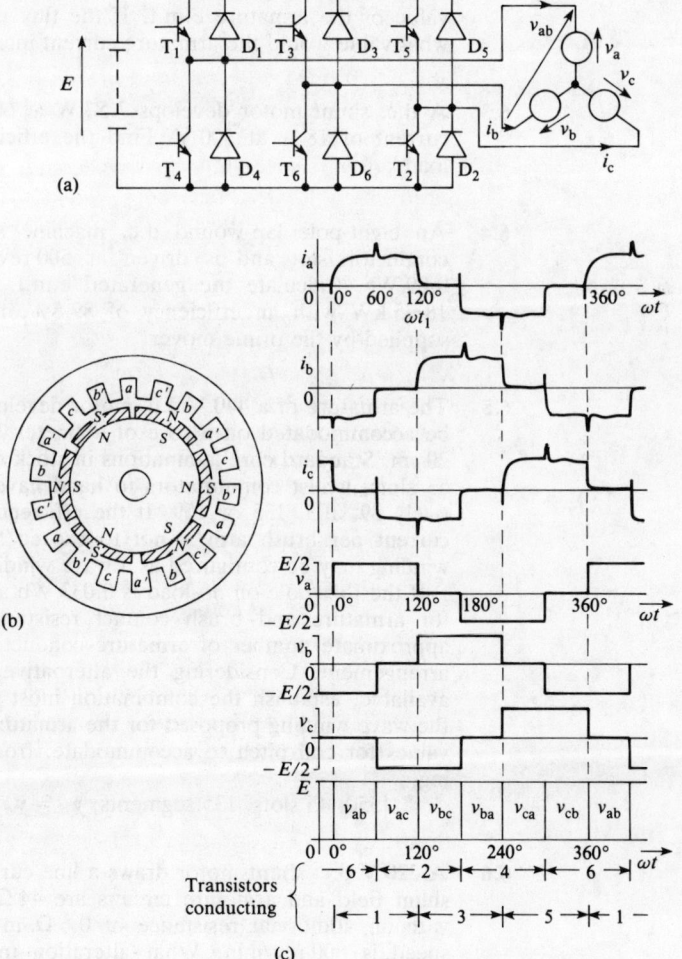

Figure 6.38 Three-phase(d) brushless d.c. motor: (a) photo-transistor bridge drive circuit; (b) section showing permanent-magnet rotor poles and stator winding with 1 slot/pole/phase; and (c) phase voltage and current waveforms.

If torque pulsations must be kept to a minimum in a brushless d.c. motor drive the magnetic circuit may be designed to induce an essentially sinusoidal e.m.f. in the armature phases. When employed in conjunction with a d.c.-derived armature supply, controlled by a high-resolution shaft encoder or 'resolver' to provide appropriately phased sinusoidal voltages for all shaft positions, such a machine may be termed a *brushless sinusoidal synchronous a.c.* machine.

6.22 Tutorial Examples

6.1 A d.c. shunt machine, driven as a generator at 700 rev/min is supplying an armature current of 60 A at a terminal voltage of 220 V. If the same machine is run as a motor with the same terminal voltage and an armature current of 40 A, calculate the speed. The armature circuit resistance is 0.2 Ω. Neglect any change in flux due to armature reaction.

[640 rev/min]

6.2 A d.c. shunt motor running off a 220 V supply takes an armature current of 15 A. The resistance of the armature circuit is 0.8 Ω. Calculate the value of the armature e.m.f. If the flux is suddenly reduced by 10%, to what value would the armature current increase momentarily?

[208 V, 41 A]

6.3 A d.c. shunt motor develops 7.5 kW at 600 rev/min when drawing a line current of 18 A at 500 V. Find the efficiency and useful torque at this load.

[83%, 119 N m]

6.4 An eight-pole lap-wound d.c. machine armature has 96 slots with six conductors/slot and is driven at 500 rev/min. The useful flux/pole is 0.09 Wb. Calculate the generated e.m.f. If the above machine delivers 18.75 kW with an efficiency of 89.5%, find the torque required to be supplied by the prime mover.

[432 V, 400 N m]

6.5 The armature of a 440 V d.c. motor developing 50 kW at 700 rev/min may be accommodated on a core of diameter 37 cm and effective axial length 20 cm. Standard core laminations in stock of this diameter have 45 slots or 53 slots, whilst commutators to hand have numbers of segments alternatively 89, 105, 135 or 159. If the efficiency is estimated at 89% and the current per brush arm is not to exceed 500 A, show that the armature winding may be configured as a wave winding with two parallel circuits.

If the flux/pole on no-load is 0.033 Wb and 17 V is allowable to account for armature and brush contact resistance on full load, calculate the approximate number of armature conductors required in a wave winding arrangement. Considering the alternative laminations and commutators available, establish the combination most appropriate for accommodating the wave winding proposed for the armature circuit. State also appropriate values for coil pitch to accommodate, from choice, a retrogressive winding.

[550; 45 slots, 135 segments; $y_b = y_c = 67$ coil sides, 540 conductors, 2 turns/coil]

6.6 A 220 V d.c. shunt motor draws a line current of 50 A. The resistances of shunt field and armature circuits are 44 Ω and 0.02 Ω, respectively and, with an additional resistance of 0.6 Ω in series with the armature, the speed is 800 rev/min. What alteration must be made in the armature circuit to raise the speed to 850 rev/min, the torque remaining constant?

[Reduction of 0.267 Ω]

6.7 A d.c. series motor of negligible resistance, operating on a linear portion of the magnetisation characteristic and running at a certain speed under a given load, draws a line current of 55 A at 420 V. If the load torque varies as the cube of the speed, find the value of additional armature resistance necessary to run at half speed.

[17.8 Ω]

6.8 A d.c. shunt machine generates 250 V on open circuit when driven at 1000 rev/min. Armature resistance, including brushes, is 0.5 Ω, and field resistance is 250 Ω. Running as a motor on no load, the machine takes 4 A at 250 V. Calculate the speed and efficiency of the machine as a motor taking 40 A at 250 V. Armature reaction weakens the field by 4% at this load.

[960 rev/min; 82.5%]

6.9 A d.c. series motor with negligible armature and field resistance, and operating on the unsaturated part of the open-circuit characteristic when driving a load, takes 40 A at 440 V. If the load torque varies as the square of the speed, find the series resistance necessary to reduce the speed by 25%.

How could the speed be increased from the base value?

[6.42 Ω]

6.10 Explain the principle of speed control of a d.c. shunt motor by variation of the field current.

A 240 V d.c. shunt motor runs at 800 rev/min with no extra resistance in the field or armature circuits on no load. Determine the resistance to be placed in series with the field winding so that the motor may run at 950 rev/min when taking an armature current of 20 A. The field-circuit resistance is 160 Ω and the armature resistance is 0.4 Ω. Neglect magnetic saturation.

[36.6 Ω]

6.11 The resistance of the armature circuit of a 250 V d.c. shunt motor is 0.3 Ω and its full-load speed is 1000 rev/min. Calculate the resistance to be inserted in series with the armature to reduce the speed at full-load torque to 800 rev/min, the full-load armature current being 50 A. If the load torque is halved, at what speed will the motor run? Neglect the effects of armature reaction, friction and windage.

[0.94 Ω, 932 rev/min]

6.12 Derive an expression for the torque of a d.c. machine in terms of the armature current, winding constants and (radial) flux/pole.

A four-pole shunt d.c. motor has its armature lap wound with 1040 conductors and runs at 1000 rev/min when taking an armature current of 50 A from a 250 V d.c. supply. The field current is 4 A, the armature circuit resistance is 0.2 Ω and the torque required to overcome windage and friction accounts for 700 W at the given running condition. Calculate

(i)   the useful flux/pole,
(ii)   the torque developed by the armature,
(iii)   the load torque available from the shaft, and
(iv)   the motor efficiency.

U. of B.   [(i) 0.0138 Wb; (ii) 114.6 N m; (iii) 107.9 N m; (iv) 83.7%]

6.13 Write brief notes on the sources of power loss in d.c. machines.

A 40 kW, 500 V shunt d.c. motor has a full-load efficiency of 0.87 and runs at 750 rev/min. If a series field winding is added to raise the speed to

800 rev/min, estimate the resulting armature (rotor) current and efficiency. The armature resistance is 0.4 Ω, the series field winding resistance is 0.1 Ω and the shunt field winding resistance 250 Ω. Assume that the gross power converted ($E_a I_a$) and the 'fixed' losses remain constant.
U. of B. [91.5 A; 85.2%]

6.14 A constant-field d.c. motor connected to a pure inertia load has a total inertia of 30×10^{-3} kg m². Armature resistance is 1.0 Ω and the motor develops an open-circuit voltage of 0.3 V at its terminals per radian per second of shaft rotation. Determine the initial acceleration of the rotor from standstill if the armature current is limited to an initial value of 5 A.
[50 rad s^{-2}]

6.15 Explain briefly why the d.c. series motor provides a suitable drive characteristic for many traction purposes.

A railway vehicle is propelled by two identical series motor each driving the wheels on one axle. The wheels on one of the axles have a diameter of 1.0 m, while those on the other have a diameter of 0.95 m. The resistance of each motor is 0.08 Ω. The characteristics of each drive when supplied at 650 V and with wheels of diameter 1.0 m are as follows.

1/speed × 10⁻² h/km	2.63	2.44	2.27	2.08	1.75
Vehicle speed (km/h)	38	41	44	48	57
Tractive effort (kN)	36	28	21	14	7
Motor current (A)	700	550	450	350	250

Determine (i) the speed of the vehicle and (ii) the mechanical output from each wheel pair, when the motors are connected *in series* to a 650 V supply and take a current of 400 A.
E.C. [(i) 21.25 km h⁻¹; (ii) 120.2 kW, 114.2 kW]

6.16 Two methods of braking a small, separately excited d.c. motor are:

(i) *dynamic braking* – where the armature supply is switched off, and the armature is short circuited; and
(ii) *plug braking* – where the polarity of the armature voltage is reversed until the speed falls to zero.

Obtain expressions for the speed and armature current as functions of time for each method, and sketch the speed/time curves. Assume that the motor is frictionless, has resistance R and total inertia J, and is initially running unloaded with armature voltage V at speed ω_0. Neglect armature inductance.

Show that in bringing the speed to zero using method (ii) the energy dissipated in the armature resistance is equal to three times the initial kinetic energy. How does this compare with the energy dissipated in method (i)?

Summarise the advantages and disadvantages of each method. How could the kinetic energy be recovered, rather than dissipated?

$$\left\{ (i)\ i_a = \frac{V}{R}\frac{\omega}{\omega_0},\ \omega = \omega_0 \exp\left(-\frac{KVt}{JR\omega_0}\right); \right.$$

$$\left. (ii)\ i_a = \frac{V(1 + \omega/\omega_0)}{R},\ \omega = \omega_0\left[-1 + 2\exp\left(-\frac{KVt}{JR\omega_0}\right)\right];\ 3\ \text{times} \right\}$$

E.C.

6.17 Show that the current flowing in the armature circuit of a separately excited d.c. machine, when motoring with continuous armature current supplied via an ideal chopper circuit from a constant voltage source V, as shown in Fig. 6.27, is given by equations of the form:

(i) during 'chopper-off' interval

$$i_a = \frac{V - E_a}{r_a} + \left(i_{a_{\min}} - \frac{V - E_a}{r_a}\right) e^{-t/\tau_a}$$

(ii) during 'chopper-on' interval

$$i_a = -\frac{E_a}{r_a} + \left(i_{a_{\max}} + \frac{E_a}{r_a}\right) e^{-t/\tau_a};$$

where $\tau_a = L/r_a$, the armature time-constant.

6.18 The armature of a separately excited d.c. motor operating at constant field current is supplied via a chopper from a 120 V battery, developing a steady load torque of 6 N m at a speed of 1000 rev/min. The chopper frequency is 1 kHz. The motor armature inductance is 0.06 H and the generated e.m.f. per rad s^{-1} is 0.6 V. This latter factor also relates mean torque (N m) to mean armature current (A). Neglecting all chopper losses, speed fluctuations and armature resistance calculate:

(i) the mean armature current; [10 A]
(ii) the mean armature terminal voltage; [62.8 V]
(iii) the 'chopper-on' and 'chopper-off' durations, [0.524 ms, 0.476 ms]
(iv) the maximum and minimum values of armature current; [10.25 A, 9.75 A]
(v) the difference between maximum and minimum energy stored in armature inductance; [0.3 J]
(vi) the mechanical energy output during each 'chopper-off' interval; [0.3 J]
(vii) the mechanical energy output during each chopper period and the mean output power; and [0.628 J, 628 W]
(viii) the electrical energy input during each 'chopper-on' interval. [0.628 J]

6.19 An ideal battery of terminal voltage 100 V supplies power to the armature circuit of a separately excited d.c. motor, with armature resistance 0.2 Ω and armature inductance 1 mH, via a (class A) thyristor chopper controller. The field current of the motor is maintained constant at its rated value. At the lowest desired speed of operation, when the main thyristor conducts for intervals of 1 ms in each overall period of 2.5 ms, the motor e.m.f. is 20 V. Sketch a circuit diagram of the arrangement and show the armature voltage waveform, assuming continuous armature current.

Calculate the average values of the armature voltage and current, and also the power entering the motor averaged over the chopper period. Ignoring friction and windage losses estimate the motor efficiency. Finally, calculate the maximum and minimum values of the armature current.

U. of B. [40 V, 100 A, 4 kW, 50%, 130.44 A, 70.74 A]

6.20 A chopper circuit configured as Fig. 6.27 is employed to control the speed of a large traction motor supplied from a 700 V d.c. source. The armature current is well smoothed at a chopping frequency of 500 Hz and peaks at 300 A.

If the main thyristor turn-off time is 35 μs and the minimum mean output voltage of the chopper circuit is to be 50 V calculate the corresponding values of the commutation capacitor and inductor, and also the charge displaced during each commutation impulse.

Estimate the corresponding current rating of the main thyristor.

[15 µF, 0.036 mH, 0.021 C, 750 A]

6.21 A single-phase rectifier bridge is energised from a (sinusoidal) 240 V supply. Calculate the steady-state average power developed in each of the following loads:

(i) 100 Ω resistance; and
(ii) 100 Ω resistance in series with a large inductance.

[(i) 576 W; (ii) 467 W]

6.22 A single-phase, fully controlled thyristor bridge supplied from a 240 V source energises a highly inductive load incorporating a 100 Ω resistance. Calculate the value of the smoothed d.c. load current when the gate-firing delay angle α is 60°, both with and without a free-wheel diode across the bridge d.c. terminals.

[1.62 A, 1.08 A]

6.23 A thyristor converter supplies a d.c. motor with armature voltage which is variable by phase control. On no-load with no gate delay ($\alpha = 0°$) the mean d.c. voltage at the converter terminals is 500 V, falling to 460 V when the motor is drawing an armature current of 100 A. This reduction in terminal voltage is due primarily to overlap effects which increase the effective source resistance of the converter supply to the motor.

The d.c. motor has an armature resistance of 0.25 Ω.

(i) Draw the equivalent circuit of the converter/machine system and calculate its parameters.
(ii) If the machine is rated at 460 V, 100 A, 1200 rev/min, calculate the machine constant $K\Phi$ which relates e.m.f. to speed in rad s^{-1}.
(iii) Determine the required firing delay angle α if the motor is to run at half rated speed with rated flux and rated armature current, and developing rated torque.
(iv) Determine the required firing delay angle α if the motor is to run at half rated speed with rated flux and one-quarter rated torque.
(v) Show that the machine constant $K\Phi$ also relates torque to armature current.

U. of B. [(ii) 3.46 V s rad^{-1}; (iii) 55.6°; (iv) 62.1°]

6.24 Draw the circuit diagram of a three-phase, single-way (half-wave), controlled converter supplied via a delta/star transformer. If the delta-connected primary winding is to be connected to a 440 V (line) supply determine, from first principles, the necessary transformer phase turns ratio to provide a *maximum* mean d.c. output voltage of 500 V. Ignore transformer leakage inductance and forward semiconductor voltage drop in your calculation, and assume also that the d.c. current is well smoothed.

Such a converter on test yielded an output voltage near to 500 V d.c. when the output current was very small but, when fully loaded at 100 A, the d.c. output voltage fell to 480 V with the firing delay angle α remaining at 0°. If the converter is connected to a d.c. motor with armature resistance 0.3 Ω determine the minimum value of delay angle necessary in order that the armature current will not exceed 120 A on starting from rest.

If the separately excited field excitation on the d.c. motor is such that its e.m.f. is related to speed through $K\Phi = 4.0$ V s rad^{-1}, what will be the maximum starting torque developed by the motor?

U. of B. [1.03:1; 83.11°, 480 N m]

6.25 A thyristor-converter d.c. machine drive system employs a three-phase bridge circuit connected via a transformer to a 400 V a.c. system. If the maximum mean d.c. voltage to be supplied by the converter on no-load is

500 V, determine the phase turns ratio of the transformer. On test, the converter mean d.c. voltage fell from 500 V on no-load to 480 V on full load, when supplying 50 A, with no gate delay. This reduction in mean d.c. voltage is primarily accounted for by overlap effects within the converter source impedance. The separately excited d.c. motor has an armature resistance of 0.3 Ω. If the motor is to be run under rated conditions at 415 V, 50 A, 1000 rev/min, establish the required thyristor firing delay angle and the motor armature constant $K\Phi$ relating e.m.f. in V to speed in rad s^{-1}.

If subsequently the machine is to *regenerate* at rated terminal voltage, current and speed, establish the corresponding firing delay angle and the necessary flux per unit of normal flux. Neglect saturation.

[1:0.925; 29.5°, 3.82 V s rad^{-1}; 142.19°, 1.075 p.u.]

6.26 Sketch the circuit configuration of a three-phase, six-pulse double-way (bridge) converter supplied via a delta/star three-phase transformer with balanced, sinusoidal a.c. voltages and having infinite inductance on the d.c. side. Deduce the corresponding waveforms of *line* current on the delta-connected primary side of the converter transformer and show that the fundamental component of the primary line current has peak value given by $6I_d/\pi$, where I_d is the value of current on the d.c. side of the transformer referred to the primary via the phase turns ratio. Check your answer by comparing a.c. input real power with d.c. load power when the conversion is uncontrolled assuming ideal transformation and for which the ratio of mean d.c. voltage to r.m.s. secondary phase voltage is given by $3\sqrt{6}/\pi$.
U. of B.

6.27 State why converter systems are generally regarded as 'sinks' for reactive volt amperes. Describe and explain how the VA$_r$ requirements of a converter supplying a resistive load change when (i) inductance is placed in series with the load resistance to smooth the d.c. current, and (ii) gate control is subsequently applied to reduce the main d.c. voltage. Finally, show how the use of a by-pass diode on a controlled converter can raise the power factor of the installation.
U. of B.

6.28 Figure 3.37 illustrates the field structure of a two-pole d.c. motor employing strontium ferrite field magnets on the stator to produce an assumed uniform radial flux density of 0.25 T in the airgap under the poles on no-load. The rotor radius is 22 mm, the airgap length 2.5 mm, the pole arc/pole pitch ratio 0.7 and the axial length of stator and rotor 50 mm.

Determine the linear current density necessary at the rotor surface to develop a torque of 0.2 N m assuming that the brushes are set to commutate coil sides located in the interpolar space. Determine also the corresponding value of radial field intensity H_n developed in the airgap at the pole tips by the armature current, ignoring the reluctance of ferrite and rotor iron.

If the armature winding has 1300 conductors set in a large number of slots calculate the armature current and the torque constant of the motor.

[7520 A m^{-1}, 72 700 A m^{-1}; 1.6 A, 0.125 N m A^{-1}]

7 Synchronous machines

7.1 Introduction

The simplest rotating a.c. machine is called a *synchronous* machine on account of the fact that it develops a steady torque at a single speed which is related simply to the frequency of the a.c. system to which it is connected. The d.c. machine of Chapter 6 is, by contrast, capable of wide variation in speed as a consequence of change in many operational parameters including torque, armature voltage and field current. The more popular a.c. alternative, the *induction* machine considered in Chapter 8, is described as *asynchronous* because its speed departs significantly from a nominal synchronous value (related to frequency by the number of pole pairs) as the torque undergoes changes in magnitude and direction. Where a robust, inexpensive machine is required in drive applications for which actual speed within a limited range of values is not important, a.c. induction motors are usually employed. If a fixed speed independent of load is required, particularly if several non-mechanically coupled drives are required to run in synchronism, an a.c. synchronous motor may be chosen. If a controlled variable-speed drive is required, either a d.c. machine or an a.c. alternative with a variable-frequency inverter supply may be installed. Since constant-voltage, constant-frequency a.c. supplies are almost universal as an available power source – except in battery-powered transport applications – the d.c. motor will require an a.c./d.c. converter, and the a.c. machine either (i) an a.c./d.c. converter followed by a d.c./a.c. converter (inverter) or (ii) a direct a.c./a.c. frequency changer (cycloconverter). Alternatively, if an induction machine has its primary (stator) winding supplied directly from a constant-voltage, constant-frequency source, some means of abstracting (and recovering) electrical energy at variable frequency from its secondary (rotor) winding will be necessary.

Alternating-current synchronous machines are important in two contrasting applications. All bulk supplies of electrical energy are presently generated using polyphase synchronous machines. Steam-turbine driven generators rotating at speeds up to 3600 rev/min, rated in excess of 1000 MW and having rotors with diameter ≈ 1 m and an axial length of several metres compare with hydroelectric plant rated to several hundred MW, having operating speeds ≈ 100 rev/min and rotors of many metres diameter and a short axial length. At the other end of the scale, small synchronous motors with ratings from 10 to 100 W feature in control systems, computer memory disc drives and audio and video tape transport systems, with quartz crystal oscillators providing alternative frequency references to the mains supply. Whilst in the larger machines considerations of efficiency and waveform require that machine design should enable the practical realisations to approach in behaviour the theoretical model, such close identification may not be necessary with small motors where impact on the supply network may be insignificant, or for which special power supplies may be individually tailored. For these machines ease and cheapness in manufacture, which are compatible with meeting a tight specification regarding performance, are more important than close proximity to a particular stereotype. Nevertheless, machines belonging to a particular family grouping share common features which are most significantly demonstrated by an idealised model which can provide the basis for theoretical analysis of type derivatives.

All a.c. synchronous machines have the following constructional features in common. On one side of the airgap, set in a high-permeability iron core, is an a.c. winding, normally polyphase, but single-phase as a special case. On the other side of the airgap is either a magnet system, permanent or electromagnetic, or a magnetic circuit of special shape which makes the airgap non-uniform, or both. The idealised model machine has sinusoidal distributions of magnetic field and flux over the airgap boundary with the armature surface and, when the airgap length is non-uniform, a particular theoretical form which can never be realised in practice. The use of such a simple model enables performance equations to be deduced, and equivalent circuits and phasor diagrams derived, from which the behaviour of practical machines may be predicted with a degree of accuracy dependent upon how closely the practical machine approaches the theoretical model. In reality, the differences may be stark. A polyphase winding may be reduced in the limit to a pair of multiturn coils set in a pair of slots separated by half a pole pitch and energised from a two-phase, square-wave voltage source, presenting an abundance of space- and time-harmonic quantities which simply do not appear in the theoretical model. Similarly, airgap profiles may be such as to maximise contributions to torque by the provision of *saliency* for which the traditional model is quite unrepresentative. Extreme departures from a standard form make custom design procedures necessary which use basic electromagnetic field theory and modern computational techniques to predict flux and mechanical stress distributions in optimising the design for particular circumstances.

During recent years, in parallel with the development of digital electronics, a new type of synchronous machine has emerged – the *stepping (stepper) motor*. The rotor of this machine moves through an increment of angle in response to a single current pulse received and may simulate continuous motion in synchronism with a train of discrete pulses. Such a machine comes within the general category of synchronous machines, although its *modus operandi* is fundamentally different from that of the a.c. machines. The stepping motor employs deep saliency on both sides of the airgap and relies on a close identification between the dynamics of its angular motion and the transient response of its electromagnetic circuits. Its response to an isolated pulse has much in common with the elementary electromechanical devices discussed in Sections 3.10 to 3.13. The *switched reluctance motor* is a development of the stepping principle which, in conjunction with its associated power electronic circuitry, can compete on rating, efficiency and cost with alternative systems which incorporate synchronous or induction motors in many drive applications.

Synchronous machines differ from asynchronous machines (including d.c. machines) in one important respect, apart from speed dependence. In an asynchronous machine the torque may well be a function of speed but is (ideally) independent of the instantaneous position of the rotor assembly of copper and iron with respect to the stator magnetic field, assuming in the d.c. machine case that the armature is on the rotor and the brushes, like the stator poles, are fixed. In a synchronous machine drive, however, the torque is dependent on the instantaneous position of the rotor – where this carries the d.c.-exciter 'field' winding – with respect to a *synchronously rotating reference frame* defined by the system voltage. Under changing load conditions, therefore, perturbations of speed about the synchronous value will occur which may give rise to stability problems in extreme cases. For this reason the dynamics of shaft motion assume particular importance in the consideration of synchronous machines.

In the present chapter we will consider initially the constructional features of conventional polyphase synchronous machines and will then proceed to develop the equivalent circuit and phasor diagram of the idealised machine with a uniform airgap. The operational features of the machine will be investigated, including a brief consideration of the dynamics of operation on constant-voltage, constant-frequency busbars in both motoring and generating modes. The unexcited salient-pole machine, the *reluctance machine*, will then be considered, followed by the more general machine with d.c. excited salient poles. Single-phase reluctance machines and stepping motors are also subjects of study, followed by permanent-magnet machines, and finally, a representative selection of power electronic supply systems appropriate for use in controlled a.c. drive applications incorporating three-phase synchronous machines.

7.2 The Synchronous Machine – Constructional Details

The simplest format of a synchronous machine resembles that of a d.c. machine, with the armature winding and commutator/brush contact replaced by a balanced, polyphase winding set in the rotor slots and connected to the external a.c. system via *slip rings* mounted on the rotor shaft in sliding contact with stationary carbon brushes. A three-phase winding may be internally connected in star or delta so that the number of slip-rings is reduced from six to three or four. The stator core would support a d.c.-excited electromagnet system with alternate N and S poles, matching in number that for which the armature is wound. Such an arrangement is not adopted for large machines, however, owing to

Synchronous machines

the fact that armature winding voltage and current may be more greatly increased if the winding is located on the rigid stator with solid, not sliding, connections to the external circuit. The rotor then carries the field pole systems, manifested either as salient-pole electromagnets bolted to a steel yoke on large-diameter low-speed machines or as provided by the solid, forged-steel rotor itself, d.c. excited by 'field-winding' conductors set in axial slots cut into the smooth rotor surface. The 'field' winding may then be distributed in such a manner that the airgap field components it produces when excited approximate closely to a sinusoidal variation with angular position. Since the armature winding is also distributed over a cylindrical surface, concentric with the rotor shaft, the resulting machine is described variously as a *round-rotor, cylindrical-rotor* or *uniform-airgap* machine, in contrast to the *salient-pole* machine whose airgap is non-uniform.

Features of these large machines are apparent from the partially completed stator armature winding shown in Fig. 7.1 and the salient-pole rotor assembly of Fig. 7.2. Figures 7.3 and 7.4 illustrate the principal components of a two-pole round-rotor synchronous generator, gas-turbine driven at 3000 rev/min for peak load duty.

Figure 7.1 Partially wound stator segments for 11 kV, 15 MVA, 0.85 p.f., 157.9 rev/min salient-pole, two-stroke diesel-driven generator. Rotor diameter is 4.02 m, and core length is 0.90 m. (*Photograph by courtesy of GEC Turbine Generators Ltd, Stafford, UK.*)

7.3 The Uniform-Airgap Synchronous Machine – Principle of Operation

The uniform-airgap synchronous machine is normally configured as a three-phase machine with a.c. armature winding on the stator, d.c.-excited distributed 'field' winding on the rotor and with a uniform airgap in between. It is thus a *doubly excited* machine,

Figure 7.2 Rotor assembly including brushless exciter for 11 kV, 30 MVA, 0.8 p.f., 100 rev/min, salient-pole,, two-stroke diesel-driven generator. Rotor diameter is 8.80 m, and core length is 0.615 m. (*Photograph by courtesy of GEC Turbine Generators Ltd, Stafford, UK.*)

having current on both sides of the airgap and developing torque at the airgap boundaries on integrating the mechanical stresses exerted on the sides of stator and rotor teeth due to the resultant field of stator and rotor currents. In modelling the machine we shall ignore the effect of slots, assuming the axially directed current to be spread over the smooth, cylindrical and concentric iron surfaces bounding the airgap, and calculating the tangential stress developed over the appropriate surfaces of the stator and the rotor due to the effective interaction of radial flux density B_n and tangential field intensity H_t. From eqn [3.22] $t_t = B_n H_t$ with, as shown formally in Section 5.4.1, B_n a function of current on *both* sides of the airgap, whereas, at a surface bounding the airgap, H_t is dependent upon *local* current only. With sinusoidal, constant amplitude distributions of B_n and H_t over the airgap surfaces, such as will be assumed in our model, a steady resultant torque is obtained on integrating the tangential stress over each complete surface only if B_n and H_t have the same pole pitch and no *relative* motion. If, therefore, the d.c. excited rotor poles rotate at a certain speed, the *resultant* radial flux density distribution must

Figure 7.3 Completed stator for 11.8 kV, 87.5 MVA, 0.8 p.f., 3000 rev/min gas-turbine driven generator which is air cooled and with core length 4.57 m. (*Photograph by courtesy of GEC Turbine Generators Ltd, Stafford, UK.*)

also rotate at the same speed and in the same direction, with the radial field contribution due to the stator (armature) current rotating similarly.

It has been proved formally in Section 5.6.1, and will be demonstrated in Worked Example 7.1, that a three-phase winding supplied with balanced, three-phase currents generates a *fundamental* field distribution which appears to move at a fixed (synchronous) speed related solely to the frequency of the current and the pole pitch of the winding. The angular motion in electrical radians per second is precisely equal to the angular frequency ω radians (elec) per second and hence, in actual revolutions per second, equals the quotient of frequency and number of pole pairs. Thus

$$n_0 = \frac{\omega}{2\pi p} = \frac{f}{p}. \qquad [7.1]$$

Therefore, if the resultant field produced by the armature winding is to appear stationary with respect to the field of the rotor pole system, it follows that the rotor poles must rotate at synchronous speed in the same direction as the field of the

Figure 7.4 Cylindrical rotor for the 11.8 kV, 87.5 MVA stator of Fig. 7.3, with rotor diameter 0.94 m. (*Photograph by courtesy of GEC Turbine-Generators Ltd, Stafford, UK.*)

armature winding. This accounts for the fact that the a.c. synchronous machine will develop a steady torque at one speed only – synchronous speed – in the appropriate direction. At other speeds the torque will vary with instantaneous rotor position and have an average value of zero over extended periods of time.

Whilst the machine is running synchronously and developing a steady torque the stationary armature conductors have constant *relative* motion at right-angles to the radial flux density crossing the airgap. Since B_n is ideally sinusoidally distributed in space with constant amplitude (eqn [5.35]), $E = B_n l v$, suggests that an e.m.f. will be developed between the terminals of the armature winding. In conjunction with the armature current, this e.m.f. accounts for the electrical power of the armature winding which, in the absence of losses, precisely matches the mechanical power identified with the product of rotor speed and torque.

Complementary torques are developed over both rotor and stator surfaces bounding the airgap. Current is necessary on both sides of a *uniform* airgap in order that resultant torque is developed. If, however, the airgap is made non-uniform by introducing salient field poles, a resultant torque may be developed in the absence of 'field-winding' current – or permanent-magnet field. *Saliency* torque or *reluctance* torque will be considered quantitatively in Sections 7.14 to 7.24, but for the present we shall be solely concerned with the phenomenon of *alignment* torque appropriate to current distributions on both sides of the airgap, and so-called because its action on the members is such as to tend to align the radial field components due to stator and rotor winding currents.

7.4 Radial Magnetic Field Developed in a Uniform Airgap by a Balanced Three-phase Winding Energised with Three-phase Currents

Worked example 7.1

(i) Calculate and sketch over a double pole pitch the distribution of radial magnetic field intensity H_n due to reference phase *a* of a three-phase, four-pole, double-layer winding set in six slots/pole with each coil short pitched by one slot, given that each coil carries instantaneous current *i*. Assume a uniform airgap of length *g*.

(ii) By considering each of the three-phase winding groups to be displaced from its neighbouring phase groups by $\frac{2}{3}$ of a double pole pitch, calculate the distribution of the resultant field due to all three phases and developed across the uniform airgap at the instant t_1 when the current in reference phase *a* is at its maximum positive value $\hat{I}_a$

and phase currents i_b and i_c are equal at $-\frac{1}{2}\hat{I}_a$. *Estimate* the relationship between the amplitudes of the resultant field and that due to phase *a* alone.

(iii) Repeat the calculation of the distribution of the resultant radial field for the three-phase winding at a later instant t_2 when the current in phase *a* has fallen to zero. The sequence of the phase currents i_a, i_b and i_c is such that, at instant t_2 $i_a = 0$, $i_b = 0.866\,\hat{I}_a$ and $i_c = -0.866\,\hat{I}_a$.

(iv) Estimate the amplitude and position of the fundamental component of the resultant radial field in case (iii) above and hence confirm that the amplitude of the resultant field is identical at both instants t_1 and t_2, appearing to move its position through $\frac{1}{4}$ double pole-pitch during the time interval of $\frac{1}{4}$ period of current supply which separates instants t_1 and t_2.

Solution The magnetic properties of a.c. machine windings are dealt with at length in Chapter 5 but the essential features are summarised in the worked solution which follows. The first requirement is the construction of a 'developed' diagram showing the airgap opened out and the conductors in cross-section. The conductors associated with each phase need to be identified. As a solution to part (i), only the conductors associated with the reference phase *a* are significant.

Since six slots in total are allocated to each pole of the magnetic field developed on exciting the winding with current then $\frac{6}{3} = 2$ slots/pole are available for each of the three phase windings. As the winding is short pitched by one slot, each coil of each phase will span $6 - 1 = 5$ slots, whilst the adjacent blocks of 'go' and 'return' currents for each phase remain separated by the full pole pitch of six slots.

Figure 7.5(b) provides a developed view of the winding with the conductors shown in Fig. 7.5(a) to be placed arbitrarily on the stator side of the uniform airgap bounded by infinitely permeable iron. Phase *a* coils and currents are identified and the consequential airgap *radial* field H_n calculated by the application of Ampère's law (eqn [2.27]), i.e. $\oint \boldsymbol{H}_n \cdot d\boldsymbol{l} = \Sigma i$ around such elemental magnetic circuits as that shown joining the points A, B, C and D. $d\boldsymbol{l}$ is measured across the gap in the radial direction, and the direction of the line integral is right-handed about the direction of the enclosed current Σi. Care is necessary when establishing the correct direction of integration as it appears in the developed diagram. Thus

$$\oint_{ABCDA} \boldsymbol{H}_n \cdot d\boldsymbol{l} = \underbrace{(H_{n\theta m1})\,g}_{A \to B} + \underbrace{0}_{B \to C} - \underbrace{(H_{n\theta m2})\,g}_{C \to D} + \underbrace{0}_{D \to A} = 4i.$$

The particular path ABCD has been chosen to embrace the maximum possible current so that considerations of symmetry would suggest that $H_{n\theta m1}$ and $H_{n\theta m2}$ will be equal in magnitude but opposite in polarity. Hence

$$H_{n\theta m1}(= -H_{n\theta m2}) = +\frac{4i}{2g} = +\frac{2i}{g}.$$

If now the path of integration is extended to make the return gap crossing E → F at θ_{m3} rather than C → D at θ_{m2}, the total current enclosed will be reduced to $3i$. The value of the integral of $\boldsymbol{H}_n \cdot d\boldsymbol{l}$ products around the complete path is correspondingly reduced by i and the effect is acknowledged by recognising that the radial field across the gap at θ_{m3} equals $(i - 2i)/g$, with the *change* in radial field value between θ_{m2} and θ_{m3} corresponding to the change in current embraced by the alternative paths of integration. Further extension of the path of integration to include the next pair of conductors carrying current $2i$ will correspond to an equivalent

Figure 7.5 Conductor current and radial field distribution in the airgap of a four-pole, double-layer, three-phase stator armature winding with two slots/pole/phase, short pitched by one slot: (a) sectional view perpendicular to the axis; (b) 'developed' view with stator/rotor and intervening airgap 'unrolled' (phase a field with $i_a = \hat{I}_a$); (c) phase b and phase c field at the same instant and the three-phase resultant; and (d) phase b field, phase c field and three-phase resultant when $i_a = 0$.

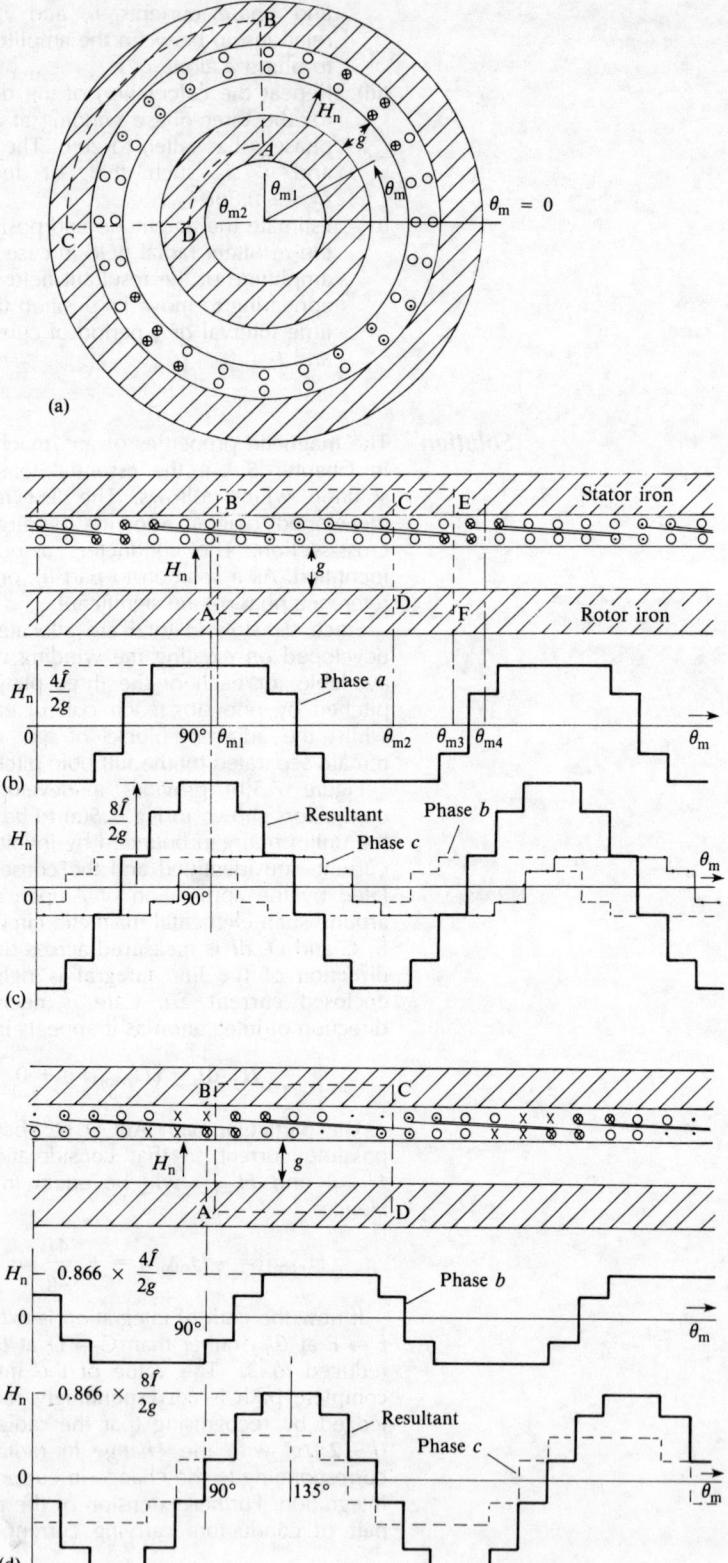

change in radial field and a consequential reversal of the direction of H_n so that $H_{n\theta m4} = i/g$.

The representation of the distribution of H_n in Fig. 7.5(b) as a progression of step changes equal to $\pm i/g$ or $\pm 2i/g$, depending upon the current content of each slot, ignores the spread of current in each slot and the locally increased airgap reluctance due to the slot itself. Such second-order effects need not concern us here. It is apparent from the figure that the use of short-pitched coils has made the transition from the plateaux to the sides of the essentially trapezoidal field distribution less abrupt and the approximation to a sinusoidal distribution more apparent.

As the phase current i alternates with time, the radial field maintains its distribution in space, pulsating about its fixed axis at $90°\varepsilon = 45°$ (mech) from the centre of each band of active coil sides.

In part (ii) the conductors of phases b and c adopt identical relationships to each other as do those of phase a. The sole distinction is that they are displaced in space by $120°\varepsilon$ and $240°\varepsilon$, respectively from those of phase a. Because the winding is chorded (assembled from short-pitch coils) some slots ($\frac{2}{3}$ the total in this case) contain coil sides of different phases. At the instant t_1, $i_a = \hat{I}_a$ whilst $i_b = i_c = -\frac{1}{2}\hat{I}_a$. Thus the radial field distributions due to phases b and c, relative to that of phase a, are then of half the amplitude, inverted and displaced in space by $120°\varepsilon$ and $240°\varepsilon$, respectively.

The phase component fields and their resultant are shown in Fig. 7.5(c). Comparison of the resultant with that due to phase a alone shows that the space-harmonic content of the former is reduced whilst the *positions* of the fundamental components are instantaneously coincident. It may readily be proved (Sect. 5.6.1) that the *amplitude* of the fundamental field is $\frac{3}{2}$ times the *peak* value of the fundamental field component of one phase. The *position* of the resultant field coincides with that of the dominant phase at an instant of peak phase current.

For parts (iii) and (iv): in moving to instant t_2 the field due to phase a becomes zero whilst that due to phase b reverses the polarity of i and increases its amplitude to 0.866 times that applicable to phase a at t_1. Simultaneously, the field due to phase c retains its original polarity but has the increased amplitude of 0.866 p.u. Figure 7.5(d) shows that the resultant field distribution has the same fundamental component *amplitude* as pertains to instant t_1 in Fig. 7.5(c) but displaced in space by $90°\varepsilon$, corresponding to the $\frac{1}{4}$-period time difference between t_1 and t_2. This demonstrates the general property of three-phase windings supplied with three-phase current, that the fundamental component of field maintains a constant amplitude equal to $\frac{3}{2}$ times the peak value per phase, appearing to move relative to the winding at a constant speed equivalent to a distance of one double pole pitch in one complete period of supply frequency. In a multipole machine the apparent rotation in mechanical measure equals the electrical (magnetic) motion divided by the number of pole pairs. Each phase of a balanced three-phase winding produces an identical spectrum of space-harmonic fields. It is characteristic of such windings that *triplen* space-harmonics (odd multiples of 3) are absent from the resultant, whilst non-triplen harmonics are increased by the same factor 1.5 as is the fundamental. The apparent difference between the resultant field waveshapes of Figs. 7.5(c) and (d) is due to relative changes in the phase of the space harmonics between instants t_1 and t_2.

7.5 Radial Airgap Field Development by Sinusoidal Current Distribution

The airgap field space-harmonics produced by a practical a.c. machine winding are generally undesirable. Their principal effect is to induce e.m.f.s in the phase windings. Such e.m.f.s are manifested at supply frequency and combine with the phase e.m.f. induced by the fundamental flux wave to balance the supply voltage. Thus sinusoidal phase currents in practical windings always give rise to sinusoidal phase e.m.f.s, regardless of the actual distribution in space of the radial flux. Torque development, however, is dependent upon the fundamental component of radial flux and hence it is desirable that the fundamental flux should constitute as large a proportion of the total flux as possible. If the phase current is sinusoidally distributed over the airgap surface – i.e. if the linear current density $(A\,m^{-1})$ is a sinusoidal function of angular position in the airgap, then space harmonics of magnetic field and flux may be completely eliminated.

Such an ideal sinusoidal distribution of current is not achievable in practice but may be approached by grading the spacing between the slots accommodating the winding conductors, and the distribution of conductors between slots. Real machines incorporate this feature only on the d.c.-excited 'field' winding of round-rotor synchronous machines; armature windings invariably have the phase conductors uniformly distributed in symmetrically spaced slots, such that the distribution of e.m.f induced therein by the fundamental radial flux is essentially unchanging as the constant amplitude airgap flux appears to move with respect to the conductors.

In our modelling of synchronous machines we shall assume sinusoidally distributed armature phase current. This inherently eliminates space-harmonic fluxes from consideration if the airgap representation is idealised and maximises the influence of the fundamental radial flux. The nature of the current distribution is such that the total current/pole/phase remains at nI (as in eqn [5.12]), where I is the current/conductor and n is the number of conductors/pole/phase. It is shown below that the corresponding current distribution must therefore be a function of airgap position θ such that the linear current density A for a two-pole winding is given by

$$A(\theta) = \frac{nI}{2b}\sin\theta \ A\,m^{-1}, \qquad [7.2]$$

where b is the rotor radius.

Proof Referring to Fig. 7.6(a) which shows axially directed current distributed sinusoidally over one pole of a two-pole winding

$$\text{current/pole} = \int_0^\pi A(\theta)\,b d\theta.$$

Substituting for current density as in eqn [7.2]

$$\text{current/pole} = \int_0^\pi \frac{nI}{2}\sin\theta\,d\theta$$

$$= -\left|\frac{nI}{2}\cos\theta\right|_0^\pi = nI,$$

as required. For the more general winding with p pole-pairs depicted in Fig. 7.6(b) the current density is necessarily expressed by

$$A(\theta_m) = \frac{pnI}{2b} \sin p\theta_m \text{ Am}^{-1} \qquad [7.3]$$

with θ_m in mechanical degrees or radians, to give

$$\text{current/pole} = \int_0^{\pi/p} A(\theta_m) b \, d\theta_m = nI,$$

as required.

Figure 7.6 (a) Sinusoidally distributed current over one pole of a two-pole winding; and (b) the same diagram for a four-pole winding.

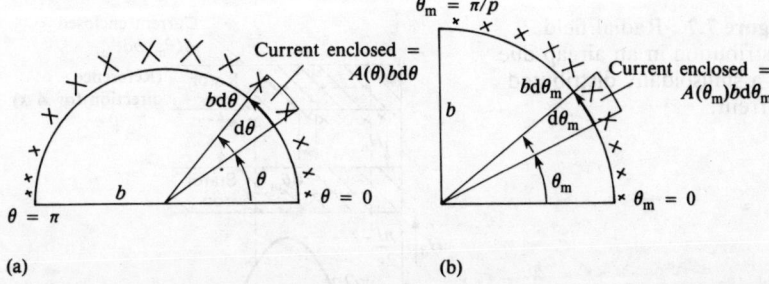

(a)   (b)

In order to calculate the radial field distribution due to a sinusoidal phase current distribution, consider the elementary path shown developed in Fig. 7.7 involving airgap crossings at points θ_{m1} and $\theta_{m1} + \pi/p$, a pole-pitch apart, which is appropriate to a general winding with $2p$ poles. Due to symmetry, the radial field at these positions will be equal in magnitude and oppositely directed, so that the line integral $\oint H_n \cdot dl$ equated to the enclosed current (about which H is right-handed) yields

$$-2H_{n\theta m1}g = \int_{\theta_{m1}}^{\theta_{m1}+\pi/p} \frac{pnIb}{2b} \sin p\theta_m \, d\theta_m$$

$$= -\frac{nI}{2} \left| \cos p\theta_m \right|_{\theta_{m1}}^{\theta_{m1}+\pi/p}$$

$$= -\frac{nI}{2}[\cos(p\theta_{m1} + \pi) - \cos p\theta_{m1}]$$

$$= nI \cos p\theta_{m1}$$

Hence

$$H_{n\theta m1} = -\frac{nI}{2g} \cos p\theta_{m1} \qquad [7.4]$$

Figure 7.7 demonstrates via eqn [7.4] that a sinusoidal current distribution of current per pole nI gives rise to a consinusoidal *radial* field distribution expressed generally as

$$H_n(\theta_m) = -\frac{nI}{2g} \cos p\theta_m \qquad [7.5]$$

with reference directions for H_n, I and θ_m as indicated in Figs. 7.6 and 7.7. Significantly the radial field distribution is out of phase with the current distribution by half a pole-pitch, peaking at those positions where the current is zero.

If we write for the current density distribution

$$A(\theta_m) = \hat{A} \sin p\theta_m \qquad [7.6]$$

then
$$\hat{A} = \frac{pnI}{2b} \text{ and } H_n(\theta_m) = -\frac{b}{pg}\hat{A}\cos p\theta_m. \quad [7.7]$$

Equation [7.7] has been derived as eqn [5.23] and demonstrates that the radial field distribution lags the responsible current density distribution by 90°, having magnitude enhanced by the factor b/pg.

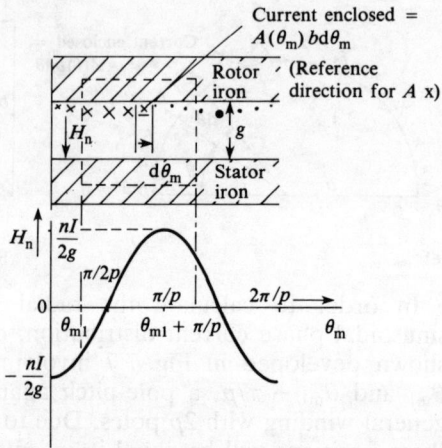

Figure 7.7 Radial field distribution in an airgap due to a sinusoidally distributed current.

7.6 Tangential Airgap Field due to Sinusoidally Distributed Current

Space harmonics are generally present in the tangential field produced in the airgap by a machine winding. The assumption of sinusoidally distributed current conveniently eliminates these harmonics and provides a simple relationship between the current density and the tangential field at the airgap bounding surface on which the current is placed. With the notation for current and field reference directions shown in Fig. 7.8 and generally adopted through this book, current on the *stator* side of the airgap develops a tangential field distribution on the *stator surface* of the airgap equal in magnitude and phase to the current density, i.e.

$$H_t(\theta_m) = \hat{A}\sin p\theta_m \text{ A m}^{-1}, \quad [7.8]$$

whilst current on the *rotor* side of the airgap, also described by $A(\theta_m) = \hat{A}\sin p\theta_m$ develops tangential field on the rotor surface which is equal and opposite to the linear current density, i.e.

$$H_t(\theta_m) = -\hat{A}\sin p\theta_m \text{ A m}^{-1}. \quad [7.9]$$

Proof Referring to Fig. 7.8(a) in which the current is spread on the rotor surface in a layer of negligible thickness, the elementary path ABCDA encloses current $A(\theta_m)b\,d\theta_m$ and is bounded on side AD by infinitely permeable iron, having sides AB and CD of negligible length, so that

$$\oint_{ABCDA} \mathbf{H}\cdot d\mathbf{l} = -H_t(\theta_m)b\,d\theta_m = A(\theta_m)b\,d\theta_m.$$

Figure 7.8 Evaluation of tangential field in an airgap due to current on (a) the rotor surface, (b) the stator surface.

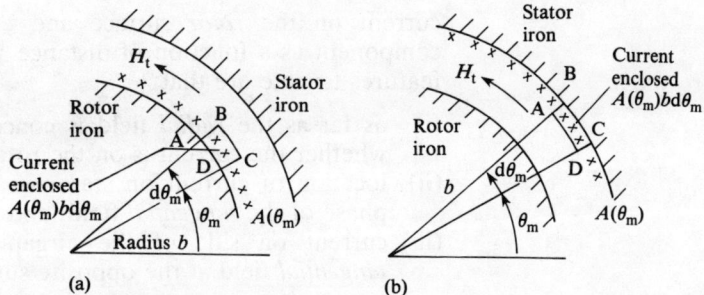

Hence, at the rotor surface

$$H_t(\theta_m) = -A(\theta_m). \qquad [7.10]$$

A complementary argument applied to the current distributed on the stator side of the airgap as in Fig. 7.8(b) yields the result

$$\oint_{ABCDA} \mathbf{H} \cdot d\mathbf{l} = H_t(\theta_m) b \, d\theta_m = A(\theta_m) b \, d\theta_m.$$

Hence, at the stator surface

$$H_t(\theta_m) = A(\theta_m). \qquad [7.11]$$

The relationships of eqns [7.10] and [7.11] are independent of the nature of the current distribution. It is formally shown in Section 5.4 that the tangential field due to current on one side of the airgap reduces linearly with radial distance across the gap, being zero-valued at the infinitely permeable bounding surface opposite.

The general relationships between sinusoidally distributed airgap/*stator* surface current density and the consequential radial and tangential components of magnetic field intensity in the airgap at the same surface are indicated in the developed diagram of Fig. 7.9. This may usefully be compared with Fig. 5.11. Both figures relate to a four-pole field pattern, but the latter figure has

Figure 7.9 Uniform airgap radial and tangential field distribution due to a sinusoidal current sheet on the stator surface.

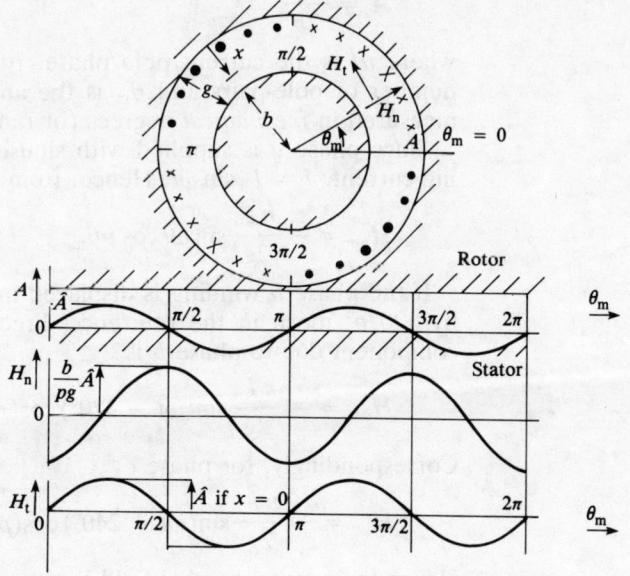

current on the *rotor* surface and expresses the tangential field component as a function of distance x from the rotor surface. Key features to note are that:

(i) as far as the *radial* field is concerned it makes no difference whether the current is on the rotor or on the stator,
(ii) location of current on stator or rotor surface determines the phase of the *tangential* field, and
(iii) current on side of the airgap contributes nothing to the *tangential* field at the opposite surface.

7.7 Determination of the Resultant Radial and Tangential Fields in a Uniform Airgap due to a Balanced, Three-Phase Winding Energised with a Balanced, Three-phase, Sinusoidally Distributed Current

The solution to Worked Example 7.1(d) indicated that the *fundamental* component of the radial airgap field due to a practical three-phase winding appeared to move at synchronous speed with respect to the winding conductors and to be of constant amplitude $\frac{3}{2}$ times that of the *peak* value of the contributing field made by each phase current. This property will now be formally proved for the radial field, free of space harmonics, produced by sinusoidally distributed phase currents, where phase windings b and c are displaced in space by $120°\varepsilon$ and $240°\varepsilon$, respectively, from reference phase a. The tangential field adopts a complementary behaviour.

Thus, for reference phase a the linear current density on, say, the *stator* surface of the airgap as depicted in Fig. 7.9 is given by

$$A(\theta_m) = \hat{A} \sin p\theta_m \qquad [7.6]$$

with

$$\hat{A} = \frac{pnI}{2b} \qquad [7.3]$$

where nI is the current/pole/phase, b is the rotor radius, p is the number of pole-pairs and θ_m is the angular position in the airgap measured in *mechanical* degrees (or radians).

Since phase a is supplied with sinusoidally time-variant alternating current, $I = \hat{I}_a \sin \omega t$. Hence, from eqn [7.5]

$$H_{na} = -\frac{n\hat{I}_a}{2g} \sin \omega t \cos p\theta_m.$$

If the phase b winding is displaced in space from that of phase a by $120/p°$ mech in the reference direction of θ_m, the radial field component due to phase b is

$$H_{nb} = -\frac{n\hat{I}_a}{2g} \sin(\omega t - 120°) \cos(p\theta_m - 120°).$$

Correspondingly, for phase c

$$H_{nc} = -\frac{n\hat{I}_a}{2g} \sin(\omega t - 240°) \cos(p\theta_m - 240°).$$

Hence the resultant radial field is given by

$$H_n = H_{na} + H_{nb} + H_{nc}$$
$$= -\frac{3}{2}\frac{n\hat{I}_a}{2g}\sin(\omega t - p\theta_m). \qquad [7.12]$$

Equation [7.12] shows that the resultant radial field at any instant is of constant peak value, varying sinusoidally with position and developing p pole pairs as θ_m varies from 0 to 2π radians. At the particular instant t_1, given by $\omega t_1 = \pi/2$, when the current in phase a has its peak positive value, the resultant field has its peak valued at

$$\frac{3}{2}\frac{n\hat{I}_a}{2g}$$

located at θ_{m1}, say, where $p\theta_{m1} = \omega t_1 - \pi/2 = 0°$. The resultant field distribution corresponds at that instant with that of the phase a component, increased in value by the factor $\frac{3}{2}$.

Equation [7.12] depicts the resultant field as a *travelling wave*, sinusoidally distributed and rotating *synchronously* with angular velocity ω/p mech rad/s in the direction of increasing θ_m, as given by the sequence of the phase winding positions a, b, c.

It is left as an exercise for the reader to show that interchange of the position in space of any pair of phase windings causes the resultant field to rotate in the opposite direction. A more general treatment of rotating fields, including the effect of space harmonics, is given in Section 5.6 and is illustrated in Fig. 5.13.

7.8 Torque Development in a Polyphase Synchronous Machine with a Uniform Airgap

A polyphase synchronous machine has its armature phase windings on one side of the airgap carrying balanced currents, which produce ideally sinusoidally distributed radial and tangential magnetic field components in the airgap, which rotate relative to the armature conductors at synchronous speed defined by the number of pole pairs and the supply frequency. At the other side of the airgap the d.c.-excited 'field' winding also produces ideally sinusoidally distributed radial and tangential components of magnetic field, which are stationary with respect to the d.c. winding and, for constant torque development, stationary with respect to the field of the armature winding. Calculation of the torque developed from an assessment of the distribution of tangential mechanical stress is simplified mathematically if the configuration adopted is such that the field winding is on the stator and the armature winding is on the rotor. Subsequent modelling of the effects of *salient* field poles is simplified if, at the outset, we adopt the centre line of a south pole on the stator, i.e. the *direct axis*, as reference position in the airgap $\theta = 0°$; or $\theta_m = 0°$ for the more general machine with p pole pairs. We shall now proceed to show that the torque developed by the machine may be simply expressed in terms of the airgap dimensions and the product of the radial field components of the armature and d.c.-excited 'field' windings and the sine of the space-phase angle between them.

Figure 7.10 shows each of three-phase windings *a*, *b* and *c* represented by single-turn coils on the rotor of a two-pole machine. When supplied with polyphase currents of sequence *abc* the ideally sinusoidally distributed resultant armature field components appear to rotate *clockwise* relative to the rotor conductors, but remain stationary in space if the rotor assembly rotates synchronously in the anticlockwise direction. The relative position of the rotor armature *radial* field component H_{nr} is maintained at an angle of lag ψ with respect to the stator 'field' winding radial field component H_{nf} such that the *resultant* radial field H_n lags H_{nf} in space by β. Figure 7.10 corresponds to a two-pole case, but in the analysis which follows the number of poles will not be restricted to a single pair, and θ_m will correspond to a *mechanical* measure of angle from the centre of an arbitrary south pole on the stator.

Figure 7.10 (a) Single-turn coil representation of a rotor-mounted three-phase winding establishing a sine-distributed radial field H_{nr} in an airgap at radius *b*. The d.c.-excited stator field winding develops a radial field H_{nf} component which is cosine-distributed about the *d*-axis. The resultant is H_n. (b) The tangential stress distribution over the rotor surface.

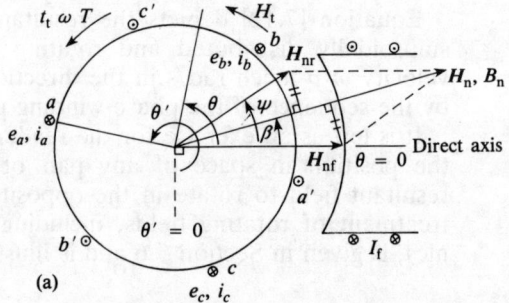

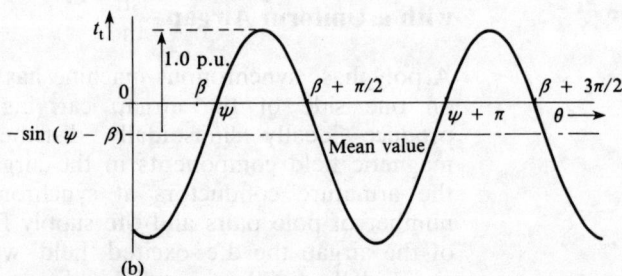

The radial field components in the airgap are expressed as general functions of angular position, as follows:
for the stator

$$H_{nf}(\theta_m) = \hat{H}_{nf} \cos p\theta_m;$$

and for the rotor,

$$H_{nr}(\theta_m) = \hat{H}_{nr} \cos p(\theta_m - \psi/p).$$

Equations [7.6], [7.7] and [7.9] provide justification that the *tangential* field H_{tr} due to the armature current on the rotor lags the radial field H_{nr} due to the same current by 90°ε and is related in amplitude by the factor pg/b, giving

$$H_{tr}(\theta_m) = \frac{pg}{b} \hat{H}_{nr} \sin p\left(\theta_m - \frac{\psi}{p}\right).$$

At the rotor surface the only tangential field component to be effective is that due to the armature current on the rotor. Both stator and rotor radial field components add, however, to give a resultant H_n which lags H_{nf} by space phase angle $\beta°\varepsilon$. Thus

$$H_n = H_{nf} + H_{nr} = H_n\angle{-\beta}$$

or

$$H_n(\theta_m) = \hat{H}_n \cos p\left(\theta_m - \frac{\beta}{p}\right)$$

with

$$H_n^2 = H_{nr}^2 + H_{nf}^2 + 2H_{nr}H_{nf}\cos\psi.$$

At any point, θ_m, in the airgap on the rotor surface the tangential mechanical stress is given by eqn [3.22]

$$\begin{aligned}t_{tr}(\theta_m) &= B_n H_{tr} = \mu_0 H_n H_{tr}\\ &= \mu_0 \frac{pg}{b} \hat{H}_n \hat{H}_{nr} \cos p\left(\theta_m - \frac{\beta}{p}\right)\sin p\left(\theta_m - \frac{\psi}{p}\right)\\ &= \mu_0 \frac{pg}{2b}\hat{H}_n\hat{H}_{nr}\left\{\sin p\left[2\theta_m - \frac{(\beta+\psi)}{p}\right]\right.\\ &\qquad\left. - \sin(\psi - \beta)\right\}. \end{aligned}\qquad[7.13]$$

Equation [7.13] shows the rotor tangential stress to have a variation with airgap location θ_m which is an offset, double-frequency sinusoid. The distribution over a double pole pitch shown normalised in Fig. 7.10(b) demonstrates that the peak stresses may be very much greater than the *mean* stress which gives rise to effective torque.

The mean value of tangential stress corresponds to a value for rotor torque given by

$$T_r = b\int_0^{2\pi} lb t_{tr_{mean}}\,d\theta_m$$

where l is the axial length of the rotor

$$\begin{aligned}&= -b^2 l\mu_0 \frac{pg}{2b}\int_0^{2\pi}\hat{H}_n\hat{H}_{nr}\sin(\psi-\beta)\,d\theta_m\\ &= -\mu_0\pi pblg\hat{H}_n\hat{H}_{nr}\sin(\psi-\beta).\end{aligned}\qquad[7.14]$$

Now, from the geometry of the parallelogram defined by the vector quantities depicting H_{nf}, H_{nr} and their resultant H_n in Fig. 7.10(a)

$$\hat{H}_{nr}\sin(\psi - \beta) = \hat{H}_{nf}\sin\beta.$$

Therefore eqn [7.14] may be rewritten

$$T_r = -\mu_0\pi pblg\hat{H}_n\hat{H}_{nf}\sin\beta. \qquad[7.15]$$

Further, since $\hat{H}_n\sin\beta = \hat{H}_{nr}\sin\psi$,

$$T_r = -\mu_0\pi pblg\hat{H}_{nr}\hat{H}_{nf}\sin\psi. \qquad[7.16]$$

Equations [7.14], [7.15] and [7.16] make the common statement that the magnitude of the torque is proportional to the product of the peak values of *any two* of the three radial field quantities and the sine of the space-phase angle between them. The torque acts in such a direction as to encourage alignment of the field components. The rotor torque is negative, opposing rotor rotation if $0 < \psi < 180°$, i.e. the rotor (armature) radial field lags in space the stator radial field by less than 180°. The rotor torque is similarly negative if $0 < \beta < 180°$. The mode of operation of the machine is thus critically dependent upon space-phase angle ψ or β.

The tangential mechanical stress distribution shown in Fig. 7.10(b) provides insight into the way the tangential stresses at different locations vary in value and direction in contributing to the effective torque. When H_{nr} and H_{nf} are co-directed, or oppositely directed, the mean stress and torque are zero, but the distributed tangential stress peaks with a value proportional to the $\hat{H}_n \hat{H}_{nr}$ product. Maximum torque for given values of $\hat{H}_n$ and $\hat{H}_{nr}$ corresponds to $(\psi - \beta) = 90°$, i.e. H_{nr} and H_n in space quadrature.

By analogy, eqn [7.16] indicates that for given values of stator and rotor radial field components maximum torque is developed at $\psi = 90°$. Equation [7.7] gives maximum torque when $\beta = 90°$ for given values of the resultant radial field $\hat{H}_n$ and (stator) 'field' winding $\hat{H}_{nf}$. Torque equations [7.14], [7.15] and [7.16] are easy to recall if it is remembered that the constant term is simply the product of the three parameters b, l and g which define the airgap, with μ_0, π and p.

7.9 Electromotive Force Equation of a Polyphase Synchronous Machine with a Uniform Airgap

The rotor of the two-pole synchronous machine depicted in Fig. 7.10(a) carries a three-phase armature winding represented by the single-turn coils aa', bb' and cc', mutually displaced in space by 120°, such that sinusoidal currents of sequence i_a, i_b, i_c give rise to fundamental rotating field components which appear to move clockwise *relative to the conductors* but which are held stationary in space due to the synchronous anti-clockwise motion of the rotor.

The motion relative to the radial airgap magnetic field induces in each rotor (armature) conductor an e.m.f. E of value

$$E = B_n l v, \qquad [5.35]$$

where v is the relative velocity of the conductor of length l at right-angles to the constant-amplitude radial flux density B_n. The relative directions of the axially directed e.m.f., motion and flux are as given by the right-hand rule, Fig. 5.15. The e.m.f. E is directed in the same sense as the electromagnetically induced component E_e of the total electric field within the conductor discussed in Section 1.3.4.

In Fig. 7.10(a) θ' is used to define the particular position in the airgap of the reference side a of the full-pitch coil aa' representing phase a of the armature winding. The position $\theta' = 0$ is chosen arbitrarily such that the radial field component in the airgap due to the stator 'field' winding may be expressed as a simple sinusoid, thus

$$H_{nf}(\theta') = \hat{H}_{nf} \sin \theta'.$$

The modelling is made more general by considering a machine with p pole pairs. With θ' then measured in mechanical degrees or radians θ'_m, we write

$$H_{nf}(\theta'_m) = \hat{H}_{nf} \sin p\theta'_m.$$

As the rotor rotates at constant synchronous speed ω_0 mech rad s^{-1} ($\equiv \omega/p$, with ω in elec rad s^{-1}) the instantaneous e.m.f. induced in conductor a by both stator and rotor components of the radial field is given by eqn [5.35] as

$$e_a(\theta'_m) = \omega_0 b l \mu_0 \left[\hat{H}_{nf} \sin p\theta'_m + \hat{H}_{nr} \sin p\left(\theta'_m - \frac{\psi}{p}\right) \right] \quad [7.17]$$

recognising that, as shown in Fig. 7.10(a), the radial field due to the rotor lags that due to the stator by electrical phase angle ψ, as seen by a rotor conductor moving in the reference direction of θ.

For a full-pitch, single-turn coil aa' the total e.m.f. will be twice the value given by eqn [7.17]. If the phase winding has many turns N, distributed in several slots *symmetrically about the locations a and a'* and possibly incorporating short-pitch coils, allowance for the difference in *phase* of the e.m.f.s induced in each contributing conductor is made by the use of *effective* turns per phase N' such that

$$N' = K_{dp1} N = K_{d1} K_{p1} N, \quad [7.18]$$

in which K_{d1} is the (fundamental) *distribution factor* for the winding and K_{p1} is the (fundamental) *pitch factor* of each constituent coil. Winding factors are discussed at length in Sections 5.3 and 5.8.2.

Hence the total e.m.f. for phase a as it rotates in the stationary radial field established by both stator 'field' winding excitation and the armature winding currents is

$$e_a(t) = 2N' \omega_0 b l \mu_0 \left[\hat{H}_{nf} \sin p\theta'_m + \hat{H}_{nr} \sin p\left(\theta'_m - \frac{\psi}{p}\right) \right] \quad [7.19]$$

where $\theta'_m = \omega_0 t$ if we choose $t = 0$ such that the reference conductor a lies then in the position $\theta'_m = 0$.

Equation [7.19] shows that $e_a(t)$ is a sinusoidally time-variant quantity with two components – those due to the stator and rotor fields. The same (reference) phase a carries current which is sinusoidally time variant and, as shown in Sections 7.4 and 7.7, has its peak positive value at the particular instant when the distribution in space of the radial field of the polyphase armature winding is coincident with that of phase a alone. Reference to Fig. 7.10(a) shows that this instant for a two-pole machine is when coil sides a and a' are located at $\theta' = p\theta'_m = \psi$ and $(\psi + \pi)$, respectively. This key property enables the *time*-dependent quantities e and i to be synchronised with the *space*-dependent quantities H_{nf} and H_{nr}.

Thus, if the current in phase a is expressed as

$$i_a(t) = \hat{I}_a \sin(\omega t + \rho) \qquad [7.20]$$

with ω the supply angular frequency and ρ an arbitrary constant, and if i_a achieves its peak positive value at instant t_1, then

$$\hat{I}_a = \hat{I}_a \sin(\omega t_1 + \rho). \qquad [7.21]$$

At the same instant t_1, phase a is positioned such that $\theta' = \theta'_1 = \psi$. Since, generally, $\theta'_m = \omega_0 t$, then at instant t_1 $\theta'_{m1} = \omega_0 t_1$, or $\theta'_1 = p\omega_0 t_1 = \omega t_1$. Thus $\omega t_1 = \psi$ and, from eqn [7.21], $\omega t_1 + \rho = 90°$, to give $\rho = (90° - \psi)$. Substituting in eqn [7.20] gives

$$i_a(t) = \hat{I}_a \sin(\omega t + 90° - \psi)$$

or

$$i_a(t) = \hat{I}_a \cos(\omega t - \psi). \qquad [7.22]$$

Now eqn [7.19] may be rewritten

$$e_a(t) = \hat{E}_f \sin \omega t + \hat{I}_a X_{ad} \sin(\omega t - \psi) \qquad [7.23]$$

where

$$\hat{E}_f = 2N'\omega_0 bl\mu_0 \hat{H}_{nf} \qquad [7.24]$$

and

$$X_{ad} = 2N'\omega_0 bl\mu_0 \left(\frac{\hat{H}_{nr}}{\hat{I}_a}\right). \qquad [7.25]$$

E_f is the e.m.f. induced by the field current alone and appears at the phase terminals when the armature current is zero. It is thus the *open-circuit e.m.f.* of the machine. Its r.m.s. value is readily shown (see Tutorial Example 5.3) to be given by an expression analogous to the *transformer e.m.f. equation* [4.14], i.e.

$$E_f = \sqrt{2}\pi f N' \Phi, \qquad [7.26]$$

where Φ is the total fundamental radial flux/pole due to the field current and f is the frequency, with

$$\Phi = B_{n_{av}} \times \text{area/pole} = \frac{2}{\pi}\mu_0 \hat{H}_{nf} \times \frac{2\pi bl}{2p}$$

and

$$f = p\omega_0/2\pi.$$

In writing eqn [7.23] the symbol of *reactance* was chosen for X_{ad}, the quantity defined in eqn [7.25] having the dimensions of impedance. It appears in eqn [7.23] as a constant of proportionality relating a component of e.m.f. induced in an armature phase winding to the current in that phase. Comparison of eqn [7.23] with [7.22], however, shows the 90° phase difference characteristic of reactance. X_{ad} is known as the *armature reaction reactance* since it accounts for the magnetic effect of the (total) armature current on the induced e.m.f. in each armature phase. It features significantly in the equivalent circuit of the synchronous machine.

7.10 Equivalent Circuit and Phasor Diagram of a Polyphase Synchronous Machine with a Uniform Airgap

Equations [7.22] and [7.23] may be interpreted in terms of the time-phasor diagram of Fig. 7.11(a), which relates r.m.s. phasor values of the components of e.m.f. E_a induced in reference armature phase a and current I_a in the same phase to a *reference phasor* instantaneously of value $\sin \omega t$.

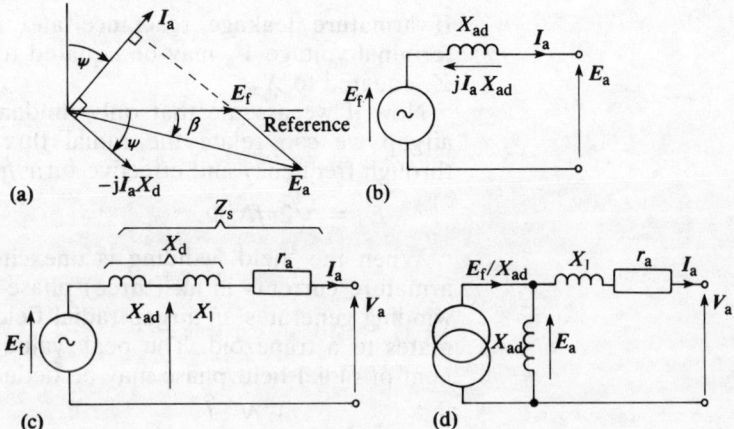

Figure 7.11 Phasor diagram and equivalent circuits for a uniform airgap synchronous machine: (a) phasor diagram, generator at leading power field: (b) voltage source equivalent circuit modelling armature reactance as source reactance X_{ad}; (c) Incorporation of armature resistance and leakage reactance/phase; and (d) current source equivalent circuit.

Thus E_f is laid horizontally from left to right with the $I_a X_{ad}$ phasor lagging by ψ. These components add to give resultant E_a while I_a lags the 90° phase advance axis by ψ. The $I_a X_{ad}$ phasor component of E_a is recognised as being proportionate in length to I_a which it lags by 90°. Since the j operator represents a 90° phase *advance* this component of E_a may be expressed as a phasor quantity $-jI_a X_{ad}$ and the equivalent circuit of Fig. 7.11(b) may be deduced as appropriate to the phasor diagram of Fig. 7.11(a). The phasor diagram and labelled equivalent circuit are consistent, as both relate the components of e.m.f. in the same way, i.e.

$$E_f = E_a + jI_a X_{ad}.$$

Thus far, the effects of armature *leakage* flux which links the individual phase windings have been disregarded. Only armature *radial* flux with respect to which the armature conductors have relative motion has been incorporated in the armature reaction reactance X_{ad}. The airgap flux produced by the armature tangential field comes into the category of leakage flux, as also does flux produced by the armature end-windings which interconnect the active conductors of the coils. Armature leakage flux effects are incorporated as a leakage reactance X_l in the phase equivalent circuit of Fig. 7.11(c), together with armature resistance per phase r_a.

$X_d = X_{ad} + X_l$ is defined as the *synchronous reactance* per phase of the uniform airgap machine. $Z_s = r_a + jX_d$ is termed the *synchronous impedance*/phase. A practical machine has $r_a \ll X_d$.

7.10.1 Evaluation of Armature Reaction Reactance X_{ad} in terms of Airgap Parameters

From the equivalent circuit of Fig. 7.11(c) it is apparent that

$$Z_s = \frac{E_f}{I_{sc}} = \frac{\text{o-c voltage/phase}}{\text{s-c current/phase}} \qquad [7.27]$$

also, that when $E_f = 0$ (unexcited machine, 'field' current zero)

$$Z_s = \frac{V_a}{I_a} = \frac{\text{terminal voltage/phase}}{\text{armature current/phase}}$$

If armature leakage reactance and resistance are ignored, the terminal voltage V_a may be equated to the 'airgap voltage' E_a and Z_s equated to X_{ad}.

Now if we assume that only fundamental flux is present in the airgap we can relate the radial flux per pole Φ to the e.m.f. through frequency and effective turns/phase via eqn [7.26]. Thus

$$E_a = \sqrt{2}\pi f N' \Phi.$$

When the 'field' winding is unexcited Φ is due entirely to the armature currents in all (three) phase windings. A *practical* phase winding generates an airgap radial field distribution which approximates to a trapezoid. The peak value of the *fundamental* component of radial field/phase may be deduced from eqn [5.12] as

$$\hat{H}_{nph} = \frac{4}{\pi} \frac{N'}{p} \frac{\hat{I}_a}{2g}$$

For a three-phase winding the resultant radial field is given by

$$\hat{H}_n = \frac{3}{2} \frac{2}{\pi} \frac{N'}{pg} \sqrt{2} I_a,$$

giving

$$\Phi = \frac{3}{\pi} \frac{N'}{p^2 g} \sqrt{2} I_a \mu_0 2bl$$

and

$$X_{ad} = \frac{E_a}{I_a} = 2\pi f \left[3 \left(\frac{N'}{p} \right)^2 \frac{\mu_0 2bl}{\pi g} \right]. \qquad [7.28]$$

Equation [7.28] is valid insofar as the radial flux distribution is adequately represented by the fundamental. A practical three-phase winding sets up space-harmonic fields of order $(2n \pm 1)$ which also induce fundamental frequency e.m.f. in the armature winding. Equation [7.28] indicates that X_{ad} is inversely proportional to airgap length g. The equivalent circuit of a uniform-airgap synchronous machine may be re-configured as the equivalent current source shown in Fig. 7.11(d). E_f (from eqn [7.24]) and X_{ad} are similarly affected by changes in airgap dimensions, so that a reduction in g makes the machine behave more nearly like a constant-current source. Increase of airgap length has the opposite effect, and synchronous machines usually have a large airgap to permit very high *electric loadings* – i.e. high current densities at the armature surface. X_{ad} is typically of the order of 1.0 p.u. – i.e. rated armature current I_a gives a value of $I_a X_{ad}$ equal to rated terminal voltage – and the voltage source equivalent circuit is usually employed.

7.11 Synchronous Machine Operation as an Alternator – Voltage Regulation

The equivalent circuit of Fig. 7.11(c) represents the synchronous machine as a voltage source with e.m.f. E_f and source impedance Z_s. E_f in the absence of magnetic saturation is proportional to the d.c. 'field' excitation current and is known as the 'e.m.f. induced by the field current alone' or as the 'e.m.f. behind synchronous impedance'. The equivalent circuit parameters of E_f and Z_s may be determined experimentally by open-circuit and short-circuit test techniques.

In the open-circuit test the machine is driven at synchronous speed by a prime-mover and a graph is obtained of no-load terminal phase voltage E_f plotted against d.c. field current. Due to magnetic saturation, particularly in the region of the armature core teeth, the graph is not a straight line but adopts the familiar shape of reducing slope at higher values of field current. In the short-circuit test the machine has its armature winding short circuited and, when rotated at synchronous speed, the relationship between armature current per phase and d.c. field current is established. The relationship is a linear one over the limited range of field current necessary to restrict the short-circuit current to values near the machine rating.

The open-circuit characteristic (o.c.c) and short-circuit characteristic (s.c.c.) are shown in Fig. 7.12 which has the latter graph extrapolated as a straight line in correspondence with the higher values of field current appropriate to the saturated part of the o.c.c. and with corresponding values for synchronous impedance being deduced using eqn [7.27]. Linear extrapolation of the s.c.c. is valid since the *resultant* magnetic field of a synchronous machine with its armature terminals short circuited never attains the higher values which exhibit saturation effects. If armature resistance and leakage reactance are ignored, it is apparent from Fig. 7.10(a) that, with the resultant radial field H_n zero under short-circuit conditions, H_{nr} and H_{nf} must sum to zero, with the space-phase angle ψ by which the armature field H_{nr} lags the 'field' winding field H_{nf} of value $\psi = 180°$. The armature radial field is seen to oppose that of the d.c. excited field winding along the *d*-axis.

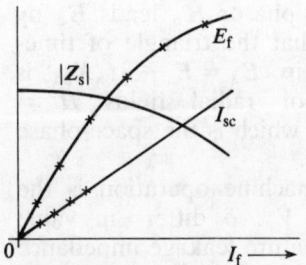

Figure 7.12 Open-circuit and short-circuit characteristics of a synchronous machine, also variation of synchronous impedance magnitude with field current on short circuit.

The same value for ψ of 180° is appropriate to all loading conditions of synchronous generators operating at zero power factor lag. A complementary zero torque condition corresponds to $\psi = 0°$ at zero power factor lead, with the armature radial field and that of the d.c. field winding similarly directed in space. Phasor diagrams relating to the limiting cases of terminal short circuit and loads of zero power factor lag and lead are shown in Fig. 7.13 with reference to the equivalent circuit of Fig. 7.11, ignoring armature resistance.

The more general loading condition for an alternator at 0.8 power factor lag is shown in Fig. 7.14(b) with armature leakage

Figure 7.13 Phasor diagram for an alternator with a uniform airgap on: (a) short circuit, neglecting armature leakage impedance, $\psi = 180°$; (b) short circuit, neglecting armature resistance, $\psi = 180°$; (c) purely inductive load, neglecting r_a, $\psi = 180°$; and (d) purely capacitive load, neglecting r_a, $\psi = 0°$.

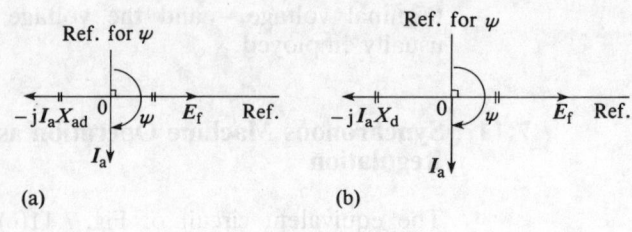

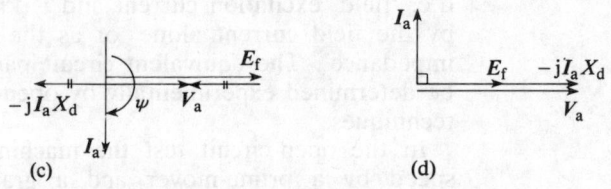

impedance retained. Figure 7.14(c) shows the case of a generator at a leading power factor of 0.8. Also shown is the induced e.m.f. phasor E_a called the 'airgap e.m.f.' since it is the consequence of the *resultant* radial airgap field H_n. The *time*-phasor diagrams of voltage and current have been extended to include the *space*-phasor radial field quantities H_{nr}, H_{nf} and their resultant H_n, correlated such that the reference vector from which ψ is measured coincides with the d.c. excitation-field H_{nf} phasor, drawn to lead the corresponding e.m.f. phasor E_f by 90°. This arrangement places the armature radial field phasor H_{nr} in phase with the current phasor I_a and the resultant field phasor H_n leads E_a by 90°. Reference to Fig. 7.14(b) confirms that the triangle of time-phasor e.m.f.s given by the relationship $E_a = E_f - jI_aX_{ad}$ is similar to the triangle of space-phasor radial fields $H_n = H_{nf} + H_{na}$. Thus E_a lags E_f in time by β, which is the space-phase angle by which H_n lags H_{nf} in Fig. 7.10(a).

Of great importance in synchronous machine operation is the time-phase angle δ by which E_f leads V_a. δ differs in value marginally from β due to the effect of armature leakage impedance $Z_L = r_a + jX_1$. δ is known as the *load angle* of the synchronous machine. For normal operation as a generator, δ is a small positive angle $< 90°$, as illustrated in Figs. 7.14(b) and (c). The role of δ will be further discussed in Section 7.15 which considers the behaviour of a synchronous machine connected to constant-voltage, constant-frequency busbars.

7.11.1 Voltage regulation – effect of magnetic saturation

A study of the phasor diagram of Fig. 7.14(b) shows that an alternator with constant d.c. 'field' excitation giving rise to constant E_f will develop a terminal voltage V_a which reduces in magnitude as the load current I_a at constant, lagging power factor is increased. The effect is more significant at lower values of power factor and is due to the demagnetising effect of the armature reaction field. Conversely, Fig. 7.14(c) shows that a capacitive load causes the terminal voltage to increase with increasing load current, the armature reaction field assisting that due to the constant

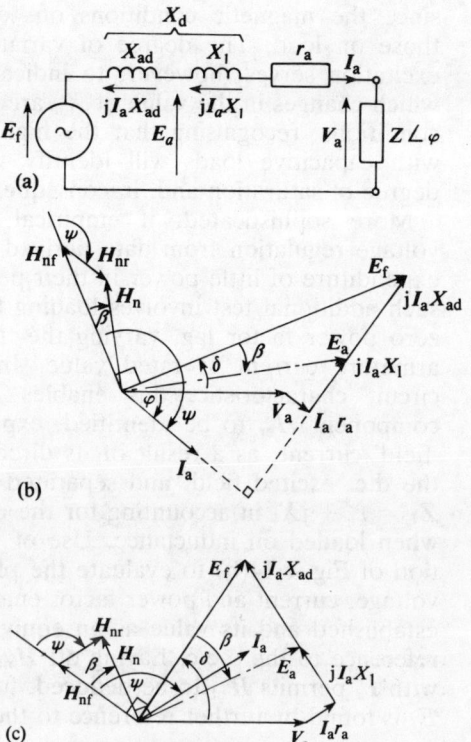

Figure 7.14 Equivalent circuit (a) with the phasor diagram for uniform airgap alternator supplying power at (b) lagging power factor (c) leading power factor.

d.c. 'field' excitation. The voltage regulation ε of a synchronous machine is defined as the *increase* in voltage appearing at the armature terminals when full-load current is thrown off, expressed p.u. of normal rated terminal voltage, i.e.

$$\varepsilon \text{ p.u.} = \frac{E_f - V_a}{V_a} \quad [7.29]$$

under conditions of rated V_a and I_a.

Clearly, voltage regulation is very much dependent upon power factor and generally has a negative value if the load is of leading power factor. Voltage regulation may be measured by carrying out direct load tests but may more conveniently be predicted using a phasor diagram construction, or a formula derived from the equivalent circuit analogous to eqn [4.13] which was derived for the transformer whose equivalent circuit is essentially identical. The fundamental difficulty with using such a predictive method for evaluating the performance of a synchronous machine is that both the parameters E_f and X_d are influenced by magnetic saturation within the machine.

If E_f is regarded as constant for a given value of field current the X_{ad} component of Z_s is a complex function of both the 'field' current and the armature current since both contribute to the *resultant* radial field in the airgap, the value of which determines the degree of saturation. It is not valid to interpret the results of the open-circuit and short-circuit tests depicted in Fig. 7.12 as displaying graphically the variation of Z_s with 'field' current I_f, since the magnetic conditions on test are not representative of those on load. The degree of variation of Z_s over the range of excitation serves, however, to indicate qualitatively the extent to which changes in the value of X_d are likely over a range of loading conditions, recognising that the high terminal voltages associated with capacitive loads will identify with the prospect of a high degree of saturation and, in consequence, a low value for X_d.

More sophisticated, if empirical, methods of predicting the voltage regulation from data derived from tests which require the expenditure of little power in their performance are available. One such additional test involves loading the machine on inductance at zero power factor lag, varying the 'field' current but maintaining armature current at rated value. In association with the open-circuit characteristic, this enables the armature reaction field component H_{nr} to be identified, expressed in terms of equivalent 'field' current, as a result of its direct demagnetising influence on the d.c. excited field, and separated from the leakage impedance $Z_L = r_a + jX_l$ in accounting for the difference between V_a and E_f when loaded on inductance. Use of the phasor diagram construction of Fig. 7.14(b) to evaluate the phasor E_a for a given terminal voltage, current and power factor enables the direction of H_n to be established and its value as an equivalent 'field' current found by reference to the o.c.c. Laying off H_{na} (in 'field-amperes') in phase with I_a permits H_{nf} to be deduced, for which the equivalent e.m.f. E_f is found by further reference to the open-circuit characteristic.

7.12 Uniform-Airgap Synchronous Machine-Operation on Constant-Voltage, Constant-Frequency Busbars

Synchronous machines seldom operate as isolated generators supplying loads such that the terminal voltage depends on the d.c. 'field' winding current – the 'field' excitation. Most commonly their armature windings are connected directly to an a.c. network of such capacity that the terminal voltage is maintained at constant voltage and frequency, regardless of the operating condition of the machine in question. The machine is then said to be connected to an *infinite busbar*. The behaviour of a synchronous machine in response to, for example, a change in 'field' excitation is markedly constrained if the terminal voltage is unable to change its magnitude. As we shall shortly establish, it is the flow of *reactive VA* between the machine and the connected system which is modified in this case, a change in field excitation having little influence upon the flow of *real* power.

With infinite busbar operation the phasor diagram and equivalent circuit construction are essentially as developed in Fig. 7.14. An abbreviated representation is usual with no recognition of the division of synchronous reactance X_d into leakage and armature reaction reactance components and no extension of the phasor diagram to include the airgap field space-phasor quantities. It is useful, however, to retain the feature of the time-phasor diagram identifying the space-phase angle ψ by which the armature winding component lags the d.c. excited 'field' winding component of the resultant airgap radial field. It will be shown that the angle ψ is significant in determining the directions of flow of real and reactive power at the terminals of the machine.

Invariably, the terminal voltage phasor V_a is adopted as the reference quantity when presenting the phasor diagram of a synchronous machine operating on an infinite busbar. The four general modes of operation which correspond to motor or generator at lagging or leading power factor are then accommodated by displaying the armature current phasor in each of the four quandrants. The four alternatives are depicted in Fig. 7.15 together with the appropriately labelled equivalent circuit. In each case, the relationship between the phasors of the voltage components displayed is such that they conform to the equation

$$E_f = V_a + I_a r_a + jI_a X_d$$

The four phasor diagrams displayed in Fig. 7.15 correspond to the different modes of operation of a synchronous machine and merit close examination. Figures 7.15(b) and (c) relate to *motoring* conditions for which the complex power $S = V_a I_a^*$ supplied to the busbar has a negative *real* part P. Where the motor operates at leading power factor as in (b), $V_a I_a^*$ has a positive *imaginary* part Q, implying reactive power flow from the machine to the system. In case (c), Q is negative and the machine absorbs reactive power. The former condition is associated with a magnitude for E_f *greater* than that of V_a (provided $I_a r_a$ is small), leading to the observation that an over-excited synchronous motor supplies VA_r to the connected system. Figure 7.15(c) demonstrates the complement of the rule. The space-phase angle ψ is a third-quadrant angle in (b) and a fourth-quadrant angle in (c). Reference to Fig. 7.10(a) confirms that in each of these cases the torque on the rotor is in the direction of rotor rotation. For both situations demonstrating motor action, the load angle δ by which E_f leads V_a is a fourth-quadrant angle, approaching 360°.

Figures 7.15(d) and (e) correspond to *generator* operation. Here, the load angle δ is a first-quadrant angle with ψ located in the first or second quadrant. Reference to Fig. 7.10(a) confirms that rotor torque in the sense to urge the stator and rotor component fields into alignment opposes rotor rotation. This figure also demonstrates that, in the condition portrayed by the phasor diagram of Fig. 7.15(d), the armature field assists the d.c. excited 'field' winding in providing the resultant radial airgap field necessary to induce the particular e.m.f required to balance the supply voltage – allowing for armature leakage impedance. The situation in (d) is analogous to the motoring alternative (c), in each case it may be concluded that the armature absorbs reactive volt amperes from the connected system in order to assist in the development of the virtually constant resultant radial airgap field required by the fixed armature terminal voltage.

In the over-excited generator of Fig. 7.15(e), however, the enhanced d.c. 'field'-winding excitation enables the armature to exert a de-magnetising influence on the airgap field by exporting VA_r to the connected system.

Figure 7.15 Equivalent circuit (a) and phasor diagrams for a uniform-airgap synchronous machine on constant-voltage busbars; (b) motoring at leading power factor and supplying VA$_r$ to system; (c) motoring at lagging power factor and absorbing VA$_r$; (d) generating at leading power factor and absorbing VA$_r$ from the system; and (e) generating at lagging power factor and supplying VA$_r$.

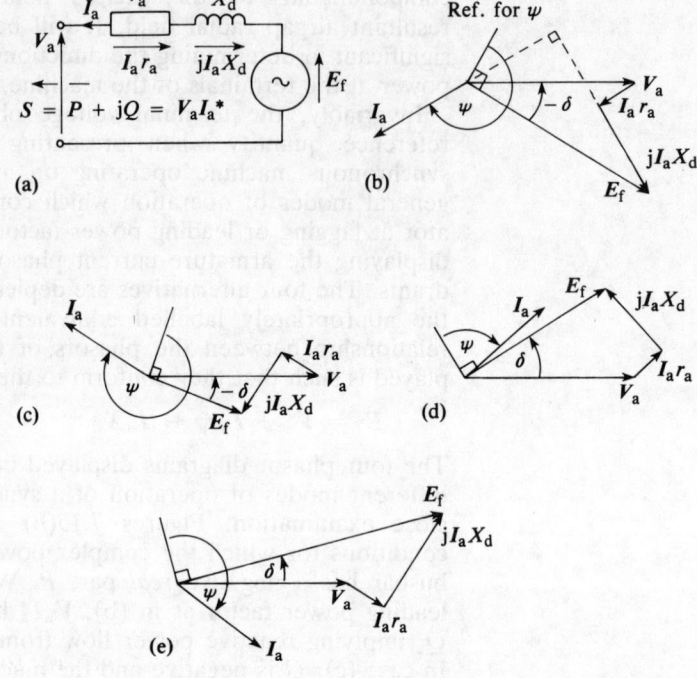

7.12.1 Starting of synchronous machines, synchronisation and putting on load

As a synchronous machine is unable to develop a steady torque at any speed other than synchronous speed, it is not self-starting on a fixed-frequency supply. Some auxiliary means is necessary to run the machine up to very near synchronous speed prior to connecting the armature winding of the d.c. field-excited machine to the supply busbar. Ideally, the connection should be made with the machine speed precisely synchronous at an instant when the armature-phase e.m.f.s induced by the 'field' winding match precisely in magnitude and phase the supply-phase voltages. The machine is then said to be *synchronised* to the supply and, if the armature e.m.f. due to the 'field' winding remains unchanged and no mechanical torque is applied to the rotor shaft, it will remain 'floating' on the busbar, with zero armature current. The subsequent behaviour of the machine is dependent upon the shaft torque applied and whether any change is made to the d.c. field current.

If the d.c. field excitation is increased but the torque is maintained at zero, current flows in the armature circuit of such phase with respect to the terminal voltage that reactive power is transferred from the machine to the system. The current is driven by the phasor difference of E_f and V_a, limited by the synchronous impedance. The phasor diagram of Fig. 7.16(a) shows the effect, relating to the equivalent circuit of Fig. 7.15(a) with armature resistance neglected for simplicity. Reduction in the d.c. field current has the opposite effect, Fig. 7.16(b), causing the machine to absorb VA$_r$.

Figure 7.16 Phasor diagrams illustrating the effect of change in field excitation and shaft torque on a synchronous machine after synchronisation: (a) increase and (b) reduction in d.c. field current; and (c) retardation and (d) acceleration of the rotor and, consequentially, of E_f with respect to V_a. The phasor quantities are defined in Fig. 7.15(a).

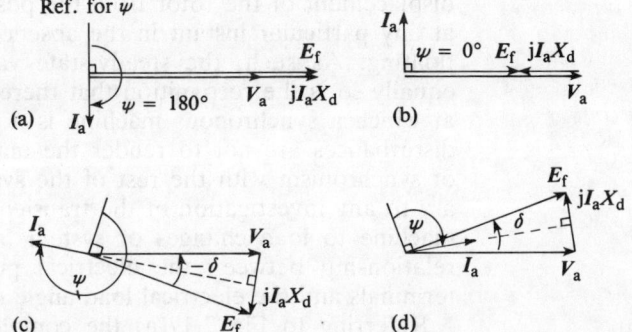

If the d.c. field excitation is kept the same as with the machine floating on the busbars and a load torque is applied to the shaft so as to oppose rotation, the rotor is momentarily slowed so that the armature windings carried on the rotor in our model are delayed in their position relative to the radial field. Consequently, the e.m.f. E_f induced in each armature phase winding is retarded in phase with respect to the fixed-frequency busbar voltage and the difference voltage accounts for the armature current flow, again limited by synchronous impedance. Immediately E_f begins to diverge from V_a, the field due to the armature current contributes to torque development in the direction of rotor motion, accelerating the rotor and enabling synchronous speed to be recovered. The situation is shown in Fig. 7.16(c) which also demonstrates that real power flows from the supply to the armature winding as $\text{Re}\{V_a I_a^*\}$ is negative. Prior to the establishment of equilibrium conditions between applied load torque and electromagnetic torque, as a consequence of the inertia of the rotating mass, oscillation of the load angle δ about the position appropriate to equilibrium will occur. Electrical power flow variations result from the changes in kinetic energy of the rotor whose speed is perturbed about the synchronous value. The dynamics relating to changes in load condition are discussed in Section 7.13.2.

Conversely, attempts to speed up the rotor by the application of torque in the direction of rotor rotation cause electromagnetic torque which opposes motion to be developed, and the steady export of real power when conditions of equilibrium are established. The relevant phasor diagram is shown in Fig. 7.16(d). Here, as also when motoring, the machine will absorb reactive volt-amperes unless the d.c. field excitation is increased.

It is apparent, therefore, that control of the d.c. excitation influences the power factor of the machine through the flow of reactive volt-amperes, whereas the flow of real power is determined by the application of mechanical torque to the rotor shaft. A moment's consideration will show that the electrical effect of applied shaft torque is the same regardless of whether the armature winding is on the rotor or on the stator – as is usual with large machines.

7.12.2 Terminal Power/Load Angle Relationships for a Uniform-Airgap Synchronous Machine on an Infinite Busbar

The electrical load angle δ is a measure of the mechanical displacement of the rotor from the position it would have adopted at any particular instant in the absence of shaft torque, i.e. when floating. As such, the steady-state value of δ is important, but equally so is the recognition that there are limits to the value of δ at which a synchronous machine is normally operated if transient disturbances are not to render the machine unstable through loss of synchronism with the rest of the system. A necessary preliminary to any investigation of the transient response of a synchronous machine to load changes or system faults is a knowledge of the relationship between the electrical power flow at the armature terminals and the electrical load angle δ.

Referring to Fig. 7.17(a) the complex terminal power supplied by one armature phase of a synchronous machine to the connected system is

$$S = P + jQ = V_a I_a^*$$

If E_f leads V_a by load angle δ and $Z_s = r_a + jX_d = Z_s \angle \rho$, taking V_a as the reference quantity and using polar co-ordinates,

$$P + jQ = V_a \left[\frac{E_f \angle \delta - V_a}{Z_s \angle \rho}\right]^* = V_a \left[\frac{E_f \angle -\delta - V_a}{Z_s \angle -\rho}\right]$$

$$= \frac{E_f V_a}{Z_s} \angle (\rho - \delta) - \frac{V_a^2 \angle \rho}{Z_s}$$

$$\therefore P = \frac{E_f V_a}{Z_s} \cos(\rho - \delta) - \frac{V_a^2}{Z_s} \cos \rho \qquad [7.30]$$

$$Q = \frac{E_f V_a}{Z_s} \sin(\rho - \delta) - \frac{V_a^2}{Z_s} \sin \rho. \qquad [7.31]$$

If V_a and E_f are line values of terminal voltage and open-circuit e.m.f. with Z_s pertaining to an equivalent star value/phase, P and Q in eqns [7.30] and [7.31] are total three-phase values. The general relationship between P and δ is shown in Fig. 7.17(b) for constant values of $E_f (> V_a)$ and $Z_s \angle \rho$.

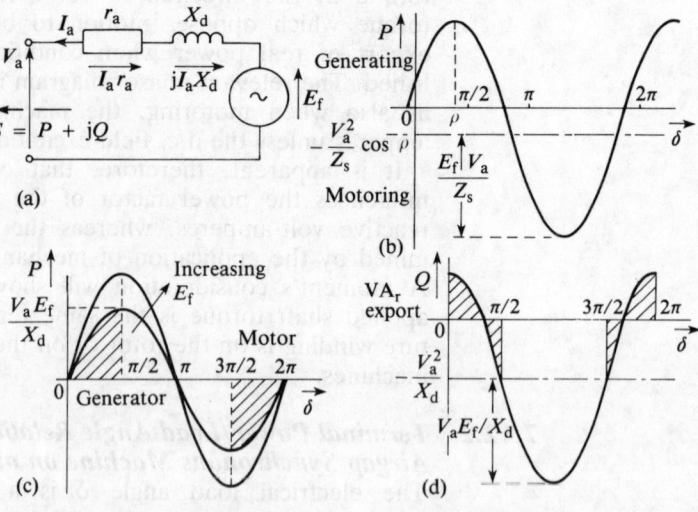

Figure 7.17 Steady-state terminal-power dependence on load angle for a uniform-airgap synchronous machine on infinite busbars; (a) equivalent circuit, (b) real power/δ, (c) as (b) with r_a neglected, and (d) reactive VA/δ with r_a neglected.

Particularly simple relationships for both P and Q apply if armature resistance is neglected, so that $Z_s = X_d \angle 90°$ to give

$$P = \frac{E_f V_a}{X_d} \sin \delta \qquad [7.32]$$

$$Q = -\frac{V_a^2}{X_d} + \frac{E_f V_a}{X_d} \cos \delta. \qquad [7.33]$$

Figure 7.17(c) shows that peak real power as a generator then occurs with $\delta = 90°$ and, as a motor, at $\delta = 270°$. With armature resistance included, maximum generator real power output occurs at $\delta = \rho$, where $\rho = \tan^{-1} X_d/r_a$. Maximum electrical input power as a *motor* occurs at $\delta = \rho + 180°$, but this is not quite coincident with maximum motoring torque as armature copper loss is dependent upon (armature current)^2. With armature losses ignored the machine torque is obtained by dividing the total terminal real power by synchronous speed expressed in mech rad s^{-1}. Then, in terms of phase quantities,

$$T = -\frac{3P}{\omega_0} = -\frac{3pE_f V_a}{\omega X_d} \sin \delta \qquad [7.34]$$

with the reference direction for rotor torque in the direction of rotor rotation.

7.13 Dynamic Response of a Synchronous Machine to Changes in Load

Consider a synchronous machine (with uniform airgap and negligible armature resistance) operating as a generator under steady load conditions with load angle δ_1 ($< 90°$) and electromagnetic and load torques in balance, such that

$$T_e = -T_m = -3p\frac{E_f V_a}{\omega X_d} \sin \delta_1.$$

If the prime-mover torque is suddenly increased by a step change ΔT_m, the electrical power output is unable to increase instantaneously through an increase in δ on account of the inertia of the rotor. The imbalance of torque will reduce as the rotor accelerates, increasing the load angle but, at the instant when the electromagnetic torque equals the prime-mover torque, the rotor is moving at above synchronous speed so δ continues to increase. This causes torque imbalance due to the excess of electromagnetic torque and exported electrical power. The rotor will thus be retarded and will return to synchronous speed when δ ceases to increase further. However, the torques are again out of balance with a deficiency of prime-mover torque so, δ begins to reduce with the rotor speed falling below the synchronous value. It is clear that δ will continue to oscillate indefinitely about the new steady-state value, out of phase with the speed changes about the synchronous value, but for *damping* effects which arise as a result of the relative motion between the armature field and the magnetic circuit of the d.c.-excited member.

It is not always the case that changes in the operating conditions of synchronous machines result simply in a damped oscillation of rotor speed about the synchronous value and an associated oscillation of load angle δ about the value appropriate to the new condition. Transient stability studies are carried out on synchronous plant to establish whether a particular disturbance might cause the machine to lose synchronism with the rest of the system, i.e. to 'fall out of step'. If a synchronous *motor* has the applied load torque *slowly* increased from zero to a value in excess of that which corresponds to a load angle modulus of 90°, the electromagnetic torque will fail to balance the load torque beyond this condition. As $|\delta|$ goes beyond 90° the electromagnetic torque falls for each increment of angle, the torque discrepancy being made up by deceleration of the rotor. As $|\delta|$ goes beyond 180° the electromagnetic torque actually reverses, causing more rapid deceleration until δ attains the value of 360° with the machine running subsynchronously. If the load torque is reduced, or the machine d.c. field excitation is increased, there is a chance that the motor will regain synchronism or 'pull into step' as $|\delta|$ increases towards 90°, the machine having slipped one pair of poles. If the machine fails to resynchronise, the rotor will eventually come to rest, although earlier disconnection from the supply is clearly advisable to avoid the large current and power swings associated with an out-of-step condition.

The steady-state stability limit for δ is therefore $\pm 90°$ for a synchronous machine with armature resistance neglected. During *transient* changes in loading conditions, however, the 90° value may be exceeded without subsequent loss of synchronism if the rotor kinetic-energy change on the first transition between the initial and ultimate equilibrium states is not excessive.

7.13.1 *Synchronising power, synchronising torque*

A measure of the ability of a synchronous machine to tolerate a transient disturbance is provided by the concept of *synchronising power*, which is defined as *the increase in electrical power transferred per unit increase in electrical load angle*. This incremental change in electrical power transferred ΔP_e in response to a change in load angle $\Delta \delta$ is dependent upon the value of δ at which the machine is operating.

From eqn [7.32], for a three-phase machine

$$P_e = \frac{3 E_f V_a}{X_d} \sin \delta$$

then

$$\left[\frac{dP_e}{d\delta}\right]_{\delta=\delta_1} = \left[\frac{\Delta P_e}{\Delta \delta}\right]_{\delta=\delta_1} = \frac{3 E_f V_a}{X_d} \cos \delta_1. \qquad [7.35]$$

The synchronising power $\Delta P_e / \Delta \delta$ is seen to be a maximum when $\delta = 0°$ and zero when $\delta = \pm 90°$. As synchronous machines feature most significantly in the generation of electrical power, stability analysis is generally presented in the context of normal electrical power flow from machine to system, and this convention will be observed in the following discussion.

7.13.2 *The natural frequency of oscillation of a synchronous machine*

Consider a synchronous machine operating under conditions of equilibrium as a generator with a constant input mechanical power P_m balanced by electrical output power P_{e1} as shown in Fig. 7.18. Assume that the rotor is given an additional impulse of torque in the direction of rotation such that the load angle is increased by a *small* amount $\Delta\delta_1$ from its equilibrium value δ_1. The electrical power exported will increase but, as the mechanical torque applied is kept constant this extra power demand $\Delta P_{e1} = P_e - P_m$ must be balanced by the kinetic energy released by a slowing down of the rotor. As the rotor slows the angle of advance $\delta_1 + \Delta\delta$ of the rotor system over the synchronously rotating reference-frame representation of the infinite busbar will fall and δ will reduce to δ_1 at which the mechanical and electrical powers are in balance. As the rotor is running subsynchronously, however, δ will continue to reduce together with P_e, the mismatch of mechanical and electromagnetic torques permitting the rotor speed to rise. The decline of δ will cease when the rotor speed is again synchronous. However, the value of P_m will then exceed P_e and the rotor will accelerate above the synchronous value and δ increase towards and beyond δ_1 once more. The load angle will then continue to oscillate about the value of δ_1 with, if damping is ignored, the *natural* frequency of oscillation.

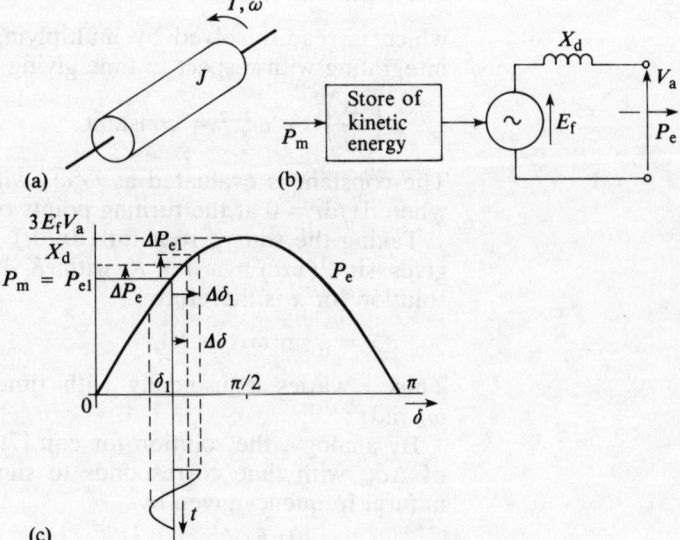

Figure 7.18 Synchronous machine with transient mechanical and electrical power mismatch: (a) mechanical energy store in rotor inertia; (b) equivalent circuit neglecting armature resistance and changes in stored energy in the system inductance which are ineffective over the time scale considered; and (c) terminal power/load angle relationship for an incremental change in accelerating power $\Delta P_e = P_{e1} - P_m$ about equilibrium, $P_{e1} = P_m$.

The quantity $\Delta P_e = P_e - P_m$ corresponds to torque imbalance of value ΔT, where

$$\Delta T = \frac{\Delta P_e}{\omega_m} \simeq \frac{\Delta P_e}{\omega_0} \qquad [7.36]$$

and ΔT is measured in N m and ω_0 is the rotor synchronous speed in mech rad s^{-1}, to which the actual speed ω_m is approximated.

ΔT accounts for retardation (negative acceleration) of the rotating system such that, with damping neglected

$$\Delta T = -J\frac{d\omega_m}{dt} = -J\frac{d^2(\Delta\delta_m)}{dt^2} \quad [7.37]$$

where J is the moment of inertia in kg m^2 of the rotor and $\Delta\delta_m$ its displacement in mech rad from a reference position appropriate to synchronous-speed running.

From eqn [7.36]

$$\Delta T = \frac{\Delta P_e}{\Delta \delta}\frac{\Delta\delta}{\omega_0} = \frac{\Delta P_e}{\Delta \delta}\left(\frac{\Delta\delta}{\Delta\delta_m}\right)\frac{\Delta\delta_m}{\omega_0} = \frac{\Delta P_e}{\Delta \delta}p\frac{\Delta\delta_m}{\omega_0}$$

for a machine with p pole-pairs. Hence, substituting eqns [7.35] and [7.37]

$$\frac{3pE_f V_a}{\omega_0 X_d}\cos\delta_1 \Delta\delta_m = -J\frac{d^2(\Delta\delta_m)}{dt^2}$$

or

$$\frac{d^2(\Delta\delta_m)}{dt^2} + \frac{3pE_f V_a}{\omega_0 X_d J}\cos\delta_1 \Delta\delta_m = 0. \quad [7.38]$$

Now the general equation for (undamped) *simple-harmonic motion* of variable $x(t)$ is

$$\frac{d^2 x}{dt^2} + \omega_n^2 x = 0,$$

which is readily solved by multiplying throughout by $2dx/dt$ and integrating with respect to time giving

$$\left(\frac{dx}{dt}\right)^2 + \omega_n^2 x^2 = \text{constant}.$$

The constant is evaluated as $\omega_n^2 c^2$, with c being the $\pm$ values of x when $dx/dt = 0$ at the turning points of x.

Taking the square root for $(dx/dt)^2$, rearranging and integrating gives $\sin^{-1}(x/c) = \omega_n t + K$ with K being another constant. The solution for x is therefore

$$x = c\sin(\omega_n t + K).$$

Thus x varies sinusoidally with time, having angular frequency ω_n rad s^{-1}.

By analogy, the solution for eqn [7.38] is such that the variation of $\Delta\delta_m$ with time corresponds to simple-harmonic motion with a natural frequency given by

$$\omega_n = \left[\frac{3pE_f V_a}{\omega_0 X_d J}\cos\delta_1\right]^{1/2}$$

or, from eqn [7.35] with *synchronising torque* equated to synchronising power/ω_0

$$f_n = \frac{1}{2\pi}\left[\frac{\text{pole pairs} \times \text{synchronising torque}}{\text{moment of inertia}}\right]^{1/2}. \quad [7.39]$$

The frequency of oscillation of the rotor position about that appropriate to constant synchronous speed is thus seen to be dependent on the value of the *synchronising torque*, which is itself dependent upon the load angle at equilibrium. Synchronising torque is also known as the 'stiffness' of a machine. A 'stiff' machine is one less likely to go unstable. For a given terminal voltage, stiffness may be enhanced by increasing the d.c. field excitation, minimising synchronous reactance and operating at a low value of load angle.

The maximum value of synchronising torque is proportional to the peak value of power transfer between machine and system and is thus related to the machine power rating. Moment of inertia J is related to the stored kinetic energy at synchronous speed $\frac{1}{2}J\omega_0^2$. Thus the ratio of machine rating to stored kinetic energy at synchronous speed is an important quantity in assessing the transient stability of a synchronous machine. The inverse ratio is called the 'inertia constant' H or 'H-constant', usually measured in MW s/MVA, or s. A value for H lies typically within the range 1–10 s, depending on machine type and pole number.

7.13.3 Dynamic response of a synchronous machine to load or system change – equal-area criterion

As stated in Section 7.13 it is important to be able to predict whether a proposed change in operating conditions will be achieved without loss of synchronism. The equation of motion describing the change in load angle throughout the transient period is derived from the basic torque balance equation which, with damping neglected, equates the imbalance between mechanical and electromagnetic torques to the angular acceleration.

Thus, for a synchronous generator subjected to a mechanical torque increase to T_{m2} from a lower value equivalent to an electrical power transference

$$p_{e1} = \frac{3E_f V_a}{X_d} \sin \delta_1$$

$$T_{m2} - T_e = J \frac{d\omega_m}{dt} = J d\left(\frac{\omega_0 + \Delta\omega_m}{dt}\right)$$

or

$$T_{m2} - \frac{3E_f V_a}{\omega_m X_d} \sin \delta = \frac{J}{P} \frac{d^2\delta}{dt^2}, \qquad [7.40]$$

where $\Delta\omega_m$ is the difference between rotor actual speed ω_m and the synchronous value ω_0 and δ is the electrical load angle.

Equation [7.40] is a non-linear, second-order differential equation, best solved for δ by step-by-step computational methods. The system is declared to be stable if the value of δ is seen to decrease after attaining a maximum value which may well exceed 90°. A graph of δ against time is called a *swing curve*.

In the plotting of swing curves knowledge of the machine inertia is a prerequisite. If the transient study being carried out is not required to provide a time scale for the development of a possible out-of-step condition, a simple alternative graphical procedure is available to judge whether a proposed change in operating conditions would occur without the transient stability limit being exceeded. Erring on the safe side, the effects of damping are ignored. The method makes use of the basic premise that, if a system is to remain stable, the change in kinetic energy of the rotor between inception of the disturbance and the first subsequent balance of mechanical and electromagnetic torque cannot exceed the change in kinetic energy between that first balance and the achievement of synchronous speed once more. The latter condition marks the instant at which the load angle δ ceases to increase.

Thus, consider again the case of a synchronous generator operating initially with electromagnetic and mechanical torques in equilibrium at load angle δ_1, subjected to an increase in mechanical torque such that the mechanical input increases from P_{m1} to P_{m2}. The net torque causes rotor acceleration until $P_e = P_{m2}$ at $\delta = \delta_2$, as shown in Fig. 7.19. During acceleration, the increase in kinetic energy of the rotor is $\tfrac{1}{2}J(\omega_2^2 - \omega_0^2)$.

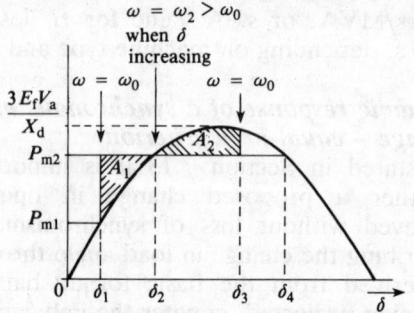

Figure 7.19 Equal-area criterion of transient stability for a synchronous generator with mechanical input power increase from P_{m1} to P_{m2}.
δ_3 = consequential maximum load angle value,
δ_4 = limiting value without loss of stability.

The area A_1 on the power/angle graph of Fig. 7.19, given by

$$\int_{\delta_1}^{\delta_2}(P_{m2} - P_e)\,d\delta,$$

has the dimensions of power and is a measure of the *increase* in energy of the rotor above that pertaining to synchronous speed. During deceleration the reduction in kinetic energy = $\tfrac{1}{2}J(\omega_2^2 - \omega_0^2)$ as the load angle increases from δ_2 to δ_3, at which the rotor regains synchronous speed. The kinetic energy given up relates similarly to area A_2, given by

$$\int_{\delta_2}^{\delta_3}(P_e - P_{m2})\,d\delta = \int_{\delta_3}^{\delta_2}(P_{m2} - P_e)\,d\delta.$$

Hence, the use of the *equal-area criterion* of transient stability involves finding graphically the area A_2 which equates to area A_1, checking that δ_3 falls short of the δ_4 beyond which further increase in δ would cause further acceleration of a rotor still running above synchronous speed. Knowledge of the machine inertia is not required.

7.13.3.1 Damping

When relative motion occurs between the sinusoidally distributed resultant radial field in the airgap and the iron structure upon which the d.c. excited field coils are mounted, e.m.f.s will be induced in any closed conducting circuits which are linked by the flux. As a consequence currents flow in these closed circuits producing a tangential magnetic field at the airgap surface. The frequency of the induced current is low when the relative motion is due to load or system changes. Torque developed by the interaction of these induced currents flowing in mainly resistive circuits and the responsible radial field is invariably such as to oppose the relative motion which gave rise to the induced e.m.f. The effect is discussed quantitatively in Chapter 8 which is concerned with induction machines. The law of conservation of energy would, however, confirm this result by requiring an absorption of mechanical energy if consequential electrical energy is accounted for by dissipation within the resistance.

7.13.4 *Transient stability improvement for a synchronous machine*

The *stability* of a power system is that property which establishes its ability to retain the synchronous running of its synchronous plant when the system is subjected to changes in its operating conditions due to a changing load profile or changes in system configuration.

System stability is a complex problem for analysis, involving the simultaneous solution of many electromechanical equations which involve system and plant impedances, plant inertias and damping factors, and a wide variation of initial operating conditions and types of disturbance. The problem of stability has been simplified in the earlier treatment by modelling the system to which a machine of interest is connected as a constant-voltage, constant-frequency busbar. The effects of a power system becoming generally unstable are serious, not simply as a result of likely loss of supply to consumers but also as a consequence of the disruption of the controlled, steady energy flows through boilers and turbines at the generating stations.

The way in which a system is operated has a significant bearing on the system stability. A machine supplying a large amount of reactive volt-amperes to the connected system will operate at a lower value of load angle than if the VA_r flows are in the opposite direction. At times of light load the overhead lines and cables of a power network are net providers of reactive volt-amperes through shunt capacitance, and any surplus must be absorbed by the synchronous plant which correspondingly requires a reduced d.c. field excitation to maintain an appropriate terminal voltage. The effect is ameliorated if the number of lines and cables in circuit is reduced. A failure in one remaining system component may then have serious consequences for the maintenance of supply through insufficient sources or routes of supply. It is no mere coincidence that catastrophic failures of power supplies generally occur at times of low demand.

Protective gear may be installed on synchronous plant to detect a developing loss of synchronism prior to actual pole slip. The change in load angle may be monitored by observing changes in rotor position relative to a synchronously rotating reference frame derived from terminal voltage. The prospects for retaining synchronism are enhanced if rapid increases in the d.c. field excitation are possible together with a close balance between heat, steam, mechanical and electrical energy flows. High values of rotating machine inertia allow more time for correcting control action to take place. Developments in machine manufacturing and operating techniques have led to a concentration of generation in fewer, larger units with relatively low values of inertia and with regard to which the rest of the system cannot be regarded as a truly infinite busbar. This trend may not be set to continue, however.

7.14 Synchronous Machines with Salient Poles – Two-reaction Theory

The analysis in Sections 7.8 to 7.13 relates strictly to a machine with a uniform airgap. Such a machine has the simplifying property that a sinusoidal distribution of current at either airgap boundary gives rise to a sinusoidal distribution of a radial flux density of proportionate magnitude (saturation neglected) and displaced in space by 90°ε. A *salient-pole* machine has a preferred axis of magnetisation, the *direct* axis, which is that axis along which the main field winding directs its radial airgap field. Thus a smaller armature winding m.m.f. would be required to develop a given total flux per pole when distributed about this axis than if the radial field produced is centred on the interpolar space – the *quadrature* axis.

The precise nature of saliency is very much an individual feature of a particular machine – the specific way in which the airgap length varies with position around the airgap, which is invariably symmetrical about the pole centres and those of the interpolar spaces. A complementary issue is that a sinusoidal distribution of current no longer produces a sinusoidal distribution of radial flux density, but a fundamental component together with odd space harmonics.

Conventional a.c. machine theory is developed from the assumption of sinusoidally distributed fields and fluxes within the airgap, giving rise to sinusoidally time-variant components of e.m.f. associated with similarly variable currents. A method of analysis for the salient-pole machine which retains the freedom to ignore space harmonics would therefore be attractive, recognising that the validity of the assumptions must be approved by checking actual machine performance against behaviour predicted via equivalent circuits or performance equations derived from analytical reasoning using experimentally determinable parameters.

The *two-reaction theory* of salient-pole synchronous machines was first proposed by Blondel in 1895. The method identifies resolved components of the sinusoidally distributed radial field due to the armature current along d- and q-axes and assigns two different values for the effective length of the airgap at all positions depending upon the particular affiliation of each flux-producing radial field component. A smaller gap length is allocated to the d-axis components, so accounting for the lower reluctance presented to radial flux crossing the airgap in the vicinity of the poles. The use of this artifice retains the concept of sinusoidal space distributions of B_n responsible for sinusoidally time-variant e.m.f.s. Generally, the machine has d.c. field winding excitation contributing to the d-axis radial field in addition to that of the armature. If the armature winding alone is excited, however, and arranged such that the radial field has components along both d- and q-axes, the resultant *radial flux density* distribution is phase shifted in space relative to the *tangential field* at the cylindrical armature/airgap boundary to provide torque. This accounts for the ability of a salient-pole synchronous machine to function as an electromechanical energy converter without field-winding excitation. This *reluctance machine* will now be considered in some detail appropriate to steady-state operation prior to incorporating the effect of d.c. field excitation which is usual with highly rated salient-pole synchronous machines.

7.14.1 *The polyphase reluctance machine – torque equation*

Consider the armature current represented in the three-phase armature windings mounted on the rotor of Fig. 7.10(a) to be the only source of airgap magnetisation. The situation is reproduced in Fig. 7.20, which shows for a two-pole machine the radial field H_{nr} resolved along d- and q-axes, the former being coincident with the axis of symmetry having minimum airgap reluctance. θ' again defines the general position in the gap of the *a* phase reference coil side rotating synchronously in the anticlockwise direction and ψ the phase angle of lag which the peak of the constant-amplitude fundamental component of the armature radial field makes with the d-axis as the balanced armature phase currents of sequence *abc* vary sinusoidally with time.

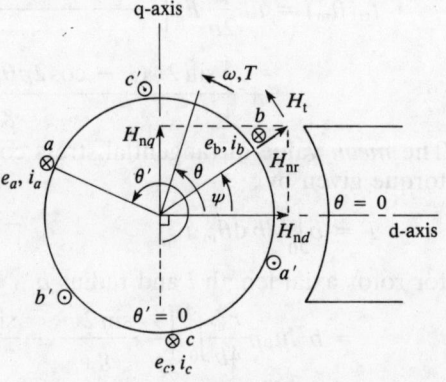

Figure 7.20 Resolution of a sine-distributed airgap field along *d*- and *q*-axes.

Accounting for saliency, the sinusoidally distributed radial-field components in the airgap may be expressed as general functions of position θ_m, thus along the d-axis

$$H_{nd}(\theta_m) = \frac{\hat{F}_{nr}}{g_d}\cos\psi \cos p\theta_m$$

along the q-axis

$$H_{nq}(\theta_m) = \frac{\hat{F}_{nr}}{g_q}\sin\psi \sin p\theta_m.$$

In the above equations, $F_{nr}(\theta_m)$ is the m.m.f. developed across the airgap by the armature current, with each resolved component, directed along d-axis or q-axis, being given by the product of the appropriate radial field component and the assigned length of corresponding gap crossing, g_d or $g_q > g_d$. p is the number of pole pairs for a multipole machine and θ_m is measured in *mechanical* degrees or radians.

The *tangential* field at the cylindrical rotor surface rotor surface is unaffected by the saliency present on the stator. From eqns [5.21] and [5.23], or eqns [7.6]. [7.7] and [7.9], and also assuming $g_q \gg g \gg g_d$

$$H_{\text{tr}}(\theta_m) = \frac{pg}{b}\hat{H}_{\text{nr}} \sin p\left(\theta_m - \frac{\psi}{p}\right)$$

$$= \frac{p}{b}\hat{F}_{\text{nr}} \sin p\left(\theta_m - \frac{\psi}{p}\right).$$

The resultant radial flux density distribution is derived from the radial field components H_{nd} and H_{nq} thus

$$B_n(\theta_m) = \mu_0 \hat{F}_{\text{nr}} \left[\frac{\cos\psi \cos p\theta_m}{g_d} + \frac{\sin\psi \sin p\theta_m}{g_q}\right].$$

Hence the tangential stress at the rotor surface is given by

$$t_{\text{tr}}(\theta_m) = B_n H_{\text{tr}}$$

$$= \mu_0 \frac{p}{b}\hat{F}_{\text{nr}}^2 \sin p\left(\theta_m - \frac{\psi}{p}\right)\left[\frac{\cos\psi \cos p\theta_m}{g_d} + \frac{\sin\psi \sin p\theta_m}{g_q}\right]$$

which may be re-written

$$t_{\text{tr}}(\theta_m) = \mu_0 \frac{p}{2b}\hat{F}_{\text{nr}}^2 \left[\frac{\cos^2\psi \sin 2p\theta_m - \frac{1}{2}\sin 2\psi(1 + \cos 2p\theta_m)}{g_d}\right.$$

$$\left. + \frac{\frac{1}{2}\sin 2\psi(1 - \cos 2p\theta_m) - \sin^2\psi \sin 2p\theta_m}{g_q}\right].$$

The *mean* value of tangential stress corresponds to a value of rotor torque given by

$$T = b\int_0^{2\pi} lb\, d\theta_m \cdot t_{\text{tr}_{\text{mean}}}$$

for rotor axial length l and radius b

$$= b^2 l\mu_0 p\frac{\hat{F}_{\text{nr}}^2}{4b}\int_0^{2\pi}\left[-\frac{\sin 2\psi}{g_d} + \frac{\sin 2\psi}{g_q}\right]d\theta_m.$$

Hence

$$T = -\mu_0\pi pbl\frac{\hat{F}_{\text{nr}}^2}{2}\left[\frac{g_q - g_d}{g_q g_d}\right]\sin 2\psi.$$

If $g_q - g_d$ is small compared wth g_d

$$\frac{\hat{F}_{\text{nr}}^2}{g_q g_d} \simeq \hat{H}_{\text{nr}}^2$$

$$T = -\mu_0\pi pbl\left[\frac{g_q - g_d}{2}\right]\hat{H}_{\text{nr}}^2 \sin 2\psi. \qquad [7.41]$$

Equation [7.41] may be compared with eqn [7.16]. Key features of eqn [7.41] show that the torque of a synchronous reluctance machine is:

(i) proportional, at constant current, to the sine of twice the

space-phase angle ψ which the armature radial field axis makes with the d-axis,
(ii) proportional to (armature current)^2 for given ψ,
(iii) proportional to the difference in effective airgap length $(g_q - g_d)$, and
(iv) directed such as to tend to align the armature radial field with the d-axis.

Such torque is described as 'saliency torque' or 'reluctance torque' as it is associated with the different reluctances associated with radial flux crossing the airgap at d- and q-axis locations.

7.14.2 *The polyphase reluctance machine – equivalent circuit and phasor diagram*

The two-reaction method of analysing the salient-pole machine is inherently dependent upon the validity of the resolution of the radial armature magnetic field into components directed along d- and q-axes. Equivalent circuit and phasor diagram representations of such machines go one step further by assigning to time-variant components of armature current the responsibility for producing these components of field directed along axes in space quadrature.

Referring to the elemental three-phase armature coil representation of Fig. 7.20, it is apparent that balanced currents of sequence *abc* produce a resultant rotor armature field H_{nr} which is stationary in space when the armature coils rotate synchronously in the anticlockwise direction. H_{nr} lags the d-axis (in the reference direction of angular measurement) by the space-phase angle ψ hence, if the armature currents were to be advanced in time by the *time*-phase angle ψ relative to their actual values the armature radial field would be directed along the d-axis. Similarly, if the armature currents were delayed in time by the equivalent of $(90° - \psi)$ then H_{nr} would lie along the q-axis. Hence, identical magnetic effects to the general case of Fig. 7.20 would be developed if each armature phase current is considered to be the resultant of fictional components I_d and I_q where, in phase *a*, I_{ad} leads I_a by ψ and I_q lags I_a by $(90° - \psi)$, each of magnitude such that $\hat{I}_{ad}/\hat{I}_a = \cos \psi$ and $\hat{I}_{aq}/\hat{I}_a = \sin \psi$.

Motion of the armature phase conductors in the radial flux distributions due to d- and q-axis components of radial magnetic field accounts for components of induced e.m.f. in each phase winding. Since these components of flux density are assumed sinusoidal, and distributed about axes in *space* quadrature, it follows that the components of e.m.f will be sinusoidal and in *time* quadrature, with that due to the d-axis flux leading the other by 90°. Reference to Fig. 7.20 shows that a rotating armature conductor encounters the d-axis 90° ahead of the q-axis. Ignoring armature leakage reactance and resistance the resultant e.m.f. induced in each phase by the motion of the conductors at right-angles to the components of radial flux will equate to the terminal voltage per phase such that, for reference phase *a*,

$$V_a = E_a = E_{ad} + E_{aq}$$

where E_{ad} leads E_{aq} by 90°.

If the machine were operated such that $\psi = 0$ the quadrature axis components of radial field and flux would disappear. Neglecting armature leakage impedance, the flux responsible for E_a would then be identified solely with the d-axis whose effective permeance is determined by minimum airgap length g_d. Invoking the concept of armature reaction reactance introduced for the uniform airgap machine in eqn [7.23] the relationship between $E_a = E_{ad}$ and $I_a = I_{ad}$ may be written

$$\frac{E_{ad}}{I_{ad}} = -jX_{ad}, \qquad [7.42]$$

where

$$X_{ad} \propto \frac{1}{g_d}.$$

The 90° phase lag of E_{ad} over I_{ad} is confirmed by reference to Fig. 7.20, recognising that $\theta' = 0$ at the instant of current maximum in phase a with the resultant armature radial field directed along the d-axis, whereas the induced e.m.f. e_a due to the same field has its peak positive value when $\theta' = 90°$ (right-hand rule) i.e. 1/4 of a period later.

Similarly, if the machine operates such that H_{nr} is directed along the q-axis with the total armature e.m.f. E_a given by E_{aq}, the reluctance relating radial flux to m.m.f. established by the armature current is determined by gap length g_q. Maximum current in phase a will then correspond to the a phase reference coil side in Fig. 7.20 being located at $\theta' = 90°$, with peak positive e.m.f. e_a when $\theta' = 180°$. Under these conditons, therefore

$$\frac{E_{aq}}{I_{aq}} = -jX_{aq}, \qquad [7.43]$$

where

$$X_{aq} \propto \frac{1}{g_q}.$$

The quantities X_{ad} and X_{aq} account for the armature reaction effect with saliency present. X_{ad} is called the *direct-axis armature reaction reactance* and X_{aq} the *quadrature-axis armature reaction reactance*.

Summarising the relationship between the 'airgap e.m.f.' E_a of phase a, the current in phase a and their 'components' we have

$$E_a = E_{ad} + E_{aq}$$

where

$$\arg\left[\frac{E_{ad}}{E_a}\right] = \beta,$$

say.

$$I_a = I_{ad} + I_{aq}$$

where

$$\arg\left[\frac{I_{ad}}{I_a}\right] = \psi$$

giving
$$I_{ad} = I_a \cos \psi, \quad I_{aq} = I_a \sin \psi.$$
Also
$$\frac{E_{ad}}{I_{ad}} = -jX_{ad}$$
and
$$\frac{E_{aq}}{I_{aq}} = -jX_{aq}.$$
Hence
$$\left. \begin{array}{c} I_a = I_{ad} + I_{aq} = -\dfrac{E_{ad}}{jX_{ad}} - \dfrac{E_{aq}}{jX_{aq}} \\ \text{with} \\ E_a = E_{ad} + E_{aq}. \end{array} \right\} \quad [7.44]$$

Equations [7.44] give rise to the phasor diagrams relating armature current I_a and resultant e.m.f. E_a shown in Fig. 7.21. Ignoring armature leakage reactance and resistance, E_a equals terminal voltage V_a. The condition specified in Fig. 7.21(a) corresponds to machine operation as a *generator* absorbing reactive power. Figure 7.21 also shows the locus of I_a as the load torque conditions are varied with constant terminal voltage. The displayed phasors are able to vary only within the constraints implied by eqns [7.44]. The terminal power exported per phase, ignoring leakage, is given by $S = P + jQ = E_a I_a^*$, and any change in real power flow is accompanied by an associated change in reactive

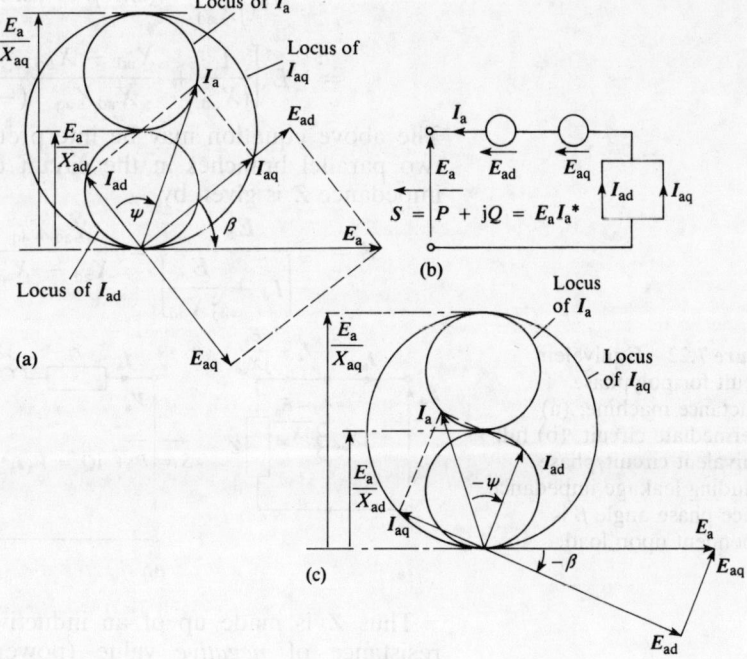

Figure 7.21 Armature current locus diagram for polyphase reluctance machine with $X_{ad}/X_{aq} = 2$ and armature leakage impedance neglected: (a) generating condition identified, (b) effective equivalent circuit, and (c) motoring condition identified.

power. Q is invariably negative, hence reactive power is always absorbed by the machine.

A motoring condition for the same machine with $X_{ad}/X_{aq} = 2$ is shown in Fig. 7.21(c). Included on the phasor diagrams are phase angles ψ and β, the latter indicating the space-phase angle by which the d-axis component of radial *flux density* leads the resultant flux density. ψ and β are equivalent only when the machine develops zero torque – at $\psi = \beta = 0°$, $180°$ or $\pm 90°$.

An equivalent circuit for a polyphase reluctance machine may be derived from eqns [7.44]. Such an equivalent circuit is of limited practical value, however, not only on account of the imperfections of the model but also because it involves the angle β which is variable with load.

From eqns [7.44], taking E_a as reference $E_{ad} = E_{ad} \angle \beta$

$$I_a = -\frac{E_{ad}}{jX_{ad}} - \frac{E_{aq}}{jX_{aq}}$$

$$= -\frac{E_a \cos\beta(\cos\beta + j\sin\beta)}{jX_{ad}}$$

$$- \frac{E_a \sin\beta[\cos(\beta - 90°) + j\sin(\beta - 90°)]}{jX_{aq}}$$

$$= -E_a \left[\frac{\cos^2\beta + \sin^2\beta - \sin^2\beta + j\cos\beta\sin\beta}{jX_{ad}} \right.$$

$$\left. + \frac{\sin^2\beta - j\sin\beta\cos\beta}{jX_{aq}} \right]$$

$$= -E_a \left\{ \frac{1}{jX_{ad}} + (\sin^2\beta - j\sin\beta\cos\beta)\left[\frac{1}{jX_{aq}} - \frac{1}{jX_{ad}}\right] \right\}$$

$$= -E_a \left[\frac{1}{jX_{ad}} + \frac{X_{ad} - X_{aq}}{X_{ad}X_{aq}} \frac{1}{(-\cot\beta + j1)} \right].$$

The above equation may be interpreted as the sum of currents in two parallel branches in the circuit of Fig. 7.22(a) for which the impedance Z is given by

$$Z = \frac{E_a}{-\left[I_a + \frac{E_a}{jX_{ad}}\right]} = \frac{X_{ad}X_{aq}}{X_{ad} - X_{aq}}(-\cot\beta + j1).$$

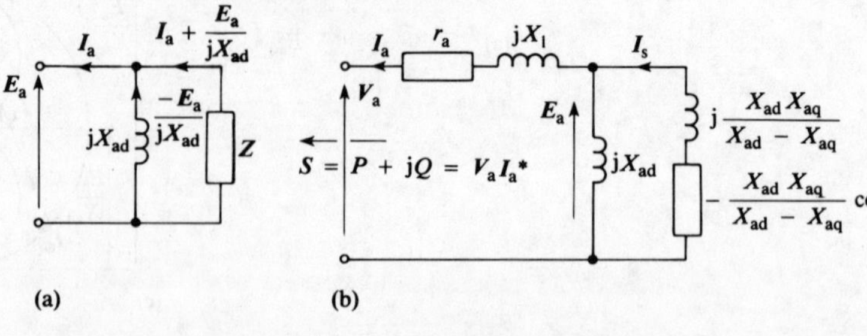

Figure 7.22 Equivalent circuit for polyphase reluctance machine: (a) intermediate circuit, (b) full equivalent circuit/phase including leakage impedance. Space-phase angle β is dependent upon load.

Thus Z is made up of an inductive reactance in series with a resistance of *negative* value (power source) if $\cot\beta > 0$. The equivalent circuit of Fig. 7.22(b) incorporates armature resistance and leakage reactance.

If leakage impedance is ignored, an expression for the terminal real power may be derived in terms of the terminal voltage and the equivalent circuit parameters. Thus, from Fig. 7.22(b),

$$P \simeq \mathrm{Re}\{E_a I_a^*\} \simeq \mathrm{Re}\{V_a I_s^*\}$$
$$\simeq \mathrm{Re}\left\{V_a^2 \left[\frac{X_{ad} - X_{aq}}{X_{ad} X_{aq}}\right](\sin\beta\cos\beta + \mathrm{j}\sin^2\beta)^*\right\}$$
$$\simeq V_a^2 \left[\frac{X_{ad} - X_{aq}}{2 X_{ad} X_{aq}}\right] \sin 2\beta.$$

If X_{ad} and X_{aq} are values per phase for a star-connected armature winding and V_a is expressed as line–line terminal voltage, the above equation gives total three-phase power. An equivalent value for torque in the direction of rotor rotation is given by

$$T = -\frac{V_a^2}{\omega_0}\left[\frac{X_{ad} - X_{aq}}{2 X_{ad} X_{aq}}\right]\sin 2\beta. \qquad [7.45]$$

Equation [7.45] may be compared with eqn [7.41]. Whereas the latter expresses torque in terms of the resultant radial magnetic field and its effective current source via a phase angle between the resultant field and its d-axis component, eqn [7.45] implies a voltage source, involving the space-phase angle between resultant radial *flux* and its d-axis component, or the corresponding time-phase difference between corresponding induced e.m.f.s. Maximum torque is developed by a reluctance machine when $\beta = \pm 45°\varepsilon$ if the terminal voltage is held constant and the leakage impedance neglected, or when $\psi = \pm 45°\varepsilon$ if the armature current is held constant.

Reluctance machines have the inherent disadvantage of operating at low power factor. If the direct axis is provided with d.c. field excitation as with the uniform-airgap synchronous machine the torque-producing capability is much enhanced and the machine made capable of supplying reactive power to the connected system. The *salient-pole synchronous machine* is the subject of detailed study in Section 7.16.

7.15 The Single-phase Reluctance Motor

A reluctance machine will develop torque as a motor or generator at synchronous speed when one phase only of its armature winding is connected to a single-phase supply. The *pulsating* radial and tangential fields developed by the winding may, if sinusoidally distributed, be resolved into sinusoidally distributed components, each of amplitude equal to one-half of the *peak* value of the resultant and *rotating* relative to the winding in opposite directions at synchronous speed determined by the frequency of the supply current and the number of pole pairs. The single-phase reluctance motor will run in either direction at synchronous speed but is not self starting. The 'forward' rotating airgap field components (radial and tangential) are responsible for the torque by a mechanism identical with that appropriate to the polyphase machine. The 'backward' rotating field components develop no average torque but cause pulsations in the instantaneous value of the total torque.

Proof Consider a two-pole radial airgap field assumed cosinusoidally distributed about $\theta = 0$ and established by sinusoidally time-variant currents i and $-i$ in the full-pitch coil shown in Fig. 7.23.

At any point θ in the airgap

$$H_n(\theta, t) = \hat{H}_n \sin \omega t \cos \theta$$

where $\hat{H}_n \propto \hat{I}_a$ and $i_a = \hat{I}_a \sin \omega t$. Therefore

$$H_n(\theta, t) = \frac{\hat{H}_n}{2}[\sin(\omega t - \theta) + \sin(\omega t + \theta)] \qquad [7.46]$$

$$= H_{n1} + H_{n2}.$$

$H_{n1} = \dfrac{\hat{H}_n}{2}\sin(\omega t - \theta)$ represents a sinusoidally distributed field of constant amplitude

$\hat{H}_n/2$ rotating in the direction of θ with velocity ω relative to the coil.

$H_{n2} = \dfrac{\hat{H}_n}{2}\sin(\omega t + \theta)$ represents a field of constant amplitude

$\hat{H}_n/2$ rotating in the direction of $-\theta$ with velocity ω.

When $\omega t = 0$, H_{n1} has the peak positive value of its distribution located at $\theta = -90°$, whereas that of H_{n2} is located at $\theta = 90°$. One quarter period later, when the current peaks at $\omega t = 90°$, both rotating field components have peak positive values at $\theta = 0°$.

The single-phase reluctance motor is considered further in Section 8.12.1.

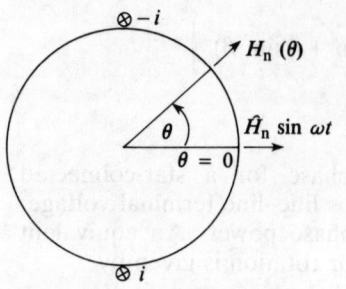

Figure 7.23 Pertaining to the resolution of a pulsating sine-distributed airgap field into equal-amplitude, contra-rotating components.

7.16 The Polyphase Salient-pole Synchronous Machine – Phasor Diagram

The procedure by which the distribution of mechanical tangential stress and consequential torque may be established for a salient-type synchronous machine follows that described in Section 7.14.1 for the polyphase reluctance machine. Saliency is accounted for in the same way – the assignation of different values of assumed uniform airgap length g_d or g_q depending upon whether a component of armature radial field is resolved along the direct axis or the quadrature axis. The additional feature of the more general machine is the provision of additional d-axis excitation derived from d.c.-energised electromagnet coils, or permanent magnets, located on the side of the airgap opposite the armature winding. The important consequence of this is that the radial field at the armature/airgap surface is modified by the d.c. field excitation and may be much enhanced to increase machine torque and power. It also offers reactive volt-ampere flow control when the machine is connected to infinite busbars, and generally combines the features of the reluctance machine with those of the uniform-airgap machine.

An equivalent circuit for the polyphase salient-pole machine may be deduced by extending that developed for the polyphase reluctance machine in Section 7.14.2 and portrayed in Fig. 7.22, to include a current source representative of the additional excitation along the d-axis. The equivalent circuit is of limited value, however, as the parameters are difficult to isolate and evaluate experimentally and one, β, has its value dependent upon the operating conditions.

In our detailed consideration of the excited polyphase salient-pole synchronous machine we shall parallel the procedures used to establish the phasor diagram of the uniform-airgap machine, Section 7.10, accounting for the components of e.m.f. induced in reference phase a and identifying their relationship to current in the same phase winding. The conductor and field configuration for the two-pole machine is therefore identical with that shown in Fig. 7.10(a), but in Fig. 7.24 the different permeance characteristics of d- and q-axes are noted. The principal consequence is that the armature radial magnetic field wave in the airgap is no longer in space quadrature with the armature m.m.f. wave responsible. The figure shows components H_{nrd} and H_{nrq} of the armature radial field H_{nr} resolved along d- and q-axes.

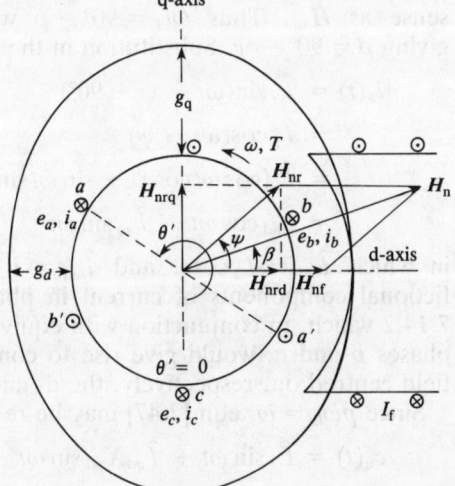

Figure 7.24 Single-turn coil representation of three-phase armature winding, rotor mounted at radius b, developing radial airgap field H_{nr} and d- and q-axis components appropriate to assumed airgap lengths g_d and g_q, respectively. Superposed stator d-axis field H_{nf} gives resultant H_n.

As the rotor of a machine with p pole pairs rotates at synchronous speed $\omega_0 = \omega/p$ in the anticlockwise direction, the reference side a of the full-pitch coil aa' which represents reference phase a of the armature winding adopts instantaneous position θ'_m in the airgap and experiences components of the assumed sinusoidally distributed components of radial field as follows:

$$H_{nf} = \hat{H}_{nf} \sin p\theta'_m$$
$$H_{nrd} = \hat{H}_{nrd} \sin p\theta'_m$$
$$H_{nrq} = \hat{H}_{nrq} \sin p(\theta'_m - 90°/p)$$

where H_{nrd} and H_{nrq} are respectively the d- and q-axis components of the radial armature field H_{nr} such that

$$H_{nrd} = H_{nr} \cos \psi$$
$$H_{nrq} = H_{nr} \sin \psi.$$

The induced e.m.f. in phase a with N' effective turns is given by eqn [5.35]

$$e_a(\theta'_m) = 2N'\omega_0 bl\mu_0\left[\hat{H}_{nf}\sin p\theta'_m + \hat{H}_{nrd}\sin p\theta'_m\right.$$

$$\left. + \hat{H}_{nrq}\sin p\left(\theta'_m - \frac{90°}{p}\right)\right]$$

$$= 2N'\omega_0 bl\mu_0[\hat{H}_{nf}\sin p\theta'_m + \hat{H}_{nr}\cos\psi\sin p\theta'_m$$

$$- \hat{H}_{nr}\sin\psi\cos p\theta'_m] \qquad [7.47]$$

where $\theta'_m = \omega_0 t$ if $\theta'_m = 0$ when $t = 0$.

If the current in phase a is expressed as the general time function

$$i_a(t) = \hat{I}_a\sin(\omega t + \rho)$$

with ω being the supply frequency in (electrical) rad s^{-1} and ρ being an arbitrary constant, then i_a will achieve its peak positive value at the instant t_1 given by $(\omega t_1 + \rho) = 90°$, when coil aa' is positioned with its radial field component directed in the same sense as $\boldsymbol{H}_{nr}$. Thus $\omega t_1 = 90° - \rho$ with $\psi = p\theta'_m = p\omega_0 t_1 = \omega t_1$ giving $\rho = 90° - \psi$. Substitution in the equation above gives

$$i_a(t) = \hat{I}_a\sin(\omega t - \psi + 90°)$$

$$= \hat{I}_a\cos(\omega t - \psi)$$

$$= \hat{I}_a(\cos\omega t\cos\psi + \sin\omega t\sin\psi)$$

$$= \hat{I}_{ad}\cos\omega t + \hat{I}_{aq}\sin\omega t \qquad [7.48]$$

in which $\hat{I}_{ad} = \hat{I}_a\cos\psi$ and $\hat{I}_{aq} = \hat{I}_a\sin\psi$. $\boldsymbol{I}_{ad}$ and $\boldsymbol{I}_{aq}$ are the fictional components of current in phase a introduced in Section 7.14.2 which, in conjunction with equivalent current components in phases b and c, would give rise to components of armature radial field centred on, respectively, the d- and q-axes.

Since $p\omega_0 = \omega$, eqn [7.47] may be re-written

$$e_a(t) = \hat{E}_f\sin\omega t + \hat{I}_{ad}X_{ad}\sin\omega t - \hat{I}_{aq}X_{aq}\cos\omega t \qquad [7.49]$$

where

$$\hat{E}_f = 2N'\omega_0 bl\mu_0\hat{H}_{nf} \qquad [7.24]$$

$$\left. \begin{array}{l} X_{ad} = 2N'\omega_0 bl\mu_0\left(\dfrac{\hat{H}_{nrd}}{\hat{I}_{ad}}\right) \\[2ex] X_{aq} = 2N'\omega_0 bl\mu_0\left(\dfrac{\hat{H}_{nrq}}{\hat{I}_{aq}}\right) \end{array} \right\} \qquad [7.50]$$

Equations [7.50] are analogous to eqn [7.25], incorporating the effective permeances identified respectively with radial flux crossing the airgap along d- and q-axes. Thus in eqn [7.49] the concept of 'reactance' has been included with the component of e.m.f. $\hat{I}_{ad}X_{ad}\sin\omega t$ recognised as lagging the component of current $\hat{I}_{ad}\cos\omega t$ in eqn [7.48] by 90°, and the e.m.f. component

Synchronous machines 313

$-\hat{I}_{aq}X_{aq}\cos\omega t$ lagging the current component $\hat{I}_{aq}\sin\omega t$ by 90°. E_f is the component of armature e.m.f. due to the field current responsible for H_{nf}.

X_{ad} is called the *direct-axis armature reaction reactance* and X_{aq} is the *quadrature-axis armature reaction reactance*. Equations [7.49] and [7.48] are interpreted by the phasor diagram of Fig. 7.25 which relates r.m.s. values of the sinusoidal components of E_a and I_a, the e.m.f. and current of phase a, to a reference phasor defined by $\sin\omega t$. The equivalent circuit diagram of Fig. 7.26 identifies the quantities displayed in Fig. 7.25 for a salient-pole machine operating as a generator at leading power factor i.e. absorbing reactive volt-amperes. Also included is armature leakage reactance X_l and resistance r_a per phase. The leakage reactance voltage may be readily incorporated within the armature reaction voltages by supplementing both X_{ad} and X_{aq}, so that

$$X_d = X_{ad} + X_l \quad [7.51]$$

$$X_q = X_{aq} + X_l \quad [7.52]$$

yielding the equivalent circuit representation of Fig. 7.27. X_d and X_q are then called the *direct-axis and quadrature-axis synchronous reactances* of the salient-pole machine. An appropriate phasor diagram for salient-pole synchronous generator operating at lagging power factor (exporting reactive volt-amperes) is shown in Fig. 7.28, with V_a as reference.

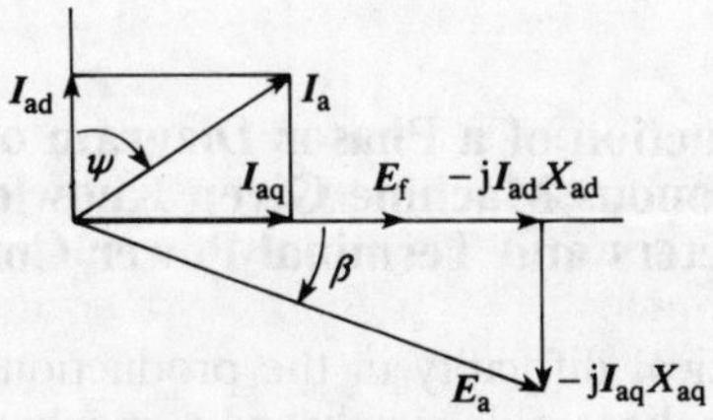

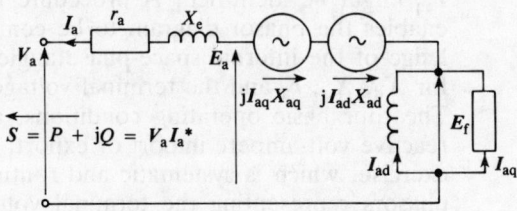

Figure 7.25 Phasor diagram of a three-phase salient-pole synchronous machine generating at leading power factor. Armature leakage impedance has been ignored. Variables are defined in Fig. 7.26.

Figure 7.26 Equivalent circuit/phase of a three-phase salient-pole synchronous machine, leakage impedance is included.

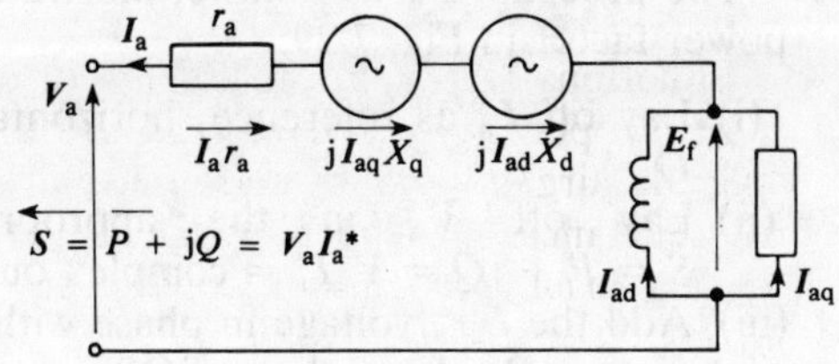

Figure 7.27 Equivalent circuit/phase of a salient-pole machine with leakage reactance subsumed within the d- and q-axis synchronous reactances X_{ad} and X_{aq}.

Figure 7.28 Phasor diagram of a three-phase salient-pole synchronous machine supplying real and reactive volt amperes to the system.

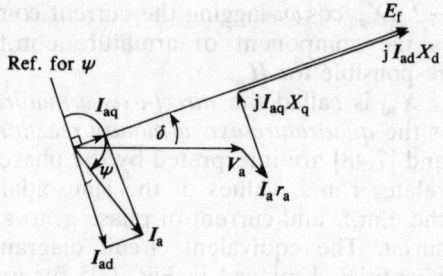

The incorporation of X_l within X_d and X_q has the consequence that E_a is no longer identifiable. The angle β is subsumed within δ, the load angle by which E_f leads V_a. Components of voltage within the equivalent circuit are related by the general equation

$$E_f = V_a + I_a r_a + jI_{aq}X_q + jI_{ad}X_d$$

which is true for all loading conditions – provided that the reference directions for voltage and current remain unchanged. Further

$$I_a = I_{ad} + I_{aq}$$

with I_{ad} and I_{aq} in quadrature and I_{aq} directed along (or in antiphase with) E_f. Also ψ, the space-phase angle by which the armature radial field lags the d.c. excitation field (the d-axis) is given by the time-phase angle of lag of I_a with respect to a reference vector 90° in advance of E_f.

7.17 Construction of a Phasor Diagram of a Salient-pole Synchronous Machine Given Knowledge of the Machine Parameters and Terminal Power Conditions

A practical difficulty in the production of the phasor diagram of a loaded salient-pole synchronous machine is that a knowledge of the value of ψ appears necessary in order that the components I_{ad} and I_{aq} might be identified. A procedure is available, however, which enables the phasor diagram to be completed without initial knowledge of the internal space-phase angle ψ. Prerequisites are values for X_d, X_q, r_a and the terminal voltage, current and power factor. The four basic operating conditions, as motor or generator with reactive volt-ampere import or export, are equally amenable to the exercise, which is systematic and routine once the positions of the phasors representing the terminal voltage and current have been defined. The current phasor I_a is taken as reference in all cases.

The procedure is as follows, illustrated for a generator at lagging power factor in Fig. 7.29.

(i) Lay off I_a as reference, horizontally to the right from origin O.
(ii) Lay off V_a in the appropriate quadrant such that $S = P + jQ = V_a I_a^* =$ complex output electrical power.
(iii) Add the $I_a r_a$ voltage in phase with I_a, to give point P.
(iv) From P draw line PQS vertically upwards such that

$PQ \equiv jI_aX_q$ and $PS \equiv jI_aX_d$.

(v) Draw perpendiculars PT and SU, where T and U are on OQ or OQ produced.

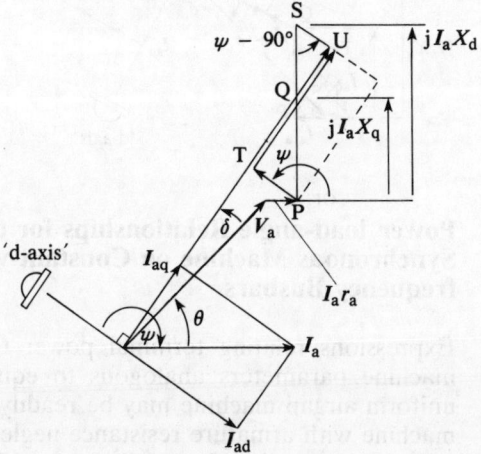

Figure 7.29 Salient-pole synchronous machine phasor diagram constructed without initial knowledge of the internal space-phase angle ψ. The machine is generating at lagging power factor.

Then OU represents E_f, the e.m.f. induced by the d.c. field excitation alone. TU represents $jI_{ad}X_d$ and PT represents $jI_{aq}X_q$, with I_{ad} and I_{aq} being so-called 'direct- and quadrature-axis components' of I_a.

Proof With reference to the geometry of Fig. 7.29 wherein ψ is identified:

$PT = jI_aX_q\cos(\psi - 90°) = jI_aX_q\sin\psi = jI_{aq}X_q$

$TU = jI_aX_d\sin(\psi - 90°) = jI_aX_d|\cos\psi| = jI_{ad}X_d$.

Figure 7.29 illustrates the case of a synchronous machine supplying both real and reactive power to the connected system. The machine is 'over-excited' with $|E_f| > |V_a|$. The reference vector for ψ measurement may be labelled the '*d-axis*', bearing in mind that such a label relates properly to a *space*-phasor diagram, not to a *time*-phasor diagram. The current component I_{ad} is seen to be directed oppositely to the 'd-axis' reference. This recognises that, for the particular load condition displayed, the armature reaction effect along the direct axis is to oppose the d.c. field excitation. That is, the armature load acts to demagnetise the machine and the d.c. field excitation has been increased to compensate, thus maintaining essentially constant airgap radial flux.

I_{aq} is identified with torque production of the machine, the q-axis field reacting with the stator and rotor components of the d-axis field to provide *alignment* torque where the radial d-axis field is due to the d.c. excited field, and *reluctance* torque where the *armature* d-axis field is involved.

The contrasting condition of an 'under-excited' synchronous motor is shown in Fig. 7.30. For every condition of load the phasor diagram on completion should satisfy the current and voltage component constraints applicable to the equivalent circuit of Fig. 7.27.

Figure 7.30 As for Fig. 7.29 for a machine motoring at lagging power factor.

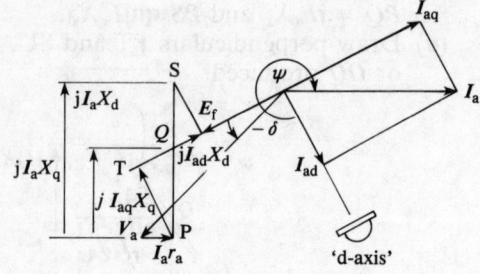

7.18 Power/load-angle Relationships for the Salient-pole Synchronous Machine on Constant-voltage, Constant-frequency Busbars

Expressions relating terminal power to the load angle δ via the machine parameters analogous to eqns [7.30] and [7.31] for the uniform airgap machine may be readily evaluated for a salient-pole machine with armature resistance neglected.

Referring to the phasor diagram of Fig. 7.29, which corresponds to operation as a generator at lagging power factor, but with $r_a = 0$, the armature real power/phase exported is given by

$$P = V_a I_a \cos\phi$$

with V_a and I_a being phase quantities also

$$(\phi + \delta) = (\psi - 90°)$$
$$\therefore \phi = (\psi - 90° - \delta)$$

and

$$I_{aq} = I_a \sin\psi, \quad I_{ad} = -I_a|\cos\psi|$$

hence

$$I_a \cos\phi = I_a \cos(\psi - \delta - 90°) = I_a \sin(\psi - \delta)$$
$$= I_a[\sin\psi \cos\delta - \cos\psi \sin\delta]$$

therefore

$$P = V_a[I_{aq} \cos\delta + I_{ad} \sin\delta] \qquad [7.53]$$

Further

$$V_a \sin\delta = I_{aq} X_q$$

and

$$V_a \cos\delta = E_f - I_{ad} X_d$$

hence

$$I_{aq} = \frac{V_a \sin\delta}{X_q}$$

and

$$I_{ad} = \frac{E_f - V_a \cos\delta}{X_d}$$

giving, on substitution in eqn [7.53],

$$P = \frac{V_a^2 \sin\delta \cos\delta}{X_q} + \frac{E_f V_a \sin\delta}{X_d} - \frac{V_a^2 \sin\delta \cos\delta}{X_d}$$

$$= \frac{E_f V_a \sin\delta}{X_d} + V_a^2 \left(\frac{X_d - X_q}{2X_d X_q}\right) \sin 2\delta. \quad [7.54]$$

The first term in eqn [7.54] is identical with eqn [7.32] which is applicable to the uniform-airgap machine, and is identifiable with the direct-axis synchronous reactance. If the d.c. field excitation is removed, the second term in eqn [7.54] remains. This is analogous to eqn [7.45] for the polyphase reluctance machine and is accounted for by the development of reluctance torque.

Figure 7.31 shows the variation of torque with load angle δ for a salient-pole machine with constant terminal voltage and d.c. field excitation. The effect of saliency is to slightly increase the maximum torque capability over the equivalent uniform-airgap machine. The salient-pole machine is the 'stiffer' of the two, i.e. the slope of the T/δ curve is steeper in the regions of normal operation as motor or generator, thus leading to improved prospects for transient stability.

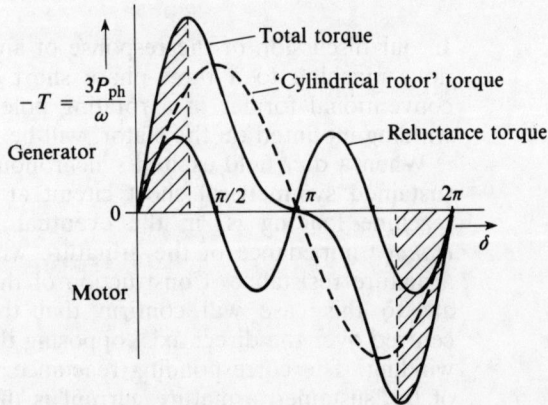

Figure 7.31 Terminal power (or torque) variation with load angle δ for a three-phase salient-pole synchronous machine, neglecting r_a.

The dependence of terminal reactive power flow on load angle, d.c. field excitation and reactance parameters may be similarly determined from Fig. 7.29. Thus

$$Q = V_a I_a \sin\phi = V_a I_a \sin(\psi - 90° - \delta)$$
$$= -V_a I_a \cos(\psi - \delta)$$
$$= V_a [I_{ad} \cos\delta - I_{aq} \sin\delta]$$
$$= -V_a^2 \left[\frac{X_q + X_d}{2X_d X_q}\right] + V_a^2 \left[\frac{X_d - X_d}{2X_d X_q}\right] \cos 2\delta$$
$$+ \frac{E_f V_a \cos\delta}{X_d}. \quad [7.55]$$

Equation [7.55] differs from eqn [7.33] derived for the uniform-airgap machine, most significantly in respect of the second term.

7.19 Measurement of Direct- and Quadrature-axis Synchronous Reactances of a Salient-pole Synchronous Machine

Approximate values for X_d and X_q may be determined experimentally by a 'slip test'. The unexcited machine is driven by an auxiliary motor at a speed slightly above or below the synchronous value, whilst normal-frequency balanced a.c. voltages and currents are applied to the armature windings. The synchronously rotating airgap field corresponding to rated armature current will then slip slowly past the pole system, the axis of the field corresponding with d- and q-axes, in sequence. Ignoring armature resistance the conditions vary sequentially between limits appropriate to the recognition of X_d and X_q as ratios of terminal phase voltage to phase current, eqns [7.42] and [7.43], [7.51] and [7.52]. Thus the r.m.s. value of phase current varies between maximum and minimum values approximating to V_a/X_q and V_a/X_d, respectively. In practice, experimentally derived values of X_d and X_q are suspect, with that for X_d being affected by the prospect of magnetic saturation.

7.20 Sudden Three-phase Short Circuit of a Synchronous Machine

In our discussion of the response of an excited polyphase synchronous machine to a three-phase short circuit at its terminals, the conventional format of a rotating pole system, with the armature winding mounted on the stator, will be assumed.

When a d.c. field-excited synchronous machine is subjected to a sustained symmetrical short circuit at its terminals, the effective machine loading is, in the eventual steady state, the inductive *leakage* impedance of the armature winding – including the small armature resistance. Construction of the phasor diagram appropriate to this case will confirm that the armature radial field is centred over the direct axis, opposing that of the d.c.-excited 'field' winding. The corresponding reactance which determines the value of the sustained armature current is the d-axis synchronous reactance X_d.

This reactance is properly described as a *positive (phase-) sequence* synchronous reactance, as the armature field rotates at synchronous speed relative to the armature conductors on the stator, such that no relative motion exists between armature field and the rotating pole system. It is a *steady-state* reactance since the amplitude of the armature field is constant, and the flux linkage which this armature field might be considered to create with closed rotor circuits is constant also – in reality, during short circuit, the radial flux created by the *total* m.m.f. is near zero.

When, however, the terminals of an excited synchronous machine are *suddenly* short circuited, a different state of affairs holds. The armature radial field is again centred over the d-axis of the machine but the amplitude of this field and the flux linkage for which it has responsibility are not steady but changing.

When applying Faraday's law, eqn [2.40], to closed circuits it is apparent that an instantaneous change in flux linkage is not possible as it implies an infinite induced e.m.f. If the proposed flux change is imposed externally the induced current will establish a magnetic field of such polarity as to oppose the attempted flux change. In a closed circuit possessing resistance the induced current will decay in calculable fashion to zero, with the time constant being inversely proportional to circuit resistance. When the induced current has ceased to flow the circuit is said to be fully *penetrated* by the external flux. It is apparent that the flux linkage of a closed superconducting circuit cannot ever change!

A synchronous machine with its armature winding short circuited has other closed circuits which need to be taken into consideration – the main field winding, closed through its low-resistance voltage source or *exciter*, and damper circuits. The damper circuits may be represented as short time-constant closed circuits located on both d- and q-axes, on the rotor adjacent to the airgap. The main field winding is of much longer time constant, and is also located on the rotor and embracing the d-axis.

In assessing the response of a synchronous machine to a sudden terminal short circuit the effect of the *theorem of constant flux linkages* is relevant to all closed windings. Prior to the application of the short circuit the machine will be considered to be excited but unloaded so that, at the instant of short circuit, the armature windings are converted from the open-circuit to the closed-circuit condition.

7.20.1 *Transient behaviour*

In the context of examining the response of a synchronous machine to step changes in the operating conditions – such as that imposed by a sudden short circuit – the *transient* period refers to that time during which the armature m.m.f is bringing about a change in the flux linked by the highly inductive d.c.-excited field circuit. In practical machines the transient period does not commence until several milliseconds after the short circuit occurs due to the effect of the closed damping circuits located near the airgap. Effectively, the evolving flux produced by the armature windings has to penetrate the low time-constant damper circuits before the armature m.m.f. can attempt to change the flux linking the field winding. When considering the so-called transient period the damper circuits are disregarded: it is assumed that they no longer have any effect on the changing flux conditions which affect the armature and main field windings.

Prior to short circuit, the flux linkages of each armature phase vary sinusoidally with time, mutually displaced by 120°. The radial flux due to the excited field winding is responsible. At the instant of short circuit the three armature phase windings become closed circuits and must retain, initially, flux 'trapped' at the corresponding value. In order to sustain flux at this level each armature phase winding has induced within it a d.c. current component of magnitude appropriate to the instant of short circuit, being different for each phase. Such d.c. components of armature current subsequently decay exponentially to zero with *armature time constant* τ_a as the trapped flux decays. τ_a is inversely proportional to the armature resistance. These d.c. currents may be considered to give rise to a stationary radial field which induces in the closed rotor field winding a complementary a.c. current of line frequency, also decaying with armature time-constant τ_a. This a.c. current induced in the heavily flux linked rotor field winding is necessary to prevent the trapped stationary armature winding flux from affecting the pre-existing field winding flux.

The stationary armature phase windings continue to have synchronous motion relative to the rotating d.c. field winding. The a.c. currents induced in the armature windings give rise to a radial field which is stationary with respect to the rotor and directed along the d-axis, attempting to reduce the flux linkage of the field winding on the rotor. This is resisted by additional d.c. current induced in the rotor winding, hence the flux produced by the armature field is obliged to flow in high-reluctance 'leakage' paths which avoid the low-reluctance field pole system. The effective *direct-axis transient reactance* X'_d which limits the *initial* value of the a.c. component of armature current is correspondingly low in comparison with the steady-state reactance X_d. Gradually, however, the armature field succeeds in penetrating the field winding and the amplitude of the a.c. component of the armature current decays exponentially with time constant τ'_d, the *direct-axis transient time constant*, from a value limited by X'_d to one limited by X_d. The additional component of d.c. current induced in the field winding also decays with time constant τ'_d. τ'_d is given by the product of the rotor field circuit time constant L_f/R_f and a flux leakage coefficient σ_{df} relating to the magnetic coupling between the armature winding and the field winding.

Analysis of the machine modelled as electromagnetically coupled circuits shows that $\sigma_{df} = X'_d/X_d = L'_d/L_d$, where L_d is the effective self-inductance of the armature winding when excited to produce radial field along the d-axis, with other windings magnetically coupled via the same axis on open circuit, whereas L'_d is the effective armature winding self-inductance including the effect of the coupled, closed field winding which, in this context, is considered to possess zero resistance and therefore to undergo no change in flux linkage. For 'tight' coupling between armature and field windings σ_{df} is small as $L'_d \ll L_d$. Consequently, the short-circuit armature currents are of high initial value compared with their eventual steady-state amplitude, and the duration of the transient period is short.

In addition to the steady and transient fundamental frequency a.c. components and the transient d.c. component, a full analysis shows that the armature current has also a transient second harmonic term which decays exponentially with time constant τ_a if the machine possesses transient saliency, i.e. has $X'_d > X'_q$, where $X'_q \simeq X_q$.

7.20.2 Subtransient behaviour

When investigating the *subtransient* behaviour of a synchronous machine the effect of damper windings modelled as closed circuits on both d- and q-axes must be considered. The subtransient occurs immediately after short circuit and is of very short duration, since the time constant of the damper circuits is deliberately made short and, due to their location adjacent to the airgap, they experience tight coupling with the armature winding.

The armature current during the subtransient period has generally the same form as during the transient period, with the initial amplitude of the fundamental frequency a.c. component limited by X_d'', the *subtransient direct-axis reactance*. X_d'' is small ($< X_d' < X_d$) corresponding to the high-reluctance paths available to the flux produced by the armature field which avoid initial linkage with the d-axis damper circuit and field winding. Rapid flux penetration of the damper circuits is achieved, however, and the fundamental-frequency a.c. component of the armature current is subsequently determined by transient behaviour considerations. The time constant associated with the change from subtransient to transient conditions is called the *subtransient direct-axis time constant* τ_d''.

Corresponding to the above a.c. component of armature current there is induced in the field winding a d.c. current component, decaying exponentially with time constant τ_d''. The field winding also carries a fundamental frequency a.c. component which balances the d.c. field current required by the armature phase windings to maintain their flux linkages at the value they had prior to short circuit. Both these currents decay with armature time constant τ_a. Additionally, if *subtransient saliency* is present, i.e. $X_d'' > X_q''$, where $X_q'' < X_q'$, the armature current includes a second-harmonic component.

Typical waveforms of armature current and field-winding current for a salient-pole machine subjected to a sudden short circuit at its terminals are shown in Fig. 7.32. The instant of short circuit is such that the d.c. component in phase *a* is near zero, i.e. the flux 'trapped' by this phase winding of three at the instant of short circuit is very small. Figure 7.33 illustrates the flux paths appropriate to the definition of various armature-winding reactances in terms of flux linkage and current. The steady-state reference conditions are shown in Fig. 7.33(a) and (b) with, on normal load, the armature radial field distribution predominantly established along the q-axis. Under short-circuit conditions, however, both transient and in the eventual steady state, the armature radial field is directed mainly along the d-axis. The sequence of evolution is from Fig. 7.33(e) to (a) via (c).

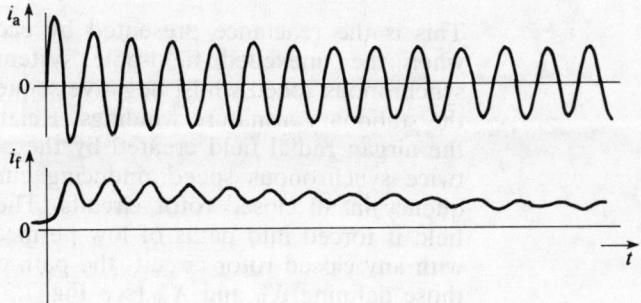

Figure 7.32 Oscillograms of armature and field-winding current for an unloaded, excited, salient-pole synchronous machine with sudden three-phase short circuit at an instant when the flux linkage of the phase displayed is near zero.

Figure 7.33 Principal flux paths relevant to definitions of salient-pole synchronous machine reactances as angular frequency and flux-linkage products per unit armature current: (a) X_d and (b) X_q, steady-state d- and q-axis synchronous reactances; (c) X'_d, d-axis transient reactance – field centred on the d-axis and of changing amplitude commencing penetration of a closed field winding on evolution to (a); (d) X'_q, q-axis transient reactance (field distribution essentially as in (b)); (e) X''_d, d-axis subtransient reactance (field centred on the d-axis of changing amplitude commencing penetration of a closed damper winding on evolution to (c)); (f) X''_q, q-axis subtransient reactance (q-axis equivalent of (e), evolving to (d)).

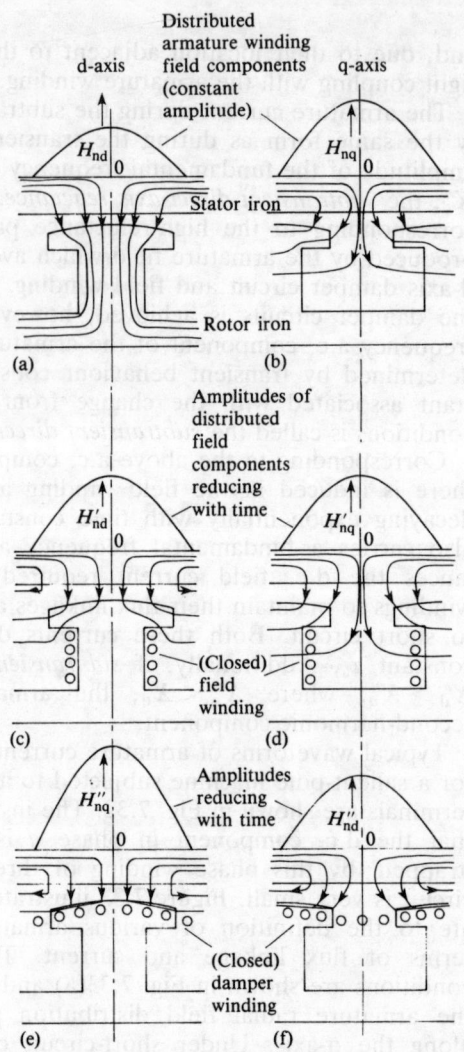

7.21 Negative (Phase-) Sequence Reactance of a Polyphase Synchronous Machine

This is the reactance presented by each armature phase winding when the unexcited field-pole system is rotated 'forwards' at synchronous speed whilst negative sequence currents are applied to the stationary armature windings. Relative to the rotor, therefore, the airgap radial field created by the armature currents rotates at twice synchronous speed, inducing currents at twice supply frequency in all closed rotor circuits. The flux due to the armature field is forced into paths of low permeance so as to avoid linkage with any closed rotor circuit, the path corresponding alternately to those defining X''_d and X''_q (see Fig. 7.33(e) and (f)). The negative sequence reactance X_2 thus lies between the values of X''_d and X''_q, which are similar.

7.22 Zero (Phase-) Sequence Reactance of a Polyphase Synchronous Machine

This is the reactive component of the impedance per phase presented by the armature winding to the flow of zero-sequence current, i.e. currents of equal magnitude and phase angle flowing in each phase of the machine. Ideally, such currents set up no fundamental field in the airgap if the currents are sinusoidally distributed and the airgap length is modelled as uniform overall or with idealised saliency. Correspondingly, zero sequence impedance is largely due to end-winding leakage reactance and winding resistance. In practice, values of zero sequence reactance are determined by the degree to which the actual winding and magnetic circuit depart from the ideal.

7.23 Stepping Motors

The synchronous motors discussed so far have been essentially high-power devices with armature voltage and current sinusoidally time variable. The armature/airgap surface is essentially cylindrical so that torque on the member carrying the armature winding is calculable via the integration of tangential stresses over that surface. Torque is fundamentally independent of rotor position and the electromechanical power converted may be accounted for by the integral of radial power density, given by the vector product $E \times H$, crossing the boundary between armature surface and airgap. Saliency, where it exists, results from projecting poles on one side only of the airgap, and, as represented by the two-reaction theory involving the separate identification of d- and q-axis components of variables, does not obscure our view of the significant happenings at the armature surface.

Stepping (stepper) motors and *switched reluctance motors* are examples of synchronous machines which are *doubly salient*, where salient poles exist on both sides of the airgap and for which classical synchronous machine theory is inappropriate. The power supply requirement, although it might originate from an a.c. source, is for neither sinusoidal voltage nor current, and the double saliency means that the stress system contributing to torque is complex on both sides of the airgap, with significant changes from instant to instant.

Such devices are indeed synchronous machines as their speed is, under normal operation, precisely related to the frequency of the energising pulse train. Additionally, the total angular displacement of a well-designed system is proportional to the number of pulses supplied. Stepping motors find many uses in low-power (< 2 kW) control applications where precise and variable speed control is required, or where accurate positioning is needed – as with numerically controlled machine tools or the read/write heads in computer memories. At extremely low power levels they are used as quartz analogue time-pieces. On the larger scale, *switched reluctance motors* compete with induction motors, synchronous motors and d.c. alternatives for industrial and domestic variable-speed drive applications.

Being capable of responding directly to unidirectional pulses of current, stepping motors are ideal devices for transforming digital data into speed or position, with rapid response and high accuracy. The market for stepping transducers is likely to parallel the growth of digital electronics over the foreseeable future.

As implied by its title, the rotational motion of the stepping motor is in terms of clearly defined steps, each step being taken in response to a pulse of input energy. At a high pulse rate the motion approximates to constant speed, proportional to the pulse repetition frequency. A full analysis of stepping motor behaviour is extremely complex, being essentially defined by repetitive transient phenomena influenced by mechanical inertia, electrical circuit time constants, magnetic properties of materials used, tooth profiles on stator and rotor, load torques, damping factors and stepping frequencies.

Stepping motors are manufactured in several formats operating on the common principle that an excited electromagnet develops forces on the iron surfaces of a composite iron/airgap magnetic circuit such that, at constant current, the total stored magnetic field energy should increase if movement occurs as a consequence of the force (eqn [3.23], Sect. 3.9). In practice, this means the development of forces which tend to reduce the airgap length. An excited salient stator pole will therefore encourage the lateral movement of a rotor salient pole into alignment. A permanent magnet exerts a similar attractive force on unmagnetised iron. In a stepping motor the polar projections on stator and rotor are organised such that a sequence of pole excitations on the stator member brings about a proportionate stepped angular movement of the rotor. Due to the torque dependency on position and rotational inertia the movement may involve a series of damped oscillations about each successive step position, if the stepping rate is low. If the stepping rate is high, the buildup of current in the exciting coils will be adversely affected by the effective time constant of the coils, influenced by the changing flux linkage due to both changing current and motion.

Modern stepping motors fall into two distinct categories – the *variable reluctance* and *hybrid* types. The former employs soft iron and d.c. excitation is necessary to produce flux. The latter incorporates a permanent magnet in conjunction with several electromagnets whose role is to steer the permanent magnet flux into appropriate paths.

7.23.1 *Multistack variable-reluctance stepping motor*

A general view of a two-stack variable–reluctance stepping motor is shown in Fig. 7.34, from which it can be seen that the projecting teeth or poles on the laminated iron rotor are displaced by a constant angle between 'stacks', whereas the excited stator poles are in line. Each stack has the same number of teeth on stator and rotor and the stator pole excitation due to the series-connected field coils *on each pole* is alternately N and S. The cross-sectional views of Fig. 7.35 show a four-pole three-stack motor with two teeth/pole, the rotor teeth being in line and the laminated stator poles being staggered. The castellation of the stator poles and the

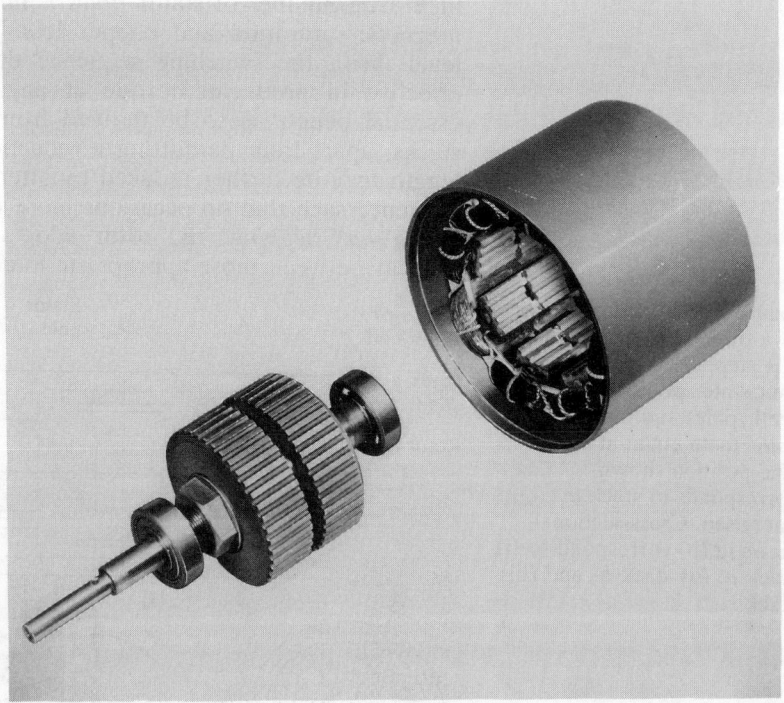

Figure 7.34 Two-stack variable-reluctance (VR) stepping motor. (*Photograph by courtesy of GEC Electromotors Ltd, Warley, UK.*)

consequential increase in the number of rotor teeth has the effect of reducing the step length. Normally, only one stack has its field coils excited at any one time and each stack is energised in sequence, giving rise to the alternative designation 'phase'. In Fig. 7.35(b) stack a is shown energised with the eight rotor and stator teeth aligned, in a position of stable equilibrium. The instantaneous direction of the current in the interconnected windings surrounding each pole is immaterial and the torque tending to return the rotor to the aligned position shown after any displacement will be proportional to (current)^2, assuming the iron to be unsaturated. Stacks b and c are observed to have their teeth out of alignment but, being unenergised, they do not contribute to any electromagnetic torque developed.

If, however, the current in the coils of stack a is reduced to zero whilst that of stack b is increased from zero, with either polarity, the rotor teeth associated with stack b will tend to align with the stator teeth in b, moving through a *step length* equal to the relative displacement of the stator poles between stacks a and b. Subsequent change of excitation to stack c produces another increment of displacement. De-energisation of c and re-energisation of a will cause rotor alignment as in Fig. 7.35(b), the rotor having moved through the equivalent distance of one tooth and one slot, i.e. one rotor tooth pitch or 1/8 revolution over three changes of excitation. Thus the step length appropriate to one change in excitation is specifically $360°/(3 \times 8) = 15°$ or generally $360°/nt_r$, where n is the number of stacks, or phases, and t_r is the number of rotor teeth.

Clearly, the torque capability is increased if the flux in the gap is high. The number of stator *poles* is immaterial. Considerations of magnetic saturation and copper loss limit the excitation current level. With the switching sequence described, only one stack is effective in producing torque at any instant, and therefore no essential benefit is to be derived from increasing the number of stacks, apart from permitting a reduction in step length. The step length may be further reduced by altering the excitation switching sequence such that on occasions more than one stack is energised. When *half-stepping*, the rotor adopts a position of equilibrium midway between those appropriate to each excited stack.

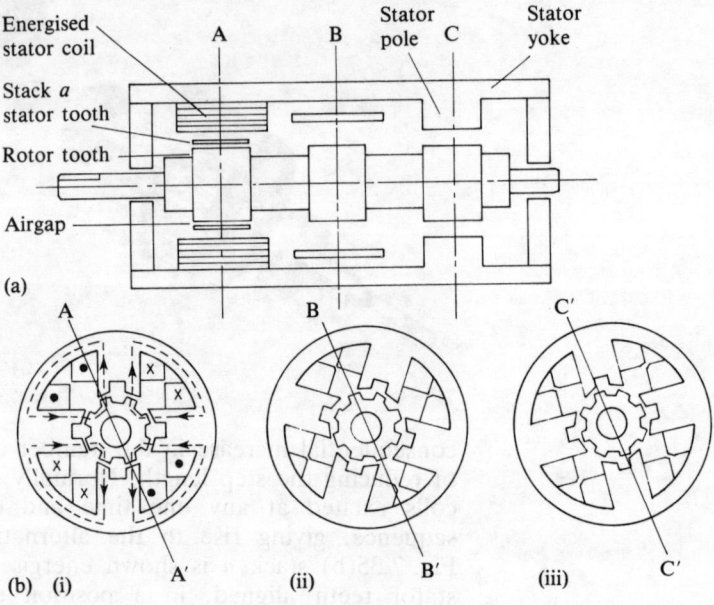

Figure 7.35 Sectional views of a three-stack (three-phase) VR stepping motor having four stator poles with two teeth/poles and number of *rotor* teeth equal at eight. The position shown corresponds to stack *a* energised. Cross-sectional views in (b) correspond to (i) stack *a*, (ii) stack *b*, and (iii) stack *c*.

7.23.2 Single-stack variable-reluctance stepping motor

This motor employs a single stack of laminated stator poles and a number of rotor poles which differs from that of the stator poles by two, typically. The machine shown in Fig. 7.36 has four stator *phases* and eight stator poles, or teeth, with six rotor poles, or teeth. Stator windings of the same phase carry current at the same time, in sequence *abcda*. Over a period when phase *a* is energised the rotor tends to adopt a position of equilibrium as shown. Most of the flux in the airgap is associated with one pair of stator and rotor poles. If the excitation is switched from phase *a* to phase *b*, the rotor will tend to move to another position of equilibrium via an anticlockwise rotation through the angular equivalent of the difference between a rotor tooth pitch and a stator tooth pitch. The step length is accordingly given by $360°/t_r - 360°/2n = 360° \times (2n - t_r)/2nt_r = 360°/nt_r$, since $2n - t_r = 2$. In the example of Fig. 7.36, the number of 'phases' $n = 4$ and $t_r = 6$ to give a step length of 15°.

Reversal of the phase sequence, giving *adcba*, changes the direction of rotor motion to clockwise. Once again, the direction of current in an energised group of stator coils comprising a phase winding is unimportant.

Figure 7.36 Sectional view of a single-stack VR stepping motor with four stator phases, eight stator poles and six rotor teeth. The equilibrium position shown corresponds to 'phase' *a* energised.

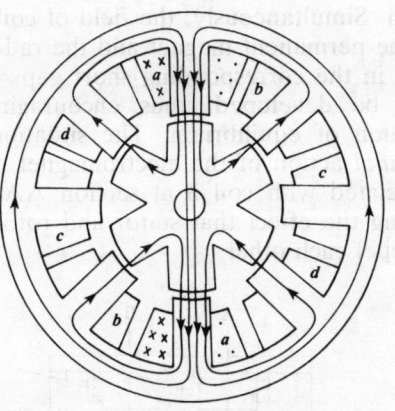

7.23.3 Hybrid Stepping Motor

A *hybrid* stepping motor incorporates a cylindrical permanent magnet with field directed axially along the rotor, the flux completing its circuit via castellated soft-iron pole caps, two airgap crossings, stator poles and stator yoke. The permanent magnet is mounted on a non-magnetic steel shaft. The working flux path is therefore basically axial in the rotor magnet, and radial and tangential across the gap. The flux path in the stator yoke is not completely axial, however, being *skewed* and having a tangential component not entirely accounted for by the fact that the castellations on the permanent magnet pole caps are staggered by half a rotor tooth pitch. The stator pole pieces complement the rotor pole caps and may be excited by extended coils which embrace pairs of axially adjacent stator poles. The number of stator poles is a multiple of two with the coils of alternate poles being connected in series to form one of two phase-winding groups. The field coils of successive stator poles in each phase grouping are wound in opposite senses so that their fields are directed alternately across the pole/airgap boundary and thus may have similar or opposite orientation to the permanent magnet field across a particular gap. Regardless of the number of stator poles, the number of *phases* is always two.

Figure 7.37(a) portrays an axial section of a hybrid stepping motor showing the permanent magnet field and a section across the upper and lower stator coils, numbered 1 and 3. Figure 7.37(b) shows transverse sections across the stator poles and rotor pole caps at AA' and BB'. The eight stator teeth contrast with nine on the rotor, which is shown in such a position that one pair of rotor teeth is aligned with the particular pair of stator teeth due to pole 1 at section AA'. The displacement of half a rotor tooth pitch between the sections at AA' and BB' means that, for the same shaft position, it is the *opposite* pole 3 which at section BB' aligns pairs of stator and rotor teeth. The position shown corresponds to coils 1 and 3 being excited with phase *a* current, of polarity such that the field of coil 1 across the airgap at section AA' acts cumulatively with the permanent magnet field there to develop a high flux density across the short gap between stator and rotor teeth. Simultaneously, the field of coil 3 at section BB' assists that of the permanent magnet and the radial rotor/stator flux density is high in the corresponding short gap. Alignment torque will therefore be developed, thus encouraging the rotor to adopt this position of equilibrium. The situation is encouraged by the *differential* action of the electromagnet and permanent magnet fields associated with coil 3 at section AA' and coil 1 at section BB', having the effect that stator and rotor teeth in these regions tend to repel each other.

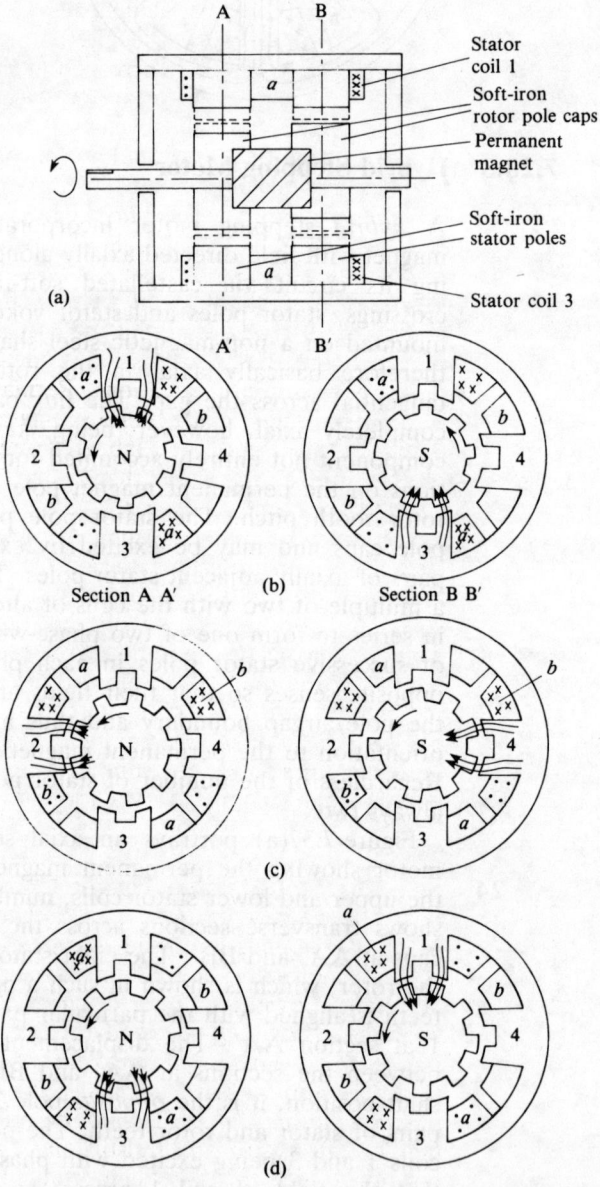

Figure 7.37 Sectional view of a two-phase hybrid stepping motor with four stator poles, eight stator teeth and ten rotor teeth. Sections for (a) and (b) correspond to phase *a* energised alone; (c) shows the rotor displaced through one quarter of a rotor tooth pitch on energising phase *b* alone; and (d) shows further increment of one quarter of a rotor tooth pitch on subsequent re-energisation of phase *a* alone.

Over the time duration when the above situation prevails the excitation of poles 2 and 4 by phase *b* current is zero but, if phase

b current is raised from zero with polarity such that the electromagnet field of coil 2 assists that of the permanent magnet at AA' and the electromagnet field of coil 4 assists the permanent magnet field at BB' whilst that of phase a is reduced to zero, the position of equilibrium will move anticlockwise through half a rotor tooth width to that shown in Fig. 7.37(c). If, subsequently, the current in phase b is reduced to zero and that in phase a is restored, but with *opposite* polarity, the rotor will make a further anticlockwise increment to the position of equilibrium shown in Fig. 7.37(d), having stator pole 3 aligned with rotor teeth at AA' and pole 1 aligned at BB'. It is evident that the exciting coils should be supplied from a two-phase alternating current supply, each complete cycle of current corresponding to movement of the rotor through one complete rotor tooth pitch. Thus for the motor of Fig. 7.37, nine complete cycles of supply current are required for one revolution of the rotor. Each successive current pulse results in an incremental movement of half a rotor *tooth width* or one quarter of a rotor *tooth pitch*. The step length is correspondingly $360°/(9 \times 4) = 10° = 90°/t_r$ in general, where t_r is the number of rotor teeth.

A feature of the hybrid stepping motor is that the permanent magnet field tends to retain the rotor in the last position prior to de-energising the windings. This 'detente' torque may be advantageous in certain applications. Permanent magnet material on the rotor contributes to a generally higher value of moment of inertia than with a VR motor. Hybrid stepping motors generally provide for smaller step lengths than variable-reluctance types and may provide a larger torque for a given size of package. A small step length may be a disadvantage, however, when large positional changes are required and the rate of excitation changes which can be achieved is limited. A popular step length is 1.8° which requires 200 stepping points per revolution. Stepping rates of $20\,000\,\text{s}^{-1}$ are commonplace, corresponding to speeds of 6000 rev/min. Stepping motors are unable to develop much torque when running at high speeds but power outputs of several kW are possible. The high-power *switched reluctance motor* considered in Section 7.26 is designed for continuous running with ratings up to 200 kW and is based on the single-stack VR format.

7.24 Stepping Motor Static Torque Characteristics

In many applications the stepping motor is selected for use because, normally, its total angular displacement may be accurately determined by the number of current pulses applied to its exciting windings. Provided the load torque, rotational inertia or rate of pulse application are not excessive, the rotor should be able to rotate in step with the applied pulse train to its final position. In practice, it may be necessary to reduce the stepping rate when moving away from standstill and when approaching the final position. Once this has been attained, however, any residual load torque will cause a displacement from its equilibrium position – the *positional error*. For a given static load torque the positional error will depend upon the rotor/stator tooth profile, the number of rotor teeth, the degree of magnetic saturation and the excitation current. Ideally, the static torque is proportional to (current)^2 for the VR motor, proportional to current for the hybrid motor. Figure 7.38 shows a family of typical static torque characteristics for a three-stack VR motor with eight rotor teeth, and a step length of 15°. The actual characteristics are often approximated to a sinusoidal variation with displacement angle. The steady-state stability limit for torque is thus achieved with a displacement of $90°/t_r = 90°/8 = 11.25°$ for the motor with eight rotor teeth. This is equivalent to a displacement of one quarter of a rotor tooth pitch or half a rotor tooth from the position of equilibrium, independent of the number of stacks.

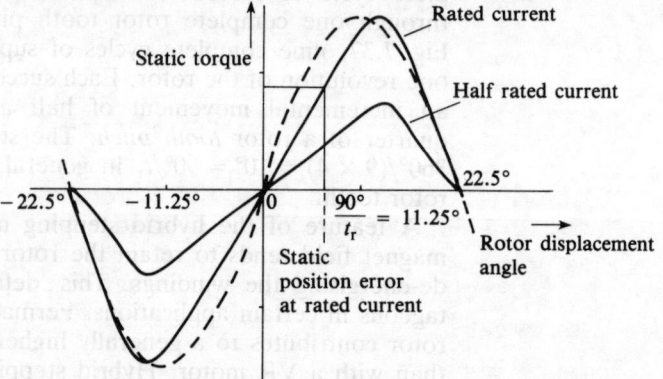

Figure 7.38 Family of static torque/displacement characteristics for a three-stack VR stepping motor at rated and half rated current, with a sine approximation for former. Step length = 45°/3 = 15°. Positional error is indicated for residual load torque at rated current.

By implication one stack (or phase) is left excited and adjacent alternative positions of rotor equilibrium are spaced by $360°/t_r$. Variable-reluctance stepping motors with the number of stacks greater than three may have a peak value of static torque increased above that appropriate to one phase alone by exciting more than one stack simultaneously. A (two-phase) hybrid motor will have its peak static torque increased by a multiplier of approximately $\sqrt{2}$ if both phases are excited.

7.25 Stepping Motor Dynamic Behaviour

The static characteristics are of interest only after the stepping motor has responded to the train of exciting current pulses, each of which should result in the rotor turning through an increment of angle equal to the step length. The torque requirement will increase if the rotor and its connected load are being accelerated and, if the developed electromagnetic torque is inadequate, the rotor will fall out of step with the pulse train. The electromagnetic torque results from stresses developed at the airgap surfaces of the iron teeth on stator and rotor in accordance with the distribution of the excitation current, switched from pole to pole in sequence. The factors influencing instantaneous torque are many and varied but torque is calculable if the instantaneous flux pattern is known. The determination of the changing flux is affected by the changing reluctance of the magnetic circuit as the relative movement of

Synchronous machines 331

stator and rotor teeth takes place, the effect of induced eddy currents in the laminated core material, the saturation and hysteresis properties of the magnetic material and the time variation of the exciting current. The phase current itself is much affected by induced e.m.f. resulting from the changing flux linkages of the exciting coils, particularly at high speeds.

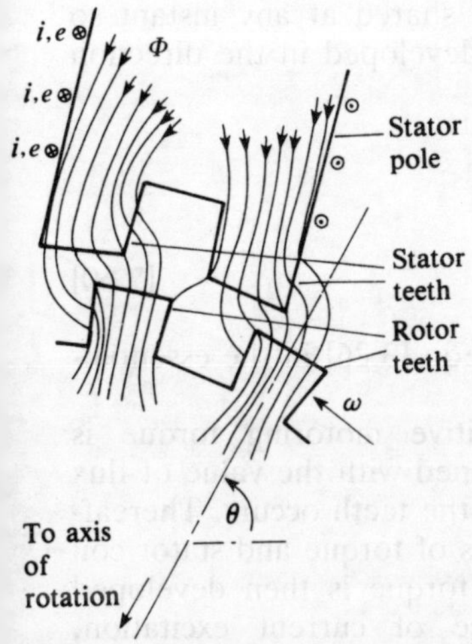

The situation for the variable-reluctance motor is illustrated in Fig. 7.39 which shows a pair of rotor teeth approaching a complementary pair of stator teeth identified with an excited stator pole. In the position shown the force on the rotor and stator teeth is clearly such as to draw the teeth into alignment. In so doing the flux linked by the exciting winding (carrying nominally constant current) will increase and an e.m.f. of self-induction e developed therein, of value given by Faraday's law, eqn [2.40], generalised in terms of the circuit flux-linkage λ, and self-inductance L. Thus, from eqn [2.41],

$$e = -\frac{d\lambda}{dt}. \quad [7.56]$$

Figure 7.39 Changing flux pattern in the vicinity of stator and rotor teeth for a VR stepping motor as teeth approach alignment.

The flux linkage λ approximates to $N\Phi$ and the reference directions of flux, e.m.f. and current are defined in the figure. For the condition shown of flux increasing with time the e.m.f. acts to oppose the flow of existing current, with the source of current being required to supply electrical power $-ei$ instantaneously. Some of that power is absorbed in raising the energy stored in the magnetic field, the remainder is the electrical equivalent of the instantaneous mechanical power output. Since the magnetic circuit configuration is changing, the inductance L relating the electric circuit flux linkage to current is also a function of time, through rotor position θ. With current also time dependent the division of electrical input power p may be accounted for by development of eqn [7.56]. Thus

$$p = -ei = \frac{i(t)d\lambda(t)}{dt} = i(t)\frac{d}{dt}[L(\theta)i(t)]$$

$$= i(t)\left[L(\theta)\frac{di(t)}{dt} + i(t)\frac{dL(\theta)}{dt}\right]. \quad [7.57]$$

The first term in eqn [7.57] represents the instantaneous rate of increase of stored magnetic field energy if, as in general, the current is changing. For the special case of constant current, however, eqn [7.57] reduces to

$$p]_{I=\text{constant}} = I^2\frac{dL(\theta)}{dt}. \quad [7.58]$$

If, over a short interval of time $t_2 - t_1$ the rotor moves through angle $\theta_2 - \theta_1$ and the circuit inductance increases from L_1 to L_2, the total energy supplied at constant current is

$$W = -\int_{t_1}^{t_2} ei\, dt = \int_{t_1}^{t_2} I^2\frac{dL(\theta)}{dt}dt$$

$$= \int_{L_1}^{L_2} I^2\, dL(\theta) = I^2(L_2 - L_1). \quad [7.59]$$

The energy represented by eqn [7.59] is recognisable as *twice* the change in stored magnetic field energy appropriate to the change in inductance, ignoring saturation. It follows, therefore, that the energy expended in developing mechanical torque in the direction of rotor rotation, during the same time interval, has the same value as the increase in field energy at constant current. The input electrical power will be correspondingly shared at any instant so that, from eqn [7.58], the rotor torque developed in the direction of rotation at ω rad s^{-1} is

$$T = \frac{p}{2\omega}\bigg]_{I=\text{constant}} = \frac{I^2}{2\omega}\frac{dL(\theta)}{d\theta}\frac{d\theta}{dt}$$

$$= \frac{1}{2}I^2\frac{dL(\theta)}{d\theta}. \qquad [7.60]$$

The above expression is encountered as eqn [3.26] in the essentially static example of Section 3.9.

Thus, referring to Fig. 7.39, a positive motoring torque is developed if the exciting current is sustained with the value of flux continuing to increase until alignment of the teeth occurs. Thereafter the flux will reduce and the directions of torque and stator coil induced e.m.f. will reverse. A *braking* torque is then developed and energy is returned to the source of current excitation, accounted for in part by a reduction in the stored field energy. Clearly, for motor action, the excitation should be controlled and timed to maximise the effect of rotor torque in the direction of rotation. Conversely, generation or regenerative braking is possible by delaying the excitation current pulses.

Such fine control requires knowledge of the instantaneous rotor position and this may be provided by means of an *encoder* attached to the shaft. Such encoders may be incremental optical types providing an output signal pulse corresponding to each step length of angular movement and incorporated in a closed-loop control system.

7.25.1 *Open-loop control*

In the absence of feedback signals regarding rotor position care must be taken in the design of the system to ensure that pole slipping is avoided, otherwise errors will occur in the integrated rotor displacement *vis-à-vis* the number of incremental movements called for. One relevant performance indicator is the *pull-out* torque/speed characteristic illustrated in Fig. 7.40. Pull-out torque is that value of load torque which, if exceeded, causes a rotating machine to stall, and thus represents the maximum torque available for synchronous operation at a given speed. In general, the speed of a stepping motor will not be constant, the rotor will frequently be accelerated or braked. At very low stepping rates the rotor, initially at rest, responds to each excitation pulse by executing a damped oscillation with natural frequency ω_n (*c.f.* eqn [7.39]) about its new position of equilibrium. If the stepping rate exceeds the *start rate* the initially stationary rotor will fail to achieve a displacement equal to its step angle between each pulse, and will either fail to move away from standstill or pole slip whilst running at too low a value of mean speed. Conversely, a stepping motor running at high speed will fail to stop within the interval between excitation changes if the stepping rate is higher than the *stop rate*, due to the inability to develop sufficient torque to decelerate in time the rotating mass. For a given system the start and stop rates have slightly different values due to friction and damping effects, friction being generally minimised to avoid wear of contacting surfaces. Viscous damping may be deliberately introduced to improve the response.

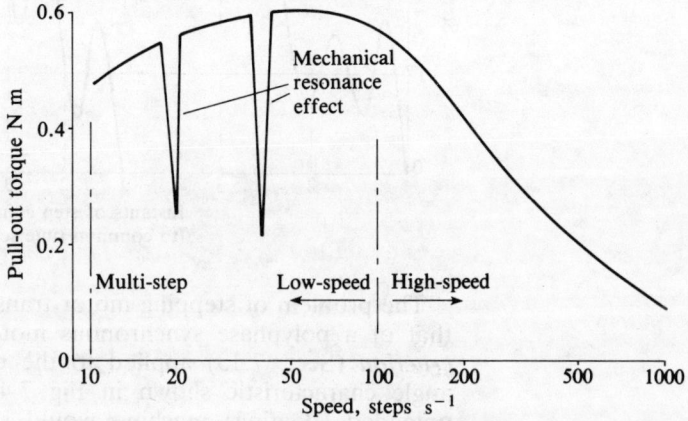

Figure 7.40 Typical pull-out torque/speed characteristic for a stepping motor. Regions identified for multistep low-speed and high-speed operation. (Reproduced with permission from *Stepping motors: a guide to modern theory and practice*, P. P. Acarnley. Published by Peter Peregrinus Ltd, Revised 2nd Edn., 1984).

The different modes of operation possible with a stepping motor are identified in the pull-out torque/speed characteristic of Fig. 7.40. At low stepping rates *multi-step* operation corresponds to oscillatory motion about each step position, the rotor periodically coming to rest and reversing direction. Figure 7.41, curve (a) illustrates this. At higher stepping rates corresponding to *low-speed* operation the rotor motion is unidirectional at all times, although variable about a mean value proportionate to the stepping rate. Such synchronous operation is described as 'slewing', which is also demonstrated in Fig. 7.41, curve (c). In the low-speed region of the pull-out torque/speed characteristic the torque is essentially constant, but in the *high-speed* region the pull-out torque falls off with increasing speed due to reduced flux development.

Mechanical resonance effects can cause instability at low stepping rates near the natural frequency of the rotor about the position of equilibrium. These are manifested as dips in the multi-step operation of the pull-out torque/stepping rate characteristic, Fig. 7.40. The rotor positional response at a low stepping rate is shown in Fig. 7.41, curve (a), to be essentially oscillatory about the position of equilibrium. At the third crossing of each successive equilibrium position the rotor has high instantaneous velocity in the direction appropriate to the next step, particularly if the system is not well damped. If the next change of excitation occurs at precisely this instant the initial velocity will encourage a larger amplitude of rotor displacement about the next position of equilibrium.

Figure 7.41 Typical plots of stepping motor rotor position against time at different stepping rates: (a) multi-step operation; (b) Multi-step operation near the natural frequency of rotor oscillation leading to loss of synchronism; and (c) low-speed operation (slewing).

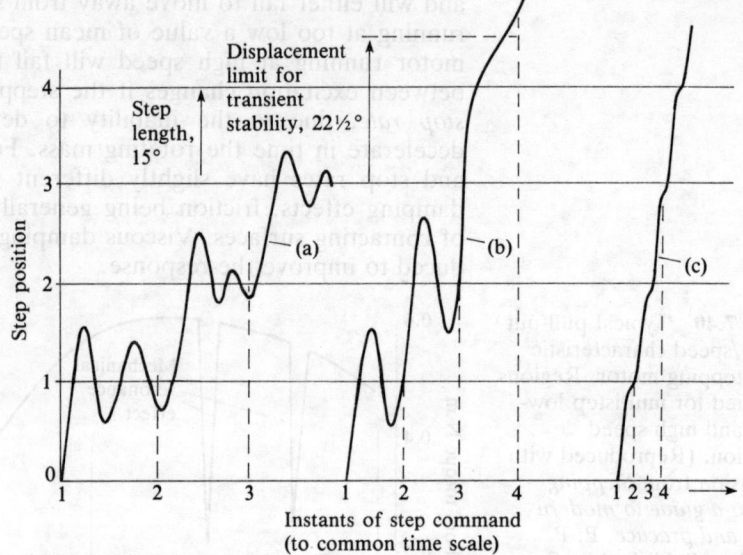

The problem of stepping-motor transient stability is analogous to that of a polyphase synchronous motor. For this, the *equal-area criterion* (Sect. 7.13) applied to the electrical power output/load angle characteristic shown in Fig. 7.42 suggests that a loss-free, unloaded (floating) machine would sustain without loss of synchronism a departure of *speed* from the synchronous value if the subsequent change in load angle δ is less than 180° (elec), twice the steady-state stability limit of 90°. In the context of a multi-stack VR stepping motor the maximum value of permissible oscillation amplitude about each step position of equilibrium is twice the steady-state displacement limit of one quarter of a rotor tooth pitch. Figure 7.41, curve (b), shows for a three-stack VR stepping motor with eight rotor teeth, a possible scenario for the development of loss of synchronism as a result of excitation changes at a rate near to the natural frequency of rotor oscillation.

Figure 7.42 Equal-area criterion of transient stability applied to an unloaded stepping motor. Plot of terminal power against load angle.

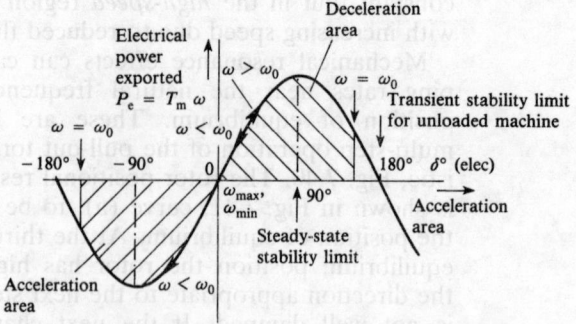

7.25.2 Low-speed operation

At higher stepping rates the rotor may no longer come to rest between successive steps but may have speed perturbations about a mean speed which is proportional to the stepping rate. This region corresponds to a stepping rate between 50 and 100 steps s^{-1} in Fig. 7.40. At higher stepping rates the duration of each excitation current pulse is proportionately reduced and the flux changes more difficult to establish as a result of opposing induced e.m.f. In the area of present interest, however, the flux/pole is assumed to be established without delay, adopting a value and distribution equivalent to single- and multi-step operation as soon as the appropriate stator poles are excited. On this basis the pull-out torque/stepping rate characteristics of Fig. 7.40 in the region of low-speed operation may be predicted from a consideration of the static torque/rotor angle characteristic of Fig. 7.38.

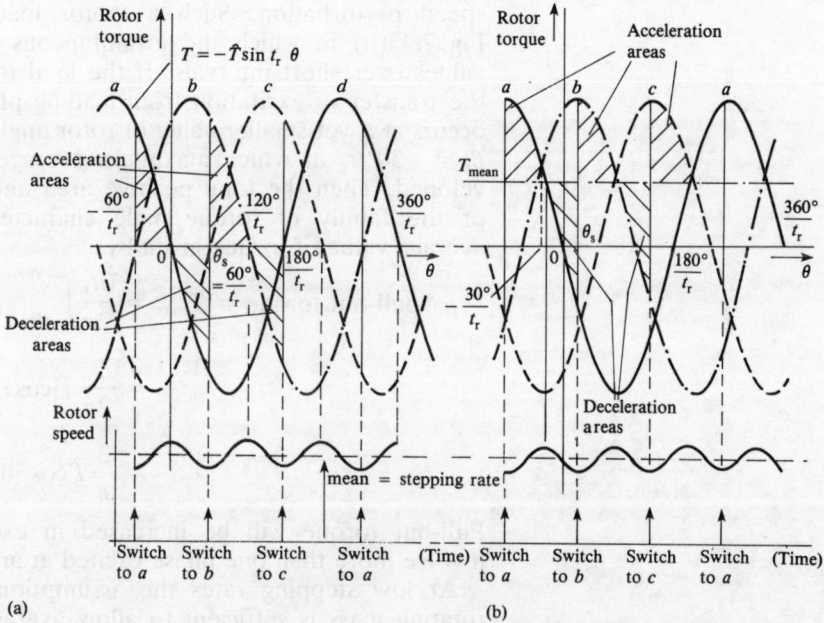

Figure 7.43 Pertaining to the prediction of pull-out torque for a VR stepping motor. Instantaneous rotor position variation with (a) zero mean load torque, and (b) motor load torque applied.

The behaviour of the motor when 'slewing' at low speed may be established by reference to Fig. 7.43 which shows the sinusoidal approximations of the static torque/rotor position curves as the rotor moves over one rotor tooth pitch with the three phases of a three-stack or three-phase single-stack VR stepping motor energised in sequence. The number of rotor teeth is t_r. The phase windings are switched sequentially at the stepping rate with sequence *abca*, the rotor motion keeping pace, on average, with the progression of the position of torque equilibrium as it increments in steps of $360°/3t_r = 120°/t_r$. Considering initially the rotor to be subjected to zero load torque and to be of sufficient inertia to ensure that speed perturbations are small, the acceleration and deceleration areas shown hatched on the torque/angle characteristics of Fig. 7.43(a) should be equal about a mean electromagnetic torque value of zero. Thus, with the rotor moving from the position of phase *a* equilibrium $\theta = 0°$ in Fig. 7.43(a), that phase being energised, the negative torque developed will tend to retard the rotor until at $\theta = 60°/t_r$ the excitation is switched to phase *b*.

Complementary rotor acceleration then occurs until the position of equilibrium with phase *b* energised is reached at $\theta = 120°/t_r$. The process continues with the excitation switched to phase *c* when $\theta = 180°/t_r$ *et seq.*

If now load torque is applied to the rotor the immediate effect is to retard the rotor motion so that it lags behind the stepped progression of the position of zero torque. As a consequence, each phase excitation naturally develops electromagnetic torque in the direction of rotor motion for a longer time, and generally of larger value than reverse-torque development, during each period of phase excitation. The *mean* torque developed is then accounted for by the load torque, the difference between instantaneous and mean values giving rise to acceleration and deceleration corresponding to speed perturbation. Such a motor load condition is shown in Fig. 7.43(b) in which the instantaneous torque still has negative values over short intervals. If the load torque is further increased the transfer of excitation from leading phase *a* to lagging phase *b* occurs at a yet smaller value of rotor angle $\theta = \theta_s$ until, ultimately, $\theta_s = -30°/t_r$ at which maximum average (pull-out) torque is developed. Then the total positive area under the positive envelope of the family of torque/angle characteristics contributes to an average value of torque given by

$$\text{pull-out torque} = T_{max} = \frac{3t_r}{2\pi} \int_{-150°/t_r}^{-30°/t_r} -\hat{T} \sin t_r \theta \, d\theta$$

$$= \frac{3}{2\pi} \hat{T} |\cos t_r \theta|_{-150°/t_r}^{-30°/t_r}$$

$$= \frac{3}{\pi} \hat{T} \cos 30° = 0.827 \hat{T}. \qquad [7.61]$$

Pull-out torque will be increased in excitation sequences which involve more than one phase excited at any instant.

At low stepping rates the assumption that the inertia of the rotating mass is sufficient to allow averaging of the instantaneous torque may not be valid. The pull-out torque may then be defined by the minimum value of the positive envelope of the torque/angle set, given by the intersection of characteristics at the instant of switching excitation. The effect is reflected in the low-speed region of Fig. 7.40.

The above analysis shows that the rotor position automatically adjusts to compensate for changes in load torque up to the pull-out value, maintaining synchronous operation. In complementary fashion, an attempt to drive the rotor at a higher speed than that appropriate to the stepping rate gives rise to net torque opposing rotation, and generator action.

7.25.3 *High-speed operation*

At higher stepping rates the excitation current can no longer be considered constant over the period when each phase winding is connected to a source of supply. As the duration of each current pulse diminishes, so the rate of change of flux linkage increases, due in part to the higher speed of the toothed rotor, and becomes more significant in affecting the establishment of exciting current. The drive circuits effectively ensure that, at lower speeds, the stator windings are current fed but, at higher speeds, the influence of e.m.f. induced by the changing flux linkage is of increasing relevance. Figure 7.44 shows the progressive distortion in excitation current waveform for a three-stack VR motor as stepping rate and speed rise. The deterioration in excitation waveform indicates a reduced ability to step a definable distribution of magnetic field and flux around the airgap. The overall effect is one of reduced pull-out torque capability, as illustrated in Fig. 7.40. The torque/speed characteristics are readily predictable at higher stepping rates if key simplifying assumptions are made in modelling the machine (Acarnley[27]).

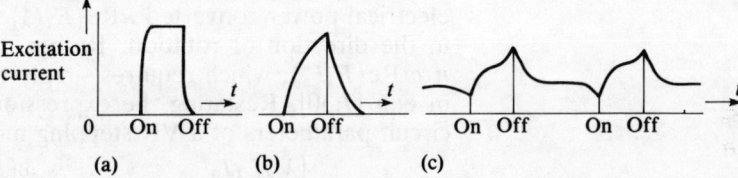

Figure 7.44 Corruption of the exciting current pulse of a VR stepping motor for progressive increases in stepping rate (a), (b) and (c) with respective time scales of order 100 : 10 : 1.

For hybrid motors the assumptions are as follows.

(i) A sinusoidal distribution of constant permanent-magnet flux, around the airgap periphery, with double pole pitch equal to one rotor tooth pitch (tooth plus space) and the flux linkage of the phase b group windings with *peak* value λ_m, out of phase by 90° (elec) with that of the phase a group.
(ii) The fundamental component alone of each alternating square wave of exciting current contributes to torque.
(iii) The sinusoidal e.m.f. induced in each stator phase due to the motion of the permanent-magnet flux is balanced by the fundamental component of the alternating square wave of the voltage applied to the phase windings and a fundamental frequency voltage drop across the phase impedance accounted for by winding resistance R and leakage reactance ωL_0.

For variable-reluctance motors the assumptions are.

(i) The distribution of total self-inductance of each phase winding has a component L_0 independent of position, together with a superposed sinusoidal variation of *amplitude* L_1 ($< L_0$) having double pole pitch equal to one rotor tooth pitch.
(ii) The unidirectional phase-current pulses have as torque-contributing element the fundamental alternating component but, in addition, a mean d.c. component I_0 which is responsible (with L_1) for an e.m.f. of fundamental frequency induced in each phase.
(iii) The fundamental component of the unidirectional phase voltage balances the e.m.f. described in (ii) above after accounting for the fundamental frequency voltage drop across phase-winding impedance $R + j\omega L_0$.

On making these assumptions it may be shown that both types of stepper motor may be modelled by the familiar equivalent circuit of Fig. 7.17(a) with source e.m.f. $\hat{E}_f$ replaced by $\omega\lambda_m$ or $\omega L_1 I_0$, as appropriate. A typical phasor diagram for a VR motor is shown in Fig. 7.45. As with the uniform-airgap polyphase synchronous machine, δ is identified as the load angle, but the stepping motor parameters are such that operation is invariably at lagging power factor and the resistive element R is comparable with ωL_0. Phase resistance R may be deliberately made high in order to preserve the excitation current drive at high speeds. A torque expression for stepping motor operation at high speed may be deduced from eqn [7.30] within the limits of validity of the equivalent circuit. Equation [7.30] gives the output power/phase at the terminals of the machine. If, as is usual with the stepping motor, phase resistance is significant, the value of terminal power $\mathrm{Re}\{V_a I_a^*\}$ cannot, after allowing for n phases, be equated to the electrical power converted $n\mathrm{Re}\{E_f I_a^*\} = -\omega T$, with rotor torque T in the direction of rotation. Rather, torque must be evaluated as $n/\omega\,\mathrm{Re}\{E_f I_a^*\}$, which requires simply the interchange of E_f and V_a in eqn [7.30]. Restating the expression in terms of the equivalent circuit parameters of a VR stepping motor gives

$$T = K\left[\frac{V_a L_1 I_0}{Z}\cos(\delta - \rho) - \frac{\omega(L_1 I_0)^2 \cos\rho}{Z}\right]. \qquad [7.62]$$

The constant K in eqn [7.62] includes a 'torque-correction factor' which accounts for the effect of the significant harmonics at phase current, giving rise to harmonic fluxes and e.m.f.s, contributing to torque.

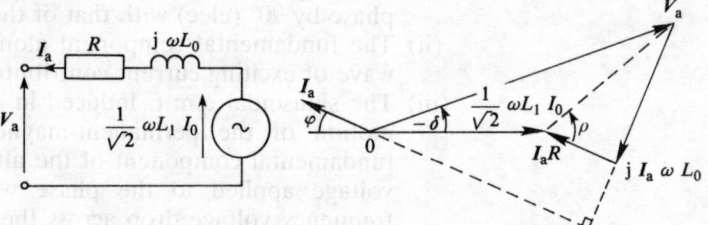

Figure 7.45 Equivalent circuit and phasor diagram representation of a VR stepping motor at high speed.

Pull-out torque corresponds to $\delta = \rho$, where $\cos\rho = R/Z$ and $Z = (R^2 + \omega^2 L_0^2)^{1/2}$, giving

$$T_{\max} = K\frac{L_1 I_0}{(R^2 + \omega^2 L_0^2)^{1/2}}\left[V_a - \frac{\omega L_1 I_0 R}{(R^2 + \omega^2 L_0^2)^{1/2}}\right]. \qquad [7.63]$$

The reduction in pull-out torque at high stepping rates is indicated in Fig. 7.40. Whether or not the pull-out torque falls to zero at a finite speed depends on the circuit parameter values. Generally, high-speed behaviour is enhanced by increasing the excitation circuit resistance R – at the expense of reduced efficiency.

The *hybrid* motor gives a similar performance, with the concept of a torque-correction factor being unnecessary as the value of the airgap flux is predominantly established by the permanent magnet.

7.25.4 *Drive circuit requirements*

The function of a stepping motor drive circuit is to provide the excitation windings with current, sequentially in accordance with the stepping rate. Hybrid motors require a two-phase supply, alternating with an ideally square waveshape. Variable-reluctance motors require a unidirectional or *unipolar* supply with, typically three to six phases. The development of 'back' e.m.f. (opposing the flow of current when motoring) is inevitable if the excited windings are to be engaged in electromechanical power conversion, but the current waveform will be modified by the induced e.m.f. unless this influence is swamped by *resistance* in the drive circuit. Such an expedient is undesirable as it is wasteful of power and also raises the necessary d.c. power supply voltage.

Figure 7.46(a) shows one phase of an elementary multiphase VR stepping motor drive circuit employing a switching transistor T_1, 'forcing' resistor R and a free-wheel diode–resistor series combination. One phase of an equivalent *bipolar* drive circuit for a hybrid stepping motor is shown in Fig. 7.46(c). Current waveforms for each circuit are shown in Fig. 7.46(b) and (d) with the effects of source resistance and rotational e.m.f.s ignored.

The bipolar bridge circuit, though more complex, constitutes the more efficient drive system as some of the field energy built up in the winding inductance is returned to the supply on each current reversal and not wasted in the forcing resistor R. The incorporation of the power supply voltage during freewheeling also has the desirable feature of reducing the duration of current decay, which role is carried out by resistor R_f in the drive circuit of Fig. 7.46(a) without energy recovery by the source. On account of their higher efficiency, bipolar drive circuits may be used in highly rated variable-reluctance stepper drives for which the direction of excitation current is immaterial. The requirement for a forcing resistor is avoided if a *chopper* drive circuit is employed as in Tutorial Example 7.23. In this arrangement a high d.c. supply voltage ensures a rapid initial rate of rise of exciting winding current until a rated value is reached at which a series-connected control transistor is turned off, obliging the winding current to freewheel via a parallel, low-resistance circuit incorporating a diode. The current slowly falls to a value at which the supply is restored, increasing the current once more to the rated value when the cycle repeats. At the end of the desired conduction period the freewheel current is re-routed via the d.c. supply to permit energy recovery. Both unipolar and bipolar chopper drive circuits are available.

Bifilar windings may be incorporated on the stator field poles of low-power hybrid stepping motors to permit the use of simpler electronic circuitry associated with unipolar drives, such that excitation current flow is unidirectional. Each field winding is duplicated and connected such that sequential current energisation creates the necessary alternating m.m.f.

7.26 Switched-reluctance Motor

The switched-reluctance motor illustrated in Fig. 7.47(a) is a high-power development of the single-stack VR stepping motor with an integrated drive-electronics package incorporating closed-loop control via rotor positional feedback. The latter facility may be

Figure 7.46 Variable reluctance stepping motor drive circuits with forcing resistor R: (a) and (b) unipolar drive circuit and waveforms; and (c) and (d) bipolar drive circuit and waveforms.

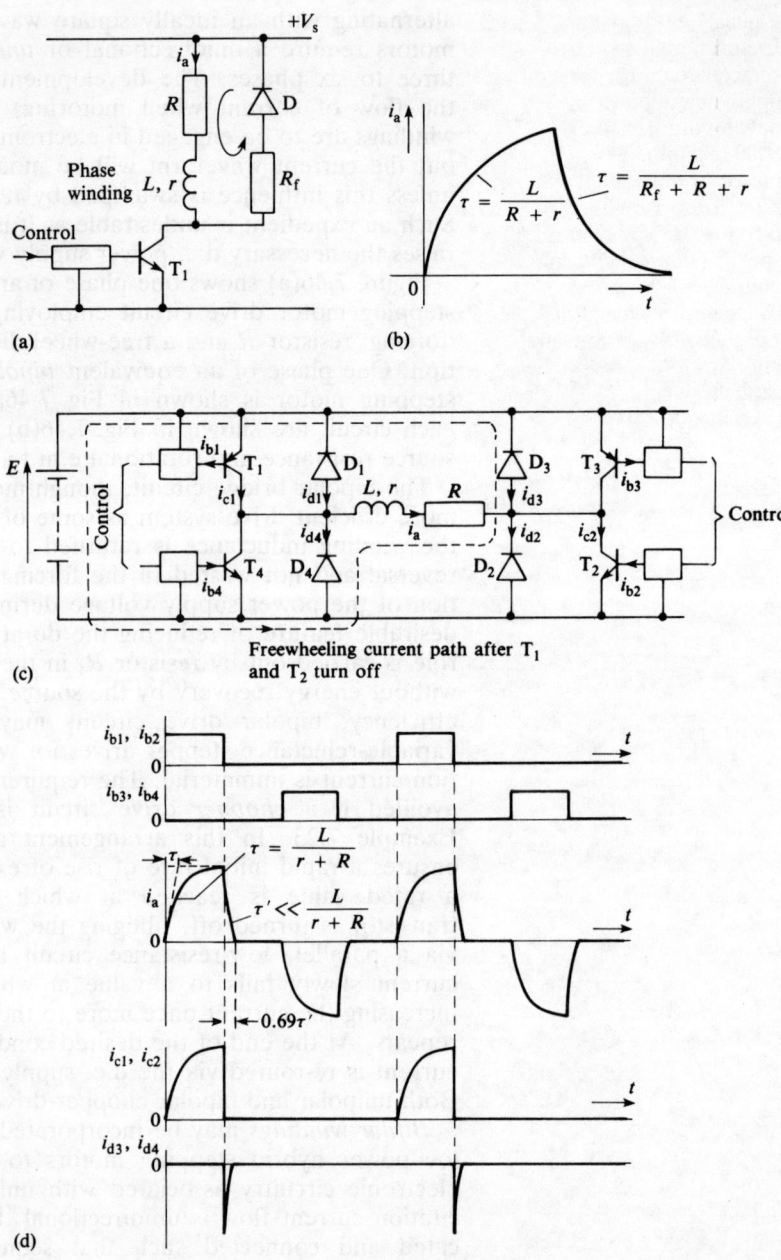

provided by optical, inductive proximity or Hall-effect sensors. The timing and duration of excitation current pulses are arranged so as to optimise performance over a wide range of speeds, providing full four-quadrant drive facility comparable in cost, efficiency and performance with alternative variable-speed drive systems supplied from constant-voltage, constant-frequency sources. The use of forcing resistors to establish a rapid buildup of exciting current and flux is inadmissible as inefficient and, for the same reason, energy recovery from the excited windings at current switch-off is necessary.

Synchronous machines 341

Figure 7.47 (a) Experimental switched-reluctance motor for domestic appliance application developing 270 W at 8000 rev/min. The *universal* motor of equivalent frame size (rotor shown left) develops 135 W at the same speed. (*Photograph by courtesy of Switched Reluctance Drives Ltd, Leeds, UK.*) (b) Bifilar winding energisation of a switched-reluctance motor, circuit and (c) waveforms.

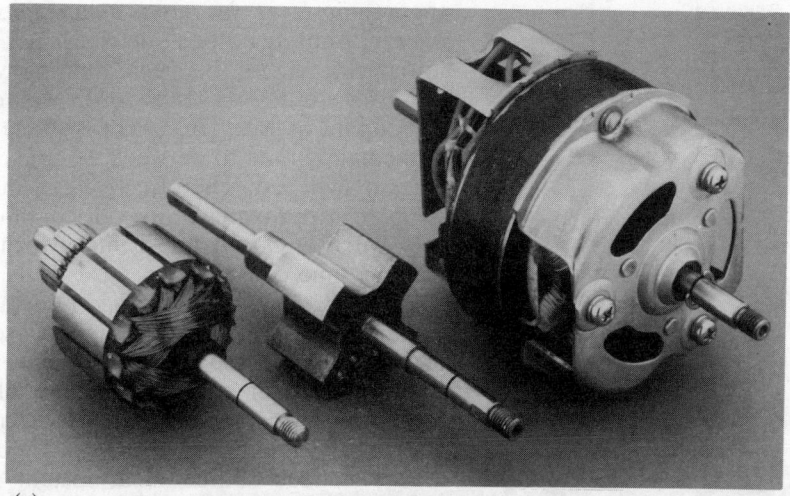

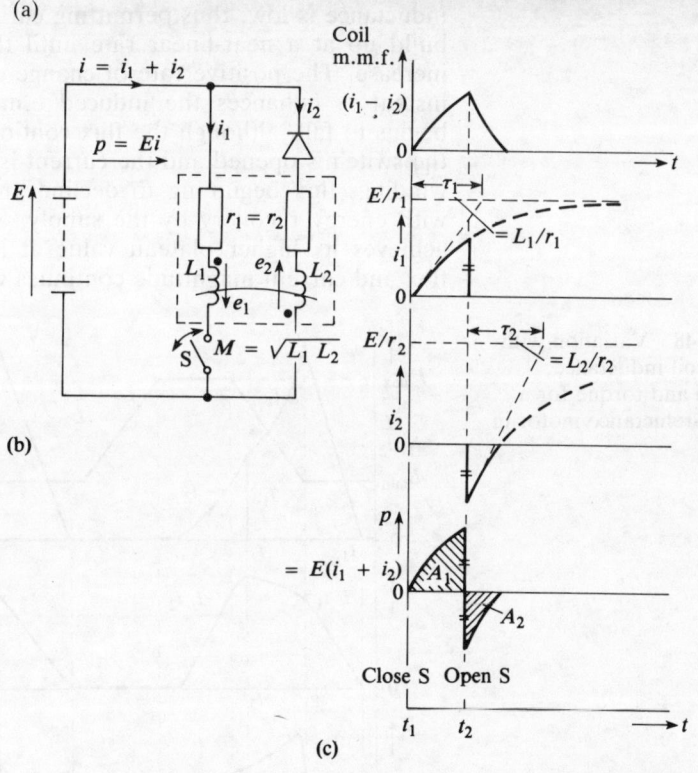

Although seldom used in practice, bifilar windings may be employed on the stator poles for this purpose, connected as in Fig. 7.47(b) and (c) which illustrate a static system. At the instant t_1 of switch S closure, current i_1 in winding 1 rises at initial rate $di_1/dt = E/L_1$ and energy is subsequently drawn from the voltage source. If the switch is opened at instant t_2 when i_1 is rising towards $I_{max} = E/r_1$ with time constant L_1/r_1, a value of current equal to i_1 is immediately established in identical winding 2, whilst i_1 falls immediately to zero – assuming that the windings are perfectly coupled with no leakage inductance. This follows from the requirement for no *instantaneous* change in stored magnetic energy. Subsequently i_2 declines, positive-going towards zero with initial rate di_2/dt between limiting values of E/L_2 and $2E/L_2$, inducing e.m.f. $e_1 = M\,di_2/dt$ in winding 1 and returning energy to the supply at rate $|Ei_2|$. The voltage across switch S on opening immediately rises to between $2E$ and $3E$ depending on the value of current switched. The effective coil m.m.f. waveform on repetitive switching approximates to a discontinuous sawtooth, provided that the switch is closed for a time which is short compared with the winding time constant.

The behaviour of a switched-reluctance motor in motion may be explained with reference to the idealised space and time waveforms of Fig. 7.48 and the bifilar winding excitation circuit of Fig. 7.47. The variation of exciting winding self-inductance L_1 or L_2 with rotor position is approximated to a trapezoid between limits L_{min} and L_{max}. The primary excitation winding is connected via switch S to the voltage source E at instant t_1 whilst the presented self inductance is low, thus permitting current i_1 and stator pole flux to build up at a near-linear rate until the coil inductance begins to increase. The positive rate of change of inductance with time from instant t_2 enhances the induced e.m.f. such that the current i_1 begins to fall, although the flux continues to increase. At instant t_3 the switch is opened and the current is transferred to the secondary winding, flux beginning to decline and current $|i_2|$ falling further with energy recovery by the supply. Subsequently, the inductance achieves its higher plateau value at instant t_4 and the decline in flux and current magnitude continues with energy recovery.

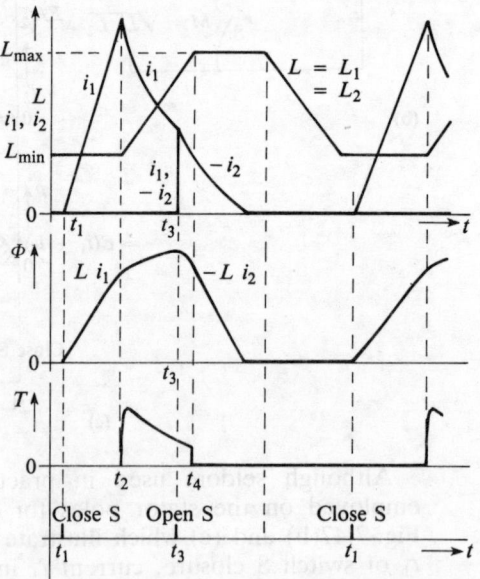

Figure 7.48 Variation with time of coil inductance, flux/pole and torque for a switched-reluctance motor in motion.

With torque defined by eqn [7.60], $T = \frac{1}{2}i^2\,dL/d\theta$, a pulse of rotor torque is developed in the direction of rotation during the time the inductance is increasing. Early closure of the switch relative to the rotor position provides for an increased value of torque as the current i_1 increases to a higher peak value and, with

$-i_2$, persists for a longer time. Torque pulsations are characteristic of the switched-reluctance drive when operated in the manner described, being particularly significant at low rotor speeds when the stored kinetic energy is low. Several alternative modes of operation are possible, however. The eminent controllability of the switched-reluctance machine makes it a contender for unconventional power generation purposes such as wind turbines, in addition to the commercial variable-speed drive market.

7.27 Permanent-magnet a.c. Synchronous Machines

The electromagnet field system of a polyphase synchronous machine may be replaced by permanent magnets in situations where the lack of field current control is acceptable. This virtually rules out large machine applications involving constant-voltage busbars where control of reactive power flows is important. The use of permanent-magnet field systems is generally restricted to small machines where, for generators, terminal voltage variation with load current is not important or, for both motors and generators, power factor is unimportant. Permanent-magnet synchronous machines may be operated more efficiently than their wound field counterparts and at a higher power factor than *induction* machines (discussed in Ch. 8) of the same rating and frame size. High-speed 400 Hz permanent-magnet generators find many applications in aerospace due to their high power/weight ratio.

The likelihood of loss of permanent magnetisation in service is great unless precautions are taken with the method of operation, or construction, or selection of magnetic materials employed. The radial field armature reaction effect on the main pole system is completely demagnetising if a generating machine is loaded on inductance or is subjected to a short circuit (Sect. 7.11). The distributions in the airgap of the radial components of armature field and 'field' winding field are then in phase-opposition, and this resultant field predominantly governs the flux in the magnetic materials bounding the airgap. If the torque limit at synchronous speed when motoring or generating is exceeded, the armature field rotates at *slip speed* with respect to the permanent-magnet structure and exerts sequentially magnetising and demagnetising influences on the permanent magnet. As noted in Section 3.17.3 the normal operating point of a permanent-magnet system incorporating an airgap is at the intersection of the *airgap line* with the *recoil line*, the latter being located in the second quadrant of the B/H plane and attached at one extremity to the demagnetisation characteristic of the permanent-magnet material. The application of an additional source of m.m.f. which is alternately magnetising and demagnetising effectively shifts the position of the constant-slope airgap line to the right and left of the origin of the B/H plane, respectively. Net demagnetisation occurs if the limit of the essentially linear recoil line is exceeded and the point of intersection moves to the steeply sloping portion of the demagnetisation curve, as illustrated in Fig. 3.35. Removal of the modulating m.m.f. then places the operating point at the airgap line intersection with a new recoil line yielding a lower value of magnet flux density B_m.

Dismantling the machine without the use of a *keeper* to 'short circuit' the permanent-magnet poles may also lead to loss of permanent magnetism by changing the slope of the airgap line.

Both these situations may be avoided by the use of a permanent-magnet material whose demagnetisation characteristic within the second quadrant element of the major hysteresis loop approximates to the ideal of a uniform slope approaching the theoretical minimum of μ_0. Rare-earth/cobalt materials approximate this condition with a high value of remanence and a correspondingly high value of coercive force. The cheaper ferrites approach this straight-line characteristic with lower values of remanence and coercive force (Fig. 3.32).

An inherent disadvantage of the synchronous motor is that it is not self-starting when run off constant-frequency supplies. *Line start* synchronous motors employ a *cage winding* adjacent to the airgap on that side carrying the field-pole system. This permits run-up to near synchronous speed as an induction motor (considered in Ch. 8), when the armature winding on the opposite side of the airgap is energised. If the 'field' is provided by an electromagnet the excitation current for this would not be supplied until the rotor has attained very near synchronous speed. The rotor should then pull into step through the development of *synchronising torque*, Section 7.13. If permanent magnets are employed, however, their field will alternately aid and oppose the armature winding field as the rotor accelerates from standstill, superimposing pulsating torque on the net value. The influence of the armature field may be reduced by arranging for the slotted, laminated core which supports the cage to provide magnetic screening for the permanent magnets during acceleration. During actual synchronisation, however, the screening effect of the cage is very much reduced and the magnets are subjected to the maximum demagnetising influence. A modern design for a four-pole line-start synchronous motor is shown in cross-section over one pair of poles in Fig. 7.49 which incorporates a computed flux plot for the machine running

Figure 7.49 Flux distribution in a loaded line-start synchronous induction motor with permanent magnet field.

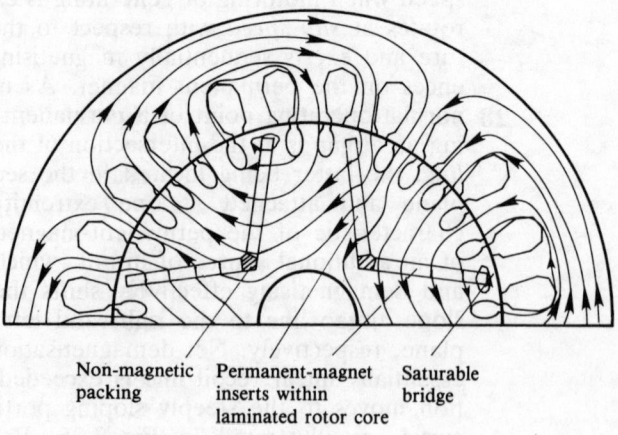

Non-magnetic packing   Permanent-magnet inserts within laminated rotor core   Saturable bridge

synchronously on load. The eight samarium-cobalt magnets set into the pack of soft-iron laminations which define the rotor pole surfaces at the airgap are noteworthy for their short *length* and small volume.

Figure 7.50(a) shows a typical torque/speed characteristic for a line-start synchronous motor when running up towards synchronous speed from rest. The rotor cage of a stator-fed machine is responsible for the induction motor torque acting in the direction of the synchronously rotating field initiated by current flowing in the stator conductors. The stator winding, however, has induced within it a component of e.m.f. and consequent current due to the relative motion between it and the rotor permanent-magnet field, the effect of which is to develop a braking torque. Provided the minimum net accelerating torque exceeds the load torque the machine will self-synchronise and will subsequently run with a load angle δ whose dependence on load torque is indicated in Fig. 7.50(b). The shape of this characteristic curve is analogous to that of the electromagnetically excited salient-pole machine presented in Fig. 7.31 and deduced from eqn 7.54.

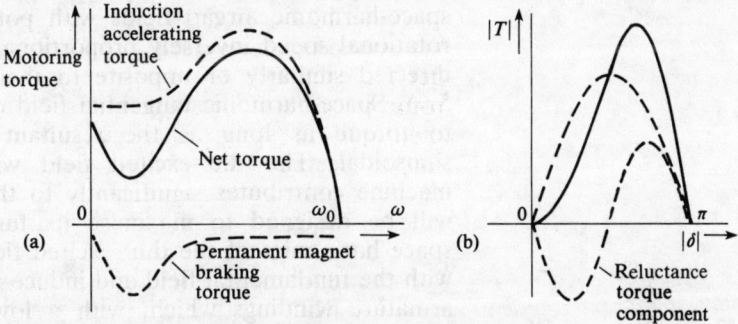

Figure 7.50 (a) Torque/speed characteristic for line-start synchronous induction motor running up from standstill, (b) torque/load angle characteristic after synchronisation.

For the permanent-magnet machine, however, the d-axis synchronous reactance X_d is smaller than the q-axis value X_q, due to the fact that the flux established along the d-axis by armature current is required to effect two traverses of permanent-magnet material in its journey within the (rotor) iron from pole to (adjacent) pole. In contrast, the q-axis rotor flux path remains in soft iron and encounters no airgap length increase characteristic of the salient pole machine considered in Section 7.14.

7.28 Power Electronic Drive Systems for Synchronous Machines

In this section we shall consider means by which appropriate power supplies may be made available to enable an a.c. synchronous machine to provide a variable-speed drive. Such a machine develops torque only at synchronous speed – which is related via the number of pole-pairs to the frequency of the supply. Varying the field excitation controls the reactive volt-ampere flow, and hence the power factor.

This identification of speed with supply frequency is a characteristic of all a.c. machines. The induction motor considered in

Chapter 8 will run at a speed whose departure from the synchronous value is dependent upon the load torque. As such it is an *asynchronous* machine with a base no-load speed dependent upon supply frequency, and relying on current induced in a secondary winding to provide m.m.f. on the side of the uniform airgap facing that excited by primary winding current. Hence, in its basic closed format, this secondary winding is not available to exert any *external* modifying influence on the radial flux within the machine. Control of the radial airgap flux must therefore be achieved by adjustment of the primary winding applied voltage.

A variable-speed a.c. drive system therefore generally requires control of both frequency and voltage. Ideally, the supply voltage should be sinusoidal, to which voltage waveshape an ideal rotating machine should present a constant impedance over each cycle of supply, assuming sinusoidal magnetic field and flux distributions, with equal instantaneous and average torques, and drawing a sinusoidal current.

In practice, a.c. machines are tolerant of imperfections in the supply voltage waveform. Fundamental frequency armature currents supplied to *practical* polyphase windings produce non-triplen space-harmonic airgap fields with pole-pitch and subsynchronous rotational speed inversely proportional to their harmonic number, directed similarly or opposite to that of the fundamental (Sect. 5.6). Space-harmonic tangential field components cannot give rise to torque as long as the resultant radial flux distribution is sinusoidal. The d.c.-excited field winding of the synchronous machine contributes significantly to the resultant radial field and will be designed to maximise its fundamental component. The space harmonics of the d.c. excited field winding maintain station with the fundamental field and induce time-harmonic e.m.f.s in the armature windings which, with a low-impedance voltage source give rise to corresponding harmonics in the armature current. Non-triplen time-harmonic components of armature current produce supersynchronous rotating-field components with pole-pitch equal to that of the fundamental, and which are able to develop torque via secondary current induced in closed circuits on the opposite side of the airgap. The torque produced is generally insignificant, however, even with *induction* machines, due to the high relative motion between field and secondary winding. The precise effect is very dependent upon winding layout and slotting. Triplen harmonic current components are identical in each phase of a three-phase armature winding and their field components should sum to zero if the winding is balanced. The synchronous machine is generally less prone to harmonic influences than the induction machine by virtue of its current-fed field winding.

Inverter drives to a.c. machines may derive their power from fixed voltage and frequency a.c. sources or variable-voltage d.c. supplies. An example of the former is the *cycloconverter* which achieves its instantaneous output by connecting each phase of the load via switches to whichever phase of the three-phase supply is appropriate to the synthesis, over successive intervals, of an output voltage waveform which has a significant fundamental component of the required frequency. Such an arrangement employs two fully controlled thyristor converters as encountered in the reversible d.c.

drive system of Section 6.17, with the gate-control signals cyclically controlled. Its inherent property of reversible power flow is of advantage with variable-speed a.c. drives, although its use is restricted to applications where the required output frequency is a small fraction of the supply frequency. Harmonic currents and voltages are substantial in this type of system but may be minimised by employing converters with a high pulse number, having a correspondingly large number of semiconductor devices.

The major concern in this book is with *d.c. link* converter systems in which a converter is used to derive an intermediate d.c. supply from a fixed-frequency, fixed-voltage a.c. supply; the d.c. power source then energises the a.c. machine via a variable-frequency inverter. The d.c. link may be maintained at a nominally constant though adjustable voltage, or current, each alternative giving rise to different drive characteristics. A feature of this type of application for converters transforming d.c. to a.c. is that the a.c.-side voltages are no longer definable by a sinusoidal source, as was the case with the *a.c. line-* or *naturally commutated* converters considered in Chapter 6. The commutation of the a.c. line currents is therefore no longer controlled by systematically changing sinusoidal source e.m.f.s. Similarly, the d.c.-side current may not be well-smoothed – which would be characteristic of a *current source* d.c. link (Sect. 7.34) – but may be rather in conformity with defined d.c. *voltages* appropriate to a low-impedance voltage-source d.c. link.

7.29 Single–phase Voltage-source Bridge Inverter

Our treatment of inverters for variable-speed drives will commence with the simple, single-phase bridge configuration, essentially identical with that of the bridge converter considered in Section 6.14.2 but now *supplied* from a constant-voltage d.c. source. We shall see that commutation problems are likely, such that special measures may be necessary to turn off previously conducting thyristors as required.

The basic circuit is shown in Fig. 7.51(a). Ignoring the antiparallel connected diodes for the moment and considering a *resistive* load, the principle of operation becomes self-evident. Simultaneous turn-on of thyristors Th_1 and Th_2 in sequence with Th_3 and Th_4 will generate square waveforms of voltage and current for the resistive load, the amplitude of the a.c. voltage ideally equalling the d.c. source e.m.f. E. It is also apparent that, whichever thyristor of each pair associated with each half of the bridge circuit is conducting, the other is forward biased to the extent of the source voltage E. In consequence, it will respond immediately to a gate-firing pulse but, unless the previously conducting thyristor is turned off by deliberately forcing its anode–cathode current to zero, a short-circuit across the supply will result. Furthermore, unless the forcing to zero of the anode–cathode current is followed by a brief period of artificially maintained reverse anode–cathode bias voltage, restoration of the normal forward bias will inhibit the turn-off mechanism and a short circuit will develop subsequently.

Figure 7.51 Single-phase bridge inverter: (a) essential circuit. Waveforms: (b) for R/L load; (c) for R/C load, load current idealised as fundamental component.

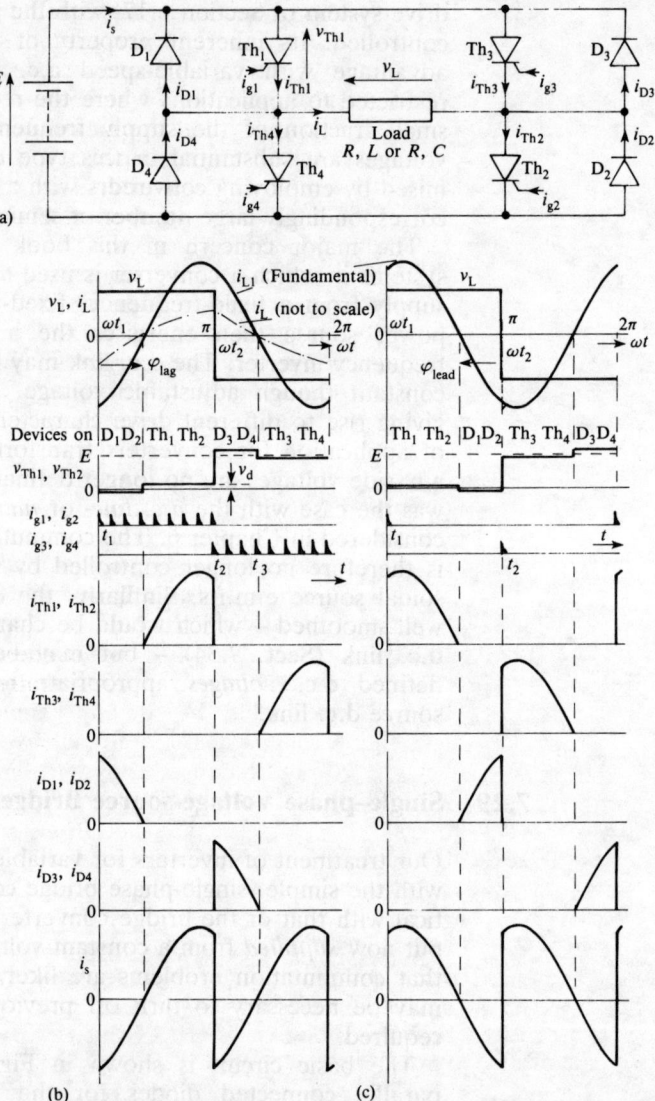

The waveforms shown in Fig. 7.51(b) correspond to an inductive load comprising R and L in series. The thyristor switching procedure is as above, generating a square-wave voltage across the load. The load current has the characteristic exponential form, non-zero valued at the instants of switching. The inductive load current cannot change in the discontinuous manner implied by the switching function so that, if Th_3 and Th_4 are gated at instant t_2, load current i_L continues to flow from left to right through the load impedance. Th_4 should have been gated for turn-on at instant t_2, and the commutation process that achieves this should also turn off Th_1, ideally at the same instant. Th_4, however, cannot conduct load current of the particular polarity required at instant t_2, therefore i_L must flow in the anit-parallel connected diode D_4 until falling naturally to zero at instant t_3. The small voltage-drop across

D_4 will reverse bias Th_4 which cannot therefore be turned on until i_L crosses zero at t_3. Over the same period and for the same reason Th_3 is unable to conduct and D_3 carries the load current which returns the energy stored in the load inductance to the source E. Provided that the gate pulses are maintained on Th_3 and Th_4, these thyristors will conduct when i_L passes through zero at t_3 and forward bias is restored. This transfer of load current from diode to thyristor occurs smoothly as the diode current falls *naturally* to zero. The waveform for source current i_s in Fig. 7.51(b) indicates the periods when the d.c. source is absorbing or supplying power through the polarity of i_s. In the figure the derived current waveforms are identified with the fundamental component of the load current only. It is apparent that *diode conduction* is associated with instantaneous power flow from load to source, whereas the power flow is from source to load when the thyristors conduct. Thus the diodes are necessary for the return of energy stored in the load inductance over the intervals of time defined by the power factor angle ϕ.

Provided the gating pulses are maintained on the thyristors, therefore, the commutation of load current from diode to thyristor is achieved naturally as it passes through zero. To commutate the current from thyristor to diode at instants when load voltage reverses, however, requires the sudden interruption of the thyristor current, i.e. its forcing to zero against a forward anode–cathode bias voltage rising to the full source voltage value. Suitable *force-commutation* circuits to achieve this end are discussed in Section 7.30 but it is now apparent that the role of such a facility is to supply a train of gate-firing pulses to the thyristor of a thyristor/diode pair connected in anti-parallel whilst also achieving the turn-off of the series-connected complementary thyristor.

Waveforms associated with *capacitive* reactive or leading power factor loads are shown in Fig. 7.51(c). Again, the fundamental components of load voltage and current are shown, together wth the current distribution throughout the circuit. The diodes again conduct when the load voltage and current are instantaneously of opposite polarity, with power flowing from load to source. This period is now *preceded* by a natural transition through zero of current commutated from thyristor to anti-parallel connected diode. A previously conducting thyristor is naturally reverse-biased by the voltage-drop across the conducting diode, and hence no special thyristor turn-off facilities are necessary. The current transition from diode to thyristor is achieved by gate-firing Th_1 and Th_2 at instant t_1, and Th_3 and Th_4 at instant t_2. The gated thyristors are forward biased by the full source voltage E and turn on immediately, reverse biasing the previously conducting series-connected diode, again by the full source voltage. No special measures are therefore required to ensure that the diodes remain off.

The capacitive leading power-factor load may therefore be commutated naturally by simply initiating gate pulses at the appropriate instant of load voltage reversal. Lagging power-factor loads, however, require forced commutation. The same rule applies to polyphase bridge inverter configurations, and hence it

7.29.1 Quasi-square Voltage Waveform Generation

It is not necessary to organise the switching of the semiconductor devices associated with the two halves of the bridge to take place simultaneously. Thyristors Th_3 and Th_2 may be gated independently of Th_1 and Th_4. This feature has particular relevance for the three-phase bridge inverters considered in Section 7.33. Whichever devices are conducting in the right-hand half of the single-phase bridge of Fig. 7.51(a) simply determine the return path for the load current, whether via the upper or the lower busbar. The complete route taken has relevance for the instantaneous power flow, as the alternative circuits will either exclude or include the source E. Use may be made of the inherent independence of device switching on each side of the load impedance to modify the waveshape of the load voltage from square to the quasi-square waveform of Fig. 7.52(a).

Most inverter applications require a.c. voltage and current waveforms which approximate to the sinusoidal. Fourier analysis shows that the quasi-square voltage waveform in Fig. 7.52(a) with $\alpha = \pi/3$ radians has triplen harmonic components eliminated, with the fifth and seventh dominant at 1/5th and 1/7th of the fundamental respectively. The figure illustrates the gate-firing sequence necessary in the circuit of Fig. 7.51(a). The instantaneous current route may be traced by observing which devices are conducting, bearing in mind that, with inductance present in the load, a force commutation circuit will be required and gate-firing pulses sustained for up to one half cycle. Immediately after each commutation the inductive load current is carried by the appropriate anti-parallel diode until current reversal occurs. During periods of zero output voltage the non-zero load current circulates between the two halves of the bridge via upper and lower busbars alternately.

7.30 Forced Commutation of a Bridge Inverter Supplied from a Voltage Source

The essential requirements of a force commutation circuit may be stated with reference to Fig. 7.53. The circuit is required to carry out the following functions sequentially in response to alternate gating of thyristors Th_1 and Th_4:

(i) With Th_1 conducting (i_L positive), to turn on Th_4 (or D_4 initially if the load is inductive) and turn off Th_1 with switch S closed in either position; and

(ii) with Th_4 conducting (i_L negative), to turn on Th_1 (or D_1 initially if the load is inductive) and turn off Th_4 with S closed in either position.

Synchronous machines 351

Figure 7.52 (a) Quasi-square voltage waveform with R/L load current and thyristor gate drive requirements. (b) Voltage and current waveforms for Worked Example 7.4 employing the circuit of Fig. 7.51(a).

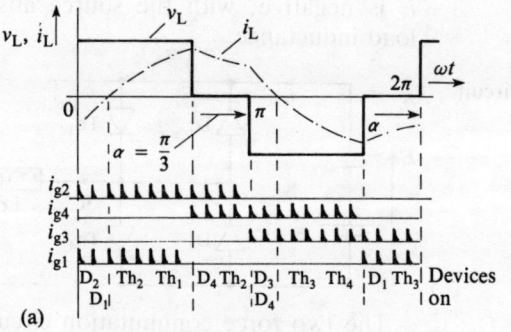

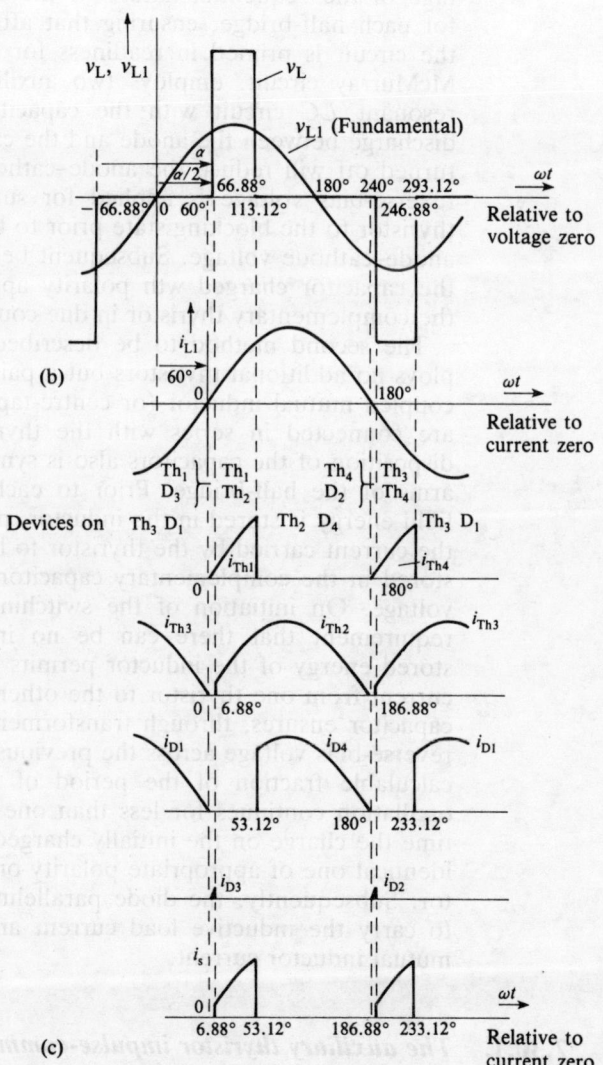

The switch S represents conditions at the other side of a single-phase bridge. With S in position 3 and D_1 conducting, say, source current i_s must be zero, but with S in position 2 and D_1 conducting

i_s is negative, with the source absorbing energy supplied by the load inductance.

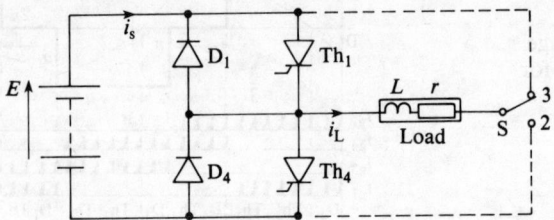

Figure 7.53 Essential circuit of one phase of a bridge inverter (half-bridge)

The two force commutation circuits to be described take advantage of the sequential nature of the switching functions required for each half-bridge, ensuring that after each switching operation the circuit is primed in readiness for the next. One method, the McMurray circuit, employs two auxiliary thyristors and a series resonant LC circuit with the capacitor initial charge such that discharge between the anode and the cathode of the thyristor to be turned off will reduce the anode–cathode current to zero. Then a reverse-bias voltage is applied for sufficient time to restore the thyristor to the blocking state prior to the re-application of forward anode–cathode voltage. Subsequent behaviour of the circuit leaves the capacitor charged wth polarity appropriate to the turn-off of the complementary thyristor in due course.

The second method to be described (McMurray–Bedford) employs no additional thyristors but a pair of capacitors and a closely coupled mutual inductor (or centre-tapped choke) whose windings are connected in series with the thyristors to be switched. The disposition of the capacitors also is symmetrical with respect to the arms of the half-bridge. Prior to each switching action magnetic field energy is stored in the inductor, proportional to the square of the current carried by the thyristor to be turned off. Energy is also stored in the complementary capacitor which is charged to supply voltage. On initiation of the switching process, the fundamental requirement that there can be no instantaneous change in the stored energy of the inductor permits an instantaneous transfer of current from one thyristor to the other, whilst the initially charged capacitor ensures, through transformer action, the maintenance of reverse-bias voltage across the previously conducting thyristor for a calculable fraction of the period of a resonant oscillation. The oscillation continues for less than one quarter-cycle, during which time the charge on the initially charged capacitor is replaced by an identical one of appropriate polarity on the complementary capacitor. Subsequently, the diode paralleling the gated thyristor is left to carry the inductive load current and to freewheel to zero the mutual inductor current.

7.30.1. *The auxiliary thyristor impulse-commutation circuit (McMurray)*

The complete circuit for one half-bridge is shown in Fig. 7.54(a) with the commutation components Th_{1A}, Th_{4A}, C and L. The capacitance has initial charge of polarity such as to turn off Th_1

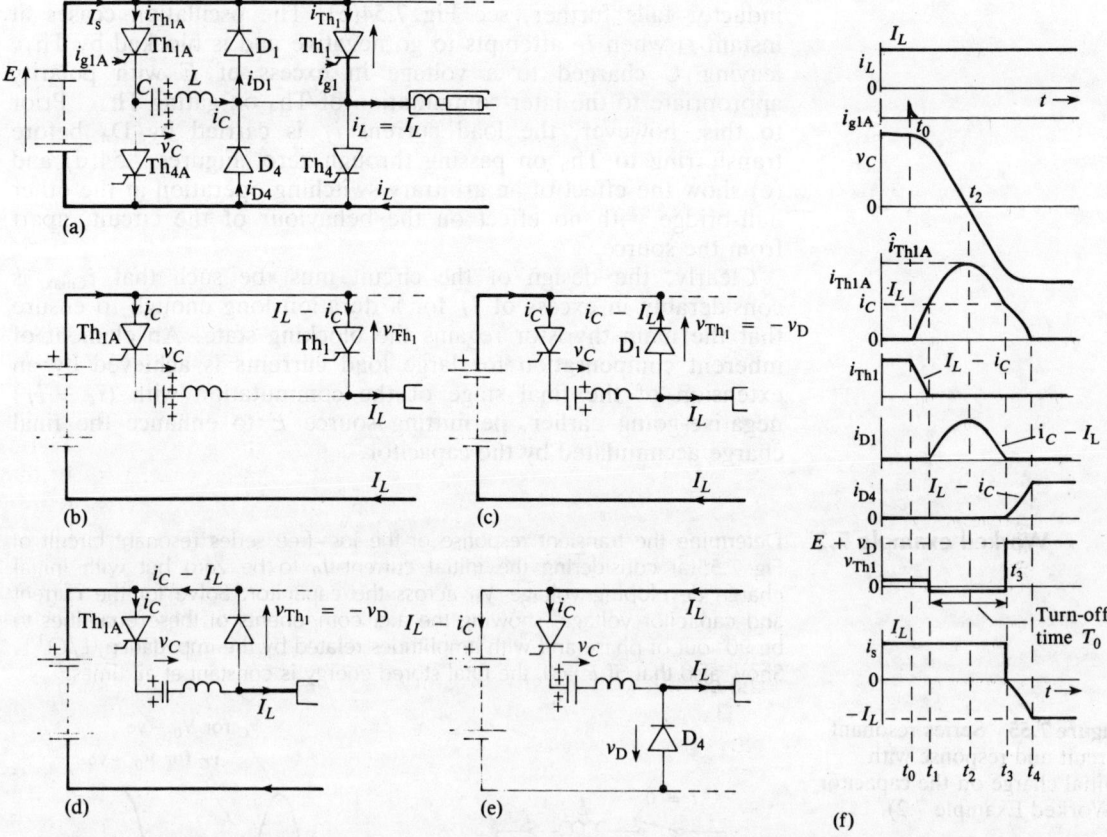

Figure 7.54 McMurray auxiliary thyristor impulse-commutation circuit: (a) essential circuit, (b)–(e) effective circuit development to achieve turn-off of Th_1 on gating Th_{1A}; (f) corresponding waveforms. (*Note*: $I_1/\hat{\imath}_{Th1A} = \cos(w_0 T_0/2)$ where T_0-turn-off time).

when Th_4 is gated. Thus, prior to turn-off, Th_1 conducts with i_L positive, continuing its circuit via the source and the lower bus. The load is inductive, therefore the expectation is that $i_L = I_L$ will be carried by diode D_4 on completion of the commutation, bypassing source E. This diode, along with D_1, also has a role to play in the commutation process. Prior to turn-off of Th_1, the auxiliary thyristor is forward biased by the voltage v_C across the commutation capacitor C, of value in excess of the source voltage E.

The effective circuit immediately after firing the auxiliary thyristor Th_{1A} at $t = t_0$ is shown in Fig. 7.54(b) with the capacitor C discharging via inductor L and thyristors Th_1 and Th_{1A}. Series resonance occurs in this circuit with i_C increasing sinusoidally as energy is transferred from the capacitor to the inductor. At instant t_1, i_C exceeds the standing current through Th_1 which turns off, with i_C transferring to D_1 whose voltage-drop reverse biases Th_1, enabling recovery to the blocking state. The resonant oscillation, demonstrated in the waveforms of Fig. 7.54(f), continues with i_C achieving peak value at instant t_2 when v_C passes through zero. Figures 7.54(c) and (d) illustrate this. Shortly afterwards at instant t_3, $(i_C - I_L)$ flowing in D_1 falls to zero and the resonant activity continues via D_4, now incorporating the source E which acts to supplement the buildup of charge on C as the stored energy in the inductor falls further, see Fig. 7.54(e). The oscillation ceases at instant t_4 when i_C attempts to go negative and is blocked by Th_{1A} leaving C charged to a voltage in excess of E with polarity appropriate to the later commutation of Th_4 on gating Th_{4A}. Prior to this, however, the load current i_L is carried by D_4 before transferring to Th_4 on passing through zero. Figures 7.54(d) and (e) show the effect of an arbitrary switching operation at the other half-bridge with no effect on the behaviour of the circuit, apart from the source.

Clearly, the design of the circuit must be such that i_{Cmax} is considerably in excess of I_L for a duration long enough to ensure that the main thyristor regains the blocking state. An element of inherent compensation for large load currents is achieved by an extension of the final stage of the commutation with $(i_C - I_L)$ negative-going earlier, permitting source E to enhance the final charge accumulated by the capacitor.

Worked example 7.2 Determine the transient response of the loss-free series resonant circuit of Fig. 7.55(a) considering the initial current I_0 to be zero but with initial charge developing voltage V_0 across the capacitor. Solve for the current and capacitor voltage, showing the a.c. components of these quantities to be 90° out of phase and with amplitudes related by the impedance $(L/C)^{1/2}$. Show also that, if $E = 0$, the total stored energy is constant at all times.

Figure 7.55 Series resonant circuit and response with initial charge on the capacitor (Worked Example 7.2).

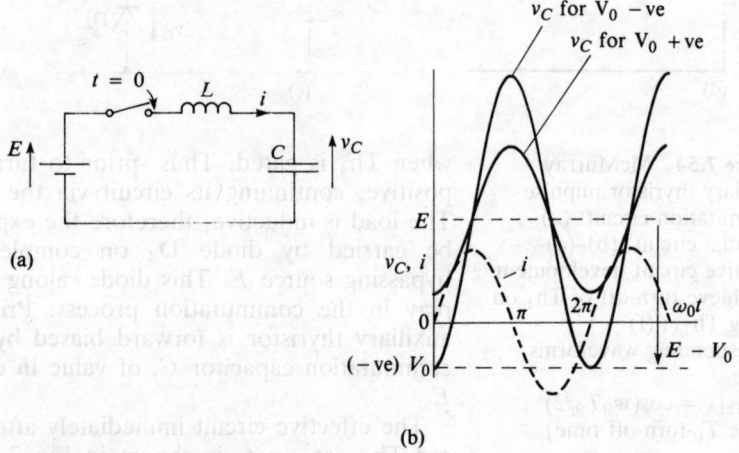

Solution Referring to Fig. 7.55(a)

$$E = L\frac{di}{dt} + \frac{1}{C}\int i\,dt.$$

Invoking the Laplace transform we get

$$\frac{E}{s} = L(s\bar{i} - I_0) + \frac{1}{C}\left(\frac{\bar{i}}{s} + \frac{V_0 C}{s}\right)$$

where I_0 is the initial current flowing at $t = 0$ and V_0 is the initial voltage across the capacitor. With I_0 zero, the solution for $\bar{i}$ gives

$$\bar{i} = \frac{E - V_0}{L(s^2 + \omega_0^2)},$$

where $\omega_0 = (LC)^{-1/2}$. Transforming back to the time domain gives

$$i(t) = \frac{E - V_0}{\omega_0 L}\sin \omega_0 t. \qquad [7.64]$$

Also

$$\bar{v}_C = \frac{1}{LC}\frac{(E-V_0)}{s(s^2+\omega_0^2)} + \frac{V_0}{s}$$

$$= \frac{E-V_0}{\omega_0^2 LC}\left(\frac{1}{s} - \frac{s}{s^2+\omega_0^2}\right) + \frac{V_0}{s},$$

giving

$$v_C(t) = E - (E-V_0)\cos \omega_0 t. \qquad [7.65]$$

Waveforms of $i(t)$ and $v_C(t)$ are shown in Fig. 7.55(b). It is apparent that $(E - V_0)$ determines the amplitude of voltage and current swing, whereas E determines the d.c. component of the capacitor voltage.

The ratio of the amplitudes of the a.c. components of v_C and i is

$$\omega_0 L = \frac{1}{\omega_0 C} = \sqrt{\frac{L}{C}}$$

and the total stored field energy with $E = 0$ is

$$\tfrac{1}{2}Cv_C^2 + \tfrac{1}{2}Li^2 = \tfrac{1}{2}CV_0^2\cos^2 \omega_0 t + \tfrac{1}{2}L\left(\frac{V_0^2}{\omega_0^2 L^2}\right)\sin^2 \omega_0 t$$

$$= \tfrac{1}{2}CV_0^2.$$

Tutorial example 7.2(A) Repeat worked example 7.2 for the general case of initial current I_0 at $t = 0$ and show that general expressions for circuit current and capacitor voltage are

$$i(t) = A\sin(\omega_0 t + \delta); \quad v_C(t) = E - B\cos(\omega_0 t + \delta)$$

where

$$A = \left[\left(\frac{E-V_0}{\omega_0 L}\right)^2 + I_0^2\right]^{1/2}, \quad B = \left[(E-V_0)^2 + \left(\frac{I_0}{\omega_0 C}\right)^2\right]^{1/2}$$

and

$$\delta = \tan^{-1}\left[\frac{\omega_0 L I_0}{E-V_0}\right]$$

7.30.2 *The complementary mutually coupled impulse-commutation circuit (McMurray–Bedford)*

The general circuit diagram incorporating the additional components of a mutual inductor and an identical capacitor pair is shown in Fig. 7.56(a). The load inpedance is assumed inductive so that load current is sustained at the value I_L during commutation. Arbitrarily, the load current is considered to return via the upper busbar. With Th_1 initially conducting, the expectation is that the load current will be transferred to diode D_4 on completion of commutation.

Referring to Fig. 7.56(a), conditions prior to gating Th_4 at instant t_0 are such that Th_1 is conducting with v_{Th1} near zero and $i_{Th1} = I_L$. All other currents are zero valued, as also is v_{C1}, but capacitor C_4 is fully charged to source voltage E. Thyristor Th_4 is

356   Electrical machines and drive systems

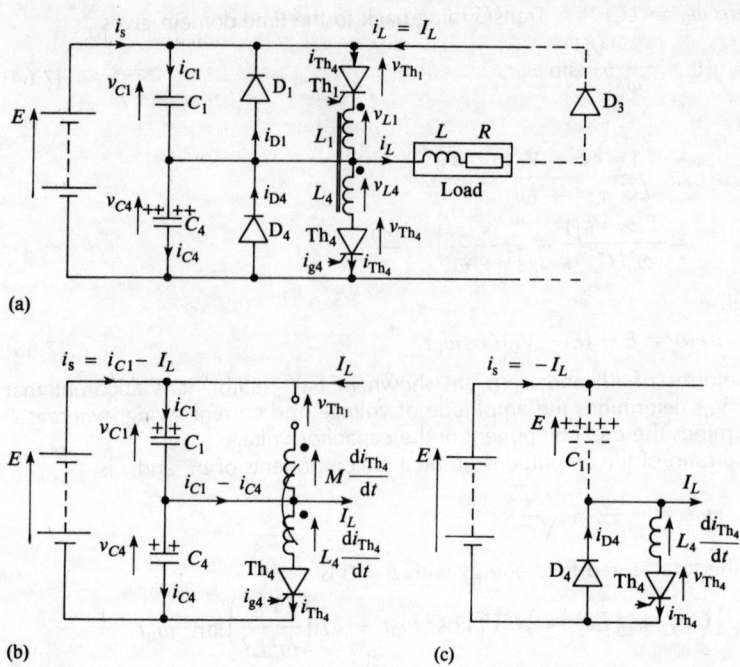

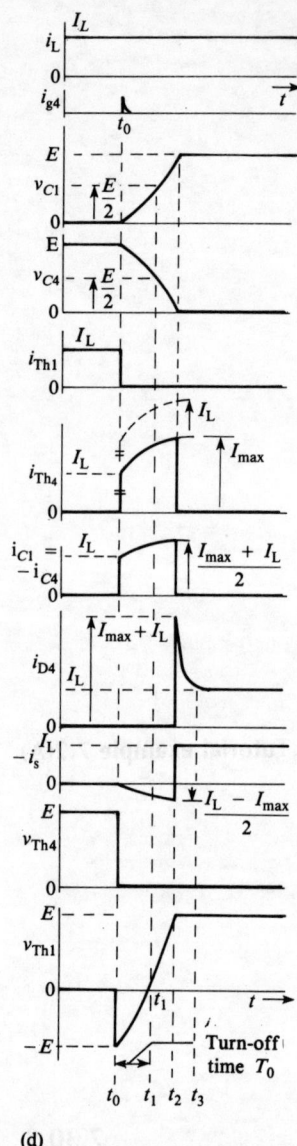

Figure 7.56 McMurray–Bedford complementary, mutually coupled impulse-commutation circuit: (a) essential circuit, (b)–(c) effective circuit development to achieve turn-off of Th_1 on gating Th_4; and (d) corresponding waveforms with load current sustained after switching.

forward biased to this voltage and responds to a gate pulse train initiated at $t = t_0$ when the effective circuit becomes as shown in Fig. 7.56(b).

The immediate effect is to place a short circuit across charged capacitor C_4 which proceeds to discharge via winding 4 and Th_4 under the influence of two constraints.

(i) There is an immediate increase in the current i_{Th_4} flowing in winding 4 from zero to I_L, accompanied by an equivalent reduction to zero of i_{Th_1} flowing in winding 1 (maintaining constant core flux), so turning off Th_1. Thereafter, i_{Th_4} must increase at a rate proportional to v_{C_4}, where $v_{C_4} = v_{L_4} = v_{L_1}$ and $(v_{L_1} - v_{C_1})$ is available to reverse bias Th_1.

(ii) C_4 loses charge at a rate such that its voltage fall is balanced by an equal rise in voltage across C_1, maintaining a constant sum equal to the source voltage E. This requires that i_{C1} must be equal and opposite to i_{C4} with the difference current $(i_{C1} - i_{C4})$ providing the continuing load current I_L and the inductor current flowing in winding 4. Thus

$$\left. \begin{array}{c} v_{C4} = L_4 \dfrac{di_{Th4}}{dt} \\ \text{where} \\ i_{Th4} = -2i_{C4} - I_L = 2i_{C1} - I_L. \end{array} \right\} \quad [7.66]$$

Since v_{C4} is positive i_{Th4} must increase from value $(2i_{C1} - I_L)$ until at $t = t_1$ its rate of change is half its initial value; v_{C4} then equals $E/2$, with an equal voltage induced in winding 1. v_{C1} at $t = t_1$ will have increased to a value $E/2$ also. Subsequently, Th_1 is forward biased, therefore the duration $(t_1 - t_0)$ marks the turn-off time for Th_1.

i_{Th4} continues to increase until it achieves a value I_{max} at $t = t_2$ when v_{C4} is zero and $v_{C1} = E$. The oscillatory interchange of energy between inductance L_4, the source and the capacitor combination should then continue with frequency $\omega_0 = (2LC)^{-1/2}$, but the negative transition of v_{C4} is prevented by diode D_4 which becomes forward biased. As a consequence, C_4 remains discharged, C_1 is left charged to voltage E and the current i_{Th4} is obliged to decay from I_{max} to zero via the freewheel path provided by D_4, as shown in Fig. 7.56(c). The commutation process is completed at $t = t_3$ with D_4 carrying the continuing load current I_L and providing a small reverse-bias voltage for Th_4.

The energy lost in the commutation process is essentially $\frac{1}{2}LI_{max}^2$. That stored in C_4 initially is transferred to C_1. This capacitor is left with a charge of polarity appropriate to subsequent commutation of Th_4 on gating Th_1. Significant waveforms are shown in Fig. 7.56(d).

Figure 7.57(b) details the variation of current in thyristor Th_4 during the turn-off of Th_1 for the alternative conditions of initial load current I_L being sustained by the load inductance and being allowed to fall instantaneously to zero as the load voltage reverses with a resistive load. In each case, the thyristor current waveform i_{Th4} is governed by a sinusoidal envelope between instants t_0 and t_2 as demonstrated by the transient analysis of the essential circuit of Fig. 7.57(a).

For the resistive load case I_0 is the value of I_L at $t = 0$, instant t_0. Subsequently, I_L is zero. With voltage E on C_4 due to the initial charge we have

$$E = L_4 \frac{di}{dt} + \frac{1}{C_1} \int i_1 \, dt$$

$$0 = L_4 \frac{di}{dt} + \frac{1}{C_4} \int (i - i_1) \, dt$$

where

$$i = i_{Th4}, \; i_1 = i_{C1}, \; C_1 = C_4 = C, \; L_4 = L.$$

Figure 7.57 Effective circuit: (a), and (b): alternative scenarios for current development in the gated thyristor Th$_4$ after achieving turn-off of Th$_1$ with (i) load current falling instantaneously from I_0 to zero, (ii) load current sustained at I_L.

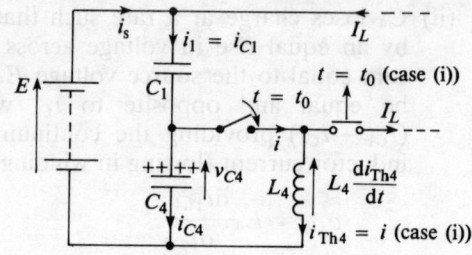

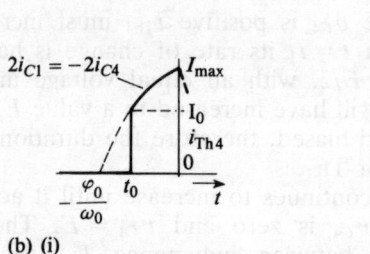

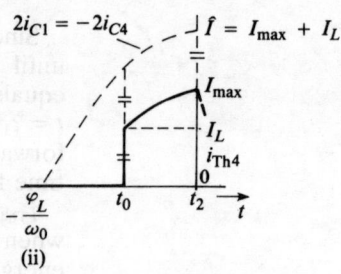

(b) (i) (ii)

Transforming to the Laplace form

$$\left. \begin{array}{l} \dfrac{E}{s} = L[s\bar{i} - I_0] + \dfrac{1}{C}\left(\dfrac{\bar{i}_1}{s} + 0\right) \\[1em] 0 = L[s\bar{i} - I_0] + \dfrac{1}{C}\left(\dfrac{\bar{i} - \bar{i}_1}{s} + \dfrac{V_0 C}{s}\right) \end{array} \right\}$$

where V_0 is the initial voltage on capacitor C_4, equal to $-E$. Solution gives

$$\bar{i}_1 = \bar{i}/2$$

and

$$\bar{i} = \dfrac{E\omega_0}{\omega_0 L(s^2 + \omega_0^2)} + \dfrac{I_0 s}{(s^2 + \omega_0^2)}$$

with

$$\omega_0 = \left(\dfrac{1}{2LC}\right)^{1/2}.$$

Thus

$$i_{C1}(t) = -i_{C4}(t) = i_{Th4}(t)/2$$

and

$$i_{Th4}(t) = I_{max}\sin(\omega_0 t + \phi_0)$$

where

$$I_{max} = \left[\left(\dfrac{E}{\omega_0 L}\right)^2 + I_0^2\right]^{1/2}$$

and

$$\phi_0 = \tan^{-1}\left[\dfrac{\omega_0 L I_0}{E}\right].$$

[7.67]

Synchronous machines 359

If the load current is sustained (due to inductance) at the value I_L the current distribution changes, with the total capacitor current increasing to offset the load current. Thus

$$2i_{C1} = -2i_{C4} = i_{Th4} + I_L = (I_{max} + I_L)\sin(\omega_0 t + \phi_L). \quad [7.68]$$

The total capacitor current waveform follows a sinusoidal waveform with v_{C4} falling to zero when, at $t = t_2$, i_{Th4} attains I_{max} and $2i_{C1} = -2i_{C4} = I_{max} + I_L$.

At $t = t_0$, $i_{Th4} = I_L$ with i_{C1} and $-i_{C4}$ of the same value. This contrasts with $i_{C1} = -i_{C4} = I_0/2$ at $t = t_0$ if the load current is not sustained. The current waveforms of Fig. 7.57(b) emphasise the sinusoidal nature of the capacitor currents. If I_0 (or I_L) is known and I_{max} is specified, bearing in mind the need to keep the commutation losses low, the commutation time may be deduced from an essentially sinusoidal di_{Th4}/dt. With turn-off time specified in microseconds for particular devices, a value for ω_0 may be deduced. An appropriate value for L may be established, either from analogous expressions to eqns [7.67] or through accounting for the additional energy supplied by the source in raising the magnetic field energy from $\frac{1}{2}LI_L^2$ (or $\frac{1}{2}LI_0^2$) to $\frac{1}{2}LI_{max}^2$ as $\int_0^{(\pi/2-\phi)/\omega_0} Ei_{C1}\,dt$. With inductive load i_{c1} is given by $i_{c1} = \frac{1}{2}(I_{max} + I_L)\sin(\omega_0 t + \phi_L)$.

Worked example 7.3 (i) Deduce appropriate values for the commutation circuit elements of a complementary mutually coupled impulse-commutation circuit (McMurray–Bedford) applied to a 6 Ω resistive load fed via a single-phase bridge inverter from a 240 V d.c. supply. Assume ideal commutation elements and a thyristor turn-off time of 30 μs, limiting the thyristor current to 150% of the steady-state value. Calculate also the commutation losses for 100 Hz operation and the maximum rate of rise of voltage across each thyristor after turn-off.
(ii) Repeat for the case where load inductance maintains the load current constant during the commutation period.

Solution (i) The circuit diagram is shown in Fig. 7.57(a). With *resistive* load

$$i_{Th4} = I_{max}\sin(\omega_0 t + \phi_0) = \left[\left(\frac{E}{\omega_0 L}\right)^2 + I_0^2\right]^{1/2}\sin(\omega_0 t + \phi_0)$$

and $I_0 = 240/6 = 40$ A. Therefore $i_{Th4} = 2i_{C1} = -2i_{C4} = 40$ A at $t = 0$, rising to $I_{max} = 1.5 \times 40 = 60$ A at $(\omega_0 t + \phi_0) = \pi/2$. Thus

$$I_{max} = 60 = \left[\left(\frac{240}{\omega_0 L}\right)^2 + 40^2\right]^{1/2},$$

giving $\omega_0 L = 5.367$ Ω. Now, $di_{Th4}/dt = I_{max}\omega_0\cos(\omega_0 t + \phi_0)$ and has initial value $I_{max}\omega_0\cos\phi_0$. ϕ_0 is such that at $t = 0$, $i_{Th4} = 40 = 60\sin\phi_0$ and therefore $\phi_0 = 0.7297$ rad. Hence, $I_{max}\omega_0\cos\phi_0 = 0.7494 I_{max}\omega_0$. di_{Th4}/dt has *half* its initial value at t_1 when $I_{max}\omega_0\cos(\omega_0 t_1 + \phi_0) = 0.3747 I_{max}\omega_0$, i.e. when $(\omega_0 t_1 + 0.7297) = \cos^{-1} 0.3747$, i.e. when $\omega_0 t_1 = 1.1867 - 0.7297$. For a turn-off time t_1 of 30 μs, $\omega_0 = (1.1867 - 0.7297)/30 \times 10^6 = 15\,230$ rad s^{-1}. Therefore $L = \omega_0 L/\omega_0 = 5.367/15\,230 = 352$ μH. Also, $C = (2\omega_0^2 L)^{-1} = 6.13$ μF. Energy dissipated in the diode per commutation = $\frac{1}{2}LI_{max}^2 = 0.636$ J. With 200 commutations per second at each side of the bridge the energy dissipated per second = $400 \times 0.636 = 254$ W. (This compares with a nominal inverter output power of $240^2/6 = 9.6$ kW). The maximum rate of rise of voltage across Th_1 after turn-off = dv_{C1}/dt at $i_{C1} = I_{max}/2$. Therefore max $dv_{Th}/dt = 30/6.13 \times 10^6 = 4.89$ V/μs.

(ii) With the inductive load sustaining $I_0 \equiv I_L$ at 40 A, $i_{Th4} = 2i_{C1} - I_L$, where $2i_{C1} = (I_{max} + I_L)\sin(\omega_0 t + \phi_L)$. For $I_{max} = 60$ A and $I_L = 40$ A, $2i_{C1} = 100\sin(\omega_0 t + \phi_L)$. At $\omega_0 t = 0$, $2i_{C1} = 100\sin\phi_L = 80$, giving $\sin\phi_L = 0.8$, $\phi_L = 0.927$ rad and $\cos\phi_L = 0.6$.

The initial rate of rise of i_{Th4}

$$= \frac{d}{dt}[100\sin(\omega_0 t + \phi_L) - 40]_{t=0} = 100\omega_0\cos\phi_L = 60\omega_0.$$

With half the initial rate of rise of $i_{Th4} = 30\omega_0$ occurring at t_1 where $100\omega_0\cos(\omega_0 t_1 + \phi_L) = 30\omega_0$, i.e. when $(\omega_0 t_1 + \phi_L) = \cos^{-1} 0.3 = 1.266$. Hence $\omega_0 t_1 = 1.266 - 0.927 = 0.339$ and $\omega_0 = 0.339/30 \times 10^6 = 11\,303$ rad s^{-1}.

By analogy with eqn [7.67] for i_{Th4}, recognising that this equation identifies with the oscillatory capacitor current,

$$2i_{C1} = (I_{max} + I_L)\sin(\omega_0 t + \phi_L)$$

where

$$(I_{max} + I_L)^2 = \left(\frac{E}{\omega_0 L}\right)^2 + (2I_L)^2$$

and

$$\phi_L = \tan^{-1}\frac{2I_L}{\left[\left(\frac{E}{\omega_0 L}\right)^2 + 2I_L^2\right]^{1/2}}.$$

Therefore

$$\left(\frac{E}{\omega_0 L}\right)^2 = (I_{max} + I_L)^2 - (2I_L)^2 = I_{max}^2 + 2I_{max}I_L - 3I_L^2.$$

$$\left(\frac{240}{\omega_0 L}\right)^2 = 60^2 + 2 \times 60 \times 40 - 3 \times 40^2 = 3600.$$

Thus $\omega_0 L = 240/60 = 4$ and $L = 4/11\,303 = 354$ μH. Also, $C = (2\omega_0^2 L)^{-1} = (2 \times 4 \times 11\,303)^{-1} = 11.06$ μF. The maximum rate of rise of voltage across Th_1 after turn-off with $i_{C1} = I_{max}/2$ is $30/11.06 \times 10^6 = 2.71$ V/μs.

Note: Dissipation of the trapped energy $\frac{1}{2}LI_{max}^2$ within a diode at each commutation may cause thermal problems and limit the operating frequency. The situation may be avoided and efficiency raised by incorporating an additional pair of diodes and a centre-tapped (feedback) transformer to allow energy recovery by the source.

7.31 Amplitude Control of Inverter Output Voltage

In many inverter applications it is desirable to control not only the *frequency* of the alternating voltage but also the *amplitude*. The most obvious method is simply to adjust the voltage level of the d.c. link supplying the inverter, either by phase control of the a.c./d.c. converter energising the d.c. link (Sects 6.14 and 6.15) or by the use of a diode rectifier followed by a d.c. chopper (Sect. 6.13). In the former case, reverse power flow through the converter is possible, should the inverter be required for a drive application in which net regeneration over extended periods of time is likely. As we have already noted in Section 7.29, inverter operation with reactive loads involves brief periods during each cycle of d.c. voltage generation when energy is fed back to the d.c. source. Such periods of inversion would be used to recharge the d.c. link capacitance associated with the d.c. source, so reducing the demand on the a.c. source energising the link.

The output voltage of an inverter may alternatively be made to vary by maintaining a constant input voltage and generating a quasi-square waveform with variable firing displacement angle α (see Worked Examples 7.4 and 7.5) or by *chopping* the output voltage in much the same manner as was described in Section 6.13 for the control of the average value of a d.c. voltage applied to the armature winding of a d.c. machine. The inverter output voltage in the latter case then becomes a series of pulses, each of uniform height equal to the d.c. source voltage and of width relative to the following space proportional to the *average* value of voltage required. If a square or quasi-square waveform is uniformly *pulse-width modulated* in this way, the average value of the a.c. voltage will be directly proportional to the mark/space ratio of the pulse train. This type of waveform is alternatively known as a *notched* waveform. In most a.c. drive applications the inverter fundamental output voltage and frequency should be of fixed ratio over most of the working range, in order to maintain a near-constant radial flux density in the machine airgap. Such a facility is automatically provided by a scheme of pulse-width modulation in which the pulse width is kept constant and the spaces of zero voltage are allowed to increase in duration as the frequency falls. The modulation is achieved by simple gating of the thyristors with a modulation frequency ≈ 20 times the fundamental of the output waveform. As with the quasi-square waveform of Section 7.29.1, the load current freewheels via a diode and a thyristor incorporated in opposite half-bridges of the single-phase converter and upper or lower busbar during the brief 'off' intervals. The commutation losses are increased with *thyristor* inverters employing pulse-width modulation. This is less of a disadvantage with lower power *transistor* inverters of maximum rating $\approx 20\,\text{kVA}$, which employ transistors in the switching mode, controlled by base drive, and which are generally less efficient than the thyristor alternative but have no commutation circuit complications.

The advantages of pulse-width modulation may be enhanced if the pulse repetition frequency is kept constant but the duration of each fixed-height pulse is made proportional to the *average* value of the appropriate part of a sinusoidal waveform between the instants which define the midpoints of adjacent 'OFF' periods. Thus in *sinusoidal* pulse-width modulation the pulse duration is modulated sinusoidally. Ideally, the pulse repetition frequency should be a large integer multiple of the modulating frequency, so that the average effect of the train of supply voltage pulses on the load is similar to that of a sine wave of voltage having the modulation frequency. Fourier analysis shows this expectation to be valid, with the lower frequency harmonics virtually eliminated, although the higher harmonics may be significant.

Worked example 7.4 Determine the Fourier series for the output voltage waveform of the single-phase bridge inverter circuit shown in Fig. 7.51, supplied from a constant voltage d.c. source E. The thyristors are gated in the sequence Th_1, Th_2, Th_4, Th_3 with the gating of Th_2 and Th_3 delaying that of Th_1 and Th_4, respectively, by the time equivalent of α radians, whereas the gating of Th_1 and Th_4 is separated by π radians. Hence show that the amplitude of the fundamental component of output voltage is given by $\hat{V}_1 = (4/\pi)E\cos v\alpha/2$, with that of the harmonic components being given generally by $\hat{V}_v = (4/v\pi)E\cos(v\alpha/2)$, where v is the harmonic number (odd harmonics only are present).

Solution Figure 7.52(b) illustrates the quasi-square waveform, drawn for $\alpha = 133.75°$, such that its fundamental component is zero-valued, positive-going at the chosen reference point $\omega t = 0°$.

Due to the *inverted symmetry* about the $\omega t = 0°$ axis, i.e. $f(\omega t) = -f(-\omega t)$, it is recognised that the Fourier series may be expressed in terms of sinusoidal quantities only, with the vth component being given generally by

$$v_v = B_v \sin v\omega t$$

where

$$B_v = \frac{1}{\pi}\int_{-\pi}^{\pi} v(\omega t)\sin v\omega t\,d\omega t = \frac{2}{\pi}\int_{-\pi/2}^{\pi/2} v(\omega t)\sin v\omega t\,d\omega t.$$

Since the function has *half-wave symmetry* i.e. $f(\omega t) = -f(\omega t + \pi)$, even harmonics will be absent. For the fundamental

$$B_1 = \frac{4}{\pi}E\left[\int_{\alpha/2}^{\pi/2}\sin\omega t\,d\omega t\right] = \frac{4}{\pi}E\cos\left(\frac{\alpha}{2}\right). \qquad [7.69]$$

For the vth harmonic (odd only)

$$B_v = \frac{4}{\pi}E\left[\int_{\alpha/2}^{\pi/2}\sin v\omega t\,d\omega t\right] = -\frac{4}{v\pi}E\left[\cos v\frac{\pi}{2} - \cos v\frac{\alpha}{2}\right].$$

Since v is odd

$$B_v = \frac{4E}{v\pi}\cos v\frac{\alpha}{2}. \qquad [7.70]$$

Equation [7.69] may be rewritten in a form which illustrates directly that the fundamental component of load voltage is proportional to the *sine* of the *pulse width*, i.e.

$$B_1 = \frac{4}{\pi}E\sin\frac{(\pi - \alpha)}{2},$$

where E is the pulse height.

Variation of the α thus provides for variation in the inverter output voltage by *pulse-width control*.

Worked example 7.5 Determine the firing displacement angle α of a quasi-square wave single-phase bridge inverter such that the amplitude of the fundamental component of output voltage is 50% of the d.c. supply voltage. List the corresponding harmonic component amplitudes of output voltage to the 13th.

Solution From eqn [7.69] $B_1 = 0.5E = (4/\pi)E\cos(\alpha/2)$. Therefore $\alpha = 2\cos^{-1}(\pi/8) = 133.75°$. From eqn [7.70], for the vth harmonic, $B_v = (4E/v\pi)\cos(133.75v°/2)$. Thus, for the harmonic series of output voltage, $B_1 = 0.5E$, $B_3 = -0.3972E$, $B_5 = 0.2296E$, $B_7 = -0.0566E$, $B_9 = -0.0667E$, $B_{11} = 0.1115E$, $B_{13} = -0.0843E$.

The lower harmonics are seen to be comparable with the fundamental component of output voltage. The harmonic content is increased further relative to the fundamental if the pulse width is reduced. *Pulse-width modulation* employed in Worked Example 7.7, achieves output voltage control with a much reduced lower harmonic content but at the expense of increased higher harmonics.

Worked example 7.6 The fundamental component amplitude of the quasi-square output voltage of a single-phase bridge thyristor inverter is reduced to 50% of the d.c. power supply voltage E by pulse-width control. The inverter is connected to a motor load which draws a sinusoidal current of 100 A peak value, lagging by 60° the fundamental component of load voltage. Deduce the waveforms of thyristor, diode and source currents. Calculate also the mean value of current supplied by the source.

Solution Using the result of Worked Example 7.5, the firing displacement angle α is calculated as 133.75° and Fig. 7.52(c) shows the corresponding load voltage waveform drawn to a time reference for which, at $\omega t = 0$, the load current i_L is zero-valued, going positive. Also shown is the device conduction sequence (incorporating by implication the thyristor gate-firing sequence) and the corresponding current waveforms. Similar devices *in each half-bridge* carry similar currents displaced by the time equivalent of 180°.

The mean value of current in the source is given by

$$i_{s\,av} = (1/\pi)\int_0^\pi i_s(\omega t)\,d\omega t = (1/\pi)\int_{6.877°}^{53.123°} 100\sin\omega t\,d\omega t$$

$$= (100/\pi)[\cos 6.877° - \cos 53.123°] = 12.5\text{ A}.$$

This value of mean source current is confirmed by comparing the average power supplied by the d.c. source $12.5E$ W with the average fundamental frequency power absorbed by the motor load expressed in r.m.s. quantities as $V_1 I_1 \cos\phi = (0.5E/\sqrt{2})(100\sqrt{2})\cos 60° = 12.5E$ W.

Worked example 7.6A Determine the mean value of current flowing in the thyristors and diodes of the single-phase bridge inverter circuit described in Worked Example 7.6.

Solution

$$i_{Th1\,av} = i_{Th4\,av} = (100/2\pi)\int_0^{53.123°} \sin\omega t\,d\omega t$$

$$= (100/2\pi)[1 - \cos 53.123°] = 6.3646\text{ A}.$$

By similar reasoning

$i_{Th2\,av} = i_{Th3\,av} = 31.7165$ A,

$i_{D1\,av} = i_{D4\,av} = 25.4664$ A,

$i_{D2\,av} = i_{D3\,av} = 0.1145$ A.

7.32 Three-phase Voltage–source Bridge Inverter

Most high-power inverter circuits are configured as three-phase circuits employing the bridge format. Three half-bridges are therefore required, with the semiconductor switching off each half-bridge arranged sequentially, such as to develop appropriate alternating line voltages between the three load terminals defined by the common semiconductor junctions of each half-bridge. Thus the general layout is similar to that of the three-phase naturally commutated bridge converter of Section 6.15. It has been shown that the natural commutation process defined by the sinusoidal a.c. source voltages applicable in that context provided for each thyristor to conduct over 120° intervals. Such an operating sequence for a three-phase system in which load line current would cease for 60° in each half-cycle of inverter output may be achieved with an artificially commutated inverter. It is more usual, however, to operate force-commutated converters in such a way that permits each anti-parallel thyristor/diode pair to conduct for 180° intervals, and the current in each phase of the three-phase load to be continuous. Force-commutation circuits such as the McMurray and McMurray–Bedford versions discussed in Section 7.30 may now be used, as one thyristor/diode pair of each half-bridge is conducting at any instant and the changeover switch characteristic of these circuits is appropriate to this mode of three-phase inverter operation.

The extended 180° conduction periods provide for a lower ratio of peak thyristor current to r.m.s. current than is the case with 120° conduction, but there is no difference, apart from phase, in the harmonic spectra of each waveform. The Fourier series for the line voltage waveform of amplitude E with 180° conduction may be shown to be

$$v_L(t) = \frac{4}{\pi}E[\sin \omega t - \tfrac{1}{5}\sin 5\omega t - \tfrac{1}{7}\sin 7\omega t + \tfrac{1}{11}\sin 11\omega t + \tfrac{1}{13}\sin 13\omega t - \tfrac{1}{17}\sin 17\omega t - \tfrac{1}{19}\sin 19\omega t + \ldots].$$

Figure 7.58(a) shows the essential routing of the switched supplies to a three-phase static load provided by a three-phase bridge inverter. Each arm of each half-bridge includes a thyristor/diode pair connected in anti-parallel. The necessary commutation circuits are not shown. Figure 7.58(b) shows the requisite thyristor gating sequences and the resulting quasi-square line voltage waveforms. The *line current* waveforms correspond to a static resistive/inductive load and have zero-crossings non-coincidental with those of the derived phase voltages. The role of the diodes in conducting reverse current during periods of inversion is clearly seen. The phase voltage and line current waveforms are predicted on the assumption that mutual coupling does not

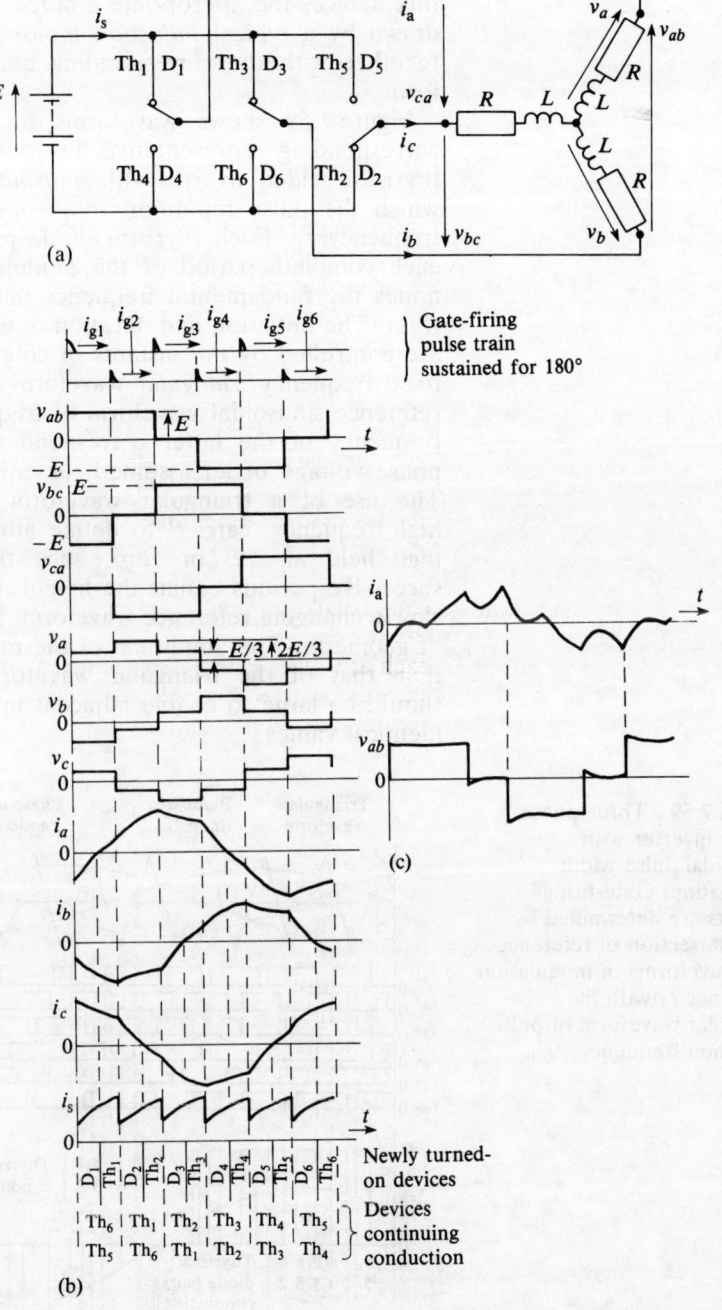

Figure 7.58 Three-phase bridge inverter: (a) essential representation by changeover switches; (b) gate current requirements and idealised waveforms for static R/L load; and (c) actual line current and voltage waveforms for an (induction) motor load.

exist between phases and would be inappropriate for a rotating machine load, although the impressed *line voltage* waveforms defined solely by the d.c. source and the switching sequence are valid. A machine load would respond differently to each harmonic contained within the line voltage wherein the 20% fifth and 14% seventh harmonics are dominant. The resulting harmonic currents are responsible for supersynchronous rotating fields whose radial flux induces the appropriate e.m.f.s. The waveform of line current drawn by a typical *induction motor* load is shown in Fig. 7.58(c) together with the corresponding quasi-square line voltage waveform.

Figure 7.59 shows waveforms for gate-firing pulse-trains and corresponding representative line output voltage waveforms for a thyristor bridge inverter with *sinusoidal pulse-width modulation* for which the pulse repetition frequency kf_1 is 10× the modulation frequency f_1. Each thyristor/diode pair is then gated 10 times in each complete period of the modulating waveform which determines the fundamental frequency and phase of the output waveform. The initiation and duration of each burst of gate-firing pulses are controlled by the instants of coincidence of a fixed-amplitude, fixed frequency *triangular* waveform having frequency kf_1, with a reference sinusoidal waveform of frequency f_1. The amplitude and frequency of the latter correspond to the required fundamental phase voltage of an assumed, or equivalent, star-connected load. The use of a triangular waveform for the constant-amplitude, high-frequency 'carrier' to define alternate periods of output voltage held at $\pm E$ or zero, such that the *average* value over successive periods equals the height above zero of an intersecting, slowly changing reference waveform, is justified as a simple matter of geometry. The amplitude of the reference sinusoid must be less than that of the triangular waveform, and the frequency ratio should be large to enable adjacent intersections to occur at nearly identical values.

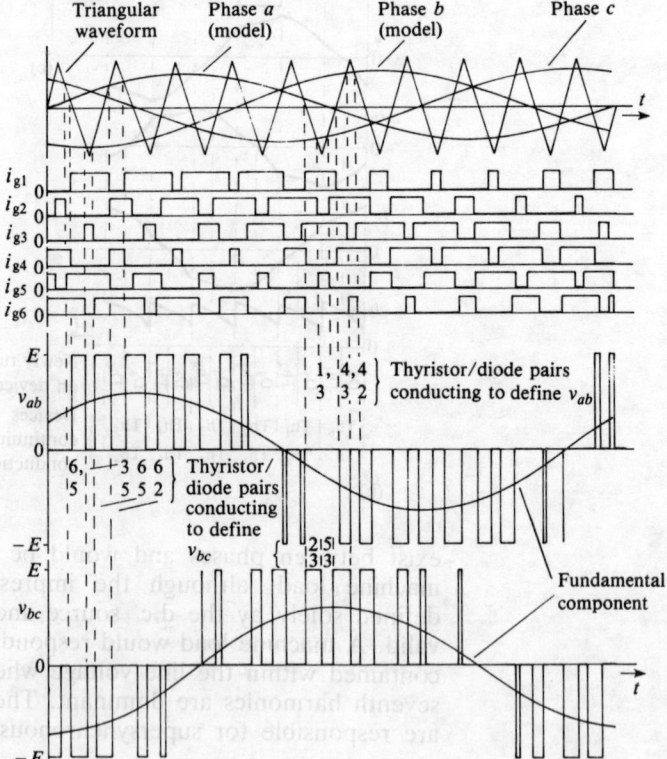

Figure 7.59 Three-phase bridge inverter with sinusoidal pulse-width modulation. Gate-firing instants are determined by the intersection of reference sine waveforms of modulation frequency f_1 with the triangular waveform of pulse repetition frequency $10f_1$.

Worked example 7.7 A single-phase, sinusoidally pulse-width modulated inverter generates five pulses of uniform height E during each half-cycle of fundamental output voltage. The switching instants are as defined in Fig. 7.60, chosen to correspond to a *peak* fundamental voltage nominally 50% of the d.c. supply voltage E. Use Fourier analysis to determine the amplitude of the harmonic components of output voltage as far as the 15th.

If the inverter supplies a motor load which draws a sinusoidal current of 100 A peak, lagging by 60° the fundamental component of supply voltage, identify the sequence of current-carrying devices and evaluate the mean value of current supplied by the d.c. source.

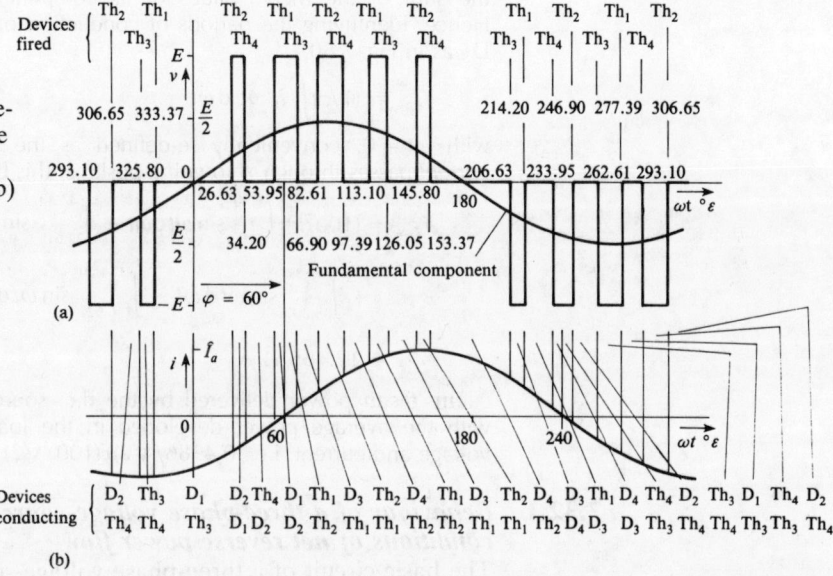

Figure 7.60 Sinusiodal pulse-width modulated single-phase inverter output voltage waveform (a) appropriate to Worked Example 7.7 with (b) the corresponding line current of fundamental frequency lagging the fundamental of the output voltage by 60°. The firing sequence and conducting devices in the circuit of Fig. 7.51(a) are also identified.

Solution Recognising that, due to symmetry, the voltage waveform of Fig. 7.60 may be expressed with respect to the chosen $\omega t = 0$ reference as a Fourier series containing only sine terms of odd harmonic frequencies,

$$v_v = B_v \sin v\omega t$$

where

$$B_v = (4/\pi)\int_0^{\pi/2} v(\omega t)\sin v\omega t \, d\omega t$$

$$= (4E/\pi)\left[\int_{26.63°}^{34.20°} \sin v\omega t \, d\omega t + \int_{53.95°}^{66.90°} \sin v\omega t \, d\omega t + \int_{82.61°}^{90°} \sin v\omega t \, d\omega t\right]$$

$$= (4E/\pi)[-\cos v34.20° + \cos v26.63° - \cos v66.90° + \cos v53.95° - \cos v90° + \cos v82.61°]$$

Substituting odd values for v gives

$B_1 = \hat{V}_1 = 0.4986E,$

$B_3 = \hat{V}_3 = 0.000\,647E,$

$B_5 = \hat{V}_5 = (-)0.000\,523E,$

$B_7 = \hat{V}_7 = (-)0.000\,525E,$

$B_9 = \hat{V}_9 = (-)0.044\,56E,$

$B_{11} = \hat{V}_{11} = (-)0.360\,72E,$

$$B_{13} = \hat{V}_{13} = 0.360\,65 E,$$

$$B_{15} = \hat{V}_{15} = 0.043\,147 E.$$

The thyristor switching sequence defining the *voltage* waveform is indicated in Fig. 7.60 with the device designation being as in Fig. 7.51. Also shown is the sequence of device *conduction* pertaining to the load current waveform described in terms of a fundamental component alone. The implicit assumption is that the motor load presents infinite impedance to harmonic applied voltages. Referring to Fig. 7.51, the source current $i_s = i_{Th1} + i_{Th3} - i_{D1} - i_{D3}$. The *mean* value of source current is equal to the sum of the mean values of its components, having regard to sign. Hence, identifying the periods of conduction of devices Th_1, Th_3, D_1 and D_3 as in Fig. 7.60,

$$i_{s\,av} = (1/\pi)\int_0^\pi i_s(\omega t)\,d\,\omega t,$$

with $\omega t = 0$, conveniently re-defined as the instant at which the *load current* passes through zero going positive, this becomes

$$i_{s\,av} = (100/\pi)\Bigg[\int_{0°}^{6.90°} \sin \omega t\,d\omega t + \int_{22.61°}^{37.39°} \sin \omega t\,d\omega t + \int_{53.10°}^{66.05°} \sin \omega t\,d\omega t$$

$$+ \int_{85.80°}^{93.37°} \sin \omega t\,d\,\omega t - \int_{146.62°}^{154.20°} \sin \omega t\,d\,\omega t - \int_{173.95°}^{180°} \sin \omega t\,d\omega t\Bigg]$$

$$= 12.465\,51\ \text{A}.$$

The mean power delivered by the d.c. source $12.465\,51 E$ W correlates with the average power developed in the load due to the fundamental voltage and current, i.e. $(0.4986/\sqrt{2})E(100/\sqrt{2})\cos 60°$ W.

7.32.1 Behaviour of a three-phase voltage-source inverter under conditions of net reverse power flow

The basic circuit of a three-phase voltage-source inverter connected to an a.c. machine is shown in Fig. 7.61. Commutation circuits have not been incorporated and the machine is simply modelled as

Figure 7.61 Reverse power flow condition for three-phase bridge inverter with synchronous motor load: (a) circuit diagram excluding commutation circuits; (b) fundamental frequency equivalent circuit/phase of machine load and phasor diagrams for all modes of operation; and (c) waveforms and device conduction sequences for reverse power flow (generation) (i) underexcited, (ii) overexcited.

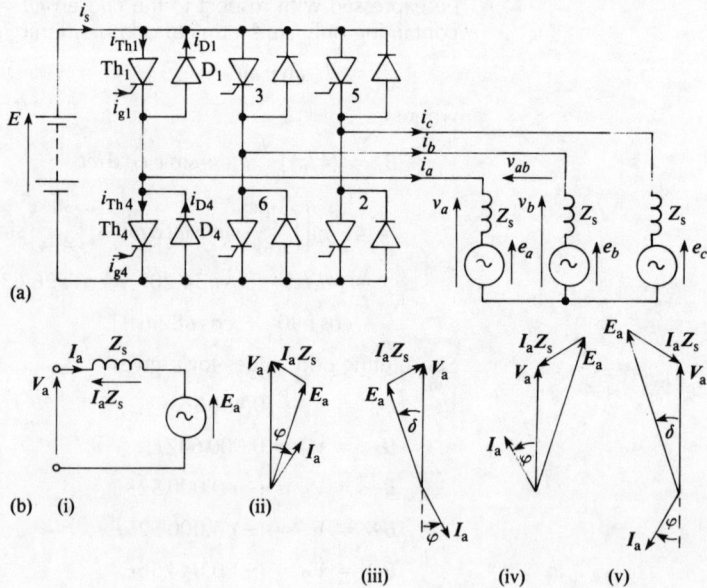

Synchronous machines

a fundamental-frequency source of e.m.f. in series with inductive impedance. There is, in consequence, an implicit assumption that the impedance presented by the machine to harmonics in the line voltage waveform is so high that the current flowing is predominantly of fundamental frequency.

Having discounted the effects of the voltage harmonics it is then appropriate to represent the line and phase voltages of the machine by their fundamental components. In Section 7.29 we considered the case of a *single-phase* inverter supplying loads at both lagging and leading power factors, represented by their fundamental-frequency current components, and established that, whereas a *leading* power factor load provided for the natural commutation of current between a conducting thyristor and its anti-parallel diode, with a *lagging* power factor load this was not so. Artificial commutation was necessary to interrupt the flow of

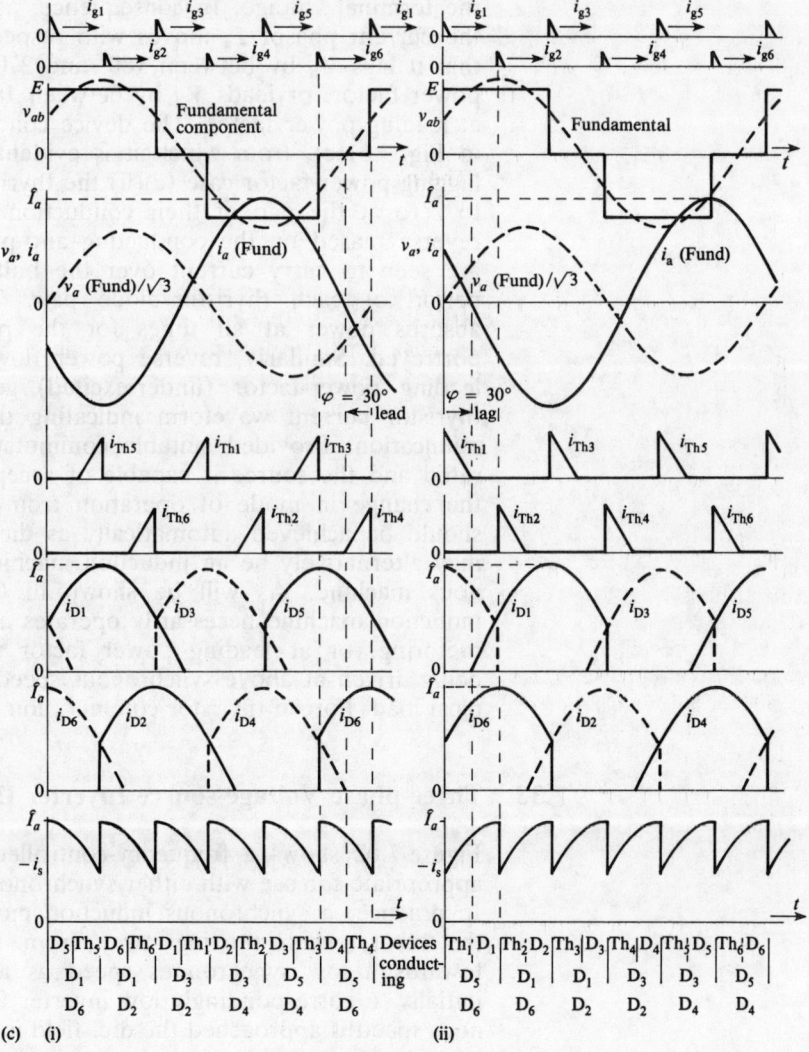

thyristor current and to establish a reverse-bias voltage for sufficient time to ensure recovery of thyristor capability to block re-application of forward bias prior to gating. The three-phase inverter with motor load behaves in essentially the same way. The waveforms of Fig. 7.61(c) relate to the operation of a three-phase inverter when the load is regenerating, i.e. providing net power flow into the d.c. source. The line voltage waveform shown corresponds to a quasi-square wave inverter, but the fundamental waveform to which it is equated would be even better justified using sinusoidal pulse-width modulation. A synchronous machine load is considered to be operating as a *generator* with respectively leading and lagging power factor, corresponding to machine d.c. field under- and over-excitation.

The transition from motor to generator operation is achieved quite simply by a reversal of rotor load torque which temporarily accelerates the rotor such that the machine armature e.m.f. phasor moves from a lagging position to a leading position with respect to the terminal voltage. In consequence, as depicted in Fig. 7.61(b), the current phasor I_a moves with respect to the V_a reference so that it lags V_a by between 180° and 270° if generating at lagging power factor, or leads V_a by between 180° and 270° if generating at leading power factor. The device conduction sequence is shown in Fig. 7.61(c), from which it is evident that in the over-excited lagging power-factor case (c)(ii) the thyristor currents fall naturally to zero at the end of their conduction periods, being thereafter reverse biased by the conducting anti-parallel diode. The diodes are seen to carry current over the bulk of the 180° conduction period of each thyristor/diode pair. The d.c. voltage source absorbs power at all times for the particular load conditions portrayed. Similarly, reverse power flow is demonstrated by the leading power-factor (under-excited) generator case (c)(i), the thyristor current waveform indicating the need for forced communication. Provided suitable commutation circuits are incorporated and the source is capable of accepting reverse power flow, the change in mode of operation from inversion to rectification should be achieved automatically as dictated by the load, which may alternatively be an induction machine rather than a synchronous machine. As will be shown in Chapter 8, however, the induction machine necessarily operates at lagging power factor, if motoring, or at leading power factor when generating through being driven at above synchronous speed by its connected mechanical load. Forced thyristor commutation is therefore a necessity.

7.33 Three-phase Voltage-source Inverter Drive Systems

Figure 7.62 shows a frequency-controlled open-loop drive system appropriate for use with either synchronous or induction machines. In practice, a synchronous induction motor may be used so that the rotor, when moving away from standstill may accelerate towards a low synchronous speed as an induction motor with, initially, a correspondingly low inverter frequency. When synchronous speed is approached the d.c. field is excited and synchronising torque is developed to pull the machine into step. Provided that the torque limits for stability are not exceeded, excitation dependent, the machine runs subsequently in synchronism with the inverter frequency, controlled by the set-speed regulator via a *voltage-to-frequency converter* which sets the frequency of the inverter firing circuit input signals. Comparison of the d.c. link voltage with the set-speed voltage provides for maintenance of a constant armature voltage/frequency ratio. The d.c. link is likely to incorporate considerable shunt capacitance to smooth the d.c. link voltage and hold it substantially constant in the short term, over a few cycles of minimum inverter frequency. If the machine becomes a net generator of electrical power the reversed power flow will cause the d.c. link voltage to rise. Suitable control of the gate-firing pulses on the naturally commutated converter should enable this to invert and return energy to the three-phase a.c. system.

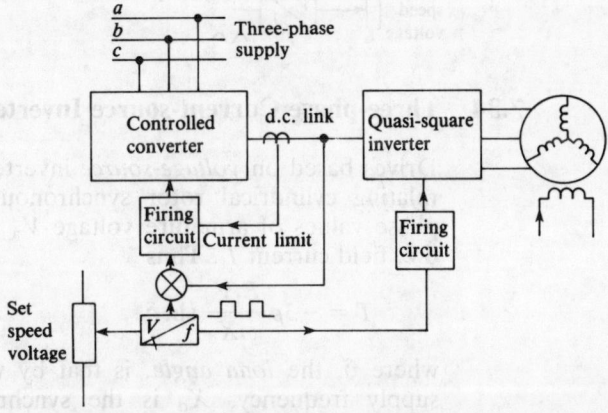

Figure 7.62 Open-loop voltage-source a.c. motor drive with quasi-square inverter and regeneration facility.

An alternative frequency-controlled open-loop drive system employing a sinusoidal pulse-width modulation inverter is shown in Fig. 7.63. The pulse-width modulation provides for automatic armature voltage/frequency compensation over a range of operating frequencies with a constant d.c. link voltage. There is, therefore, no need for a controlled converter to provide the d.c. link. If, however, dynamic braking is likely on occasions, a braking resistor may be installed in parallel with the smoothing capacitor for the d.c. link. The effective value of a fixed braking resistor may be controlled by means of a series-connected thyristor. Pulse-width modulated inverter systems are preferred to the quasi-square alternatives at low frequencies where the commutation losses are relatively low and the low harmonic content is a distinct advantage.

Closed-loop control is seldom necessary with synchronous machine drives unless the torque stability limit is likely to be exceeded. To reduce the likelihood of this, constraints may be incorporated to limit the permissible rate of change of the controlling frequency in open-loop drives. Schemes incorporating closed-loop control make use of a rotor shaft position indicator to define the instants at which the inverter thyristors are gated, ensuring stability under all conditions of load. Whereas in open-loop control schemes for synchronous motors incorporating voltage-source inverters the load angle δ changes in accordance with the torque demand, in closed-loop schemes δ may be fixed, and fixed at a high value to maximise the torque for given field current and link voltage.

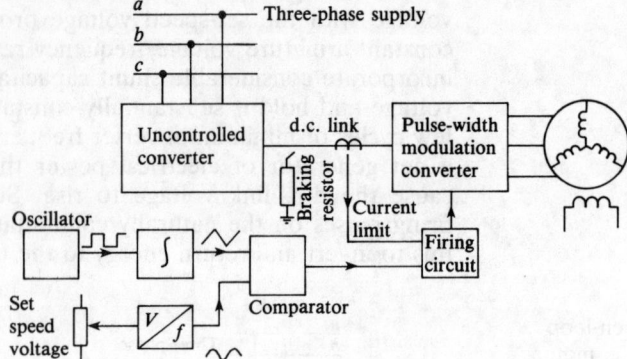

Figure 7.63 Open-loop voltage-source a.c. motor drive with sinusoidal pulse width modulation.

7.34 Three-phase Current-source Inverter Drive Systems

Drives based on *voltage-source* inverters make use of eqn [7.34] in relating cylindrical rotor synchronous machine torque to r.m.s. phase values of armature voltage V_a and e.m.f. E_f established by d.c. field current I_f. Thus

$$T = -3p \frac{E_f V_a}{\omega X_d} \sin \delta \qquad [7.34]$$

where δ, the *load angle*, is that by which E_f leads V_a, ω is the supply frequency, X_d is the synchronous reactance per phase (proportional to frequency) and p is the number of pole pairs. Armature resistance is neglected.

E_f is also proportional to frequency and hence, if the field current is maintained constant,

$$|T| \propto \frac{V_a I_f}{\omega} \sin \delta.$$

Thus, if the d.c. link voltage is made proportional to ω

$$|T| \propto I_f \sin \delta. \qquad [7.71]$$

In complementary fashion, drives based on *current-source* inverters make use of eqn [7.16] which relates torque to the sinusoidally distributed *radial* field components with peak values $\hat{H}_{nr}$ due to the armature current and $\hat{H}_{nf}$ due to the d.c. excited field winding.

$$T = -\mu_0 \pi p \, blg \hat{H}_{nr} \hat{H}_{nf} \sin \psi, \qquad [7.16]$$

where ψ is the angle by which the armature winding field lags in space the d.c. 'field' winding field.

The fact that, in practice, the d.c. field will be on the rotor side of the airgap, and the armature winding on the stator is irrelevant. Hence, for a given machine

$$|T| \propto I_f I_a \sin \psi, \qquad [7.72]$$

where I_a is the r.m.s. value of the fundamental component of armature current per phase, proportional to the d.c. link current.

The *torque angle* ψ is a space-phase angle, by definition, but it also appears on the time-phasor diagram of the loaded machine, Fig. 7.14. In an inverter-fed machine with shaft position indication, the torque angle may be controlled directly by relating the instants of gating the thyristors which control the stator armature winding to the known position of the d.c. field poles carried on the rotor. Thus a closed-loop drive system with ψ fixed and constant field current has machine torque (at any speed) proportional to the d.c. link current. The varying power demands of constant torque at varying speeds are accommodated by changes in the d.c. link voltage.

Neglecting armature resistance, machine torque for given values of armature and field current is maximised by maintaining $|\psi|$ at 90°, which is the inherent torque angle of a conventional d.c. machine with brushes set to commutate coils having sides located on the quadrature axis. Inverter-fed synchronous motors with closed-loop control fixing the torque angle at 90° are often described as *brushless d.c. machines*, a label particularly appropriate if permanent magnets are used to provide the rotor field. Such machines are described in Section 6.21. The thyristor switching procedures and flux distribution may be modified on small machines with few armature coils to minimise the effect of torque variations with time through rotor position.

The locus of V_a for variable load current I_a with a torque angle set to $(-)90°$, and fixed speed and field current, is shown in Fig. 7.64(b) to correspond to motor operation at lagging power factor, which is likely to worsen the prospects for inverter commutation. In practice, torque angle and/or field current are likely to be controlled such that the motor operates at *leading* power factor, requiring that $E_f > V_a$. Under such over-excited conditions, illustrated in Fig. 7.64(c), force-commutation circuits are unnecessary for the *current-source* inverter employed.

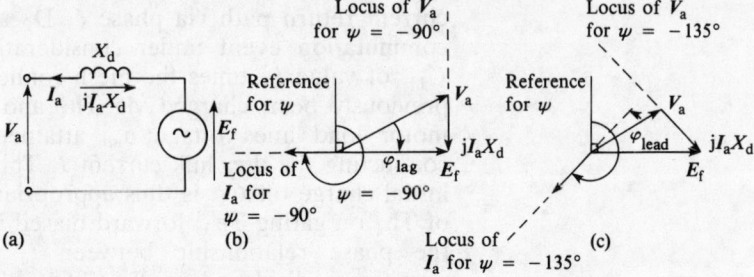

Figure 7.64 Equivalent circuit (a) and phasor diagram for a synchronous motor showing terminal voltage and armature current loci for constant excitation and internal space-phase angle (torque angle) ψ maintained, (b) at maximum value $(-)$ 90° and lagging power factor, and (c) at $-135°$. Armature resistance is neglected.

7.34.1 The current-source inverter

In contrast to the voltage-source inverter considered in Sections 7.28 to 7.33 the current-source inverter is supplied from a d.c. link having substantially constant current over a time-scale of several cycles of inverter output, rather than having constant voltage. The d.c. link is characterised by possessing large series *inductance* which permits the *instantaneous* values of converter d.c. voltage and inverter d.c. voltage to diverge significantly, the difference voltage being developed across the inductor, accounting for changes of stored energy therein.

Thus the *current* waveform in a current-source inverter load is defined in terms of current blocks of amplitude fixed in the short term. The load *voltage* is dependent upon the load impedance values appropriate to the current fundamental and harmonic components. The nature of a rotating machine load is such that the developed line voltage waveform is reasonably sinusoidal.

In the current-source inverter the load voltage may be made responsible for facilitating commutation, and this feature of self-commutation may be implemented by essentially simple commutation circuits incorporating only diodes and capacitors. Since the inverter load voltage is isolated from the converter voltage supplying the d.c. link by the link inductance which maintains unidirectional current flow at all times, reversals of power flow in which the load machine regenerates are achieved simply by permitting reversal of the mean d.c. link voltage polarity. This may be achieved at the machine terminals by the inherent change in phase relationship between the fundamental voltage and the current accompanying torque reversals whilst, at the a.c./d.c. converter, mean d.c. voltage magnitude and polarity are determined by the commutation delay angle α. In practice, natural commutation of the current-source inverter by the load voltage is facilitated best with leading power-factor motor loads. At low motor speeds or when generating, the excitation may be inadequate to ensure appropriate charging of the commutation capacitors. Forced commutation of the inverter is one of several techniques available to circumvent this difficulty.

The three-phase current-source converter circuit is shown in Fig. 7.65(a) and (b), emphasising the current path through the inverter when current is about to be transferred from phase a to phase b by the turn-on of Th_3 and the turn-off of Th_1. The link current return path via phase c, D_2 and Th_2 is unaffected by the commutation event under consideration. Commutation capacitor C_{13}, of value $1\frac{1}{2}$ times the mesh-connected capacitor value/ph, has previously been charged via Th_1 and D_3 to the highest value of motor load line voltage v_{ab} attained since Th_1 and D_1 began conducting the d.c. link current I. The polarity of v_{C13o} due to the initial charge on C_{13} is thus appropriate to the subsequent turn-off of Th_1 on gating Th_3, forward biased by the capacitor voltage. For the phase relationship between i_a and v_{ab} pertaining to the waveforms displayed in Fig. 7.65(e), $v_{C13o} = \hat{V}_{ab}$, capacitor discharge as v_{ab} falls below this peak value is blocked by D_3.

At instant t_1, defined in Fig. 7.65(e), Th_3 is fired to turn off Th_1 and supply i_a at constant current I via Th_3, C_{13} and D_1. v_{ab} continues to change, in conformity with the load, but at a lower rate than v_{C13} which now proceeds to reduce its charge due to injection of I. The effective circuit is shown in Fig. 7.65(c) with $i_a = -i_{C13} = I$.

Synchronous machines 375

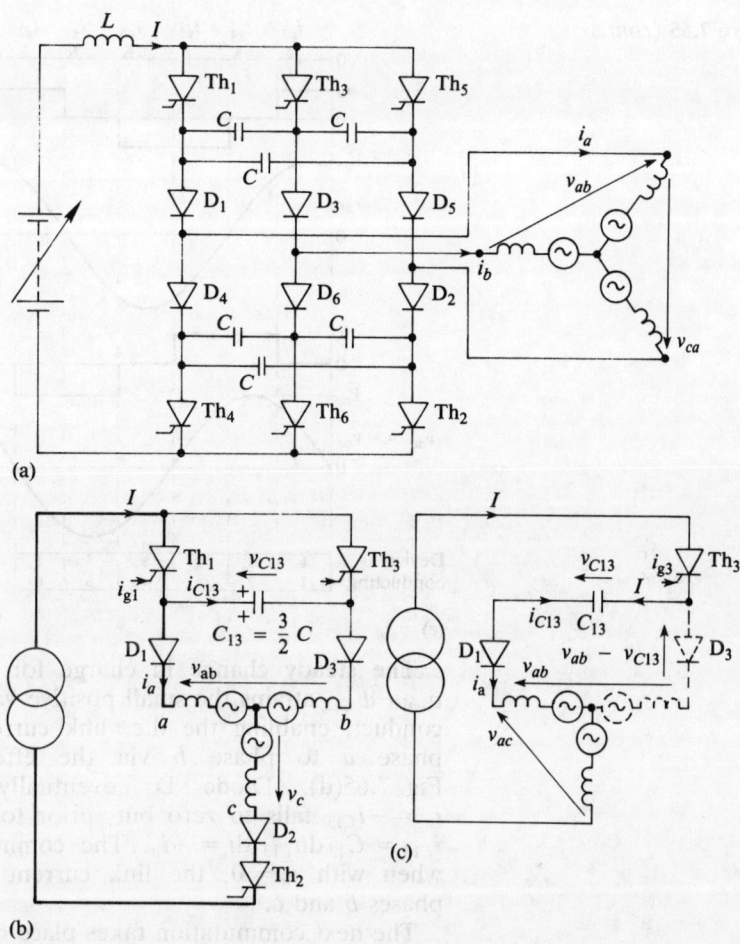

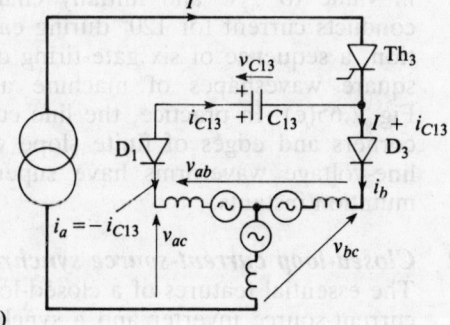

Figure 7.65 Three-phase current-source inverter: (a) circuit diagram including commutation capacitors and motor load; (b)–(d) effective circuitry as current derived from a constant-current source commutates from phase *a* to phase *b*; (e) idealised waveforms, thyristor switching and device conduction sequences for leading power factor load, $\phi_L = 30°$.

Figure 7.65 (*cont.*)

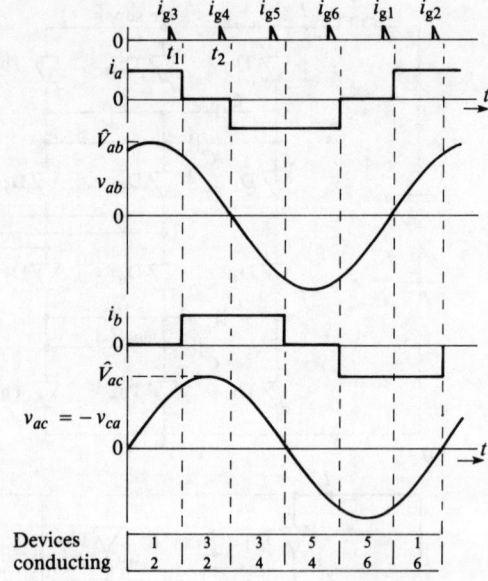

(e)

The steady change in charge for C_{13} means that, ultimately, $v_{ab} - v_{C13}$ attains the small positive value necessary to cause D_3 to conduct, enabling the d.c. link current to initiate transfer from phase a to phase b via the effective equivalent circuit of Fig. 7.65(d). Diode D_1 eventually ceases conduction when $i_a = -i_{C13}$ falls to zero but, prior to this instant, $v_{C13} \simeq v_{ab}$ and $i_{C13} = C_{13}\,dv_{C13}/dt = -i_a$. The commutation process is complete when with $i_a = 0$, the link current flows in the load between phases b and c.

The next commutation takes place 60° later at instant t_2 between phases c and a, with the turn-off of Th_2 being achieved through the turn-on of Th_4 via effective commutating capacitor C_{24}, equal in value to $1\tfrac{1}{2}C$ and initially charged to $\hat{V}_{ac}$. Each phase thus conducts current for 120° during each half-cycle of inverter operation, a sequence of six gate-firing operations generating the quasi-square waveshapes of machine armature current illustrated in Fig. 7.65(e). In practice, the line current waveforms have rounded corners and edges of finite slope whilst the essentially sinusoidal line-voltage waveforms have superimposed 'spikes' at the commutation instants.

7.34.2 Closed-loop current-source synchronous motor drive

The essential features of a closed-loop drive system incorporating a current-source inverter and a synchronous motor are illustrated in Fig. 7.66. For set values of field current and torque angle the torque developed is proportional to the d.c. link current I. Divergence of speed from the set-speed value alters the d.c. link current and hence the motor torque. Torque angle and field current are also controlled as functions of I to maintain the over-excited conditions which facilitate self-commutation of the inverter.

Figure 7.66 Closed-loop, current-source synchronous motor drive system.

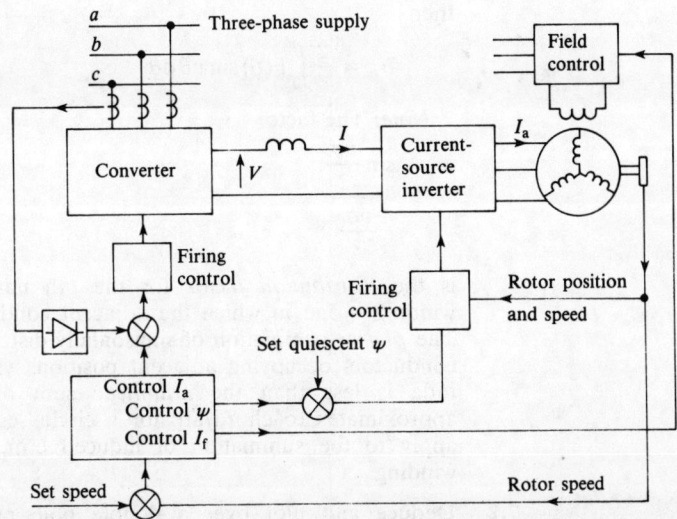

Natural commutation becomes difficult at low speeds when the terminal voltage is low. As an alternative to providing forced commutation at low link voltages amendments may be made to the constant-current source by a.c./d.c. converter control.

7.35 Tutorial Examples

7.1 Each phase winding of an a.c. machine armature generates, when carrying current, a radial airgap field component which approximates to a trapezoidal distribution with each side of the trapezoid corresponding to a group of conductors carrying current of like polarity.

Considering such a winding to be set up in q slots/pole/phase with Z_s conductors per slot and with I_c the r.m.s. current per conductor (sinusoidally time-variant) show that:

(i) the height of the trapezium representing the radial magnetic-field intensity distribution at the instant of maximum phase current is given by

$$\hat{H}_n = \frac{q}{2g} Z_s \sqrt{2} I_c$$

where g is the airgap length;

(ii) the harmonic series representing the radial field distribution at the same instant may be expressed as follows

$$H_n(\theta) = \frac{4}{\pi} \hat{H}_n \left[\sum_{v=1}^{v=\infty} \frac{1}{v} \left(\frac{\sin \frac{v\sigma}{2}}{\frac{v\sigma}{2}} \right) \sin v\theta \right]$$

where v is the harmonic number (odd only) and σ is the *phase spread*, i.e. the angle (elec) subtended by the band of adjacent conductors in each phase winding.

Hint: If the harmonic series can be restricted to

$$F(\theta) = \sum_{v=1}^{v=\infty} b_v \sin v\theta$$

then

$$b_v = \frac{2}{\pi}\int_0^\pi F(\theta)\sin v\theta\, d\theta$$

Note: The factor

$$\frac{\sin\dfrac{v\sigma}{2}}{\dfrac{v\sigma}{2}}$$

is the *distribution factor* for the vth harmonic of a finely distributed winding – one in which the adjacent conductors are very close together. The *phasor* summation of sinusoidally distributed field components due to conductors occupying adjacent positions yields a resultant whose magnitude is less than the *arithmetic* sum of the components. The ratio approximates to chord/arc for a circle segment. Identical considerations apply to the summation of induced e.m.f.s within a distributed phase winding.

7.2 Deduce and plot over a double pole pitch the radial magnetic field distribution established in the airgap of a balanced, three-phase machine having a single-layer armature winding set in 12 slots/pole, considering one phase only to be carrying current. Ignore slot effects and relate the peak value of the radial field to the current/coil side, I, and the uniform airgap length, g.

Each phase winding has four full-pitch coils per pole pair, distributed over adjacent slots.

Repeat for a condition in which one phase carries current as above, the other two phases carrying one-half the current of the original phase, of opposite polarity.

U. of B. $\qquad [2I/g,\ 4I/g]$

7.3 An 11 kV, three-phase, round rotor synchronous machine has its armature winding connected in star to 11 kV infinite busbars. Armature resistance and synchronous reactance/phase are 1.0 Ω and 12.0 Ω, respectively. Determine the value of the e.m.f./phase induced by the field current alone when the machine motors, taking a line current of 100 A at 0.8 power factor lag.

By what percentage would the field current need to be increased if the machine were required to generate at the same current and power factor? What would be the corresponding change in load angle?
Neglect saturation.

$\qquad [5.6\text{ kV};\ 28\%,\ 16.5°\varepsilon]$

7.4 An 11 kV, three-phase, four-pole, 50 Hz, star-connected synchronous generator has synchronous impedance/phase of j12.0 Ω. Determine the e.m.f./phase induced by the field current and the load angle when the machine supplies 3 MW at 0.8 power-factor lag and rated voltage to its busbar. Ignore saturation.

If the excitation remains constant, evaluate the pull-out torque – maximum torque before synchronism is lost.

$\qquad [8.0\angle 13.9°\text{ kV},\ 81\text{ kN m}]$

7.5 A 2 kV, three-phase, four-pole, 16.67 Hz star-connected cylindrical-rotor synchronous machine has synchronous impedance 0.2 + j10 Ω/phase and motors at rated voltage, absorbing 80 kW from the supply. Calculate the armature open-circuit voltage/phase and the rotor displacement angle (° mech) when operated at (i) 0.8 power factor lag, (ii) 0.8 power factor lead.

$\qquad$ [(i) $1.0\text{ kV}\angle -6.6°$ mech; (ii) $1.33\text{ kV}\angle -5.0°$ mech]

Synchronous machines 379

7.6 A 750 kW, 6000 V, three-phase, star-connected synchronous motor has a synchronous impedance of 1.5 + j16.0 Ω/phase. It is excited to develop an open-circuit e.m.f. of 5000 V (line). Draw the locus of the armature current phasor for motor loads from zero up to 1 MW with this excitation and estimate the maximum value of the power factor.

[0.89]

7.7 Show that the complex terminal power output of a round-rotor polyphase synchronous machine with negligible armature resistance is given by an expression of the form

$$S = P + jQ = \frac{E_f V_a}{X_d} \sin \delta - j\left(\frac{V_a^2}{X_d} - \frac{E_f V_a}{X_d} \cos \delta\right).$$

Plot the locus of E_f on a phasor diagram relating to a three-phase, star-connected synchronous machine motoring at 16 MW total power and varying power factor, connected to 11 kV (line) busbars. The direct-axis synchronous reactance/phase is 4.5 Ω. Evaluate the open-circuit voltage/phase at unity power factor.

[7.4 kV]

7.8 A 20 MVA, 11 kV, three-phase, 50 Hz, star-connected synchronous machine has synchronous reactance $X_d = 4.5$ Ω/phase and negligible armature resistance. It operates as a generator supplying rated load at rated voltage and 0.8 power factor lag. Ignoring saturation effects, determine graphically or otherwise

 (i) the load angle, i.e. the time-phase angle by which the armature e.m.f. E_f leads the terminal voltage V_a,
 (ii) the space-phase angle by which the armature field lags the 'field' winding field,
 (iii) the modulus of E_f, and
 (iv) the maximum real power output of the machine before synchronism is lost, if this value of excitation is maintained and the input shaft power increased.

Repeat with the same initial real power output for a power factor of 0.9 lead and, by comparing results for (iv), deduce which of the two alternatives is the preferred operating condition.
U. of B.

[(i) 22.4° 40° (ii) 149° 104° (iii) 9.93 kV$_{ph}$ 5.89 kV$_{ph}$ (iv) 42 MW (preferred), 25 MW]

7.9 An 11.5 kV, three-phase, star-connected, cylindrical-rotor synchronous machine has an open-circuit curve as follows.

Field current	60	80	100	120	140	160	180	A
Armature e.m.f./phase	5.3	6.2	6.8	7.2	7.5	7.7	7.9	kV

On short circuit a field current of 80 A gave an armature current of 350 A. Determine the corresponding value of synchronous impedance/phase.
Use a phasor diagram construction to evaluate the variation of armature current with field current for the machine running as a motor off 11.5 kV busbars and developing 900 kW. Neglect armature winding resistance and consider what other assumptions are implied.
[*Note:* The resulting curve is commonly known as a 'Vee curve'.]

[17.7 Ω]

7.10 Why do modern turbogenerators have 'large' airgaps? An ideal cylindrical-rotor synchronous generator has a synchronous reactance of 2.0 p.u. and is designed to operate at a power factor of 0.87 lagging. On open circuit the unsaturated e.m.f. is 2.8 p.u. when the rotor excitation is 1.0 p.u. Calculate the load angle at rated output.

If the airgap length is doubled, estimate (i) the excitation for rated voltage on open circuit, (ii) the synchronous reactance, and (iii) the excitation for rated current on short circuit.

Use a phasor diagram to show that the power output will be limited by rotor heating to about 0.6 p.u. but that the machine has a higher steady-state stability.

E.C. [41°; (i) 0.714 p.u., (ii) 1.0 p.u., (iii) 0.714 p.u.]

7.11 A synchronous motor with constant excitation initially receives 0.25 p.u. of the maximum power that it is capable of receiving from an infinite busbar. If the mechanical load on the motor is suddenly doubled to 0.5 p.u., calculate the maximum value of load angle developed during swinging about the new equilibrium value. Ignore losses.

[46.5°]

7.12 Draw a simple per-phase equivalent circuit for the steady-state operation of a non-salient synchronous machine, identifying any symbols used. Comment on the limitations of the model and indicate briefly how the synchronous reactance might be measured.

Sketch the phasor diagram for the situation where the machine is generating at lagging power factor and show that, if the stator resistance is neglected, the output power/phase is given by

$$P = (EV/X_s)\sin\delta$$

where the symbols have their usual meaning.

The excitation of 600 MVA, two-pole, 50 Hz turboalternator operating on infinite busbars is adjusted so that the steady-state stability limit is 400 MW. The inertia constant (stored kinetic energy/VA) is 5 s. Estimate the frequency of rotor oscillations following a small load change from an initial steady load angle of 30°. Neglect the influence of damping currents in the rotor and state any other assumptions.

E.C. [4.254 rad s^{-1}]

7.13 What is the 'equal-area criterion' in synchronous-machine technology, and to what conditions does it apply?

(i) Briefly explain the sequence of events initiated in a generating station by the sudden disconnection from the busbars of a loaded turbogenerator.

(ii) A 400 V, 50 Hz, three-phase, star-connected, eight-pole synchronous motor operates at unity power factor with a line current of 40 A. The instantaneous rotor position is then 4° (mech) behind that for no load. Estimate the frequency at which the rotor will tend to oscillate, given that the total inertia is 8 kg m^2.

E.C. [(ii) 3.95 Hz]

7.14 A 400 V, six-pole, 50 Hz, three-phase reluctance motor with armature leakage flux ignored has a direct-axis synchronous reactance of 15 Ω/phase and a quadrature-axis synchronous reactance of 6 Ω/phase. Neglecting losses determine

(i) the maximum shaft torque developed before pull-out; and
(ii) the maximum power factor obtainable and the corresponding power output.

[(i) 76.4 Nm; (ii) 0.42, 7 kW]

7.15 State how salient-pole synchronous machine theory accommodates the variable permeance of the machine airgap and explain how the concepts of synchronous reactances X_d and X_q arise.

A 625 kVA star-connected salient-pole synchronous machine is connected to a three-phase 2400 V (line) system and has unsaturated synchronous reactances $X_d = 6\,\Omega$/phase, $X_q = 4\,\Omega$/phase and armature resistance 0.5 Ω/phase. Determine the open-circuit voltage/phase required in order that the machine may operate as a motor absorbing 500 kW from the supply at power factors of (i) unity, (ii) 0.8 lead, (iii) 0.8 lag.
Determine also, in each case the load angle.
U. of B. [(i) 1.49∠−20° kV, (ii) 2.01∠−17° kV, (iii) 1.00∠−25° kV]

7.16 Justify the construction used to derive the phasor diagram of a salient-pole polyphase synchronous machine for which data is provided in terms of its direct- and quadrature-axis synchronous reactances, armature resistance, terminal voltage and current and power factor. A 6000 kVA, 2400 V, star-connected, three-phase, 50 Hz salient-pole synchronous machine operates as a generator supplying rated load at 0.9 power factor lag. Machine parameters X_d and X_q are 1.0 Ω/phase and 0.667 Ω/phase, respectively. Neglecting saturation and armature resistance, determine the value of the e.m.f./phase induced by the field current, the load angle and the space-phase angle between armature and field m.m.f.s. Calculate the torque from an expression involving the machine reactances and the load angle, and compare with a value derived directly from consideration of real power at the machine terminals.
U. of B. [2.37 kV∠25.5°, 142°, 5.4 sync. MW]

7.17 A single-stack variable-reluctance stepping motor has four rotor teeth and six stator poles energised in a three-phase sequence *abca*. Sketch the polar arrangement in cross-section and determine the step length.
[30°]

7.18 Determine the step length for a single-stack variable reluctance stepping motor having eight rotor teeth and 10 stator poles excited in a five-phase sequence *abcdea*.
[9°]

7.19 A three-stack variable-reluctance stepping motor is energised in a *half-stepping* sequence represented as $a, a+b, b, b+c, c, c+a, a$. Determine the step length appropriate to a 12-tooth rotor.
[5°]

7.20 A hybrid stepping motor employs an 18-tooth rotor and is provided with eight-stator poles, with two teeth per pole. Adjacent stator poles are energised from alternate phases of a two-phase, bipolar source of sequence $+a, +b, -a, -b, +a$. Deduce the step length.
[5°]

7.21 The single-stack variable-reluctance stepping motor illustrated in Fig. 7.36 has stator-phase inductance variation with rotor position which may be approximated to sinusoidal between limits of 3 and 7 mH. Derive an expression for the torque dependence on rotor angle displacement from the position of equilibrium for a steady current of 1 A flowing in one phase only.
$[T = 0.006\sin 6\theta\,\text{N m}]$
Determine the maximum and minimum values of the *instantaneous* torque developed by the motor when running at low speed with sequentially switched phase currents of 1 A and load torque increased to the pull-out value.
[0.006 N m, 0.0042 N m]

Assuming a total rotor inertia of 5×10^{-6} kg m^2 and ignoring damping, calculate the frequency of oscillation of the rotor about each new position when responding to single steps of excitation at 1.0 A.

[13.5 Hz]

7.22 A variable-reluctance stepping motor employs the transistor bridge bipolar drive circuit illustrated in Fig. 7.46(c). With ideal devices, turn-on of transistors T_1 and T_2 gives rise to an eventual steady-state phase-winding current of $E/(R + r)$. Show that subsequent turn-off of transistors T_1 and T_2 at instant $t = 0$ gives rise to the following expression for phase current.

$$i_a = \frac{-E}{R + r}(1 - 2e^{-t/\tau})$$

where $\tau = L/(R + r)$ with L, the phase winding inductance, assumed constant.

Show also that the phase current falls to zero in a time equal to 0.69315τ and that 61.37% of the initial stored magnetic field energy is returned to the d.c. supply.

7.23 Figure 7.67 shows a unipolar chopper drive circuit for one phase of a variable-reluctance stepping motor. Establish the switching sequence necessary for the transistors T_1 and T_2 in order that the value of phase-winding current i may be controlled within defined maximum and minimum values after the initial simultaneous turn-on of T_1 and T_2. Plot typical waveforms of winding current and terminal voltage assuming constant inductance L and demonstrate that stored magnetic field energy is restored to the voltage source on turn-off of both T_1 and T_2.

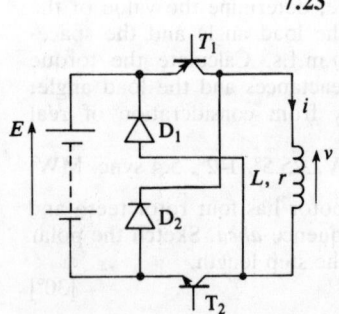

Figure 7.67 Figure for 7.23.

7.24 A switched-reluctance motor, bi-filar wound as in Fig. 7.47(b), is configured in such a way that the self-inductance of each stator phase winding varied trapezoidally with time and rotor position between limits defined by minimum and maximum values of 2 mH and 8 mH as the rotor rotates at constant speed. The several stator phase windings are energised in sequence from a 150 V d.c. supply.

Each inductance transition occurs over a rotor displacement of 20° which corresponds to a duration of 6 ms when the motor is rotating at 556 rev/min. With rotation at that speed each phase winding is connected to the source at an instant t_1, 4 ms before the instant t_2 at which the inductance begins to increase linearly from its lower level. Calculate (i) the energy absorbed from the source by a stator phase winding during the 4 ms interval which elapses whilst the self-inductance remains at 2 mH, and (ii) the value to which the primary phase current i_1 rises (from zero) during this time.

[(i) 90 J; (ii) 300 A]

Each primary switch is sequentially opened at an instant t_3, 6 ms after t_2 and coincident with that at which the self-inductance achieves its maximum value of 8 mH. Calculate (iii) the value of primary current i_1 at the instant t_3 and *estimate* (iv) the energy absorbed from the source over the period $t_2 \to t_3$.

[(iii) 187.5 A; (iv) 198 J]

Calculate also (v) the energy returned to the supply during the freewheeling period after t_3 when $-i_2$ falls linearly to zero and (vi) the duration of this freewheeling period over which the winding self-inductance is assumed constant at the higher value of 8 mH.

[(v) 140.63 J; (vi) 10 ms]

Hence *estimate* (vii) the net energy absorbed from the supply during each cycle of stator phase energisation and (viii) the *average* torque contribution made over one revolution by each stator phase that this energy represents.

[(vii) 148 J; (viii) 24 N m]

Finally, using an estimate of the mean-square value of i_1 over the period $t_2 \to t_3$ estimate (ix) the value of the torque *pulse* effective over 20° of rotor motion using the relationship $T = \frac{1}{2} i_1^2 dL/d\theta$, eqn [7.60]. Hence derive (x) an average torque contribution value for comparison with (viii) above.

[(ix) 425 N m; (x) 24 N m]

[*Notes:*

(i) The induced e.m.f. developed across a varying inductance is given by

$$v = \frac{d(Li)}{dt} = L\frac{di}{dt} + i\frac{dL}{dt}.$$

(ii) With practical machines the assumption of current-independent inductance is invalid as the magnetic circuit operates highly saturated. Non-linear computation is necessary to predict realistic performance characteristics.

(iii) A preferred alternative to a bifilar winding is depicted in Fig. 7.67 which associates the monofilar field winding with a pair of switches and a pair of diodes. Closure of S_1 and S_2, represented by transistors T_1 and T_2 respectively, enables energy to be transferred from the source to the field of the coil. Subsequent opening of S_1 followed by S_2 permits energy recovery by the source after a brief period of freewheeling.]

7.25 A notched waveform inverter provides an output voltage having 10 pulses of equal width and equal separation in each cycle. The switching sequence is ordered as for the pulse-width modulated (p.w.m.) inverter of Worked Example 7.7 and during alternate half-cycles of fundamental frequency the output voltage is switched between zero and $\pm E$. Determine the angular equivalence of the duration of each pulse necessary to provide a *fundamental* component of output voltage equal in amplitude to 50% of the d.c. source voltage E.

Determine also the amplitude of each component of output voltage to the 15th harmonic.

[13.94° ≡ 0.2433 rad; $B_3 = \hat{V}_3 = 0.187E$, $B_5 = \hat{V}_5 = 0.146E$,
$B_7 = \hat{V}_7 = 0.169E$, $B_9 = \hat{V}_9 = 0.407E$,
$B_{11} = \hat{V}_{11} = (-) 0.364E$, $B_{13} = \hat{V}_{13} = (-) 0.121E$,
$B_{15} = \hat{V}_{15} = (-) 0.082E$]

7.26 Determine the values of the commutating circuit components L and C required for an inverter bridge circuit employing auxiliary thyristor impulse commutation (McMurray). The maximum load current to be commutated is 250 A and is maintained constant during the commutation period. The main thyristor must be reverse biased for a minimum duration of 30 μs at turn-off. Neglect commutation losses, assume that the capacitor charges to $1\frac{1}{2}$ times the d.c. supply voltage of 200 V and limit the auxiliary thyristor current to 400 A.

[12.56 μH, 22.33 μF]

7.27 Show that the relationship between $\hat{I}_1$, the peak value of the oscillatory current, and I_L, the load current assumed constant over the duration of commutation in an auxiliary thyristor impulse-commutated bridge inverter (McMurray) operated with minimum commutation energy ($=\frac{1}{2}CV_{co}^2 = \frac{1}{2}L\hat{I}_1^2$) is given by the expression

$$\hat{I}_1 = 1.533 I_L.$$

[*Note*:

$$\frac{d}{dx}\cos^{-1}\left(\frac{1}{x}\right) = \frac{1}{x(x^2-1)^{1/2}}.]$$

Thus evaluate for minimum commutation energy at maximum load the commutation components L and C for a three-phase induction motor drive application in which the r.m.s. load current should not exceed 300 A. The main thyristor turn-off time is 25 μs. Ignore commutation losses and assume that the capacitor charges to 400 V.

[8.94 μH, 23.6 μF]

7.28 A single-phase bridge inverter circuit employing the complementary mutually coupled impulse commutation circuit (McMurray–Bedford) illustrated in Fig. 7.56 is required to commutate a motor load current which will not exceed 37.5 A. The load current may be assumed constant over the commutation period. Deduce appropriate values for the commutation circuit components if the d.c. supply voltage is 250 V and the thyristor current is limited to 1.6 times the maximum load current value. Assume ideal circuit elements and a thyristor turn-off time of 50 μs.

Calculate also the energy loss associated with each commutation.

[0.545 mH, 16.9 μF; 0.981 J]

7.29 Sketch the equivalent circuit of a single-phase current-source inverter supplied from a voltage source with source inductor L. Such an inverter supplies a resistive load with a nominal square wave of current having 12.5 A amplitude from a 200 V d.c. source. Determine the source inductance necessary to limit the rate of change of source current to 20 A s^{-1}.

[10 H]

7.30 A load resistor $R\ \Omega$ is supplied from a *single-phase* current-source inverter supplied at I A. Identical commutation capacitors C are employed, charged with appropriate polarity to E V prior to each commutation. Show from first principles that the load and capacitor currents may be expressed as functions of time t measured from the instants of gate-firing as follows.

$$i(t) = I\left[1 - \exp\left(\frac{-t}{2CR}\right)\right] + \frac{E}{R}\exp\left(\frac{-t}{2CR}\right)$$

$$i_C(t) = \frac{1}{2}\left(I - \frac{E}{R}\right)\exp\left(\frac{-t}{2CR}\right)$$

Sketch the waveform of voltage developed across each previously conducting thyristor as the capacitors recharge and indicate how the equation for load current might be used to predict the relationship between circuit parameters and thyristor turn-off time.

8 Induction machines

8.1 Introduction and Principle of Operation

The polyphase a.c. induction motor is the most popular machine employed in basic industrial drives, on account of its fundamental simplicity and rugged construction which leads to an economical first cost and reliable service with easy maintenance.

The induction machine has speed/torque characteristics which, when the supply is from constant-voltage, constant-frequency busbars, matches that of a shunt-field d.c. motor at constant armature voltage, with speed reducing almost linearly at a controllable rate with increasing load torque from a no-load value. The no-load speed of an induction motor approximates very closely to the synchronous value n_o rev/s defined by the number of armature winding pole pairs and the frequency of the supply, such that $n_o = f/p = \omega/2\pi p$.

The polyphase induction machine was invented around 1889 by Tesla and, in its simplest form, comprises a balanced, polyphase armature winding set in slots cut in a laminated iron core on the stator and a balanced, closed, polyphase winding on the rotor, also set in slots in a laminated iron core. The airgap is nominally uniform. There is no requirement that the number of phase windings on both sides of the airgap should be the same and advantage is taken of this fact in common use of *cage* rotor windings as illustrated in Fig. 8.1. Here the number of rotor phases is given by the number of axially directed conducting rotor *bars* divided by the number of pole pairs. Each bar is connected at its ends to short-circuiting *end-rings* which, if of zero impedance, compel the sum of the induced e.m.f.s and voltage drops developed between the ends of each bar to equal zero.

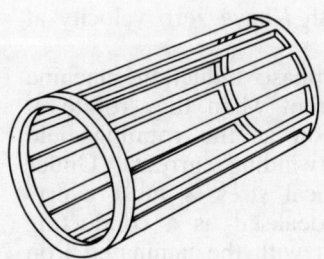

Figure 8.1 Rotor cage for polyphase induction motor.

Each rotor bar is subjected to the radial magnetic field set up in the adjacent airgap region by the current energised conductors on both sides of the airgap and hence has a component of induced axially directed e.m.f. accounted for by the relative motion of the rotor conductor of length l with respect to the radial component of airgap flux density. Equation [5.35] defines this e.m.f. to be

$$e = B_n l v \qquad [8.1]$$

where v is the velocity of the conductor relative to a frame of reference with respect to which the flux linkage appears to be unchanging with time, and where relative directions of e.m.f., flux density and motion are defined by the right-hand rule (Fig. 5.15).

If the rotor conductors carry zero current then B_n is due solely to stator current but, generally, B_n is established by the resultant radial field of both stator and rotor conductors. In practice, constraints are placed on the value of B_n by the stator winding terminal voltage which is generally sinusoidally time variant, of constant amplitude and frequency, or with amplitude and frequency proportional to each other. Induction machines are usually operated with virtually constant radial flux over a range of load conditions, to maximise performance whilst avoiding saturation of the magnetic circuit, although this rule may beneficially be relaxed under light load conditions to reduce iron losses and raise power factor.

The radial field distribution set up in the airgap by polyphase stator currents flowing in a balanced polyphase winding may, by the arguments advanced for the armature winding of the polyphase synchronous machine in Section 7.7, be represented as a sinusoidal distribution which appears to rotate at synchronous speed with respect to the stator conductors. If, then, we look for a condition in which the *current* in the rotor conductors must be zero, the net e.m.f. induced therein must also be zero as the voltage drop across the inevitable conductor resistance will be zero. In addition to e.m.f. induced by the *radial* airgap flux accounted for by eqn [8.1] any *tangential* field changes would also account for a component of axially directed rotor conductor e.m.f., but the responsible m.m.f. is then due to the rotor conductors themselves and we have proposed that, for the present, these currents should be zero.

With zero current in the closed rotor conductor circuit, therefore, the only source of electromagnetically induced e.m.f. is the radial flux density B_n, and eqn [8.1] shows this e.m.f. to be zero when the rotor conductors of axial length l have zero velocity at right-angles to the finite radial flux.

Thus the rotor conductors of a polyphase induction machine have zero induced e.m.f. and zero current when they rotate at synchronous speed in the same direction as the rotating field established in the airgap by the stator winding currents. Under these conditions the tangential mechanical stresses must everywhere be zero on the rotor surface, idealised as a conducting cylinder of negligible thickness in contact with the laminated iron rotor core.

The induction machine at synchronous speed thus behaves in a fashion identical to that of a cylindrical-rotor synchronous machine without d.c. field excitation. The *primary* armature winding currents establish a rotating field in the airgap, the radial component of which cuts the armature conductors to induce 'armature-reaction' e.m.f. in the primary winding but, being stationary with respect to the synchronously rotating *secondary* winding conductors, induces no e.m.f. therein. The primary winding, almost invariably mounted on the *stator* in induction machines, is also responsible for a *tangential* component of magnetic field in the airgap as detailed in Sections 5.4 and 7.6. The value of this tangential component varies linearly to zero across the airgap from a maximum value at the stator surface equal to the linear current density (A/m). Tangential mechanical stresses are developed on

Induction machines

the stator iron/airgap interface but the relationship between field components at this surface is such that the resultant tangential stress integrated over a double pole pitch is zero.

The induction machine differs from the uniform-airgap synchronous machine with d.c. excitation in the manner by which current is caused to flow in the secondary (rotor) winding conductors. Equation [8.1] indicates that, if the rotor conductors rotate at speeds other than synchronous speed, e.m.f. will be induced in each rotor conductor which is proportional to the product of the instantaneous value of the radial flux density and the peripheral speed of the rotor assembly with respect to the synchronous motion of the radial flux. If the latter is sinusoidally distributed and the motion of the rotor conductors in space is allowed to *slip* behind the synchronously rotating flux established (provisionally) by the stator currents then, *for a two-pole machine*, the e.m.f. induced in a rotor conductor will be of the form

$$e_2(t) = ls\omega b \hat{B}_n \sin(s\omega t + \alpha) \qquad [8.2]$$

where ω is the angular velocity of the rotating flux density in space, $(1 - s)\omega$ is that of the rotor conductors at radius b, and α indicates the spatial position of each particular rotor conductor. The above relationship is formally derived as eqn [8.12] in the general analysis of Section 8.3.

Thus each rotor conductor has sinusoidal induced e.m.f. of equal magnitude and frequency, features which are proportional to the speed discrepancy $s\omega$, and of phase related to the particular position of each with respect to the common radial flux-density wave. However, the end-rings short circuiting the symmetrically disposed rotor bars compel the *terminal voltages* of each conductor to be zero, as illustrated in Fig. 8.2(a). Thus the e.m.f. of eqn [8.2] must be responsible for a short-circuit current in each conductor of value limited by the effective impedance of that conductor, such that the voltage drop across the impedance equals the responsible e.m.f. The impedance of each rotor conductor has a resistive component due to ohmic effects and an inductive reactive component due to its own current, but excluding its contribution to the *radial* flux density B_n. The reactive component of the rotor conductor self-impedance is therefore identified with *leakage* inductance, including the effect of a tangential airgap field component.

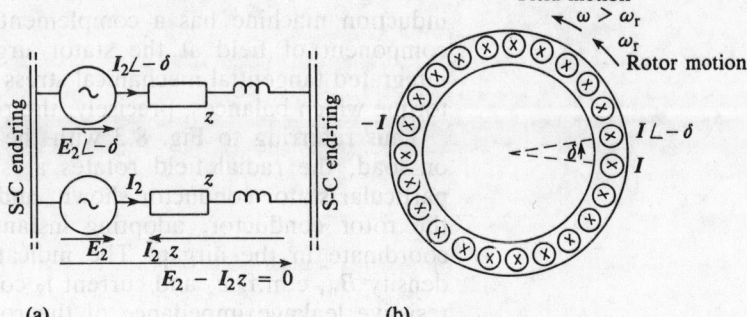

Figure 8.2 (a) Axial voltage distribution in a lumped circuit equivalent of conducting cage rotor elements. (b) Axial current distribution, two-pole rotor. δ is the angular separation (°ε) of adjacent conducting bars.

Due to symmetry, each rotor conductor has the same impedance, and therefore the current flowing therein will be sinusoidally time variant with the same frequency as the induced e.m.f. due to the radial flux, lagging this e.m.f. by the leakage impedance phase angle ϕ which is of value dependent upon rotor frequency. Thus the current in each rotor conductor is of the form

$$i_2(t) = ls\omega \frac{\hat{B}_n}{z(s)} \sin[s\omega t + \alpha - \phi(s)] \qquad [8.3]$$

where $z(s)\angle\phi(s)$ is the leakage impedance of each conductor, a function of *slip s*. The distributed rotor conductors set up a near-sinusoidal distribution of axially directed, sinusoidally time-variant current on the rotor iron/airgap boundary to influence the magnetic field conditions there. Figure 8.2(b) illustrates this for a phasor representation of the current. The e.m.f.-inducing capability of the resulting tangential field H_{tr} has already been accounted for, as contributing to the reactive component of the rotor conductor leakage impedance but, in conjunction with the radial flux density B_n, the tangential field H_{tr} also develops tangential mechanical stress on the rotor surface. Integration of such stress over the entire rotor surface gives rise to net torque which invariably acts in a direction such as to encourage the rotor conductors to run synchronously with the responsible rotating field. Thus the torque developed on the rotor of an induction motor running subsynchronously is in the direction of the synchronously rotating field.

As noted above the rotor currents also influence the radial field in the airgap. It may be recalled from the summary at the end of Section 7.6, or from Section 5.7, that in determining the *radial* airgap field it is immaterial whether the responsible current is on the rotor or stator side of the airgap. Hence the induced rotor current will modify the radial flux density B_n, but within the constraints imposed by the normal connection of the *stator* winding to a supply voltage busbar of constant voltage/frequency ratio. Hence, after making allowance for stator winding leakage impedance, the *magnitude* of the radial airgap flux density must remain virtually unchanged for all conditions of load reflected by changes in the rotor conductor current. The machine automatically compensates for the influence of rotor current by changing the *stator* conductor current such as to maintain virtually constant radial flux, in much the same manner as the primary current of a static transformer changes to compensate for the effect of the secondary winding load current. The change in stator (primary) current of the induction machine has a complementary effect on the tangential component of field at the stator/airgap boundary such that the integrated tangential mechanical stress on load gives rise to a stator torque which balances, precisely, the rotor torque.

Thus referring to Fig. 8.3 with the two-pole machine motoring on load, the radial field rotates at speed ω with respect to the particular stator conductor shown, and at speed $s\omega$ with respect to the rotor conductor, adopting instantaneously the same angular coordinate in the airgap. The indicated directions of radial flux density B_n, e.m.f. e_2 and current i_2 correspond to a predominantly resistive leakage impedance of the rotor conductor. If this is the case, the space phasor H_t due to all rotor conductors will coincide with that of the inducing radial flux density B_n whereas, generally, H_t will lag B_n by a *space*-phase angle equal to the *time*-phase angle of the rotor leakage impedance, with a value dependent upon slip frequency sf. If the current i_2 flowing in a particular rotor conductor is not to modify the radial flux established by the stator winding alone when i_2 is zero, then the current in the stator conductor facing the rotor conductor across the airgap should be increased in the direction shown by an equivalent current i_2' such that $i_2' = i_2$. This identity assumes an equal number of stator and rotor conductors. The total current enclosed within the elementary magnetic circuit shown by the broken lines to include two radial crossings of the airgap is therefore unchanged, as required for an unchanged distribution of radial airgap flux. It is significant that the magnetic effect of *load* current within the airgap is to develop *tangential* magnetic field components which act in the same direction locally, whereas the *radial* field components cancel. Superimposing the tangential and radial field components within the airgap in the region of interest displayed demonstrates that the resultant flux lines crossing the gap are redirected such that the concept of flux lines as 'stretched elastic strings pressing each other apart' confirms that the tangential mechanical stress on the rotor is in the direction of its motion, with that on the stator acting in the opposite direction.

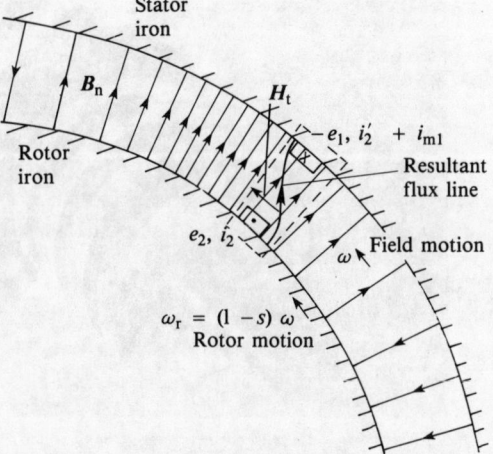

Figure 8.3 Electromotive force e_2 and current i_2 induced in an axial rotor conductor by sine-distributed rotating radial flux due to polyphase stator current sheet – of which the elemental conductor shown facing across airgap has e.m.f. e_1 and current components i_2' and i_{m1}; the latter is required to establish radial field and flux.

8.2 Construction of Polyphase Induction Machines

The stator winding and laminated core assembly of the polyphase induction machine are essentially identical with those of the polyphase synchronous machine illustrated in Fig. 7.1. On the other side of the uniform airgap, however, the rotor winding may take several alternative forms. For maximum flexibility in operation, a *wound* rotor will be employed, having a distributed polyphase winding – usually three-phase – connected to external terminals via slip-rings and brushes. The rotor windings are usually internally connected in star or delta to reduce the number of slip-rings required. The rotor circuit may then be connected to external resistance, or a more general *source* or *sink* of slip-frequency power, for control purposes, or simply short circuited for normal asynchronous running at just below synchronous speed.

Figure 8.4(a) shows such a wound rotor with its three shaft-mounted slip-rings. Practical machines with the rotor windings set in slots often have their active conductors 'skewed' from the true axial direction in order to minimise the effect of harmonic fields on torque developed as the machine runs up from standstill. Figure 8.4(b) shows a cage rotor with the aluminium conductors die-cast integrally with the end-rings and fins or 'wafters' which aid the circulation of air for cooling purposes. The ruggedness and low cost of manufacture of the cast aluminium cage rotor accounts for much of the popularity of the induction motor for industrial drives. Like the stator, the iron circuit of the rotor is made from laminated steel to reduce rotor eddy-current loss. As is also the case for the *hysteresis* component of rotor iron loss, rotor eddy-

Figure 8.4 Polyphase induction motor rotors: (a) wound rotor with three shaft-mounted slip-rings: (b) die-cast aluminium cage rotor incorporated in exploded view of totally enclosed, fan-cooled machine.
(*Photographs by courtesy of GEC Electromotors Ltd, Warley, UK.*)

current loss is not significant under normal running conditions at low slip, as the rotor frequency is then low. Stator iron losses, however, are independent of slip and are substantially constant over the working range of the machine when operated at a constant value of radial flux/pole.

8.3 Electromotive Force and Torque Evaluation in Polyphase Induction Machines

An ideal polyphase induction machine has its 'primary' winding normally located on the stator side of a uniform airgap and connected to a constant voltage supply of constant frequency, or one of constant voltage/frequency ratio. The flow of balanced sinusoidal currents in the stator phase windings sets up radial and tangential magnetic field components in the airgap. As established by each individual phase the component fields pulsate about fixed axes defined by the winding locations but, when considered collectively, they produce a harmonic series of sinusoidally distributed rotating field components of which the fundamental is predominant, rotating through a double pole pitch during each complete period of exciting current.

A detailed study of the airgap field components set up in a uniform airgap of length g by a three-phase winding is presented in Sections 7.4 to 7.7 with further analysis in Section 5.6. Polyphase stator windings for induction machines are essentially identical with those of synchronous machines and it is sufficient here to quote the relationship given in eqns [7.6], [7.7] and [7.11] for the fundamental components of radial and tangential field at the stator surface produced by sinusoidally distributed phase a current $i_a = \hat{I}_a \sin \omega t$.

$$H_{na}(\theta_m, t) = -\frac{n\hat{I}_a}{2g} \sin \omega t \cos p\theta_m \qquad [8.4]$$

$$H_{ta}(\theta_m, t) = \frac{pg}{b} \frac{n\hat{I}_a}{2g} \sin \omega t \sin p\theta_m \qquad [8.5]$$

where n is the number of phase conductors per pole, p is the number of pole pairs, b is the radius of the stator bore and θ_m is the angular position in the airgap measured in mechanical degrees of radians.

With phase windings b and c identical with the phase a winding, but displaced in the direction of θ_m measurement by 120° ε and 240° ε respectively, and supplied with currents $i_b = \hat{I}_a \sin(\omega t - 120°)$, $i_c = \hat{I}_a \sin(\omega t - 240°)$, the rotating radial airgap field expression analogous to eqn [7.12] and due to *stator* current alone is

$$H_{ns}(\theta_m, t) = -\frac{3}{2} \frac{n\hat{I}_a}{2g} \sin(\omega t - p\theta_m) \qquad [8.6]$$

with the rotating tangential field at the stator surface due to stator current leading H_{ns} in space by 90°ε

$$H_{ts}(\theta_m, t) = \frac{3}{2} \frac{pg}{b} \frac{n\hat{I}_a}{2g} \cos(\omega t - p\theta_m). \qquad [8.7]$$

The reference directions for current flow, H_n, H_t, stator winding disposition and θ_m measurement are as illustrated for a two-pole machine in Fig. 8.5(a), which is deliberately configured so as to establish rotating airgap field components which appear to move in the anticlockwise reference direction of rotor rotation and angular measurement when the stator windings are excited with current of phase sequence *abc*.

Consider a *stator* conductor of length l located at θ_m. The e.m.f. induced by the radial rotating field is given by eqn [8.1] as

$$e_s(\theta_m, t) = B_n(\theta_m, t)\, lv.$$

The reference direction for e_s is shown in Fig. 8.5(a) appropriate to anticlockwise motion of the conductor relative to the radial flux. Since the radial flux, as derived from eqn [8.6] when the rotor currents are zero, is actually rotating at synchronous speed ω/p mechanical radians/second in the *anticlockwise* direction relative to a fixed *stator* conductor

$$B_n(\theta_m, t) = \hat{B}_n \sin(\omega t - p\theta_m) \qquad [8.8]$$

and $v = -\omega b/p$ to give

$$e_s(\theta_m, t) = -\frac{\omega}{p} bl\hat{B}_n \sin(\omega t - p\theta_m). \qquad [8.9]$$

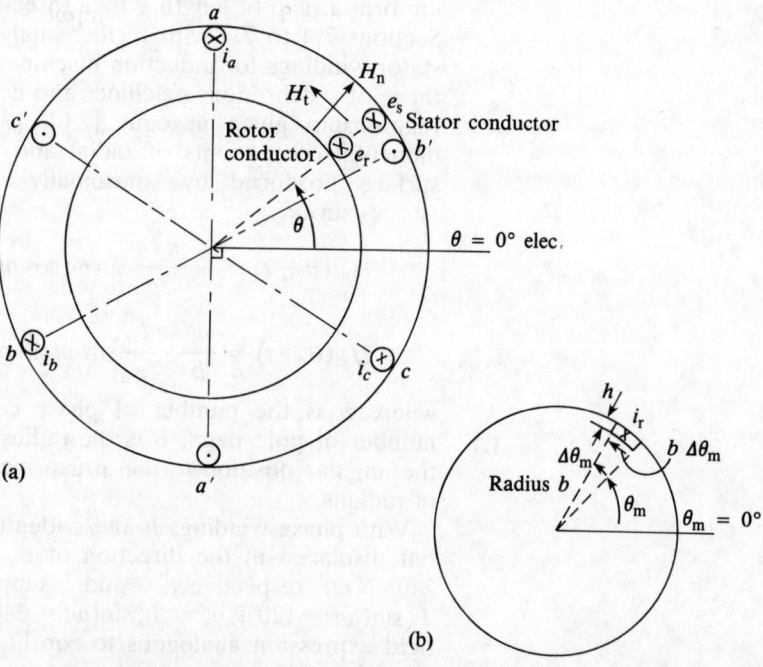

Figure 8.5 Conductor and airgap field configuration for model three-phase induction machine. (a) Representative stator and rotor conductors for a two-pole machine, where $\theta = \theta_m$ defines position with reference to radial field axis of stator phase *a*. (b) Elemental rotor conductor of $2p$ pole machine, located at $\theta_m = p\theta_{\text{elec}}$ and carrying current i_r.

A *rotor* conductor also of length l, located instantaneously at the same angular position θ_m in the airgap but mounted on a rotor assembly moving at angular velocity ω_r mech rad/s has motion relative to the same synchronously rotating radial flux such that $v = (\omega_r - \omega/p)b$ and a corresponding induced e.m.f. of value

$$e_r(\theta_m, t) = \left(\omega_r - \frac{\omega}{p}\right) bl\hat{B}_n \sin(\omega t - p\theta_m). \qquad [8.10]$$

Equation [8.10] quantifies the e.m.f. induced by the radial rotating field in a rotor conductor occupying any airgap position θ_m at any instant t. It reflects an axially directed electric field $\boldsymbol{E}_r(\theta_m, t) \equiv e_r/l$ at the rotor surface bounding the airgap, this field rotating synchronously with the radial flux density wave, in phase quadrature.

A synchronously rotating axial electric field $\boldsymbol{E}_s(\theta_m, t)$ also exists at the stator airgap boundary, indicated by eqn [8.9]. The magnitudes of the axial electric fields at the airgap bounding surfaces of rotor and stator are in proportion to the apparent speeds of the rotating radial flux-density wave with respect to the conductors located on these surfaces.

Equation [8.10] does not purport to show the e.m.f. variation with time *alone* for a rotor conductor as it moves at right-angles to the rotating radial magnetic flux produced generally by both stator and rotor currents. Such a conductor has position given by

$$\theta_m = \omega_r t + \theta_{mo} \qquad [8.11]$$

where θ_{mo} is its location at $t = 0$ when the rotating radial flux has its maximum value at a position in quadrature with the axis of phase a – if the stator windings alone carry current. From eqns [8.10] and [8.11]. the rotor conductor e.m.f. is

$$e_r(t) = \left(\omega_r - \frac{\omega}{p}\right) b l \hat{B}_n \sin[\omega t - p(\omega_r t + \theta_{mo})]$$

$$= -\frac{s\omega}{p} b l \hat{B}_n \sin(s\omega t - p\theta_{mo}) \qquad [8.12]$$

where

$$s = \frac{\omega/p - \omega_r}{\omega/p} = \frac{\text{synchronous speed} - \text{rotor speed}}{\text{synchronous speed}}. \qquad [8.13]$$

The quantity s is known as the *slip* and represents the speed *reduction* from the synchronous value per unit of synchronous speed. Equation [8.12] demonstrates that the *rotor* conductor e.m.f. due to its motion at speed ω_r in a radial field rotating at synchronous speed ω/p is at *slip frequency sf*.

8.3.1 Rotor induced current and magnetic field

The e.m.f. induced in each rotor conductor gives rise to conductor current of value determined by the conductor leakage impedance. Thus, from eqn [8.12], the rotor conductor current is

$$i_r(t) = -s \frac{\omega}{p} \frac{bl}{z} \hat{B}_n \sin[s\omega t - p\theta_{mo} - \phi_2] \qquad [8.14]$$

where $z \angle \phi_2$ is the leakage impedance of the conductor at *slip frequency*. Equation [8.14] shows that the induced current in an individual rotor conductor is at slip frequency. The rotor current *distribution* over the rotor surface is obtained by substituting from eqn [8.11] for θ_{mo} in eqn [8.14] to give

$$i_r(\theta_m, t) = -s \frac{\omega}{p} \frac{bl}{z} \hat{B}_n[s\omega t - p(\theta_m - \omega_r t) - \phi_2].$$

Substituting eqn [8.13]

$$i_{\mathrm{r}}(\theta_{\mathrm{m}}, t) = -s\frac{\omega}{p}\frac{bl}{z}\hat{B}_{\mathrm{n}}(\omega t - p\theta_{\mathrm{m}} - \phi_2). \qquad [8.15]$$

Equation [8.15] shows that the rotor-induced current distribution is essentially sinusoidally time and space variant, corresponding to a *travelling wave* of current. It will therefore establish rotating tangential and radial magnetic field components in the airgap.

If the current in a rotor conductor located at θ_{m} is spread over an element $b\Delta\theta_{\mathrm{m}}$ of rotor periphery, the corresponding distribution of *tangential* field over the rotor surface, given by eqn [7.10] with A the linear current density in A m^{-1}, is

$$H_{\mathrm{tr}}(\theta_{\mathrm{m}}, t) = -A(\theta_{\mathrm{m}}, t) = -\frac{i_{\mathrm{r}}(\theta_{\mathrm{m}}, t)}{b\Delta\theta_{\mathrm{m}}}. \qquad [8.16]$$

Thus, from eqns [8.15] and [8.16], the tangential field at the rotor surface due to the distributed rotor current is

$$H_{\mathrm{tr}}(\theta_{\mathrm{m}}, t) = \frac{s\omega}{p}\frac{l\hat{B}_{\mathrm{n}}}{z\Delta\theta_{\mathrm{m}}}\sin(\omega t - p\theta_{\mathrm{m}} - \phi_2). \qquad [8.17]$$

Now z is the leakage impedance at slip frequency of the rotor conductor shown in Fig. 8.5(b), whose current is spread over element $b\Delta\theta_{\mathrm{m}}$ of rotor periphery. Assuming a rectangular section of (small) depth h the 'impedivity' η of the rotor conductor at slip frequency is given by

$$\eta = z\frac{hb}{l}\Delta\theta_{\mathrm{m}}. \qquad [8.18]$$

Thus *impedivity* is a property of a conducting medium analogous to *resistivity* but incorporating inductive reactance effects at the appropriate frequency. Substitution for z in eqn [8.17] gives the tangential field at the rotor surface as

$$H_{\mathrm{tr}}(\theta_{\mathrm{m}}, t) = \frac{s\omega}{p}\frac{hb}{\eta}\hat{B}_{\mathrm{n}}\sin(\omega t - p\theta_{\mathrm{m}} - \phi_2). \qquad [8.19]$$

By analogy with the relationships between tangential and radial field components established by stator current and presented in eqns [8.6] and [8.7], the *radial* field across the airgap developed by the induced rotor current relates to that expressed in eqn [8.19] for the tangential field by a multiplying factor b/pg and a space-phase advance of $90°\varepsilon$.

Thus

$$H_{\mathrm{nr}}(\theta_{\mathrm{m}}, t) = -\frac{s\omega hb^2}{p^2 g\eta}\hat{B}_{\mathrm{n}}\cos(\omega t - p\theta_{\mathrm{m}} - \phi_2). \qquad [8.20]$$

8.3.2 Rotor tangential mechanical stress and torque

The tangential mechanical stress t_{tr} established at the rotor surface over the elemental peripheral area occupied by one rotor conductor $lb\Delta\theta_{\mathrm{m}}$ at radius b is given by eqns [3.22], [8.8] and [8.19].

$$t_{\mathrm{tr}}(\theta_{\mathrm{m}}, t) = B_{\mathrm{n}} H_{\mathrm{tr}}$$

$$= s\frac{\omega}{p}\frac{hb}{\eta}\hat{B}_{\mathrm{n}}^2 \sin(\omega t - p\theta_{\mathrm{m}})\sin(\omega t - p\theta_{\mathrm{m}} - \phi_2)$$

$$= s\frac{\omega}{p}\frac{hb}{\eta}\frac{\hat{B}_n^2}{2}\{\cos\phi_2 - \cos[2(\omega t - p\theta_m) - \phi_2]\}.$$

[8.21]

This result is significant in showing that the tangential stress over each elemental area of rotor surface ideally takes the form of a sinusoidally distributed, double-frequency travelling wave of stress superimposed upon a mean value proportional to the power factor of the rotor conductor. The time-dependent component, integrated over the entire rotor surface is zero at all times. The average value is the same for *all* such elementary areas of rotor surface and gives rise to a resultant *instantaneous* rotor torque of value

$$T_r = b^2 l \int_0^{2\pi} t_{tr}(\theta_m)_{\text{mean}} \, d\theta_m$$

$$= \pi s \frac{\omega}{p} \frac{b^3 lh}{\eta} \hat{B}_n^2 \cos\phi_2.$$

[8.22]

Equation [8.22] indicates that the rotor torque in the direction of rotor rotation is proportional to the product of slip, rotor-circuit power factor and (radial flux density)^2. Further algebraic manipulation identifies the motoring torque with the power absorbed by the rotor-circuit resistance.

Consider the elemental rotor conductor identified in Fig. 8.5(b), located at θ_m and having cross-sectional area $hb\Delta\theta_m$ to axial current flow over length l. With impedivity η and power factor $\cos\phi_2$ the *resistance* of the conductor is given by

$$r_c = \eta \frac{l}{hb\Delta\theta_m} \cos\phi_2.$$

The current i_r through the elemental rotor conductor is given by eqn [8.14] and has r.m.s. value

$$I_r = \frac{1}{\sqrt{2}} s \frac{\omega}{p} \frac{bl}{z} \hat{B}_n$$

Since $r_c = z\cos\phi_2$ the average power dissipated in the elemental rotor conductor resistance is

$$I_r^2 r_c = \frac{1}{2}\left[s\frac{\omega}{p}\frac{bl}{z}\hat{B}_n\right]^2 \frac{l}{hb\Delta\theta_m} \eta \cos\phi_2.$$

The *total* average power dissipated in all such elemental rotor conductors is given by

$$\sum I_2^2 r_c = \frac{2\pi}{\Delta\theta_m} I_r^2 r_c = s\frac{\omega}{p}\left[\pi s \frac{\omega}{p} \frac{b^3 lh}{\eta} \hat{B}_n^2 \cos\phi_2\right].$$

The term within the square brackets on the right-hand side of the above equation may be recognised from eqn [8.22] for torque, whilst the ratio ω/p is synchronous speed ω_o measured in mech rad s^{-1}. The product $\omega_o T$ with torque T measured in N m is known as the *torque in synchronous watts*. Since $\sum I_r^2 r_c$ is the *total* rotor-circuit copper loss, which is equal to the copper loss per rotor phase multiplied by the number of rotor phases, we get the important result for the polyphase induction machine

$$\text{Torque (sync W)} = \frac{\text{total rotor circuit copper loss (W)}}{\text{slip}}. \quad [8.23]$$

8.3.3 Maximum torque condition at constant radial flux

Equation [8.22] shows that torque at constant flux is proportional to the product of slip and rotor-circuit power factor. If, with a slight change in notation we designate each closed rotor circuit to have resistance R_2 and *standstill* reactance X_2, the impedance of each rotor circuit or phase will be $(R_2^2 + s^2 X_2^2)^{1/2}$ at slip s. When an induction machine is loaded, its speed, and hence the slip, will change until the electromagnetic torque matches the load torque. Maximum torque at constant flux will be developed at the value of slip s for which

$$\frac{d}{ds}(s \cos \phi_2) = \frac{d}{ds}\left[\frac{sR_2}{(R_2^2 + s^2 X_2^2)^{1/2}}\right] = 0,$$

i.e. when

$$(R_2^2 + s^2 X_2^2)^{1/2} R_2 - sR_2(R_2^2 + s^2 X_2^2)^{-1/2} 2sX_2^2 = 0$$

giving

$$R_2^2 = s^2 X_2^2$$

or

$$s = \pm R_2/X_2. \quad [8.24]$$

Equation [8.24] shows that maximum torque is achieved when the rotor leakage reactance per phase equals the rotor resistance per phase. The $\pm$ values for slip correspond to motoring and generating conditions. The corresponding value of rotor circuit power factor angle ϕ_2 at maximum torque is 45°.

An alternative expression for torque in terms of the resultant radial flux density and the radial field may be deduced from eqn [8.21] by substitution of *one* $\hat{B}_n$ factor with the equivalent quantity deduced from eqn [8.20], i.e.

$$\hat{B}_n = \hat{H}_{nr} \frac{p^2 g \eta}{s \omega h b^2}$$

giving

$$t_{tr}(\theta_m)_{mean} = \frac{pg}{b} \hat{H}_{nr} \frac{\hat{B}_n}{2} \cos \phi_2.$$

Integration of t_{tr} over 2π radians and multiplication by radius b gives rotor torque

$$T_r = \pi pblg \hat{B}_n \hat{H}_{nr} \cos \phi_2$$
$$= \mu_0 \pi pblg \hat{H}_n \hat{H}_{nr} \sin(90° + \phi_2). \quad [8.25]$$

Equation [8.25] is identical with eqn [7.14] for the torque developed on the rotor of a synchronous machine in terms of the rotor radial field component, the resultant radial field and the space-phase angle between them. The torque is described as *alignment* torque as it acts in such a direction as to encourage alignment of the field phasors in space. A positive value for rotor torque with $0 < \phi_2 < 90°$ confirms rotor torque in the direction of rotation, i.e. motor action.

The space-phase relationship between H_n and H_{nr} when the machine is motoring is shown in Fig. 8.6, corresponding to eqns [8.8] and [8.20]. For completeness the *stator* radial field component H_{ns} has been included, in accordance with the requirement that $H_{ns} + H_{nr} = H_n$.

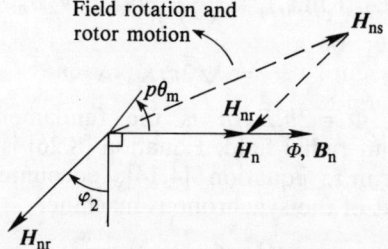

Figure 8.6 Space-phase relationship between stator and rotor components and resultant radial field phasors for induction machine motoring. Positions are depicted at $t = \pi/2\omega$. H_{nr} leads H_n by $(90° + \phi_2)$

8.4 Equivalent Circuit and Phasor Diagram of a Polyphase Induction Machine

8.4.1 *Rotor circuit representation*

Each phase of the rotor windings has induced e.m.f. at slip frequency absorbed by the voltage drop across its impedance which is made up of resistance and leakage reactance. Equation [8.12] shows that the magnitude of the induced e.m.f. is proportional to slip; each rotor phase may therefore be represented by an equivalent circuit of the form shown in Fig. 8.7(a), in which E_2 and X_2 correspond to induced e.m.f. and leakage reactance at $s = 1$, when the rotor is at standstill. In a cage rotor with c conductors per pole pair there will be cp such circuits, with e.m.f.s and currents in adjacent circuits being mutually separated by $360/c°\varepsilon$. If the rotor winding is configured as a three-phase winding then each phase will be modelled similarly using phase-circuit values X_2 and R_2 and a value for E_2 equivalent to the phasor sum of the component e.m.f.s, allowing for the fundamental frequency distribution and pitch factors of the winding $K_{w2} = K_{dp1} = K_{dl}K_{pl}$. Winding factors are discussed in Sections 5.3 and 5.8.2. The currents in each phase group of rotor conductors will be in phase with each other, but will be mutually displaced in time between phases by one third of a period at slip frequency, in sympathy with the phase e.m.f.s.

As stated in eqn [8.23], torque in synchronous watts equals the rotor copper loss represented by $I_2{}^2R_2$ in Fig. 8.7(a) divided by the slip s and multiplied by the number of rotor phases. If the rotor phase equivalent circuit is replaced by one shown in Fig. 8.7(b) in which the impedances are increased in value by the factor $1/s$ and the source e.m.f. by the same factor, rotor current per phase I_2 will be unchanged and the same magnetic field effect will be retained, but the real power absorbed by the circuit resistance R_2/s now represents the phase contribution to *torque* measured in synchronous watts. The source e.m.f. E_2 is now independent of slip and related to the radial flux density B_n via eqn [8.12] with $s = 1$. Thus

$$E_2(\text{r.m.s.}) = E_r(\text{r.m.s.}) = \frac{1}{\sqrt{2}}\frac{\omega}{p}bl\hat{B}_n.$$

For a rotor phase winding with N_2 turns and fundamental winding factor K_{w2}

$$E_2(\text{r.m.s.}) = \frac{2}{\sqrt{2}}\frac{\omega}{p}blK_{w2}N_2\hat{B}_n$$

$$= \sqrt{2}\pi f K_{w2} N_2 \Phi \qquad [8.26]$$

where $\Phi = 2bl\hat{B}_n/p$ is the fundamental flux/pole due to the resultant radial field. Equation [8.26] is identical with the transformer e.m.f. equation [4.14], encountered as eqn [7.26] in the context of the synchronous machine.

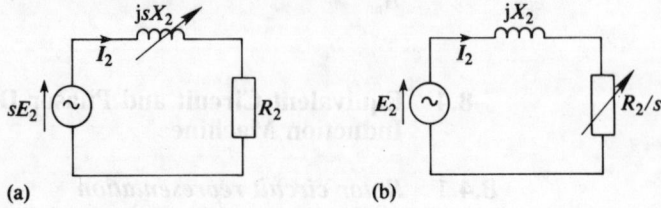

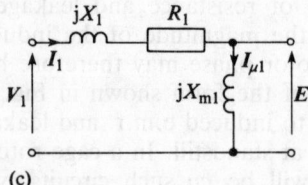

Figure 8.7 Alternative equivalent circuits (a) and (b) for the rotor circuit of a polyphase induction motor with rotor current invariance. (c) Stator (primary) equivalent circuit.

8.4.2 Stator circuit representation

An equivalent circuit for one phase of the stator winding of the induction machine is shown in Fig. 8.7(c), where R_1 and X_1 are the resistance and leakage reactance/phase and E_1 is the effective e.m.f./phase due to the summation of the e.m.f.s induced in the individual stator phase conductors by the rotating *radial* airgap flux. The instantaneous value of conductor e.m.f. is given in eqn [8.9], from which it follows that the r.m.s. value of E_1 for a stator phase winding of N_1 turns with fundamental winding factor $K_{w1} = K_{dp1} = K_{d1}K_{p1}$ is

$$E_1(\text{r.m.s.}) = \frac{2}{\sqrt{2}}\frac{\omega}{p}blK_{w1}N_1\hat{B}_n$$

$$= \sqrt{2}\pi f K_{w1} N_1 \Phi. \qquad [8.27]$$

Equation [8.26] and [8.27] show that E_1 and E_2, induced by the common radial flux, have magnitudes related by the effective stator/rotor phase turns ratio.

Under no-load conditions B_n is established by the stator current alone, with $\hat{H}_{ns}$ related to the peak value of no-load stator current/phase via eqn [8.6] for a *sinusoidal* distribution of current/pole/phase

$$\frac{\hat{H}_{ns}}{\hat{I}_a} = \frac{3}{2}\frac{n}{2g} \qquad [8.28]$$

A practical stator phase winding with N_1 turns gives rise to a *trapezoidal* distribution of radial field. Appropriate Fourier analysis giving rise to eqn [5.12] shows that the following substitution for n, the number of phase conductors per pole, should then be made in eqn [8.28] to establish the fundamental component

$$n = \frac{4}{\pi}K_{wl}\frac{N_1}{p}. \qquad [8.29]$$

The fundamental radial flux/pole established by the stator current under these conditions of zero rotor current is correspondingly

$$\Phi = \mu_0\hat{H}_{ns}\frac{2}{\pi}\frac{2\pi bl}{2p}.$$

Substitution for Φ in eqn [8.27] shows that the magnetising reactance X_{ml} per phase referred to the stator winding, which is equal to the ratio of r.m.s. quantities E_1/I_a, is given by the relationship

$$X_{ml} = \frac{E_1}{I_{ao}} = 2\left[\frac{\omega}{p}\right]K_{wl}N_1\mu_0 bl\left[\frac{\hat{H}_{ns}}{\hat{I}_{ao}}\right]$$

$$= 3\omega\left[\frac{K_{wl}N_1}{p}\right]^2\mu_0\frac{2bl}{\pi g}. \qquad [8.30]$$

The expression given in eqn [8.30] for the magnetising reactance of a three-phase induction machine is identical with that given in eqn [7.28] as the armature reaction reactance of a three-phase synchronous machine. Whereas, however, synchronous machines usually have large airgap length g to keep the synchronous reactance relatively low, induction machines have a short airgap length to keep the magnetising reactance high and the *magnetising current*, $I_{\mu l}$ in Fig. 8.7(c), small. Nevertheless, in small machines, the magnetising current may approach 50% of the full-load stator current value.

The stator (primary) and rotor (secondary) equivalent circuits of Fig. 8.7(b) and (c) are linked in such a way that the e.m.f.s E_1 and E_2 are related by turns ratio, whereas the currents I_1, I_2 and $I_{\mu l}$ satisfy the constraints imposed by the space-phase relationship between the components of radial field. Thus, as illustrated in Fig. 8.6,

$$H_{ns} + H_{nr} = H_n$$

where H_n is established by stator magnetising current/phase $I_{\mu l}$ when $H_{nr} = 0$.

The magnetic effect of the rotor current on the *radial* field is therefore such as to require that due to the stator current to modify its magnitude and position so that the *resultant* radial field is unchanged, retaining its synchronous motion relative to the stator conductors. The stator phase windings have therefore to carry additional components of current which, in combination, generate a resultant synchronously rotating radial component which balances precisely, in magnitude and position, the radial field of the rotor currents. Such a compensating field is shown as $-H_{nr}$ in the space-phasor diagram of Fig. 8.8(a). It is apparent that the magnitude and *space*-phase relationships between H_n and $-H_{nr}$ displayed here must be identical with the magnitude and *time*-phase relationships between $I_{\mu 1}$ and $-I'_2$ on the time-phasor diagram of Fig. 8.8(b), where $-I'_2$ is the additional stator current required in one phase to generate the compensating field component required. The process is analogous to the 'referring' of the secondary winding current of a loaded transformer to the primary winding and likewise involves a change in the number of turns.

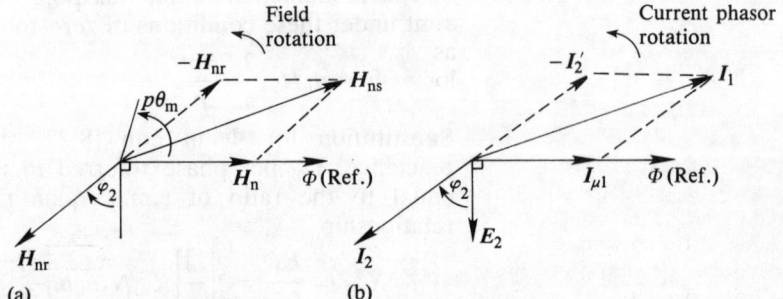

Figure 8.8 (a) Space- and (b) time-phasor diagram of a polyphase induction machine related such that resultant radial flux/pole Φ identifies with the magnetising current of reference stator phase.

Thus $I'_2 = I_2 K_{w2} N_2 / K_{w1} N_1$ and the rotor e.m.f./phase E_2 may similarly be referred to the stator number of turns/phase as $E'_2 = E_2 K_{w1} N_1 / K_{w2} N_2$. Equations [8.26] and [8.27] demonstrate that $E'_2 \equiv E_1$, enabling the stator and rotor equivalent circuits per phase to be linked as shown in Fig. 8.9. This equivalent circuit retains the isolating properties of an ideal transformer, with primary/secondary phase turns ratio $K_{w1} N_1 / K_{w2} N_2$. In most applications of the equivalent circuit it is convenient to eliminate the ideal transformer by *referring* the rotor circuit impedances to the stator number of turns, such that

$$\frac{R'_2}{s} = \left[\frac{K_{w1} N_1}{K_{w2} N_2}\right]^2 \frac{R_2}{s}$$

and

$$X'_2 = \left[\frac{K_{w1} N_1}{K_{w2} N_2}\right]^2 X_2. \qquad [8.31]$$

Further simplification may be achieved by moving the magnetising branch to the stator supply terminals, but such an approximation is less valid than was the case with the static transformer of Chapter 4, due to the relatively low value of magnetising reactance X_{m1} when compared with $R'_2/s + jX'_2$ at low values of slip s. The equivalent circuit of Fig. 8.10(a) retains the correct location for the magnetising branch, within which is incorporated a resistance element R_{m1} to account for the losses in the iron core. Representation of iron loss by an equivalent fixed resistor in parallel with X_{m1} validly identifies dependence on (radial flux density)^2 but does not

Figure 8.9 (a) Equivalent circuit and (b) time-phasor diagram for a motoring polyphase induction machine.

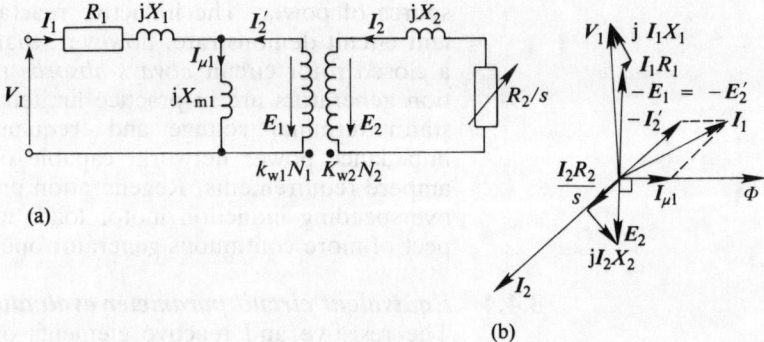

adequately reflect the effect of the much reduced *rotor* iron loss at low values of slip. Frequently, for simplicity, the rotational *mechanical* losses due to windage and friction are lumped with iron losses as a shunt resistor equivalent. Since the mechanical rotational losses are higher at the higher speeds associated with lower slip values, this artifice provides an element of compensation for the lower iron-loss element then applicable.

Figure 8.10 (a) Equivalent circuit and (b) time-phasor diagram for a polyphase induction machine referred to the stator winding.

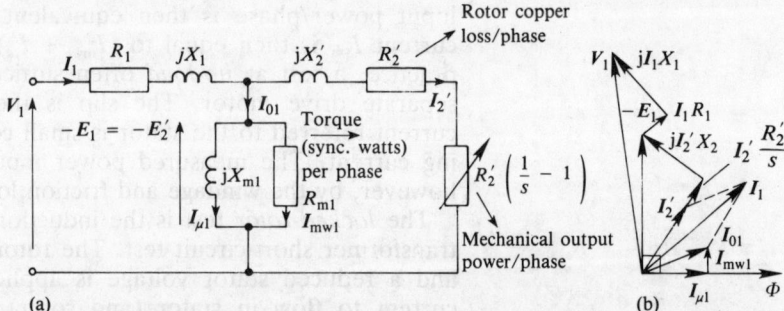

Figure 8.10(a) shows the effective rotor resistance/slip element split into components R'_2 and $R'_2(1/s - 1)$. This recognises that the input power to the rotor circuit, transferred across the airgap, equates to the sum of the rotor copper loss and the electrical equivalent of the mechanical output power. Equation [8.23] has shown that the sum of these components of electrical power equals the developed torque in synchronous watts.

8.4.3 Induction generator action

The reference directions for the electrical variables of current and voltage shown in Fig. 8.10(a) are those adopted for the transformer, as illustrated in Fig. 4.8. The phasor diagrams of Fig. 8.10(b) and Fig. 4.9 are comparable in form, corresponding in each case to power flow, with current at lagging power factor, absorbed from the primary supply for transmission to a secondary circuit load. As with the static transformer, power flow in the induction machine is essentially reversible. Generator action is achievable when the rotor is driven at above synchronous speed so that the slip s is of negative value. The effective representation of the mechanical load R'_2/s then becomes negative resistance and, correspondingly, a source of power. The inductive reactance elements of the equivalent circuit demonstrate, however, that the induction machine with a closed rotor circuit *always absorbs* reactive volt-amperes. Induction generators are in practice limited to applications in which the stator terminal voltage and frequency are defined by a low-impedance power network, capable of meeting the reactive volt-ampere requirements. Regeneration provides braking torque for an overspeeding induction motor load; wind turbines offer the prospect of more continuous generator operation.

8.4.4 *Equivalent circuit parameter evaluation by test*

The resistive and reactive elements of the equivalent circuit of a polyphase induction machine are readily determined by practical test procedures which are analogous to the open-circuit and short-circuit tests applied to a transformer as discussed in Section 4.9. The equivalent of the open-circuit test is strictly a test carried out at *synchronous* speed and rated stator supply voltage with the rotor current zero. The input impedance per stator phase would then correspond to the relatively low stator winding impedance $R_1 + jX_1$ in series with the magnetising branch comprising R_{m1} in parallel with jX_{m1} (Fig. 8.10(a)). Ignoring stator impedance the input power/phase is then equivalent to V_{ph}^2/R_{m1} and the input current I_{ph} is then equal to $(I_{mwl}^2 + I_{\mu l}^2)^{1/2}$, with $X_{m1} = V_{ph}/I_{\mu 1}$. In practice, a test at *no-load* often suffices, avoiding the need for a separate drive motor. The slip is then very low and the rotor current referred to the stator is small compared with the magnetising current. The measured power input is significantly increased, however, by the windage and friction losses of the rotor.

The *locked-rotor* test is the induction machine equivalent of the transformer short-circuit test. The rotor is prevented from rotating and a reduced stator voltage is applied such as to permit rated current to flow in stator (and rotor) windings. Measurement of stator input power, voltage and current/phase enables values of equivalent resistance and leakage reactance/phase, referred to the stator winding, to be deduced, with the effect of magnetising current being ignored. Under these conditions, iron losses are low in consequence of the reduced radial flux. Mechanical losses are non-existent but magnetising current is a not-insignificant part of the stator current. If the 'accurate' equivalent circuit is to be modelled, it is usual practice to assign equal values to stator impedance/phase and that of the rotor *when referred to the stator*.

Worked example 8.1 A 12-pole, 420 V, 10 kW, 50 Hz, three-phase induction motor yielded the following test results.

Open-circuit test 420 V 6.7 A 500 W

(including 230 W mechanical loss)

Locked-rotor test 99 V 14.0 A 980 W

Calculate the parameters of the equivalent circuit per phase with an assumed star-connected stator winding, assigning the copper losses and leakage reactance VA$_r$ equally between stator and rotor windings. Determine also the torque, mechanical output power, input line current, power factor and efficiency if the machine operates with a slip of 0.03, using the approximate equivalent circuit.

If the stator phase windings are actually connected in delta, estimate the value of resistance/phase likely to result from a direct measurement.

Solution Appreciating that the open-circuit test data relate to line values of voltage and current and total three-phase power, on an equivalent star basis

$$R_{m1} = \frac{V_{ph}^2}{P_{ph}} = \frac{420^2 \times 3}{3 \times 270} = 653\ \Omega,\ I_{m1} = \frac{420}{\sqrt{3} \times 653} = 0.371\ \text{A}.$$

Therefore

$$I_{\mu 1} = (6.7^2 - 0.371^2)^{1/2} = 6.689\ \text{A},\ X_{m1} = \frac{420}{\sqrt{3} \times 6.689}$$

$$= 36.25\ \Omega.$$

From the short-circuit test data

$$Z_{eq_1} = \frac{99}{\sqrt{3} \times 14} = 4.083\ \Omega,\ R_{eq_1} = \frac{980}{3 \times 14^2} = 1.667\ \Omega.$$

Therefore

$$X_{eq_1} = (4.083^2 - 1.667^2)^{1/2} = 3.727\ \Omega.$$

Allocating the equivalent leakage impedance Z_{eq_1} equally between stator and referred rotor circuits, the equivalent circuit is as shown in Fig. 8.11(a). The approximate equivalent circuit with the magnetising branch moved to the input terminals is shown in Fig. 8.11(b).

Figure 8.11 Equivalent circuits per phase of an equivalent star-connected stator winding for the polyphase induction motor of Worked Example 8.1: (a) 'exact', (b) approximate.

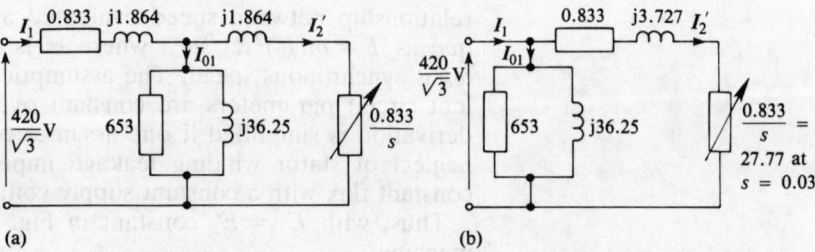

From the approximate equivalent circuit with $s = 0.03$.

$$I_2' = \frac{420}{\sqrt{3}}(27.77 + 0.833 + j3.727)^{-1} = 8.337 - j1.086\ \text{A}$$

$$= 8.41\ \angle -7.42°\ \text{A}.$$

Torque $= 3(I_2')^2 R_2'/s = 3 \times (8.41)^2 \times 27.77 = 5892$ sync. W.

Synchronous speed $\omega_o = 2\pi f/p = 100\pi/6$, hence

$$\text{torque} = 5892/\omega_o = \frac{5892 \times 6}{100\pi} = 112.53\ \text{N m}.$$

Gross output power $= 5892(1-s) = 5892 \times 0.97 = 5715$ W.

After allowing for windage and friction losses, output power = $5715 - 230 = 5485$ W.

Total copper losses $= 3(I_2')^2(R_1 + R_2') = 3 \times (8.41)^2 \times 1.667$

$$= 354\ \text{W}.$$

Including core and mechanical losses, total loss = 354 + 500
$$= 854 \text{ W}.$$

$$\text{Efficiency} = \text{output}/(\text{output} + \text{losses}) = \frac{5485}{5485 + 854} = 86.5\%$$

Input line current $I_1 = I'_2 + I_{01} = 8.337 - j1.086 + 0.371 - j6.689$
$$= 8.708 - j7.775 = 11.67 \angle 41.76°$$
$$= 11.67 \text{ A at } 0.746 \text{ power/factor lag}.$$

$R_1 = 0.833\ \Omega$ is the value of stator resistance per phase of equivalent star-connected winding. For the same power developed in delta configuration with same line voltage and current, impedance/phase must be increased by a factor of three. Due to skin effect the resistance of a machine winding under a.c. conditions exceeds that for d.c., so the value of stator phase resistance measured with d.c. current will be somewhat less than 2.5 Ω.

8.5 Speed/torque Characteristics of Polyphase Induction Machines

The variation of shaft speed with load torque is an important performance characteristic of any motor drive. Actual behaviour is dependent upon the characteristics of the source, and therefore speed/torque characteristics are usually presented on the basis of a constant-voltage, constant-frequency supply. For the induction machine, the equivalent circuit may be used to establish the relationship between speed evaluated as $\omega_r = (1 - s)\omega_o$ and torque as $T = m(I'_2)^2 R'_2/s\omega_o$, where m is the number of phases and ω_o is synchronous speed. The assumption is made that the equivalent circuit parameters are constant over the working range. The derivation is simplified if one assumes constant radial flux; further neglect of stator winding leakage impedance then associates the constant flux with a constant supply voltage if at fixed frequency.

Thus, with $E_1 = E'_2$ constant in Fig. 8.10(a), for a three-phase machine

$$\text{torque in synchronous watts} = 3(I'_2)^2 R'_2/s$$

$$= \frac{3(E'_2)^2}{\left(\dfrac{R'_2}{s}\right)^2 + (X'_2)^2} \frac{R'_2}{s} \qquad [8.32]$$

$$= \frac{3(E'_2)^2}{\left[1 + s^2\left(\dfrac{X'_2}{R'_2}\right)^2\right]} \frac{s}{R'_2}.$$

At low values of slip, such that $sX'_2 \ll R'_2$, the torque at constant flux is proportional to slip; at high values of slip, such that $sX'_2 \gg R'_2$ the torque is inversely proportional to slip. Maximum torque occurs when

$$\frac{d}{ds}\left(\frac{s/R'_2}{1 + s^2(X'_2/R'_2)^2}\right) = 0,$$

i.e. when

$$\left[1 + s^2\left(\frac{X'_2}{R'_2}\right)^2\right]\frac{1}{R'_2} = \frac{s}{R'_2}\left[2s\left(\frac{X'_2}{R'_2}\right)^2\right]$$

or when

$$s^2\left[\frac{X'_2}{R'_2}\right]^2 = 1 \text{ or } sX'_2 = \pm R'_2. \qquad [8.33]$$

Equation [8.33] has been earlier derived more fundamentally as eqn [8.24] for the maximum torque condition at constant radial flux. Speed/torque characteristics for various $X'_2/R'_2{}'$ ratios are presented in Fig. 8.12. It is apparent that, at constant flux, the value of maximum torque is independent of the standstill reactance/resistance ratio. This is evidenced by substitution of eqn [8.33] in eqn [8.32] to give

$$T_{\max} = \frac{3}{2}\left(\frac{E'_2}{X'_2}\right)^2 X'_2 = \frac{3}{2}\frac{(E'_2)^2}{X'_2}\text{sync W}. \qquad [8.34]$$

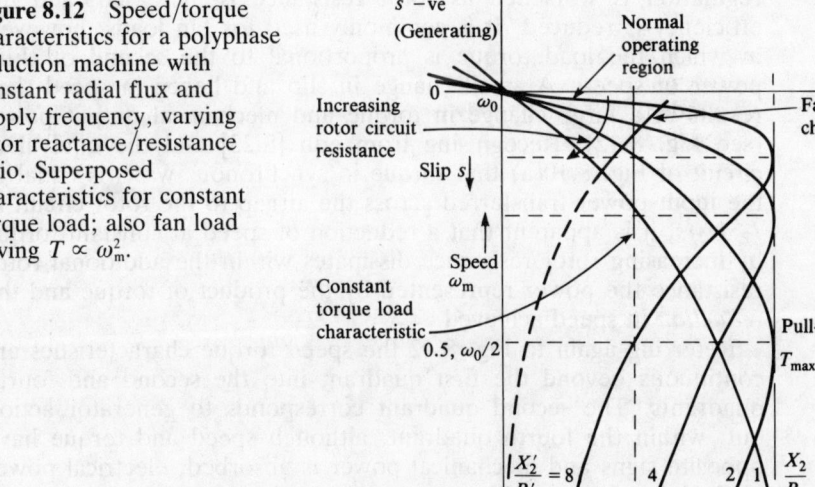

Figure 8.12 Speed/torque characteristics for a polyphase induction machine with constant radial flux and supply frequency, varying rotor reactance/resistance ratio. Superposed characteristics for constant torque load; also fan load having $T \propto \omega_m^2$.

Equation [8.34] yields the important result that the peak torque at constant radial flux is independent of slip and is inversely proportional to rotor leakage reactance. The 'pull-out' torque of an induction machine is thus primarily controlled by the rotor circuit *leakage reactance*. The rotor circuit *resistance* defines the value of slip at which a particular machine develops electromagnetic torque to balance the load torque, and the speed at which the motor will stall if the torque demand exceeds the pull-out value.

8.6 Starting and Speed Control of Polyphase Induction Motors

8.6.1 Wound rotor

The representative speed/torque curves with the constant radial flux assumption shown in Fig. 8.12 for several R_2/X_2 ratios extend beyond the first quadrant within which induction motors are normally operated, with R_2/X_2 low in order to develop a low value of slip when on load. Equation [8.23] shows that the rotor copper loss equals the product of torque (sync. W) and slip. Figure 8.12 demonstrates that low R_2/X_2 also provides good speed regulation, i.e. a small change in speed for a large change in torque. A low value of R_2/X_2 is accompanied by a low starting torque, however, on account of the fact that at $s = 1$ the rotor circuit power factor is low with the rotor current at supply frequency. For improved starting torque the rotor circuit resistance may be increased by including external resistance if a wound rotor with brush/slip-ring connections is employed. An external resistance can also provide speed control if appropriately rated, but this method of speed control is of limited value because the speed regulation is worsened as rotor resistance is increased, and the efficiency is reduced. It is commonly used for fan loads, however, in which the load torque is proportional to the second or third power of speed. A small change in slip and hence in speed then results in a large change in torque and mechanical output power (see Fig. 8.12). Recognising from eqn [8.23] and the equivalent circuit of Fig. 8.10(a) that torque in synchronous watts is equal to the input power transferred across the airgap to the rotor circuit at $I_2^2 R_2/s$, it is apparent that a reduction of speed at constant torque by increasing rotor resistance dissipates within the additional rotor resistance the power represented by the product of torque and the *reduction* in speed achieved.

Referring again to Fig. 8.12 the speed/torque characteristics are continuous beyond the first quadrant into the second and fourth quadrants. The second quadrant corresponds to generator action but, within the fourth quadrant, although speed and torque have opposite signs and mechanical power is absorbed, electrical power is also absorbed at the stator terminals. In this context $s > 1$ and positive, and therefore R_2/s is positive and the rotor circuit absorbs real power. The machine in this mode is undergoing 'plug-braking' or *plugging*. Energy derived from the mechanical load being braked and from the electrical supply is dumped in the rotor circuit resistance. Plug-braking torque may be increased by including additional resistance in the rotor circuit.

8.6.2 Cage rotor

The principal disadvantage of the cage rotor is that good performance on load is compromised by poor starting performance manifested as low torque at low power factor, and hence high current. Cage rotors may be designed with inherently low leakage reactance and hence high peak torque. High starting currents may be reduced by reducing the stator phase voltage during acceleration from standstill – e.g. by the use of a star/delta starter which connects the stator windings in star during starting but changes to

Induction machines   407

mesh for running. Consequential division of supply phase voltage by $\sqrt{3}$ reduces the *line* current by the factor 3, but reduces torque by the factor 3 also, with the power factor unchanged. Alternatively, a short-time rated autotransformer may be used to reduce the supply voltage during starting.

The benefits of a high-resistance rotor winding for improved motor starting performance, combined with a low-resistance rotor winding for efficient running are achieved in the *double-cage* rotor introduced by Boucherôt. This construction employs a high-resistance cage accommodated in small slots near the rotor surface together with low-resistance cage employing larger section conductors set in deep slots, as illustrated in Fig. 8.13(a). The lower cage has inherently a much higher leakage reactance than the upper cage and, with reactance the predominant cage impedance parameter at high slip frequencies, the higher power factor *starting* cage has a higher induced current during starting. When the machine has accelerated to near synchronous speed the predominant rotor impedance feature is resistance, and the low-resistance *running* cage is preferred to the electrically parallel starting cage.

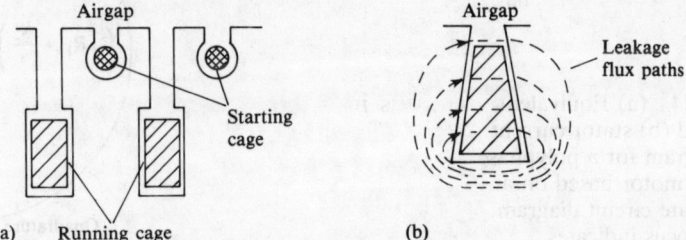

Figure 8.13 (a) Double-cage rotor section. (b) Deep slot cage rotor conductor.

An alternative to the double-cage rotor employs conductors proportioned so that the slot is deep. Illustrated in Fig. 8.13(b), the self-inductance component of the *deep-bar* conductor due to leakage flux which links only part of the conductor current is greatest at the bottom of the slot. This alters the distribution of current along the slot so that, when the slip frequency is high, the current is disposed to concentrate in the upper portion of the conductor, giving rise to an effective increase in conductor resistance. At low slip frequencies the current density is more uniform and conductor resistance is reduced.

8.7 Stator Current Locus of Polyphase Induction Machine

If the polyphase induction machine is represented by the approximate equivalent circuit illustrated in Fig. 8.14(a), I'_2, the rotor current referred to the stator winding, traces a circular locus with respect to the fixed stator voltage phasor as the slip is varied. Proof of this is established by noting that the voltage across the slip-variable quantity is proportional to I'_2 and, when added to a voltage component in quadrature, it yields a constant sum. That the point P in Fig. 8.14(b) lies on a circle follows from the geometric property that the diameter of a circle subtends an angle of 90° at any point on the circumference. A change of origin to O'

such that $O'O$ equals the magnetising current phasor I_{01} gives the locus of the stator current appropriate to constant supply voltage. The locus shows clearly regions of motor, generator and brake operation. Plug-braking is achieved by reversing the direction of the rotating airgap field through interchanging two stator supply leads. A maximum value of slip approaching 2.0 is achieved as the rotor torque reverses direction. The stator supply must be removed at the instant when the rotor comes to rest, or the rotor will accelerate in the reverse direction. The locus diagram of current shows that the stator current during plugging exceeds the standstill value.

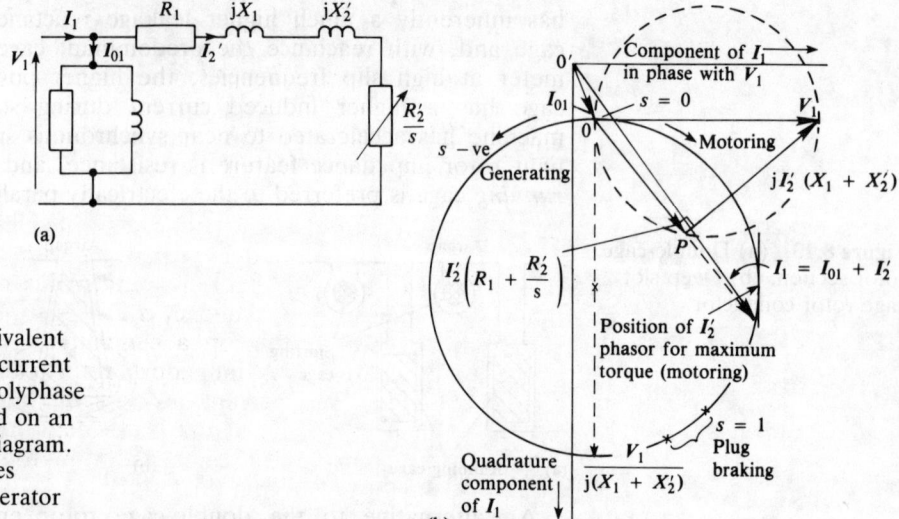

Figure 8.14 (a) Equivalent circuit and (b) stator current locus diagram for a polyphase induction motor based on an approximate circuit diagram. Circular locus indicates regions of motor, generator and brake operation.

8.8 Doubly Fed Induction Machines

The incorporation of external resistance in the rotor circuit of an induction machine is electrically equivalent to the insertion within the rotor circuit of a source of e.m.f. E_j, of slip frequency and having value $I_2 R_{2\text{ext}}$. As such, it represents a special case of a rotor circuit connected to a low-impedance voltage source. The general *doubly fed* induction machine admits no phase or magnitude restrictions on the source of rotor-injected e.m.f. E_j. The sole constraint is that the rotor e.m.f. must be at slip frequency for electromechanical energy conversion to occur.

The incorporation of a slip frequency source of e.m.f. then permits control of speed above and below the synchronous value, and also control of stator supply power factor, by the adjustment of the magnitude and phase of E_j. For supersynchronous operation E_j in Fig. 8.15 requires (in each rotor phase) a component in anit-phase with the rotor current. If, further E_j has a component lagging I_2 by 90° the magnetising volt-ampere requirements of the machine magnetic circuit may be met by this source, improving the power factor of the stator supply.

Figure 8.15 Equivalent circuit of a doubly fed polyphase induction machine with slip-frequency e.m.f. E_j.

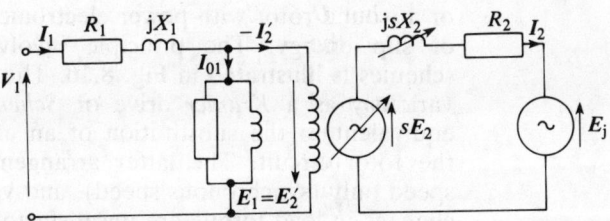

Thus Fig. 8.15 illustrates that the particular requirement I_2 for a given torque and flux might be met by an enhanced rotor-induced e.m.f. sE_2, offset by the injected e.m.f. E_j such that the difference $sE_2 - E_j$ equates to the rotor impedance voltage $I_2(R_2 + jsX_2)$. With sE_2 and E_j in phase, the larger $|E_j|$ is for a given torque, the larger s becomes, and the speed falls. As the rotor current and airgap flux are basically unchanged, the stator *power* is unchanged, and the surplus rotor-circuit power is absorbed by the injected e.m.f. source. The need for a slip-frequency source of controlled phase limits the versatility of the technique. Historically, however, many successful variable-speed induction-motor drives have been built which incorporate the frequency-changing property of the commutator, demonstrated by eqn [6.2] with the brush angle ψ proportional to time. Thus a radial airgap flux rotating at synchronous speed induces line frequency e.m.f.s as they appear between stationary brushes bearing on a commutator connected to rotor conductors, whereas the individual rotor conductors have slip frequency e.m.f. and current. A variable magnitude e.m.f. at supply frequency injected at the brush may then be used to vary speed. Alternating-current commutator motors provided with a voltage regulator operate on this principle.

Alternatively, the *Schrage* motor employs its primary winding on the rotor with the secondary winding on the stator. Slip-frequency e.m.f.s induced in the secondary circuits correspond to an airgap field rotating at slip frequency, the rotor moving with speed $(1-s)\omega_0$ in the direction opposite to that of the rotating field. The latter rotates at synchronous speed relative to the rotor conductors. An auxiliary d.c.-type closed winding, also located in the rotor slots, has coil connections to a commutator, such that the constant-amplitude supply frequency e.m.f.s induced in the conductors develop slip frequency e.m.f.s between stationary brush pairs bearing on the commutator. Three pairs of brushes for each pair of poles are placed symmetrically around the commutator to develop phase e.m.f.s at $120°\varepsilon$ separation, but of magnitude and phase relative to the primary winding e.m.f. determined by the number and position of commutator segments spanned by each brush pair. Each brush circuit is completed via one of the three stator phase windings for which it provides the injection e.m.f. Such machines, now effectively obsolete, found application as a.c. traction motors with a speed range ~ 3:1, including both sub- and super-synchronous operation.

The high initial cost and limitations on power and voltage rating imposed by mechanical commutation made this type of induction machine generally uncompetitive with d.c.-machine variable-speed drives, or alternative induction-motor drives based on a cage rotor, or a wound rotor with power electronic circuitry to enable recovery of slip energy. The principle involved in *slip-energy recovery* schemes is illustrated in Fig. 8.16. The drive system shown, known variously as a *Kramer* drive or *Scherbius cascade*, is not simply equivalent to the substitution of an active 'sink' for resistance in the rotor circuit. The latter arrangement provides for one base speed only (synchronous speed), and values of speed regulation on changes in load torque are unsatisfactory at high values of slip, i.e. low speed. These features are apparent from the speed/torque curves of Fig. 8.12 with high R_2'/X_2'.

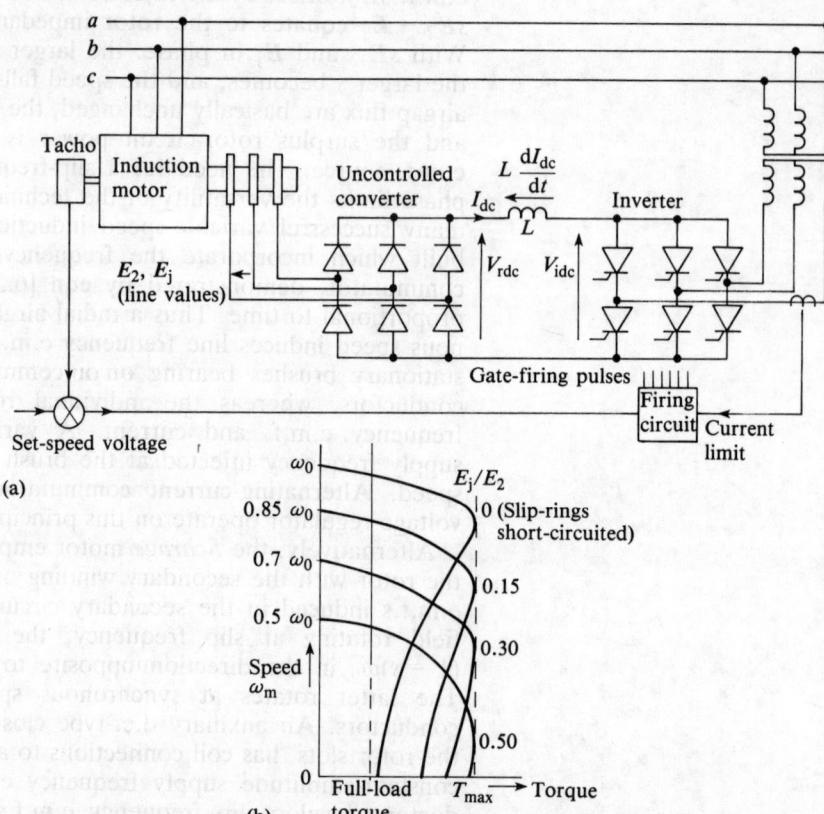

Figure 8.16 Slip-energy recovery scheme (a) for a slip-ring induction motor incorporating closed-loop control via feedback of rotor speed to compensate for the speed drop on increased load torque. An error signal provides advance or retard of gate-firing pulses about the preset value of firing angle α. (b) Speed/torque characteristics for a basic drive with E_j/E_2 equalling the ratio of the slip-ring voltage equivalent d.c. link voltage V_{rdc} to the open-circuit slip-ring voltage induced by radial flux; constant (radial) flux is assumed.

The Kramer drive, however, is normally operated such that the slip-ring voltage is maintained substantially constant at a fixed, though adjustable, value. With constant supply voltage at constant frequency maintaining essentially constant radial flux, changes in load torque are accommodated by changes in load current. When the load current is small, the rotor-induced e.m.f. roughly balances the impressed slip-ring voltage, obliging the no-load speed to be a proportion of that appropriate to short-circuited rings, i.e. synchronous speed. Where the applied torque is within the range to full load, the speed/torque curves at different slip-ring voltages are substantially parallel, and the same maximum torque is applicable as for constant flux.

The slip-ring voltage is effectively set by the d.c. link voltage.

This has identical *mean* values at both ends of the a.c. link under steady conditions: the value set by the firing angle of the inverter thyristors and the a.c. source voltage. Connected to a well-defined a.c. supply of constant voltage and frequency, a naturally commutated inverter is employed. The inverter is thus essentially a converter of the type considered in Section 6.15, operating with firing delay angle α in excess of 90°, and with high values of d.c. link voltage corresponding to α approaching 180°.

The rotor circuit of the induction motor, the d.c. link and the d.c. side of the inverter circuit are effectively in series, obliging the current flowing therein to be instantaneously the same. A feature of the circuit of Fig. 8.16(a) is the large inductance in series with the d.c. link. One function of this is to smooth the link current, developing, in the steady state, identical *mean* voltages at the two ends of the link whilst permitting large *instantaneous* voltage differences. As the *rectification* is uncontrolled, the fundamental component of the slip-ring current, the induction-motor rotor current, is in phase with the substantially sinusoidal slip-ring voltage. Harmonic current components due to the smoothing effect of the link inductance increase the VA_r demand of the induction motor stator above that required by fundamental-frequency magnetising and leakage inductances.

Similarly, the harmonics present in the inverter line currents increase the VA_r absorbed from the connected system, the demand for which is further increased due to the fact that the fundamental component of inverter a.c. line current lags the a.c. supply voltage by the supplement of the delay angle α. If the induction machine functions at constant radial flux, a constant torque load will require a constant rotor current, and hence a constant link current and inverter a.c. line current. Therefore the VA_r demand with a constant torque load is most significant at low values of slip for which the rotor voltage would also be low and comparable with the rectifier diode conduction voltage drops. External rotor resistance may be used to facilitate limited speed control in this region near synchronous speed, in addition to starting duties.

The role of the d.c. link inductance is not simply to minimise current ripple but also to provide a reservoir of stored energy which plays an important role when the operating conditions change. Suppose that the induction motor is supplying a constant torque load at constant speed when a call for reduced speed is made. The speed reduction is initiated by increasing the inverter delay angle α to nearer 180°. This raises the mean d.c. voltage V_{idc} at the inverter end of the d.c. link. Initially, there is no change in the mean d.c. voltage at the rectifier end of the link and a negative mean voltage difference equivalent to $L\ dI_{dc}/dt$ is established across the link inductance, thus obliging the link current to fall and reducing its stored energy. The induction-motor rotor current falls in sympathy and reduces the electromagnetic torque. The deficiency in torque provision to the load is met by deceleration of the rotating masses. As speed falls slip increases, as does the rotor e.m.f. proportionately. Thus the mean d.c. voltage V_{rdc} at the rectifier end of the link increases to reduce the mean d.c. voltage difference across the link inductance. The rate of fall of link current reduces in consequence and, in due course, will reverse. Evidently, the rotor speed will exhibit damped oscillations about a new, lower value whilst the d.c. link current settles down once more at the initial value appropriate to the constant torque load.

8.9 Variable-speed Induction-motor Drives Employing Cage Rotors

Wound-rotor induction motors are inherently more expensive to construct than the cage-rotor alternatives, and the necessity to mount slip-rings on the rotor shaft is demanding of space. As an alternative to control of slip by variation of rotor-circuit resistance, the slip at a given load torque may be increased by reducing the supply voltage. Equation [8.32] shows that the peak torque developed at constant supply frequency is proportional to (supply voltage)^2 and occurs at a fixed value of slip. Reduction of supply voltage when the machine is on light load reduces the magnetising current and core losses, raising the motor power factor and efficiency. Anti-parallel connected thyristors placed in series with each supply line may be used to control the effective value of voltage applied to the stator windings. Operation at increased slip then means increased rotor copper loss and reduced efficiency, whilst a progressively distorted stator current waveform gives rise to parasitic torques and further increased losses.

Variation in speed may more satisfactorily be achieved by varying the *synchronous* speed value, i.e. that at which the airgap field moves with respect to the stator (primary) conductors. Equal to the ratio ω/p, synchronous speed is proportional to the supply frequency $f = \omega/2\pi$, or is inversely proportional to the number of pole pairs. Synchronous speed change for a constant-frequency supply is therefore achieved by altering the number of magnetic poles established by the stator winding currents.

This may be accomplished by several means, including single windings with changed interconnections or completely separate windings identified with different numbers of poles. The *pitch* of a coil may be fixed in terms of the number of slots embraced, but the *pitch factor* (Sect. 5.3.1) is determined by the relationship between the individual coil pitch and the separation between poles which is established by the windings of one phase taken together. A particular coil may appear short pitched for a low pole number but long pitched for a higher pole number. Common pole-number combinations are 4 and 8 with a 2:1 synchronous speed range. Overpitched coils are disadvantageous from the point of view of conductor resistance and leakage reactance, in contrast to the short-pitch coils usually employed from choice in a.c. windings.

An alternative to reconfiguring the stator winding for pole changing such that each phase produces a fundamental field distribution of the required pole number, is provided by *pole-amplitude modulation*. This facility extends the range of the pole combinations obtainable with a single compact stator winding. Each stator phase develops distributed fields with two dominant pole numbers, e.g. 6 and 10. When the three appropriately spaced phase windings are excited together with polyphase currents, the resulting rotating field has one pole number dominant with the other suppressed. Other space-harmonic fields are much in evidence. Machines incorporating p.a.m. windings may offer two or three synchronous speeds of close or wide ratio from a fixed-frequency supply, with ratings to several MW, commonly for fan or pump loads.

8.9.1 *Supply frequency control*

Variable-frequency supplies derived from a fixed-frequency, constant-voltage source via controlled converters are applicable to polyphase induction machines in much the same way as discussed primarily for synchronous machines in Section 7.33 *et seq*. Whereas an overexcited synchronous machine supplying VA_r to the system is able to commutate satisfactorily a naturally commutated voltage-source inverter, the inherent reactive power demand of the *singly excited* induction machine requires forced commutation. Gate turn-off thyristors and MOSFETS are increasingly employed in this application. Figure 8.17 shows schematically an induction-motor drive incorporationg slip compensation by feedback of the rotor speed such that, under steady-state conditions, the inverter frequency as defined by the set-speed voltage is enhanced by the factor $(1 + s)$. Speed regulation on load is thus reduced to a very small amount.

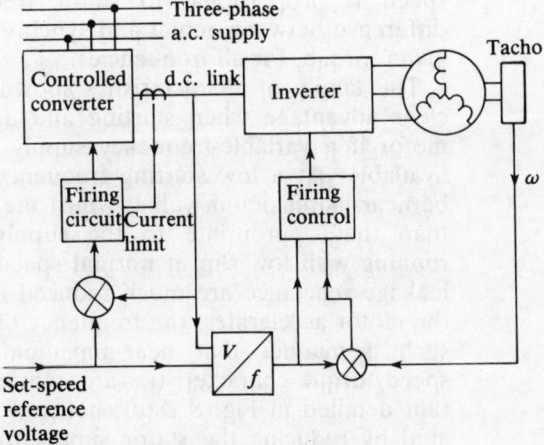

Figure 8.17 Variable-frequency induction-motor drive.

Speed/torque characteristics for an inverter-fed induction motor load at constant radial flux over a range of sinusoidal supply frequencies are shown in Fig. 8.18(c). The characteristics are essentially identical and displaced along the speed axis such that the torque is zero at synchronous speed for a particular frequency. It is readily demonstrated that the form of each speed/torque characteristic is independent of stator frequency as torque at constant radial flux is identified with the product of rotor current and power factor. For a given machine the latter quantities are then dependent solely on rotor frequency.

Figure 8.18 Speed/torque characteristics of a polyphase induction motor with variable-frequency sinusoidal supply voltage. Constant radial flux is assumed, and hence constant ratio E_1/f. X_2 = rotor leakage reactance/phase at frequency f_a. (a) Rotor equivalent circuit referred to stator at slip s_a and frequency f_a. (b) As for (a) at slip s_b and frequency f_b. (c) Speed/torque and stator current/torque for $f = 100$, 67 and 33 Hz.

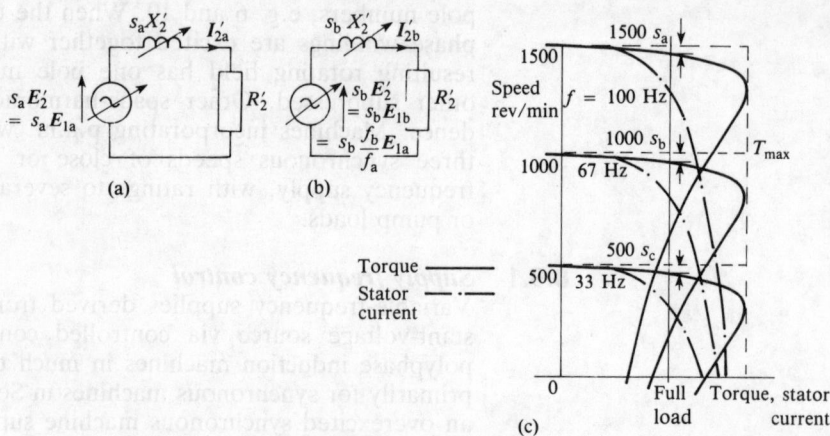

This is confirmed by reference to Fig. 8.18(a) and (b) in which the rotor equivalent circuits relate to that of Fig. 8.7(a) at different supply frequencies f_a and f_b. To maintain constant radial flux the *stator* winding e.m.f.s are related by the frequency ratio. Referred rotor currents I'_{2a} and I'_{2b}, identical in magnitude and phase, are established if $s_b f_b / f_a = s_a$ or $s_b f_b = s_a f_a$, i.e. the *rotor* frequencies are identical.

For each constant-frequency characteristic of Fig. 8.18(c), the divergence in speed from the synchronous value is given by the product of slip and synchronous speed. Since each synchronous speed is proportional to stator frequency it follows that the difference between actual and synchronous speeds is the same at a given torque, for all frequencies.

The family of characteristics shown in Fig. 8.18(c) illustrates a clear advantage when starting and accelerating a cage induction motor if a variable-frequency supply maintaining constant flux is available. At a low starting frequency the torque developed may be near its maximum value, whilst the starting current is much less than that appropriate to the supply frequency best suited to running with low slip at normal speed. The undesirable effects of leakage reactance are much reduced at the lower frequencies. As the motor accelerates, the frequency of the supply may be raised in such a manner that near-maximum torque is sustained. The speed/torque characteristics are continuous beyond the first quadrant detailed in Fig. 8.18(c), hence a motor load may be decelerated by reducing the stator supply frequency such that its speed exceeds synchronous speed. Slip is then negative, and regenerative braking occurs – provided that the inverter supply is capable of accepting the reverse power flow.

Referring to the arrangement of Fig. 8.17, the procedures corresponding to running up an induction motor and its connected load are as follows. The d.c. link voltage will initially be zero and the set-speed voltage raised to correspond to a selected speed. Gate-firing delay on the controlled converter will be reduced, raising the d.c. voltage until the d.c. link current limit is reached. Since the link voltage is low the inverter frequency is also low. If inverter voltage and frequency are held at constant ratio and a constant d.c. link current is sustained, the drive will accelerate with substantially constant torque near maximum for the given flux until the current limit becomes ineffective as the link voltage approaches the set-speed value. Inverter frequency is then determined primarily by the set-speed voltage, modified by the measured slip.

A recent development in a.c. machine drives known as *flux-vector* or *field-orientation* control seeks, when on steady load or undergoing transient changes, to maintain a space-quadrature relationship between specific field components analogous to that relating stator and rotor radial-field components for the normal d.c. machine. The object of such a facility is to provide independent control of developed electromagnetic torque at all speeds.

Reference to the equivalent circuit of the polyphase induction machine, Fig. 8.10(a), illustrates that any required torque at any speed is theoretically achievable by control of the stator current and the slip, via rotor frequency. Airgap radial flux then becomes a dependent variable and problems of magnetic saturation may occur unless the magnetising component of current is controlled. Operation at high slip values implies a high ratio of rotor copper loss to mechanical output power. One control strategy applicable to steady-state operation is to select a value of slip at which maximum overall efficiency occurs.

Further reference to Fig. 8.10(a) shows that, if the stator supply current and frequency are held constant, and slip is considered to decrease from unity towards zero, torque evaluated as the power developed in resistance R_2'/s (synchronous watts) will increase with speed. This is an unstable operating condition for a motor drive as an increase in load torque, although met initially by inertial torque derived from a reduction in rotational speed, is not subsequently balanced by an increase in electromagnetic torque – unless the source current is increased or the source frequency is lowered to reduce the slip. Hence a current-fed induction-motor drive requires closed-loop control. At lower slip values a high proportion of stator current is accounted for as magnetising current, thus increasing the prospect of magnetic saturation.

8.10 Harmonic Effects on Polyphase Induction-motor Performance

Practical induction motors may behave quite differently from that predicted from simple theory, due to harmonic fields that are invariably present to a greater or lesser degree. Harmonic fields develop net torque only when components with the same pole pitch are present on both sides of the airgap. The stator winding, fed with assumed sinusoidal current and voltage, establishes a spectrum of non-triplen space-harmonic fields and fluxes in addition to the dominant fundamental components. Relative amplitude is dependent upon the harmonic number and appropriate winding factor, and each harmonic field may appear to rotate in one direction or the other at subsynchronous speeds inversely proportional to the harmonic number (see Sect. 5.6). If the machine has a cage rotor with a large number of axial conductors, the rotor-induced current can respond to any stator-imposed pole pitch

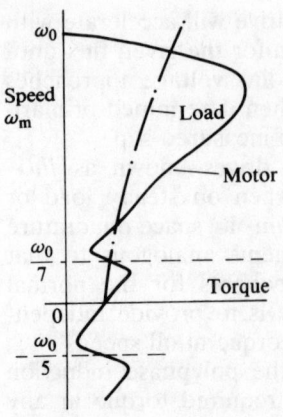

Figure 8.19 Speed/torque characteristic of polyphase induction motor incorporating effects of fifth and seventh space-harmonic fields. For the laod characteristic shown motor crawls at $\omega_0/7$.

not too far removed from the fundamental. Thus the predominant fifth and seventh harmonic fields may develop torque/speed characteristics analogous to that of the fundamental but identified with synchronous speeds respectively at $-1/5$ and $+1/7$ that of the fundamental. The negative torque component due to the seventh harmonic developed when an accelerating induction motor has just passed through $1/7$ of the fundamental synchronous speed may cause sufficient reduction in total torque that the motor fails to accelerate further, if the load torque is relatively high. Under these conditions, the motor is said to *crawl*. Figure 8.19 illustrates these conditions and also demonstrates that the effect of the fifth space harmonic is to modify the behaviour of the machine when plug-braking.

The fact that rotor and stator windings are set in slots exacerbates the practical difficulties. With the numbers of slots on both sides of the airgap simply related, crawling effects at certain critical speeds may be apparent, or 'cogging' may occur with the machine failing to move away from standstill. The local variations in airgap length due to the slots complicate the field distribution whilst the relative motion at an increasing rate as the motor accelerates further complicates the issue. In order to minimise slotting effects, most machines incorporate 'skewing' on stator or rotor, with the active conductors displaced from the true axial direction by a small amount – typically one or two slot pitches over the conductor length. Skewing has the general beneficial effects of reducing torque pulsations and vibration as the rotor and stator teeth alignment changes.

Induction motors energised from non-sinusoidal sources, whether voltage- or current-fed, are subject to *time* harmonics of current and voltage. It is usual to account for the contribution of each harmonic to the mean torque by use of the conventional equivalent circuit with parameters appropriate to the harmonic frequency. Supply voltage and current harmonics are then related by the effective impedance referred to the stator. Such time harmonics contribute small forward or reverse torque components.

Taking into account the $\pm 120°$ space-phase relationship for the phase windings of the three-phase machine, the principal field and flux waves due to the fifth time harmonic rotate at five times the fundamental synchronous speed, but in the opposite direction to that of the fundamental, whereas the principal field and flux due to the seventh time harmonic rotate at seven times the fundamental synchronous speed in the same direction as the fundamental. For an induction motor running with fundamental slip s the value per unit of slip appropriate to the fifth time-harmonic field is $(6 - s)/5$ and for the seventh time harmonic $(6 + s)/7$. On account of the high leakage reactance values at harmonic frequencies, and harmonic slip values ~ 1.0, harmonic contributions to the net torque tend to be small with, for example, the braking torque of the fifth harmonic partially cancelled by the motoring torque of the seventh harmonic.

Rotating-field and radial-flux components due to time harmonics of differing frequencies interact to develop torque components which pulsate about an average value of zero. The most significant of these is due to the fifth and seventh current harmonics which interact with the fundamental radial flux to generate a resultant pulsating torque varying at six times the supply frequency. Induction machines fed from non-sinusoidal sources are noisier and less efficient than those provided with sinusoidal voltage and current.

8.11 The Single-phase Induction Machine

In many domestic and light-industrial applications there is a need for a robust and cheap motor of nominally constant speed, with rating ~ 1 kW and for which a single-phase supply of fixed voltage and frequency is readily available. A washing-machine pump motor provides such an example, others include a circulating pump motor for domestic central heating and a refrigerator compressor motor. At these ratings the single-phase induction motor, with invariably a cage rotor construction, is broadly comparable in cost and performance with the three-phase alternative. The basic machine possesses the fundamental inability to develop starting torque, but several alternative techniques are available to overcome this deficiency.

The cage rotor of the single-phase machine is theoretically indistinguishable from that employed with the polyphase induction machine. Electromotive force and current are induced therein by changes in airgap radial flux defined by the stator supply voltage, if stator winding impedance is neglected. Induced rotor currents modify the stator current and produce tangential field components in the airgap which interact with radial flux density to develop tangential mechanical stress. Integrated over the rotor surface, such tangential stresses give rise to resultant torque in the direction of motion for a range of speeds from almost standstill to just below synchronous speed. When in motion in either direction at general speed ω_m within this range, the rotor torque is naturally in the direction of rotor rotation, appropriate to motor action. If the machine is driven above synchronous speed, the net torque reverses direction and generator action results, with the machine absorbing VA_r.

8.11.1 *Principle of operation*

A simple view of the single-phase induction motor recognises that, when supplied with a.c. current from a sinusoidal voltage source, the stator winding will ideally develop sinusoidally distributed magnetic field components and flux which pulsate in the airgap about fixed axes. Currents induced in the rotor conductors will modify the airgap field but, with the rotor at rest, the symmetry of the system will permit no net torque to be developed in either direction. If, however, the rotor is given initial motion in one direction or the other, the change in rotor-conductor induced e.m.f. is such as to cause an imbalance in the current distribution which modifies the tangential mechanical stress distribution and gives rise to net torque in the direction of initial impetus. The effect is cumulative and the machine accelerates towards a maximum speed in that direction.

One can consider the single-phase induction motor as responding to the influence of a pair of *contra-rotating* airgap field and flux systems, sinusoidally distributed and moving relative to the stator at synchronous speed defined by supply frequency and the number of pole pairs. The cage rotor responds to each of these rotating-field components in much the same way as does the rotor of a polyphase machine. The rotor currents of the single-phase machine modify the behaviour such that the 'forward' and 'reverse' fields and fluxes are equal only at standstill. On other occasions, the forward field dominates, with the plug-braking torque developed by the reverse field always present to reduce output power and increase losses – but not to an excessive extent.

8.11.2 Torque evaluation for the single-phase induction machine

Consider a single-phase induction machine with uniform airgap and p pole pairs to be motoring on load, and *assume* that the radial flux responsible for the e.m.f. induced in stator and rotor conductors is cosinusoidally distributed and pulsating sinusoidally with frequency $\omega = 2\pi f$ about an axis defined by $\theta_m = 0$ in Fig. 8.20, which illustrates the two-pole case with $\theta = \theta_m$. Rotor motion at ω_r is in the anticlockwise direction of θ_m measurement. Then

$$B_n(\theta_m, t) = \hat{B}_n \sin \omega t \cos p\theta_m.$$

A rotor conductor instantaneously at θ_m will experience a changing radial magnetic flux density given by

$$B_n(\theta_m, t) = \hat{B}_n \sin \omega t \cos p(\omega_r t + \theta_{mo}),$$

where θ_{mo} defines the rotor conductor position at $t = 0$ when the radial flux density is zero. Thus

$$\begin{aligned} B_n(\theta_m, t) &= \frac{\hat{B}_n}{2}[\sin(\omega t + p\omega_r t + p\theta_{mo}) \\ &\quad + \sin(\omega t - p\omega_r t - p\theta_{mo})] \\ &= \frac{\hat{B}_n}{2}\{\sin[(2-s)\omega t + p\theta_{mo}] + \sin(s\omega t - p\theta_{mo})\} \end{aligned}$$

[8.35]

where

$$s = \left(\frac{\omega - p\omega_r}{\omega}\right) = \frac{\omega/p - \omega_r}{\omega/p}$$

$$= \frac{\text{synchronous speed} - \text{rotor speed}}{\text{synchronous speed}} \quad [8.13]$$

Equation [8.35] shows that the rotor conductors experience the *assumed* radial flux as equivalent to a pair of travelling waves, each of amplitude $\hat{B}_n/2$. The first appears to rotate in the *clockwise* direction with angular velocity $(2-s)\omega/p$, the second to rotate in the *anticlockwise* direction with angular velocity $s\omega/p$. Under normal motoring conditions $s < 1$. At standstill, $s = 1$ and the component fields are seen to rotate in opposite directions at synchronous speed.

An axially directed rotor conductor of length l, located instantaneously at θ_m but moving anticlockwise at radius b with angular

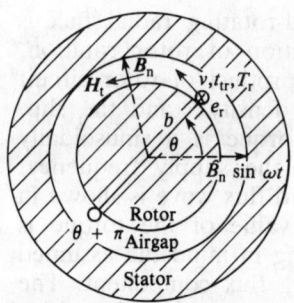

Figure 8.20 Two-pole single-phase induction motor. Reference directions for quantities at rotor/airgap boundary; radius b, axial length l, and angular velocity ω_r. Radial flux density distributed cosinusoidally about horizontal axis $\theta = 0$ (defined by single-phase stator winding).

velocity ω_r, has anticlockwise motion *relative* to the clockwise rotating radial flux component of $(2-s)\omega/p$ and anticlockwise motion *relative to* the anticlockwise-rotating flux of $-s\omega/p$. Invoking eqn [8.1], $e_r = B_n lv$, with reference directions as defined in Fig. 8.20, the e.m.f. induced in the rotor conductor is

$$e_r(t) = -\frac{\hat{B}_n}{2}bl\left\{s\frac{\omega}{p}\sin(s\omega t - p\theta_{mo})\right.$$
$$\left. - (2-s)\frac{\omega}{p}[\sin(2-s)\omega t + p\theta_{mo}]\right\}. \quad [8.36]$$

The first term in eqn [8.36] is identical with eqn [8.12] except for the factor $1/2$ and represents a slip frequency e.m.f. which, by reasoning similar to that developed in Section 8.3.1, gives rise to a distribution of rotor surface current analogous to eqn [8.15]. Developing the parallel argument further, a *tangential field* at the rotor/airgap boundary analogous to eqn [8.19] may be deduced, involving the concept of impedivity.

Identical procedures may be applied to the second term in eqn [8.36] so that, with both forward and reverse rotating flux components present, the total tangential field at the rotor surface is given by

$$H_{tr}(\theta_m, t) = \frac{\hat{B}_n}{2}\frac{hb}{p}\left[\frac{s\omega}{\eta_s}\sin(\omega t - p\theta_m - \phi_s)\right.$$
$$\left. - \frac{(2-s)}{\eta_{2-s}}\sin(\omega t + p\theta_m - \phi_{2-s})\right] \quad [8.37]$$

where $\eta_s \angle \phi_s$ is the impedivity of the rotor conductors at slip frequency, related to rotor conductor leakage impedance at slip frequency by eqn [8.18], and $\eta_{2-s} \angle \phi_{2-s}$ is the impedivity at angular frequency $(2-s)\omega$.

The tangential mechanical stress t_{tr} established at the rotor surface over the elemental area $l\,b\,\Delta\theta_m$ at radius b is

$$t_{tr}(\theta_m, t) = B_n H_{tr}$$
$$= \frac{\hat{B}_n^2 hb}{2p}\sin\omega t \cos p\theta_m\left[\frac{s\omega}{\eta_s}\sin(\omega t - p\theta_m - \phi_s)\right.$$
$$\left. - \frac{(2-s)\omega}{\eta_{2-s}}\sin(\omega t + p\theta_m - \phi_{2-s})\right]. \quad [8.38]$$

The contribution to rotor torque made by the first term in eqn [8.38], corresponding to the forward-rotating flux component is

$$T_{rf} = b^2 l \int_0^{2\pi} t_{trf}(\theta_m)_{mean} d\theta_m.$$

Now

$$\sin\omega t \cos p\theta_m \sin(\omega t - p\theta_m - \phi_s)$$
$$= \tfrac{1}{4}[\cos(2p\theta_m + \phi_s) - \cos(2\omega t - \phi_s) + \cos\phi_s$$
$$- \cos(2\omega t - 2p\theta_m - \phi_s)].$$

Hence

$$T_{rf} = \pi s\frac{\omega}{p}\frac{b^3 lh}{\eta_s}\left(\frac{\hat{B}_n}{2}\right)^2[\cos\phi_s - \cos(2\omega t - \phi_s)]. \quad [8.39]$$

Equation [8.39] shows that the forward-rotating radial flux is responsible for rotor torque in the direction of rotor rotation, having a mean value proportional to the product of rotor circuit power factor at slip s, (peak radial flux density)^2 and the slip. Superimposed upon this mean value of torque is a sinusoidally time-variant quantity alternating at twice the supply frequency. The instantaneous torque due to the forward flux wave is shown in Fig. 8.21. The expression for the *mean* value of the torque is directly comparable with eqn [8.22], bearing in mind the assumed amplitude of $\hat{B}_n/2$ for the forward-rotating flux component. The latter equation was derived for the polyphase machine, for which *instantaneous* and *average* values of rotor torque are identical.

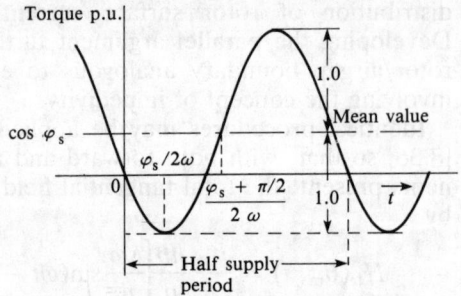

Figure 8.21 Variation with time of instantaneous torque due to a forward-rotating flux wave of a single-phase induction machine with $1 > \text{slip} > 0$.

By an argument similar to that pursued in Section 8.3.2 the mean torque developed by the forward-rotating flux of the single-phase machine, expressed in synchronous watts, may be equated to the total rotor circuit copper loss divided by the slip. Thus, analogous to eqn [8.23],

$$T_{rf}\frac{\omega}{p} = \frac{\text{total rotor copper loss (at slip frequency } sf)}{\text{slip}}$$

$$= I_{rs}^2 \frac{R_{rs}}{s} \qquad [8.40]$$

where I_{rs} is the r.m.s. value of slip-frequency current flowing in an equivalent rotor circuit impedance with resistance R_{rs} and power factor $\cos\phi_s$.

Considering now the contribution to rotor torque made by the reverse or 'backward' rotating flux component

$$T_{rb} = b^2 l \int_0^{2\pi} t_{trb}(\theta_m)_{\text{mean}} d\theta_m.$$

Referring to the second term of eqn [8.38] and invoking the identity

$$\sin\omega t \cos p\theta_m \sin(\omega t + p\theta_m - \phi_{2-s})$$
$$= \tfrac{1}{4}[\cos\phi_{2-s} - \cos(2\omega t + 2p\theta_m - \phi_{2-s}) + \cos(-2p\theta_m + \phi_{2-s}) - \cos(2\omega t - \phi_{2-s})]$$

$$T_{rb} = -\pi(2-s)\frac{\omega}{p}\frac{b^3 lh}{\eta_{2-s}}\left(\frac{\hat{B}_n}{2}\right)^2 [\cos\phi_{2-s} - \cos(2\omega t - \phi_{2-s})]. \qquad [8.41]$$

Equation [8.41] shows that the backward-rotating flux develops rotor torque in the direction *opposing* rotation, having a mean value proportional to the product of 'backward slip' $(2 - s)$ and the rotor circuit power factor at the corresponding frequency $(2 - s)f$. The backward torque alternates about the mean value at twice supply frequency. The expression for torque may also be compared with eqn [8.22] and, paralleling the presentation of Section 8.3.2, developed to yield an equation of the mean backward torque in synchronous watts with the rotor circuit copper loss at frequency $(2 - s)f$ divided by the backward slip factor $(2 - s)$. Thus

$$T_{rb}\frac{w}{p} = \frac{\text{total rotor copper loss [at backward slip frequency } (2-s)f]}{\text{backward slip}}$$

$$= -I_{r(2-s)}^2 \frac{R_{r(2-s)}}{2-s}. \qquad [8.42]$$

Neglecting skin effect, which would raise the resistance of the rotor circuit at the higher frequency, $R_{rs} = R_{r(2-s)} = R_r$, say, and the addition of eqns [8.40] and [8.42] gives for the total *mean* torque of a single-phase induction machine, in synchronous watts

$$T_r\omega_0 = (T_{rf} + T_{rb})\frac{\omega}{p} = I_{rs}^2 \frac{R_r}{s} - I_{r(2-s)}^2 \frac{R_r}{2-s} \qquad [8.43]$$

Equation [8.43] is generally true and is not qualified by the initial assumption that the amplitudes of the forward- and backward-rotating radial flux-density components are equal to each other at $\hat{B}_n/2$. In practice, the contra-rotating components of flux density are equal only at standstill, whilst the corresponding current components always have r.m.s. values proportional to the relevant inducing flux density. The forward and backward flux components must, however, be such that together they induce in the *stator* winding an e.m.f. which balances the stator supply voltage, allowing for stator winding leakage impedance.

8.11.3 *Equivalent circuit derivation*

Accepting that rotor current components I_{rs} and $I_{r(2-s)}$ may be expressed as ratios of rotor-circuit induced e.m.f. and impedance, with e.m.f. and leakage reactance proportional to rotor frequency, eqn [8.43] may be rewritten

$$T\omega_0 = \frac{s^2 E_{2f}^2}{R_r^2 + (s\omega L_r)^2}\frac{R_r}{s} - \frac{(2-s)^2 E_{2b}^2}{R_r^2 + (2-s)^2(\omega L_r)^2}\frac{R_r}{(2-s)} \qquad [8.44]$$

where E_{2f} and E_{2b} are r.m.s. values of equivalent rotor-circuit e.m.f. induced by forward and reverse rotating radial-flux components at slip $s = 1$.

Equation [8.44] may be rewritten

$$T\omega_0 = \frac{E_{2f}^2}{\left(\frac{R_r}{s}\right)^2 + \omega^2 L_r^2}\frac{R_r}{s} - \frac{E_{2b}^2}{\left(\frac{R_r}{2-s}\right)^2 + \omega^2 L_r^2}\frac{R_r}{(2-s)}. \qquad [8.45]$$

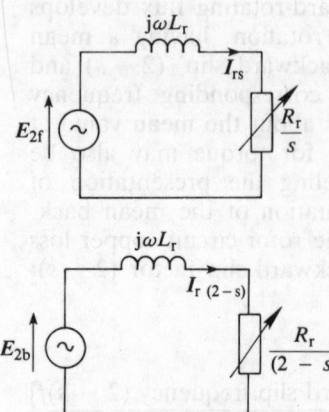

Figure 8.22 Equivalent networks for the rotor circuit of a single-phase induction machine.

Figure 8.23 Equivalent circuit of the rotor of a single-phase induction machine referred to the stator winding.

Equation [8.45] may be interpreted as the *difference* of the real powers dissipated in the equivalent networks of Fig. 8.22, which compares directly with the rotor equivalent circuit derived for the polyphase induction machine as Fig. 8.7(b). The currents I_{rs} and $I_{r(2-s)}$ in the equivalent networks of Fig. 8.22 are identified with the r.m.s. phasor current components induced in the rotor cage conductors by the forward and reverse travelling waves of radial flux. The presence of the rotor currents modifies the *stator* current in such a way that the *total* radial flux across the airgap is essentially unchanged, with the tangential field at the stator/airgap boundary attempting to balance that at the rotor/airgap boundary.

The machine is effectively 'field driven' from the stator side as the stator magnetic field intensity established by the winding current is obliged to pulsate along an axis fixed by the stator winding and, assuming a sinusoidal distribution of this field, its resolution into equal contra-rotating components is in order (*c.f.* Eqn [7.46]). Hence the values of I_{rs} and $I_{r(2-s)}$ when referred to the stator winding are obliged to be equal and the rotor equivalent circuits of Fig. 8.22, *when referred to the stator winding*, appear as Fig. 8.23.

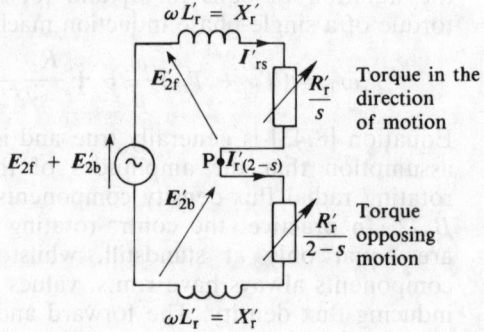

The equivalent circuit presented to the stator winding terminals may be completed by the addition of stator leakage impedance and magnetising impedance referred to the stator as shown in Fig. 8.24. In practice, however, the series connection of the stator winding turns constrains the distribution of stator current so that the balancing of stator and rotor current is not possible at all points in the airgap. Some accommodation of this effect in an equivalent circuit is afforded by dividing the magnetising impedance into two equal parts and connecting the mid-point to point P identified in Fig. 8.23. Typical speed/torque characteristics are shown in Fig. 8.25.

Figure 8.24 Equivalent circuit of a single-phase induction machine referred to the stator winding.

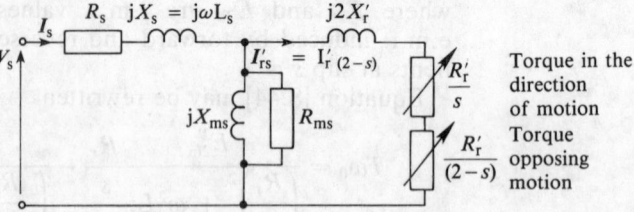

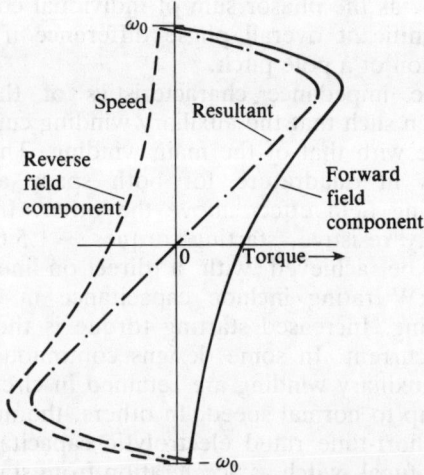

Figure 8.25 Speed/torque characteristic for a single-phase induction machine.

In practice, the equivalent-circuit parameters of a single-phase induction machine are difficult to evaluate and unreliable. The existence of current at relatively high frequency in the rotor circuit gives rise to a significant influence of *skin effect* on current distribution, making R'_r dependent upon slip. As a further consequence, the value of R'_r, where it appears as R'_r/s, is different from the value in $R'_r/(2-s)$. Taking a nominal supply frequency value for R'_r as a base it is common practice to reduce the value of R'_r/s by an empirical factor of about 1.2 and increase the value of $R'_r/(2-s)$ by 1.8 times, where s is small, to correspond with normal running conditions. The supply frequency value of R'_r may be deduced from a locked-rotor test, after allowing for R_s – the stator d.c. resistance value multiplied by the factor 1.2 to establish an effective a.c. value. Locked-rotor test results at rated current also give rise to measured values of leakage reactance which may be allocated equally between X_s and $2X'_r$.

A no-load test at normal voltage and low slip may be used to estimate the values of the magnetising branch parameters X_{ms} and R_{ms} but the presence of both forward and reverse torques even on light load means that, from choice, an auxiliary driving motor is used to drive the machine under test at precisely synchronous speed.

8.11.4 Provision of starting winding

As noted above, the single-phase induction motor is not self-starting. It may be made capable of developing torque at standstill in a preferred direction by modifying the flux conditions to enhance the 'forward' synchronously rotating component. This is often achieved through the provision of an auxiliary *starting* winding placed in stator slots left vacant by the main single-phase winding and arranged such that the magnetic axes of the radial fields of starting and main windings are at right-angles to each other. With single-phase windings it is uneconomic to wind more than about 2/3 of the available slots due to the low *distribution factor* of single-phase windings with a wide spread. As detailed in Section 5.8.2 the distribution factor accounts for the total winding e.m.f. as the phasor sum of individual coil-side e.m.f.s which have a significant overall phase difference if distributed over a large fraction of a pole pitch.

The impedance characteristics of the auxiliary winding are chosen such that the auxiliary winding current is significantly out of phase with that of the main winding. The winding m.m.f. components in quadrature for both space and time emphasise one rotating field effect above the other. If the auxiliary winding is mainly resistive, starting torques ~ 1.5 times the full-load torque may be achieved with a direct-on-line start. Larger machines $\sim 1\,\text{kW}$ rating include capacitance in series with the auxiliary winding. Increased starting torque is then available with reduced line current. In some designs continuously rated capacitance and the auxiliary winding are retained in circuit when the machine has run up to normal speed. In others, the auxiliary circuit incorporating short-time rated electrolytic capacitance is disconnected by a centrifugal switch as acceleration from standstill gets underway.

8.11.5 The shaded-pole single-phase induction motor

Small, single-phase induction motors may employ a laminated salient-pole stator field system in which the trailing edge of each pole is embraced by a closed conducting loop called a *shading ring*. But for the shading ring, the flux emanating from the salient pole would be everywhere of the same phase at the same instant. Flux which links the shading band, however, induces e.m.f. and current therein, the magnetic effect of the latter being generally such as to oppose the change in the enclosed flux. The overall effect is to reduce and delay in phase the portion of the pole flux which is linked by the shading ring in comparison with that of the remainder of the pole. The combination of these flux components may be considered to give rise to a rotating radial flux of sorts, moving in the direction towards the shaded portion, and able to develop net torque in a stationary cage rotor whose presence also modifies the airgap field.

The efficiency of a shaded-pole motor is low and the torque capability limited, but such machines are cheap to manufacture, robust and reliable. They find domestic application in fans and washing-machine pumps. A shaded-pole stator is illustrated in Fig. 8.26. Alternative asymmetries deliberately introduced on the stator-pole system may be used to provide a small motor with starting torque in a preferred direction.

8.12 Other Single-phase Motors

8.12.1 The reluctance motor; the hysteresis motor

The induction principle is only one of several which may be utilised to provide a single-phase a.c. motor drive. The single-phase *reluctance motor*, commonly employed in clocks and time-switches, and briefly described in Section 7.15, employs a simple salient-pole rotor. Through representation of coil inductance variation with rotor position by an expression of the form $L(\theta) = L_0 + L_2 \cos 2\theta$, use of eqn 3.26 shows that, provided the

Figure 8.26 Unwound stator lamination pack for a four-pole, shaded-pole, single-phase induction motor. The copper shading strip is formed into a closed loop by soldering. (*Photograph by courtesy of GEC Electromotors Ltd, Warley, UK.*)

rotor rotates in either direction at synchronous speed, the torque developed has a unidirectional mean value – about which it varies in time as twice supply frequency. Increase in load torque causes the angular displacement between rotor and stator magnetic axes at the instants of peak current to increase, the operating principle being summarised in the observation that the rotor tends to align itself in the position of minimum reluctance, equivalent to maximum stored field energy at constant current. Such machines are not self-starting and may be provided with a *split-phase* stator winding and cage rotor winding for running up towards synchronous speed as an induction motor. Very small reluctance motors may be started manually by spinning the rotor with the fingers to a speed above synchronism in the required direction with the stator winding excited. The rotor should then lock into synchronism as the speed subsequently falls.

The *hysteresis motor* is an alternative synchronous motor in which the rotor comprises a smooth cylinder of permanent-magnet material having a wide hysteresis loop. The stator winding is required to supply a rotating radial flux, ideally sinusoidally distributed. The quality of the field distribution is important to keep the iron losses low. In responding to the *fundamental* component of the rotating field set up by the stator currents, the flux emanating from or entering the rotor surface lags (in space) behind the axis of radial field intensity H_n due to the stator winding. The angle of lag is constant provided the peak flux density is constant, and each elementary part of the hysteretic material is subjected to cyclic magnetisation. If the rotor runs at below synchronous speed the hysteresis loop of the material is traversed at slip frequency, with respect to which the area of the loop and consequently the angle of lag is independent. Hysteresis loss is then proportional to slip and the motor exhibits the unique feature of developed torque independence of slip from standstill up to synchronous speed. At synchronous speed the lag angle adjusts to match the load torque demand. The attractive run-up performance characteristic of the hysteresis motor is somewhat offset by the high cost and machining difficulties of hysteretic material.

8.12.2 The single-phase a.c. commutator motor (universal motor)

The *d.c. machine* discussed in Chapter 6 possesses a commutator armature winding on the rotor with fixed brushes placed to commutate armature coils whose sides lie instantaneously on the q-axis. If the field flux is constant the armature coils short circuited during commutation are subject to no induced e.m.f. from this source. The radial field of the *armature winding* due to load current flowing in all armature coils is directed along the q-axis, in space quadrature with the radial field due to the stator *field* winding, and providing maximum torque for given values of armature and field current.

In the *series* d.c. machine the armature and stator field windings are connected in series, making their currents equal or proportionate so that, from eqn [6.11], torque is proportional to (armature current)². If, therefore, the series machine is supplied with *alternating* current, the stator and rotor radial field components will pulsate along axes at right-angles to each other, providing a unidirectional torque which varies at twice the supply frequency, having a mean value equal to half the peak value.

$$T = K\hat{I}_a^2 \sin^2\omega t = K\left(\frac{\hat{I}_a}{\sqrt{2}}\right)^2 [1 - \cos 2\omega t]. \qquad [8.46]$$

It appears, therefore, that a series commutator motor will develop torque when connected to an a.c. supply. The rotation of the armature in the pulsating radial flux develops an e.m.f. between brushes which alternates at supply frequency and is capable of balancing the supply voltage after allowance is made for the alternating voltage developed across the series field winding. Commutation problems are likely, however, as the d-axis radial field due to the field current pulsates. A *transformer* e.m.f. will be induced in the particular armature coils short circuited by the brushes as, with coil sides placed instantaneously on the q-axis but linking the d-axis flux, they undergo commutation.

An expression for the armature e.m.f. of a single-phase a.c. commutator motor is directly comparable with the e.m.f. equation for a two-pole d.c. machine, eqn [6.4]. The e.m.f. is in time phase with the radial flux and is proportional to the product of speed and flux.

The pulsating d-axis flux also induces an e.m.f. in the stator field winding, in time quadrature with the flux. Laminated cores are essential on both stator and rotor. The series connection of field and armature windings requires that the phasor sum of armature and field e.m.f.s balance the alternating supply voltage, after allowing for resistive voltage drops and inductive effects not yet accounted for.

Alternating current flowing in the armature winding produces a pulsating radial field distribution about the q-axis. This is responsible for an armature reaction e.m.f. appearing between the q-axis brushes, in time quadrature with the armature current. Unless a *compensating winding* is fitted the armature reaction voltage will be very high, thus limiting the armature current and hence the machine torque and power factor for a given supply voltage. The compensating winding is installed about the quadrature axis, accommodated in stator slots, either connected in series with armature and field windings or closed upon itself. Such compensated machines have been used for single-phase a.c. traction purposes, usually in conjunction with low-frequency supplies ~ 16.67 Hz. Uncompensated a.c. commutator motors are used in domestic appliances where the requirement is for low power at high speed (perhaps adjustable).

Single-phase *induction* motors are limited in speed to around 3000 rev/min with a 50 Hz supply. Single-phase commutator motors may exceed this, and will also operate on d.c. as universal machines. Speed control is by supply voltage, with the speed/torque characteristic generally as for the series motor, Fig. 6.22(iii). With a.c. supplies the reduction in speed at high-load currents ameliorates the commutation difficulties, eased further by the use of high-resistance brushes, a minimal number of turns for each armature coil and a large number of poles.

8.13 The Polyphase Linear-induction Motor

All the electrical machines considered in detail heretofore have been described in a format which provides for mechanical power as a product of angular velocity and torque. In expounding the theory of such energy conversion devices it has been convenient on occasion to represent the machine graphically in *developed* form i.e. as though the rotating machine was sectioned along a radius and was then unrolled or opened out. Such a linear representation of a rotating machine is valid as it has been recognised that the ends of the development are arbitrary and do not represent an actual discontinuity, a strict equivalent being dependent upon the dimensional ratio b/pg being large, i.e. there being a large ratio of pole pitch to airgap length.

Electrical energy conversion devices may generally be realised with motion either translational (linear) or rotational. The latter configuration is the more common and is well suited to many applications. Symmetry about the axis of rotation makes the device easier to design, whilst testing can be carried out at a fixed location without stationary and moving portions tending to separate. Direct-current and a.c. synchronous motors may be constructed as linear equivalents of the rotating version but it is the polyphase linear induction machine which is most commonly encountered, because of the fact that a cage or conducting-sheet secondary winding requires no external power supply and little maintenance.

Linear motors promise advantages over rotating machines when the object is to provide linear motion or thrust. Some applications require significant force development at low speed over short distances, e.g. actuators. Others require the rapid acceleration over short distances of a projectile, e.g. an amusement-park ride. Other examples include textile looms, sliding doors, curtain pullers, conveyor systems, travelling cranes, liquid-metal pumps and tracked vehicles, including trains. In traction applications the excited primary winding will normally be located on the moving member whilst the secondary winding or conducting sheet constitutes the fixed track. Since the track is usually much longer than the primary winding, such a configuration is described as 'short-stator', by analogy with the rotating-field machine. Clearly the electromagnetic effects of thrust development will be expressed only in the vicinity of the primary winding, and significant end-effects will modify behaviour in those regions where the secondary approaches and leaves the primary. Responses representative of that experienced by the rotary machine will be realised only in the central region of the primary winding.

A *single-sided*, *short-stator* linear-induction motor with a conducting sheet secondary is shown in Fig. 8.27(a). Unlike the rotating machine, the number of poles is not required to be an even number. Currents are induced in the conducting sheet at slip frequency as the travelling flux progresses synchronously in the direction of motion. The induced currents are directed primarily in the plane of the sheet at right-angles to the direction of motion, except at the edges which overlap the electromagnetic coils, completing the circuit with flow along the length of the sheet in this region. Force on the conducting sheet in or against the direction of motion is developed by the interaction of the 'radial' field at right-angles to the ideally parallel primary and secondary surfaces, with the 'tangential' field directed as the motion. The general effect of the slip-dependent secondary current is to enhance the tangential field in the space between primary and secondary conductors and to reduce the 'radial' field.

Mechanical stresses in the effective area of the conducting strip are developed in the direction of motion, distributed in nature and locally variable in both magnitude and direction. The *mean* value is influenced by secondary-circuit power factor with the thrust proportional to the secondary 'copper' loss. Thus far, the linear-induction machine mirrors the behaviour of the rotating machine. Two significant differences however exist. The first relates to the discontinuities present at the extremities of the primary where the conducting secondary enters and leaves. The phenomenon observed may be explained by reference to the *theorem of constant flux linkages*, Section 7.20, which states that the flux linkage of a closed conducting circuit cannot change instantaneously.

Consider an element of secondary conducting sheet just before it enters the region influenced by the primary field. Such an element would seem to be subjected to a sudden increase in flux. A rapid increase in the flux linked will be opposed by the development of a

Figure 8.27 (a) Single-sided and (b) double-sided, short-stator, three-phase linear-induction motors.

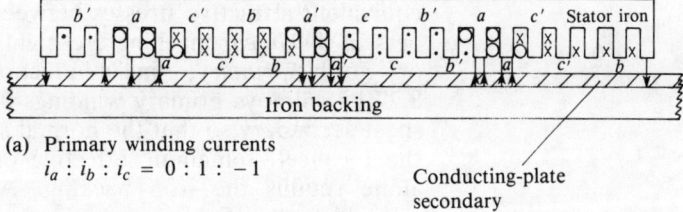

(a) Primary winding currents
$i_a : i_b : i_c = 0 : 1 : -1$

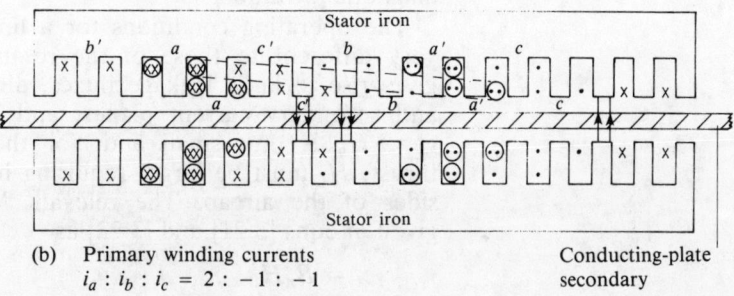

(b) Primary winding currents
$i_a : i_b : i_c = 2 : -1 : -1$

large secondary current flowing in an appropriate path. The effect is to increase the secondary-circuit 'copper' loss in this region and also to reduce the resultant field. It is usual practice to connect in series the primary phase windings over several pole pitches so as to ensure that the primary currents in the entry region are limited.

A complementary effect occurs at the exit end of the primary winding. Here the secondary conducting sheet emerges with closed circuits linking finite flux. The embraced flux is sustained by a secondary current flow for a duration dependent upon the time constant of the secondary circuit. Entry- and exit-edge effects are dependent upon speed, which is itself dependent upon supply frequency and pole pitch when the poles are arranged in line as in so-called 'axial-flux' machines.

The second significant feature contrasting the linear-induction motor with the rotating machine is in the effective airgap between the iron backing primary and the secondary conductors – at least an order of magnitude greater with the linear machine. Some linear-motor designs dispense altogether with iron backing to the secondary winding. The effect is a much increased magnetising current and a reduced 'radial' flux density directed across the gap. This reduces the linear thrust for a given value of secondary current or, conversely, for a given thrust requires the secondary current and its referred value in the primary circuit to increase. The reduced 'radial' flux density and the increased 'tangential' field have important consequences for the forces of attraction between primary and secondary members. In rotating machines these radial forces, ideally sinusoidally distributed, have little practical effect if the rotor axis, ideally concentric with the stator bore, is securely fixed via the shaft bearings. These radial stresses are very much greater than the tangential stresses which give rise to torque, but ideally sum to zero over the stator and rotor surfaces. The equivalent attractive stresses between primary and secondary members in the *linear* machine are highly significant. The *double-sided*, or *double-primary*, linear-induction motor illustrated in Fig. 8.27(b) employs primary windings on both sides of the conducting sheet secondary so that the normal stresses on opposite sides of the sheet cancel, remaining effective on the primary structure which alone retains the iron backing. Alternatively, in a single-sided assembly, the field components within the gap may be proportioned in such a way as to provide net repulsion between primary and secondary members, offering the prospect of electromagnetic levitation.

The operating conditions for a linear-induction motor are thus very different to those of the rotating machine. A large airgap gives rise to large leakage fluxes, directed parallel to the primary and secondary current planes, and low flux densities across the gap. H_n is thus small and, for the development of significant thrust, H_t must be large, requiring high current densities on both sides of the airgap. The relevant Maxwell stress equations are given in eqns [3.21] and [3.22] as

$$t_t = B_n H_t \qquad [8.47]$$

$$t_n = \tfrac{1}{2}[B_n H_n - B_t H_t] = \tfrac{1}{2}\mu_0[H_n^2 - H_t^2]. \qquad [8.48]$$

Equation [8.48] indicates that over the particular areas of the secondary conducting sheet where $H_t > H_n$ the net stress is directed such as to repel the plate. The same travelling-field system which accounts for linear motion may thus also provide lift for a high-speed transport application. Such a suspension system would have a large demand for reactive volt-amperes, if little power.

The velocity/thrust characteristic for a linear-induction machine is essentially similar to the $(1 - \text{slip})$/torque characteristic of the rotating machine shown in Fig. 8.12, although the large airgap and high leakage fluxes mean that any assumption of constant 'radial' flux is inappropriate. Representation by the traditional equivalent circuit is less valid with the flux distribution along the length of the stator influenced by the end effects. Linear motors operate with higher values of slip and lower efficiencies than do rotating machines, with maximum speed similarly limited by supply frequency to the synchronous value at two pole pitches per cycle – assuming the poles are arranged in line with the direction of motion. High speeds with this configuration demand a large pole pitch, which increases the length of secondary current paths parallel with the edges of the conducting sheet. This has the effect of increasing the secondary circuit resistance. As an alternative to compensation by increasing the width of the secondary conducting plate, a 'transverse flux' configuration may be employed in which the flux crossing the airgap completes its path through the iron portion of the magnetic circuit at right-angles to the direction of motion, the flux retaining its variation with time along this major axis.

A Japanese National Railways project achieved levitation and linear motion using a *synchronous* motor drive having a powered polyphase track and a superconducting magnet located in the vehicle. The d.c. excited magnet provided lift against an aluminium plate installed in the track at speeds in excess of 40 km/h. It also provided excitation for the field poles of the synchronous linear motor. Such an arrangement is clearly expensive in track costs, the polyphase winding being energised in sections as the locomotive progresses, with variable frequency providing speed control. The track costs might be justifiable in regions of high traffic density with speeds ~ 500 km/h in prospect.

8.14 Tutorial Examples

8.1 The input power supplied to a three-phase, 50 Hz, six-pole induction motor is 60 kW. The stator copper losses total 1 kW and the stator iron losses are 500 W. Establish the total mechanical power developed (including the friction and windage losses), the rotor copper loss/phase and the speed at a slip of 0.03 p.u.

[56 745 W, 585 W, 970 rev/min]

8.2 The power input to a 500 V, 50 Hz, six-pole, three-phase induction motor running at 975 rev/min is 40 kW. The stator copper and iron losses total 1 kW and the friction and windage losses total 2 kW. Calculate (i) the mechanical output power, (ii) the shaft torque, (iii) the rotor copper loss and (iv) the efficiency.

[(i) 36 025 W; (ii) 352.8 N m; (iii) 975 W; (iv) 90.1%]

8.3 A three-phase induction motor with a star-connected rotor winding develops 60 V between slip-rings at standstill on open circuit. The rotor impedance/phase at standstill is $0.6 + j4.0\ \Omega$. Estimate the rotor current when the motor runs at 0.03 p.u. slip with the rotor winding short circuited.

[1.70 A]

8.4 A 400 V, 50 Hz, six-pole, three-phase induction motor has a delta-connected stator winding and a star-connected rotor winding. The rotor impedance/phase at standstill is $0.1 + j0.55\ \Omega$ and the stator/rotor ratio of effective turns/phase is 2.2.

Rated load conditions are appropriate to a rotor speed of 950 rev/min. Calculate the corresponding rotor current and the value of external rotor resistance/phase required to reduce the speed to half the synchronous value at the same value of torque.

[87.7 A, 0.9 Ω]

8.5 (i) What relatively simple tests could be done to estimate the rated voltage and current of a cage induction motor from which the rating plate has been lost? Comment on the effect of insulation quality on the results.
(ii) What voltage would you recommend to allow a 12 kW, 500 V, 60 Hz induction motor to be used without modification from a 50 Hz supply? Estimate the corresponding output power.
(iii) What is the purpose of no-load (open-circuit) and locked-rotor (short-circuit) tests for induction machines? Why are these tests particularly important for extremely large machines?
(iv) Using an equivalent circuit, explain why the overall efficiency of an induction motor must be less than $(1 - s) \times 100\%$, where s is the slip.

E.C.
[(ii) 417 V, 10 kW]

8.6 A 150 kW, 1000 V, 25 Hz three-phase induction motor with a star-connected stator winding has a star-connected rotor winding with effective stator/rotor turns ratio of 3.6. Phase values of rotor resistance and leakage inductance are 0.01 Ω and 0.64 mH.

Neglecting stator impedance, evaluate (i) the stator line current and power factor when starting on normal voltage with slip-rings short circuited, (ii) the stator current and power factor at 0.03 p.u. slip and (iii) the necessary external resistance/phase configured in delta to reduce the stator current on starting to 300 A.

By what factor does the provision of this value of external rotor resistance raise the starting torque?

[(i) 444 A, 0.1; (ii) 128 A, 0.96; (iii) 0.30 Ω; 5 times]

8.7 A four-pole, three-phase induction motor has a star-connected stator winding and is supplied at 220 V, 50 Hz. The equivalent-star rotor leakage impedance/phase at standstill is $0.1 + j0.9$ Ω. The ratio of stator/rotor effective turns is 1.75 and the slip at full load is 0.04 p.u. Calculate the corresponding torque and mechanical output power, stating the assumptions made. Determine also the pull-out torque and the speed at which this occurs.

[3.63 kg m, 5.37 kW; 5.7 kg m, 1330 rev/min]

8.8 (i) The full-load slip of a three-phase delta-connected induction motor is 5%. When started direct-on-line the current is 6 p.u. Using the approximate equivalent circuit, and neglecting stator resistance and magnetising current, estimate (a) the p.u. leakage reactance and rotor resistance, and (b) the p.u. starting torque.

(ii) The pull-out torque of a squirrel-cage motor driving a boiler feed pump in a small isolated power station is 210% of the full-load torque, and the motor is called on to deliver up to 150% torque during peak periods.

Show that there is a danger of total shutdown of the plant if the system voltage is allowed to fall by more than 15%.

E.C.   [(i) (a) 0.159 p.u., 0.05 p.u., (b) 1.8 p.u.]

8.9 Sketch typical torque/speed and output power/speed curves for a cage-rotor induction machine, covering the range of slip between ±1.5 p.u. Indicate the motor, generator and brake regions, and make clear where stable operation can be expected.

A slip-ring induction motor having a rotor phase resistance r drives a constant-torque load at a slip s_0 when the slip-rings are short circuited. Show that the motor will be brought to rest if a resistor $R = r(1 - s_0)/s_0$ is inserted in series with each rotor phase. Compare the input current and power at standstill with the corresponding values for the running condition.

E.C.   [equality]

8.10 (i) Draw a sketch which illustrates the influence of rotor resistance on the torque/speed curve of a slip-ring induction motor. Indicate one method by which stable speed control of a constant-torque load may be achieved, and mention any shortcomings of the method. Explain why the locked-rotor (starting) torque can be increased by increasing the rotor resistance from its low initial value, despite the fact that this causes a reduction in the current in the rotor winding.

(ii) The total standstill leakage reactance/phase of a four-pole, 415 V, three-phase, 50 Hz star-connected induction motor is 1 Ω (referred to the stator), and the referred rotor resistance is 0.1 Ω. The motor drives a winch of 0.3 m diameter via a 4:1 step-down gearbox.

Neglecting stator resistance, magnetising current, iron losses and all friction losses, calculate the steady speed of the motor when lifting a mass

of 1000 kg ($g = 9.81$ m/s^2).
E.C. [1442 rev/min]

8.11 Explain why a polyphase induction motor cannot run at synchronous speed. Account for the variation in magnitude and phase of the stator (primary) winding supply current for speed changes above and below the synchronous value made possible by the use of an auxiliary machine, mechanically coupled.

A three-phase, 3000 kW, 4 kV, 50 Hz, six-pole induction motor gave the following test results with locked rotor (line values):

2000 V, 500 A, 0.415 p.f.

Assuming equal division of leakage impedance voltage drop between stator and rotor windings, deduce the phase equivalent circuit of the machine referred to the stator winding, assuming a star connection. Suggest a likely value for the magnetising reactance in ohms/phase, referred to the stator, if magnetising current is typically 5% of rated current.

Ignoring magnetising current, calculate the torque developed by the machine when run at rated voltage and 0.02 p.u. slip. Express the result in (i) synchronous kW, (ii) N m, (iii) per unit of full load torque deduced from rating plate details. What value of slip is implied when running at full load torque?
U. of B. [$\simeq 100\ \Omega$; (i) 638.4 kW; (ii) 6090 N m; (iii) 0.21 p.u.; 10%]

8.12 Show that, with constant radial airgap flux, the ratio of polyphase induction-motor torque T at slip s p.u. to maximum torque T_{max} is given by the expression

$$\frac{T}{T_{max}} = \frac{2}{s/\alpha + \alpha/s}$$

where α is the ratio of rotor resistance to standstill leakage reactance.

Thus show that the run-up time of a polyphase induction motor with pure inertia load from standstill to slip s, with the same assumption, is given by

$$\tau = \frac{J\omega_0}{2T_{max}} \left(\frac{1 - s^2}{2\alpha} + \alpha \log_e \frac{1}{s} \right)$$

where ω_0 is synchronous speed in rad s^{-1} and J is the moment of inertia in kg m^2.

8.13 Show that, with constant radial airgap flux, the rotor current/phase I_2 of a polyphase induction motor at any slip s may be related to its value at standstill I_{2sc} as follows

$$I_2 \simeq I_{2sc} \left[\left(\frac{\alpha}{s} \right)^2 + 1 \right]^{-1/2}$$

provided that α, the ratio of rotor resistance r_2 to standstill leakage reactance x_{12}, is very much less than 1.

Hence, by use of the identity (proved in the solution of **8.12**)

$$dt = -\frac{J\omega_0}{2T_{max}} \left[\frac{s}{\alpha} + \frac{\alpha}{s} \right] ds$$

where ω_0 is synchronous speed and J is the moment of inertia, show that the energy loss/phase over the period during which the rotor accelerates a pure inertia load from standstill to synchronous speed is given by

$$W_{ph} = \int I_2^2 r_2 dt = \frac{I_{2sc}^2 r_2 J \omega_0}{4 T_{max} \alpha}.$$

Hence show further that the total energy dissipated in the rotor phases during run up is equal to the kinetic energy of the rotor at synchronous speed.

8.14 A four-pole, 440 V, three-phase, 50 Hz induction motor gave the following test results:

No-load: 440 V, 22 A, 0.5 kW/phase

Standstill: 110 V, 40 A, 1.0 kW/phase.

Derive the approximate equivalent circuit parameters referred to the stator, assigning to the rotor 45% of the total copper loss at standstill.

Hence deduce the torque developed (in synchronous kW and N m) the line current, efficiency and power factor at a slip of 0.022 p.u.

[14.2 kW, 90 N m, 32 A, 87%, 0.66]

8.15 A 400 V, three-phase, 50 Hz induction motor has negligible stator impedance and a star-connected rotor winding with resistance 0.3 Ω/phase and leakage inductance 8 mH/phase. The ratio of stator/rotor turns/phase is 2:1.

Determine the voltage developed between rotor slip-rings on open circuit if the stator winding is (i) star-connected, (ii) delta-connected.

Calculate also the stator line current for (iii) the rotor at standstill with the stator winding star-connected, and (iv) the motor running with a slip of 0.03 and the stator winding delta-connected.

[(i) 200 V; (ii) 346 V; (iii) 22.8 A; (iv) 16.8 A]

8.16 A six-pole, three-phase 50 Hz induction motor with negligible stator leakage impedance has a value of standstill rotor leakage impedance/phase of the equivalent-star connection referred to the stator given by $0.5 + j8\,\Omega$. Considering the rotor-circuit resistance to be the only source of power loss, compare the efficiencies of the following methods of speed control when the motor is required to develop maximum (pull-out) torque at 600 rev/min:

(i) incorporation of external rotor resistance; and
(ii) reduction of the supply frequency with proportionate reduction in supply voltage to maintain constant radial flux.

[*Note*: The torque developed will be of the same value in each case.]

[(i) 0.60 p.u.; (ii) 0.906 p.u.]

8.17 A 440 V, two-pole, three-phase, 50 Hz induction motor with negligible stator leakage impedance has a standstill rotor leakage impedance/phase of the equivalent star, when referred to the stator, of $0.8 + j12\,\Omega$.

Plot the torque/speed characteristics corresponding to inverter supply with sine-wave voltage at constant voltage/frequency ratio for frequencies of value (i) 50 Hz, (ii) 30 Hz, and (iii) 10 Hz.

[Each characteristic has $T_{\max}$ of 25.67 N m occurring at a speed 20.94 rad s^{-1} below the appropriate synchronous value]

8.18 A three-phase, four-pole, 50 Hz, 600 V wound-rotor induction motor has rotor resistance and standstill reactance of 1.2 and 10 Ω/phase, respectively, referred to the stator winding. The stator leakage impedance and core loss are negligible.

Calculate the maximum torque in synchronous watts and N m with the rotor slip-rings short circuited. What external rotor resistance is required to give maximum torque at half-speed? (The stator winding has twice as many turns as the rotor winding; both windings are star connected.)

The motor is now energised by an inverter with a constant voltage/frequency ratio. Determine the maximum torque and the speed at which it occurs when the output frequency is 25 Hz. The inverter has a near-sinusoidal output voltage.

Sketch the torque/slip curves for the three modes of operation.

U. of B.   [18 sync kW, 115 N m; 0.95 Ω/phase; 115 N m, 570 rev/min]

8.19 A 220 V, four-pole, single-phase, 50 Hz induction motor yielded the following stator input test data: (i) with locked rotor, 60 V, 20 A, 700 W; and (ii) when driven at synchronous speed, 220 V, 5 A, 320 W. The d.c. resistance of the stator winding is 0.54 Ω.

Determine the parameters of an equivalent circuit appropriate to an estimate of performance at low values of slip by incorporating multiplying factors of 1.2 and 1.8 at 50 Hz and 100 Hz, respectively to account for the increase in winding a.c. resistance over the d.c. value.

Hence estimate the gross torque developed at a slip of 0.04 p.u., ignoring the effect of the magnetising branch elements. Estimate also the net mechanical power output and efficiency of the motor under these conditions, including core loss and allowing 220 W for friction and windage losses.

[20.8 N m, 2920 W, 77%]

Bibliography

1. Kraus, J. O. *Electromagnetics*, 3rd Edn, McGraw–Hill, New York, 1984.
2. Carter, G. W. *The Electromagnetic Field in its Engineering Aspects*, 2nd Edn, Longman, 1967.
3. Hammond, P. *Applied Electromagnetism*, Pergamon Press, 1971.
4. Solymar, L. *Lectures on Electromagnetic Theory*, 2nd Edn, Oxford University Press, 1984.
5. Solymar, L. and Walsh D. *Lectures on the Electrical Properties of Materials*, 3rd Edn, Oxford University Press, 1976.
6. Bleaney, B. I. and Bleaney B. *Electricity and Magnetism*, 3rd Edn, Oxford University Press, 1976.
7. Hague, B. *Principles of Electromagnetism applied to Electrical Machines*, Dover, New York, 1962.
8. Maxwell, J. C. *Treatise on Electricity and Magnetism*, 3rd Edn, Dover, New York, 1954. (Original Oxford 3rd Edn, 1891.)
9. Carpenter, C. J. Surface integral methods of calculating forces on magnetised iron parts, *Proc. IEE*, **107C**, 1960, pp. 19–28.
10. Rotors H. C. *Electromagnetic Devices*, Wiley, 1941.
11. Woodson, H. H. and Melcher, J. R. *Electromechanical Dynamics, Part 2 – Fields, Forces and Motion*, Wiley, 1968.
12. Tustin, A. *DC Machines for Control Systems*, Spon, 1952.
13. Edwards, J. D. *Electrical Machines: an introduction to principles and characteristics*, 2nd Edn, Macmillan, 1986.
14. Richardson, B. Transformer core losses, *Electronics and Power*, **32** (5), May 1986, pp. 365–8.
15. Slepian, J. The Flow of Power in Electrical Machines, *The Electric Journal*, **XVI**, Pt 5, 1919, pp. 303–11.
16. Hawthorne, E. I. Flow of Energy in D.C. Machines, *AIEE Trans.*, **72**, Pt 1, September 1953, pp. 438–44.
17. Hawthorne, E. I. Flow of Energy in Synchronous Machines, *AIEE Trans.*, **73**, Pt 1, March 1954, pp. 1–9.
18. White, D. C. and Woodson, H. H. *Electromechanical Energy Conversion*, Wiley, 1959.
19. Alger, P. L. and Erdelyi E. Electromechanical Energy Conversion *Electro-Technology*, **Art 33,** Science and Engineering Series, September 1961, pp. 96–120.
20. Seely, S. *Electromechanical Energy Conversion*, McGraw–Hill, New York, 1962.
21. Gray, C. B. Airgap power flow and torque development in electrical machines – can we teach the fundamentals?, *IEEE Trans.*, **PAS-103**, No. 4, April 1984, pp. 874–9.
22. Say, M. G. and Taylor E. O. *Direct Current Machines*, Pitman, 1980.
23. Hindmarsh, J. *Electrical Machines and their Applications*, 4th Edn, Pergamon Press, 1984.

24. Slemon G. R. and Straughan, A. *Electric Machines*, Addison–Wesley, 1980.
25. Say, M. G. *Alternating Current Machines*, 4th Edn, Pitman, 1976.
26. Hindmarsh, J. *Worked Examples in Electrical Machines and Drives*, Pergamon Press, 1982.
27. Acarnley, P. P. *Stepping Motors: a guide to modern theory and practice*, Revised 2nd Edn, Peter Peregrinus, 1984.
28. Kenjo, T. *Stepping Motors and their Microprocessor Controls*, Oxford University Press, 1984.
29. Kenjo, T. and Nagamori S. *Permanent Magnet and Brushless DC Motors*, Oxford University Press, 1985.
30. Howe, D. and Low. W. F. Performance calculations for devices with permanent magnets, *Proc. Polymodel 6 Conf.*, 1983, pp. 21–32.
31. Birch, T. S., Howe, D. and Mitchell, J. K. Design and analysis of brushless d.c. motors, *Proc. ICEMADS '86*, Romania, September 1986.
32. Chalmers, B. J., Devgan, S. K., Howe, D. and Low, W. F. Synchronous performance prediction for high-field permanent magnet synchronous motors, *Proc. ICEM*, Munich, 1986, pp. 1067–70.
33. Laithwaite, E. R. and Freris, L. L. *Electric Energy: its generation, transmission and use*, McGraw–Hill (UK), 1980.
34. Lander, C. W. *Power Electronics*, 2nd Edn, McGraw–Hill (UK), 1987.
35. Bradley, D. A. *Power Electronics*, Van Nostrand Reinhold (UK), 1987.
36. Shepherd, W. and Hulley, L. N. *Power Electronics and Motor Control*, Cambridge University Press, 1987.
37. Rashid, M. H. *Power Electronics: circuits, devices and applications*, Prentice–Hall, 1988.
38. Lipo, T. A. and Novotny, D. W. Vector control and field orientation, *Conf. on Dynamics and Control of AC Drives*, Univ. of Wisconsin, Madison, USA, 1985.
39. Laithwaite, E. R. Linear induction motors – a new species takes root, *Electronics and Power*, **32** (5), May 1986, pp. 355–9.
40. Oshima, K. Superconducting magnetic levitation train project in Japan, *IEEE Trans. Magn.*, **MAG-17**, (5) September 1981, pp. 2338–42.

Index

a.c. supply current harmonics, 226, 235, 244–6, 252, 411
actuator, 74–81, 83–6, 96–7, 428
airgap line, 93–5, 254, 343
alignment principle, 56
Ampère, André Marie, 55, 57, 69
Ampère's law, 35, 37–40, 46, 51, 65, 92, 102, 137, 141, 153, 197–8, 271
anisotropy, 60–1, 96
applications,
  aerospace, 96, 343
  disk drives, 96, 265
  DIY hand tools, 238
  domestic appliances, 323, 417, 424, 427
  fan motor, 406, 413, 424
  industrial drives, 251, 323, 385, 413
  pen drive, 86
  precise positioning, 75, 323
  robotic systems, 2, 75
  tape transport, 265
  traction motor, 226, 409, 428
  wind generator, 343, 402

battery supply, 226, 235
barrier surface, 37–9
Biot–Savart law, 51
Bloch wall, 61–2
Blondel, André Eugene, 302
Boucherôt, Paul, 407
boundary relationships, magnetic field, 64–6

castellation, 82, 324, 327
chopper drive, 225–34, 260–1
  commutation, 226–7, 229, 232, 261
  free-wheel diode, 229–30
  light motor load, 230
  regeneration, 230–2, 235

chopper drive (*cont.*)
  repetition rate, 227, 233–5
  turn-off time (interval), 229
cisoidal, 164
coercive force, (coercivity), 62, 91, 96, 344
cogging, 257, 416
concentrated coil, 42, 178, 199, 205
conductivity, 5, 6, 20–1, 27–8
conjugate phasor, 15
constant current source, 214, 285–6, 347
constant flux linkages, theorem of, 319, 428
constant voltage source, 214, 264, 285–7, 347
converter, line-(naturally)
  commutated, 225, 234–53, 261–2
  single-phase, half-wave, 236–8
  single-phase, half-controlled bridge, 241–3
  single-phase, fully-controlled bridge, 238–41
  three-phase, half-controlled bridge, 252
  three-phase, fully-controlled bridge, 244–52
  instant of natural commutation, 244, 247
  commutating reactance, 250
  overlap, 244–6, 247–50
conventions, 11, 15–16, 33–4
Coulomb, Charles Augustin de, 69
cross product, (see *vector product*)
Curie effect, 60
Curie temperature, 60, 96
curl, 32–7
current sheet, 82, 154–8, 191, 197
current density, A m^{-2}, (see *linear current density*)
cycloconverter, 264, 346–7

d.c./d.c. converter, (see *chopper*)
d.c. current transformer, 253

d.c. link, 347, 371–3, 410–12
direct current machines, 178–225
  acceleration and braking, 223–5, 235
  airgap field of armature current sheet, 154–6, 195–201,
    radial component, 154, 195–201
    tangential component, 154, 198–9, 201
  armature e.m.f., 179, 187, 195, 211
  armature inductance, 211, 223, 226, 229–30, 236, 238, 251
  armature reaction, 195, 200, 211, 213
  armature winding, 178, 185, 189, 192, 194, 255
  back pitch, 192
  bipolar drive circuit, 256
  braking of series motor, 220
  brush shift, 181, 183, 185, 188, 192, 203
  brush width, 189
  brushless commutation, (see *electronic commutation*)
  brushless permanent magnet machines, 211, 254, 373
  chopper drive, 226–35, 260–1
  closed loop control, 253
  commutating poles, (see *interpoles*)
  commutation, 179, 191, 200, 205–211, 226–9, 238
  commutation time, 181, 189, 205
  commutation diode, (see *free-wheel diode*)
  commutation thyristor, 229
  compensating winding, 200, 207–8
  compound field winding, 212, 215, 220
  constructional features, 135
  current feedback, 253
  direct axis, 181, 185
  diverter resistance, 212, 215
  dynamic braking, (see *rheostatic braking*)
  dynamic response, 214, 222–5, 235, 253
  electronically controlled drives, 225–54
  electronic communication, 208, 255, 373
  e.m.f. equation, 188, 194, 216
  equalising connections, 194
  external characteristics, (see *machine performance*)

direct current machines (*cont.*)
  free-wheel diode, 226–30, 238–243
  front pitch, 192
  generator, 183, 202, 207–8, 214–215, 230–5, 242, 250–2
  instant of natural communication, 246, 247
  interpoles, 179, 208–10
  iron loss, 216
  lap-wound armature, 189–94
  linear current density, 191, 197, 202
  loss torque, 216
  machine constant, K, $K\Phi$, 203, 217, 233
  machine performance, 213–220
  mechanical stress, tangential, 195–6, 201–4
  open-circuit characteristic, 213–5
  output coefficient, 204
  permanent magnet machines, 254–8
    brushless, 211, 255, 373
  phase-controlled a.c./d.c. converter drives, 225, 235–54, 262–3
  plug braking (plugging), 225, 260
  quadrature axis, 181, 185, 188–9, 207–8, 211
  regenerative braking, 226, 230–5, 241, 251–2, 256
  rheostatic braking, 220, 225
  separate excitation, 221, 218–22, 224–6
  series field, 212, 219, 225, 426
  shunt field, 212–3
  single time-constant electromechanical system, 223
  specific electric loading, 205
  specific magnetic loading, 205
  starting resistance, 223
  straight line communication, 206
  switched armature, 251
  torque development, 194–5, 201–4
  torque equation, 202, 216
  torque pulsation, 227, 236
  Ward–Leonard drive, 221, 226
  wave-wound armature, 192–4
  windage and friction torque, 194, 216
delay angle, 240, 244, 411
demagnetisation characteristic, 90–5, 98, 254, 343
demagnetisation by linked current, 95, 98, 254
diamagnetism, 59

dipole (magnetic), 57, 60
disc motor, 89, 94
discontinuous supply current, 236–40, 241–3
divergence, 32–3, 36
domain (magnetic), 60–2, 118
domain of closure, 61
dot convention, 44, 106
dot product, (see *scalar product*)
doubly excited system, 156
drift velocity, 3, 5, 18
drive, two quadrant, 220, 230–1
drive, four quadrant, 221, 226, 250–2, 340
duality, 39
duty cycle, 117, 235, 251

eddy current loss, 63
efficiency, 1, 3, 116, 217, 226, 232, 251, 254, 266, 406
electric field of conductor, 4–9, 18
electromagnetic induction, 2, 4, 6, 8–9, 31, 42–5, *passim*
electromechanical energy conversion, 1–2, 17, 171, 211, 236, 339, 408
electron spin, 57, 59, 61
electrostatic field, 5, 7–9, 18, 20–31
  boundary relationships, 5, 9, 27–8
  capacitance, 9, 14, 16, 23, 28–31
    of coaxial cable, 27, 50
    of parallel-plate capacitor, 28–31, 50–51
  charge, 3–9, 18, 20–33, 42, 50
  charge density, surface, 4, 28
  charge density, volume, 5, 18, 33
  dielectric, 5, 20–2, 30
  energy storage, 1, 28–31
  equipotential surface, 25–7
  field of coaxial cable, 26–7
  field of twin conductor line, 50
  flux, 20–2, 32
  flux density, 20–2, 26, 28–33
  fringing flux, 31
  Gauss' law, 21–3, 26, 28, 33
  ionisation, 1, 21, 49
  mechanical (Maxwell's) stresses, 21, 24, 29–31
  permittivity, 1, 20–3, 26–7, 29–31
  potential (scalar) 21, 23–5
  potential difference, 21, 24–5
  voltage gradient, 25–6
e.m.f., 6–8, 43, 89, 101, 113, 135, 194
encoder, 211, 258, 332, 371

energy flow, (transfer), (see *power flow*)
energy product, (density), 93, 95–6
energy storage in magnetic field, 1, 45, 58, 66, 71, 76–8, 173, 210, 226–36, 250–2, 324, 331, 411
equivalent star winding, 17, 124
error signal, 253

Faraday, Michael, 69, 86
Faraday's disc, 87–9, 97, 134
Faraday's law, 43, 318, 331
Faraday's rotator, 86–7, 134
FET, 211
ferromagnetism, 57, 60–4
field intensity (magnetic), 1, 21, 32, 34–5
  tangential component, 65, 74, 83, 88
  due to current in long, straight wire, 38
  due to twin conductor transmission line, 38–41
Fleming's right-hand rule, 167, 180, 202, 282, 385
flux (magnetic), 21, 32
flux density, 1, 8, 21, 36–7, 42, 53–6
  normal, (radial) component, 64–6, 74, 83
flux cutting, 44, 169
flux, leakage, 78, 90, 94, 102, 105–7, 120–1, 156
flux-linkage, 41–5, 105, 238, 318–22
flux-linkages, theorem of constant-, 318, 428
flux penetration, 114, 319
force on current loop, 55–6
Fourier analysis, 141–4, 154, 158–9, 244–6, 350, 361–8
fringing flux (magnetic), 68–9, 81

Gauss' law (for magnetic field), 142
generator, turbine-, 2, 264
GTO thyristor, 210, 413

Hall-effect devices, 211, 340
heteropolar, 89, 94, 97, 135, 178, 205
homopolar, 87–9, 97–8
hydroelectric generator, 264
hysteresis, 58, 61, 112–4, 213, 331
  loop, 61–3, 90–2, 113
  minor loop, 91
  motor, 425–6

impedance, 7, 12, 15

Index

impedivity, 394–5, 419–20
impulse commutation, 352–60, 383–4
inductance, 41–2
  leakage, 42, 44, 165, 247
  loop, 47, 52
  mutual, 42–4, 100–3
  self, 42–4, 102, 191
inductance machines (polyphase), 264, 343, 385–417
  cage rotor, 344, 385, 390, 407
    double cage rotor, 407
    deep bar, 407
  closed loop control, 413, 415
  cogging, 257, 416
  commutator machine, 409
  constructional details, 384, 389–90
  crawling, 416
  current-fed induction motor drive, 415
  doubly fed machine, 390, 408–12
  efficiency, 404, 406, 412
  e.m.f. evaluation for rotor circuit, 387, 393
  end rings (rotor), 385, 387, 390
  equivalent circuit, 397–401
  field orientation (flux-vector) control, 415
  frequency changing property of commutator, 409
  harmonic effects on performance, 412, 415–17
  impedivity, 394–5, 419–20
  induction generator, 401
  iron loss, 386, 390, 400–1
  Kramer drive, 410–12
  linear induction machines (polyphase), 427–30
    levitation, 432
    short stator single-sided, 428–9
    short stator double-sided, 430
    velocity/thrust characteristic, 432
  locked rotor test, 402–3, 431
  magnetising current, 399
  magnetising reactance, 399–400
  mechanical stress, tangential, 386, 388–9, 394–5
  mechanical windage and friction losses, 401–3
  no-load test, 402–403
  phasor diagram, 397–401, 408
  plug braking, 406, 408
  pole amplitude modulation, 412
  pole changing, 412
  power factor improvement, 386, 412

induction machines (*cont.*)
  pull-out torque, 396, 405, 432
  radial airgap field distribution due to stator winding, 386, 391, 399
  regenerative braking of inverter fed machine, 414
  rotating airgap field, 386, 391–4
  rotor circuit impedance referred to stator winding, 399–400
  rotor induced current and field, 387–9, 393–4
  rotor leakage impedance, 387–8, 393, 414
  rotor standstill reactance, 396–7, 405
  Schrage motor, 409
  slip, 176, 387–8, 393, 401, 415–6
  slip compensation, 413
  slip energy recovery, 410
  slip frequency, 389, 393
  speed control, cage rotor, 406–7, 412–15, 433–4
  speed control, wound rotor, 406, 408–12
  speed regulation, 406, 410
  speed/torque characteristics, 404–5, 432–3
    for inverter fed machine, 413–4, 434
  starting, 406–7
    autotransformer-, 407
    star/delta-, 406
  stator current locus, 407–8
  supply frequency control, 413
  supply voltage control, 412
  torque development, 391–6
  torque in synchronous watts, 395–6, 401
  torque, maximum, (see *pull-out torque*)
  winding factors, 397–8, 400
induction machines (single-phase), 417–24, 427
  cage rotor, 417, 425
  contra-rotating fields and fluxes, 309, 418, 421–2
  equivalent circuit, 421–3, 435
  impedivity, 419–20
  mechanical stress, tangential, 417, 419
  principle of operation, 417–8
  shaded pole motor, 424
  slip, 418–423
    backward, 421
  speed/torque characteristic, 423
  starting winding, 423–4
  start/run capacitance, 424

induction machines (*cont.*)
   torque evaluation, 418–21
infinite busbar, 264, 290
interference, electromagnetic, 232
inversion, 241–6
inverters, 346–77
   amplitude control of output voltage, 360–3
      pulse-width modulation, 361
      sinusoidal pwm, 366–8
      notched waveforms, 361, 383
   commutation, forced, 349–50
   commutation, natural, 347, 349
   current-source inverter, 373–6, 384
   impulse commutation, McMurray, 352–4, 383–4
   impulse commutation, McMurray–Bedford, 352, 355–60, 384
   quasi-square waveform, 350, 361–4
   voltage-source inverter, 347–70
      single-phase bridge, 347
      three-phase bridge, 364–70
         reverse power flow, 368–70
isotropy, 96

keeper (magnet), 344
Kirchhoff's voltage law, 7
Kramer drive, 410–12

lammellar field type, 32
Laws relay, 83–5
leakage factor, 94, 98
leakage flux, (see *flux, leakage*)
leakage inductance, (see *inductance, leakage*)
Lenz's law, 43, 103
levitation, 430
line integral, 6–7, 20–1, 25–7, 37–8, 165–8
linear current density, A m^{-1}, 65, 138, 152–60, 164, 191, 201–2, 274–8, 394
linear induction motor, 427–30
linear motion transducer, 2, 75, 85–6
Lorentz equation, 8, 42, 52–3, 165, 168–9
loss of permanent magnetism, 89, 254, 257, 343
loudspeaker transducer, 55

magnetic dipole moment, 56, 59
magnetic saturation, 62, 90, 94, 104, 106, 138, 199, 213–15, 287–90

magnetic scalar potential, 32, 37–9
magnetic susceptibility, 59–60
magnetic vector potential, 8, 21, 26, 36–7, 42–5, 51, 66, 165–9
magnetisation, 59–60
magnetomotive force, (m.m.f.), 21, 38, 57, 67, 94, 102–4, 115, 120, 130, 151, 303
Maxwell, James Clerk, 70
Maxwell's (principal) stresses in magnetic field, 39–41, 69–73, 79–81, 87
mechanical stress at iron/air boundary,
   normal (radial) component, 68, 72–7, 83, 169, 195–6, 429–30
   tangential component, 72–4, 82–3, 169, 173–4, 195–6, 429–30
mobility, 5, 18
moment of inertia, 223–5, 298–301
MOSFET, 413
moving coil instrument, 55, 85
moving iron instrument, 78–9

Oersted, Hans Christian, 55
orthogonality, 39
overlap, 242–4, 247–50

paramagnetism, 57–60
permanent magnets, 83–9, 209–11, 310, 324, 327–9
permanent magnetism, 59–61, 89–96, 254–8, 263, 343–5, 425–6
permanent magnet materials, 62, 95–6, 425
permeance, 68, 102, 105
permeability, 1, 20–1, 57–8, 61–2, 64–6
   relative, 58
Poisson's equation, 36
polarity, 33–4, 87
pole piece, pole shoe, 68, 90, 92
polyphase a.c./d.c. converter, (see *converter*)
power (electrical), 3, 10–17
   apparent power (complex power), 10, 15
   reactive power (see *reactive volt-amperes*)
   real power (average power), 3, 10, 13–17, 19
power factor, 3, 15, 19, 241, 246, 309, 345, 349
power flow, 10, 17, 156, 171–5, 301, 401–2
   over coaxial cable, 48–9
   over twin conductor transmission line, 49–50

power transfer at surface of conductor, 45–7, 135, 151, 171–3
power transfer between transformer windings, 120–1
power (mechanical), 2, 173–4, 230, 297–300, 331–2
Poynting vector, 46–50, 97, 120–1, 156, 172–6
pulse number, 244–6
pulse-width modulation, 361, 366–8
pulsating field, 273, 309–10, 417

quartz crystal oscillator, 265, 323
quasi-square waveform, 350, 362–6, 370–1

rating, 100, 118
reactive volt-amperes, 3, 10, 13–17, 19, 290, 301, 309–10, 317, 402
recoil line, 91–5, 254, 343
recoil permeability, 91, 95
rectification, 100, 181, 215, 411, (see also *converter*)
reference frame, 8, 44, 53, 165–8, 266
reluctance motor, polyphase, 302–9, 380
reluctance motor, single-phase, 309–10, 424–5
residual flux density, (remanence), 61–2, 89–91, 95–6, 215, 344
residual m.m.f., 133, 252
resistivity, 9, 63, 114, 207, 394
reversing drive, (4-quadrant), 221, 226, 250–2, 340
right-handedness, right-hand screw rule, 33–5, 43, 46, 55–6, 137, 156
right-hand rule, Fleming's, 167, 180, 202, 282, 385
ripple, 139, 183–5, 243–6, 411
rotary actuator, 79–86
rotating field, 161–4, 174–6, 273, 279–80, 386
rotating-shaft encoder, (see *encoder*)

saliency, 80–2, 135, 265–7, 302–18, 424
  double, 323–45
salient pole, 82, 134, 178, 199
saturation, (see *magnetic saturation*)
scalar product, 7, 21–2, 24, 33–4, 137
Schrage motor, 409

shaded-pole motor, 424
single-phase converter, (see *converter, line-commutated*)
single-phase commutator motor, 426–7
single-phase induction motor, 417–24, 427
skewing, 257, 327, 390, 416
skin effect, 423
sliprings, 266, 389–90
slot effects, 83, 136
snubber circuit, 210
solenoid, 75–8, 96–7
solenoidal field type, 32, 64, 138, 142
spontaneous magnetism, 60
stabilisation (of permanent magnet), 91, 254
Stokes' theorem, 35
Steinmetz index, 113, 133
Stepping (stepper) motors, 265, 323–43
  bifilar winding, 339, 341–2
  braking torque, 332
  closed loop control, 332
  damped oscillation, 324, 332–3
  detente torque, 329
  drive circuit, unipolar, 339
  drive circuit, bipolar bridge, 339, 382
  drive circuit, chopper, 339, 382
  dynamic behaviour, 330–8
  dynamic torque, 332, 337–8, 342–3
  energy recovery, 339, 342
  equal area criterion, 334
  forcing resistor, 339
  generation, 332, 336
  half-stepping, 326, 332
  high-speed operation, 333, 336–8
  hybrid type, 324, 327–9, 337–9
  low-speed operation, 333–6
  mechanical resonance, 333
  multistep operation, 333
  natural frequency, 332, 382
  open-loop control, 332–8
  positional error, 329
  pull-out torque, 332–3
  start rate/stop rate, 332–3
  static torque characteristics, 329–30
  step length, 325–6, 329, 381
  slewing, 333
  switched reluctance motor, 266, 323, 329, 339–43, 382–3
  tooth pitch, 329
  torque pulsation, 343
  variable reluctance type, single-stack, 326, 337–8

stepping motors (*cont.*)
  variable reluctance type, multi-stack, 324–6, 334, 337–8, 339
stress (mechanical), (see *mechanical stress at iron/air boundary*)
susceptibility (magnetic), 59
switch mode power supply, 225
switched reluctance motor, 339–43, 382–3
symmetrical components, 124, 322–3
synchronous induction motor, 370
synchronous machines (polyphase), 264–309, 310–23
  'airgap' e.m.f., 283–5, 288
  airgap field due to sinusoidally distributed armature current, 270–9
  (see also *windings for rotating machines*)
    radial component, 273–6, 278–9
    tangential component, 276–8
  airgap radial field distribution (trapezoidal) due to armature phase winding, 270–3, 286, 377
  airgap harmonic field components, 265, 273–4, 286, 346
  alternator, 287
  armature leakage flux, 285
  armature leakage reactance, 285, 309, 313, 323
  armature reaction reactance, d-axis, 284–6, 313
  armature reaction reactance, q-axis, 313
  armature resistance, 285
  armature time constant, 319–20
  armature winding, 265, 267
  closed loop control, 371
  constructional features, 265–6
  current-source inverter drive, 372–3, 376–7, 415
  damping, 295, 300, 319
  direct axis, 279, 302, 318–21
  dynamic behaviour, 16, 293, 295–302
  electric loading, 286
  e.m.f. equation, 282–4, 287
  equal area criterion, 299–300, 334, 380
  equivalent circuit of reluctance machine, 305–9
  equivalent circuit of round rotor machine, 285–7

synchronous machines (*cont.*)
  equivalent circuit of salient-pole machine, 310, 313
  field excitation control, 290–3, 310, 315, 345, 373, 379
  field winding (d.c. excited), 265, 267, 274, 279
  flux-leakage coefficient, 320
  flux penetration, 319–21
  inertia constant, H, 299, 380
  infinite busbar operation, 290–302, 307, 310–17
  line-start synchronous motor, 344
  load angle, 288, 293–300, 314–17, 338, 372
  magnetic saturation, 287, 289, 318
  maximum real power, 295
  mechanical stress, tangential, 268, 279–81, 304
  natural frequency of oscillation, 297, 380
  open-circuit test, 287
  permanent magnet field machines, 343–5
  phase sequence reactances, 318, 322–3
  phasor diagram of reluctance machine, 305–8
  phasor diagram of round rotor machine, 285, 287–93
  phasor diagram of salient-pole machine, 310–16
  power electronic drive systems, 345–77
  power factor, 287–95, 309, 313–15
  power swings, (see *dynamic behaviour*)
  quadrature axis, 302
  reluctance machines, 302–9, 380
  reluctance (saliency) torque, 270, 305, 317
  rotating field, 273, 279–80, 282
  saliency, 265–7, 310, 321
  short-circuit characteristic, 287
  short-circuit test, 287
  slip test, 318
  space-phase angle, ψ, 280–2, 287, 290–1, 303–5, 311–15, 372–3
  starting, (see *synchronisation*)
  steady-state stability limit, 296
  stiffness, (see *synchronising torque*)
  sudden three-phase short circuit, 318–21

Index

synchronous machines (*cont.*)
  swing curve, 299
  synchronisation, 292, 344
  synchronising power, 296–9
  synchronising torque, 298–9, 317, 344
  synchronous impedance, 285, 287
  synchronous reactance, d-axis, 285, 313–18, 345
  synchronous reactance, q-axis, 313, 318, 345
  synchronous speed, 264, 269, 344–5
  synchronously rotating reference frame, 266, 297
  terminal power/load angle relationship, 293–5, 316–17, 379
  torque angle, (see *space-phase angle*, ψ)
  torque development, 270, 279–82, 303–5
  transient stability, (see *dynamic behaviour*)
  transient and sub-transient reactances and time constants, 319–21
  trapped flux, 319–21
  two-reaction theory, 302, 323, 381
  voltage-source inverter drive, 370–2, 413
  winding distribution factor, 283, 378
  winding pitch factor, 283
synchronous speed, 161, 269, 387, 393, 395

tangential stress, (see *mechanical stress, tangential component*)
teeth, 83, 134, 169, 195–6
Tesla, Nikola, 385
three-phase bridge converter, (see *converter, line-commutated*)
torque, 2–3, 55–6, 79–89, 140, 152, 169, 194–5, 202–5, 216, 227, 236, 258, 268–70, 323, 329–32
  alignment, 270, 281–2, 315–7
  harmonic, 161, 338, 346
  reluctance (saliency), 81–2, 303–5, 315–17
torque motor, 82, 134
transistor, 211, 256–7, 361
transient response of series resonant circuit, 354–5
transformer e.m.f. equation, 112–3, 175, 284

transformer, 100–31
  ampere-turn balance, 104–5, 116, 122, 126–7, 133, 254, 389
  autotransformer, 126–7
  back-to-back test, (see *load test*)
  capitalised losses, 118
  core construction, core-type polyphase, 121–3, 133
  core construction, shell-type, 119, 122
  core losses, 106–7, 109–10, 112–14, 116–19, 122, 128–9, 131
  core saturation, effect of, 104, 116, 128, 130
  current transformer, 128–30, 133
  delta tertiary winding, 126
  eddy current loss, 113–14
  efficiency, 116–18, 126–7, 131
  e.m.f. equation, 112–13, 130, 175, 284
  equivalent circuit, 106–7, 110, 128–9
  equivalent leakage impedance, 110–12
  flux, core (mutual), 100–10, 112–17, 119–24, 128–30
  flux, leakage, 102, 105–7, 120–1
  flux-linkages, partial, 105, 120
  flux, residual, 115–6
  harmonic e.m.f.s, 114–15, 122–4, 133
  harmonic flux, 115, 122–4
  high frequency transformer, 128
  hysteresis, 112–14
  ideal transformer, 105–9
  inductance, magnetising, 107–10, 130
  inductance, mutual, 100–2, 130
  inductance, self, 102, 130
  interconnected star winding, (see *zig-zag*)
  laminated core, 63, 106, 114, 129
  laser-scribed laminations, 118
  leakage reactance, 106–12, 119–20, 126–7
  load factor, 118
  load test, 119, 131–2
  magnetising ampere-turns, 104
  magnetising current, 102, 104–5, 109–10, 113–16, 122–4
    harmonic components, 114–15, 122–4
    inrush, 115–16
  mechanical force on windings, 116, 119
  mitred joint, 106
  neutral point oscillation, 123

transformer (cont.)
　no-load (open-circuit) test, 119, 131
　phasor diagram, 108–12, 131
　power transfer, 120–1
　protection (differential), 116
　pulse transformer, 132
　referred parameters, 108–9
　resistance, 107–12, 119–20, 129
　short-circuit test, 119, 131
　Steinmetz index, 113, 133
　Sumpner test, (see *load test*)
　three-phase transformers, 115, 121–26
　　unbalanced operation, 124–6
　variable ratio autotransformer, 127
　voltage regulation, 110–12, 117, 126
　winding capacitance, 107, 128
　winding resistance, (see *resistance*)
　zero sequence current flow, 124–6
　zig-zag winding, 126, 132

universal motor, 426

vector product, 8, 55–6, 167, 172

Weiss, Pierre, 60
Weiss constant, 60
windings for rotating machines, 134–77
　airgap field of armature winding, 136–64
　　radial component, 137, 140–5, 149–50, 152–64
　　tangential component, 137–8, 140, 151–61, 164
　angular position, 141–2, 155
　armature winding, 135–177

windings (cont.)
　chorded coil, (see *short-pitch coil*)
　coil pitch, 142
　coil-pitch factor (coil-span), 143, 171, 176, 412
　coil sides, 139, 144, 165, 169
　coil interconnection (effect on e.m.f.), 153, 169
　concentrated winding, 42, 178, 199, 205
　continuous winding, 139, 255
　developed winding layout, 137
　distributed winding, 138, 145–51
　distribution of winding current,
　　for d.c. armature (uniform sheet), 154, 191, 197
　　for sinusoidal current sheet, 155–8
　distribution factor (coil-spread), 145–51, 169–71, 378, 423
　double-layer winding, 144, 148, 169, 175, 185
　e.m.f. (induced), 140, 164–71
　end windings, 139, 189, 285
　field windings, 135, 174, 179
　fractional-slot windings, 150
　integral-slot windings, 150
　phase spread, 150, 171, 377
　polyphase winding, resultant field, 161, 174, 273
　power flows, 171–6
　practical winding, 159, 164, 270, 286, 346
　short-pitch coil, 141–5, 169–71, 273, 283
　slot angle, 146–7, 165, 171
　space harmonics, 138, 140–5, 149–53, 157–64, 170–1, 174–7, 265, 273–4, 286, 346
　　of short pitch coil, 142–5
　winding factor, 149, 165, 169, 174–5